ELECTROKINETIC SEPARATION METHODS

ELECTROKINETIC SEPARATION METHODS

Editors
P.G. RIGHETTI
Università di Milano, Cattedra di Biochimica Generale, Via Celoria 2, 20133 Milan, Italy
C.J. VAN OSS
Immunochemistry Laboratory, Department of Microbiology, State University of New York at Buffalo School of Medicine, 207 Sherman Hall, Buffalo, NY 14214, U.S.A.
and
J.W. VANDERHOFF
Center for Surface and Coatings Research, Lehigh University, Bethlehem, PA 18015, U.S.A.

1979

ELSEVIER/NORTH HOLLAND BIOMEDICAL PRESS
AMSTERDAM · NEW YORK · OXFORD

ISBN Elsevier/North-Holland: 0-444-80113-8

Published by:
Elsevier/North-Holland Biomedical Press
335 Jan van Galenstraat, P.O. Box 211
Amsterdam, The Netherlands

Sole distributors for the USA and Canada:
Elsevier North-Holland Inc.
52 Vanderbilt Avenue
New York, N.Y. 10017

Library of Congress Cataloging in Publication Data

Main entry under title:

Electrokinetic separation methods.

Bibliography: p.
Includes index.
1. Electrophoresis. 2. Biological chemistry--
Technique. I. Righetti, P. G. II. Van Oss, Carel J.
III. Vanderhoff, John William, 1925-
QD79.E44E43 543'.087 79-11502
ISBN 0-444-80113-8

Printed in The Netherlands.

CONTENTS

INTRODUCTION - Brief historical survey

PIER GIORGIO RIGHETTI, CAREL J. VAN OSS° and JOHN W. VANDERHOFF∞
Department of Biochemistry, University of Milano, Milano 20133, Italy.
° Department of Microbiology, State University of New York at Buffalo, Buffalo, New York 14214 (USA) and ∞ Center for Surface and Coatings Research, Building 7, Lehigh University, Bethlehem, Pennsylvania 18015 (USA).

A few months after the call for a moratorium on electrophoresis books (in that particular case, though, it was for the nth book on isoelectric focusing)[1] an additional *opus* on electrophoretic techniques might appear preposterous. Yet, after the classical two-volume set edited by Bier[2], electrokinetic separation methods have gained new impetus, through many important refinements of pre-existing techniques, the introduction of new ones, e.g. isoelectric focusing, and the "rediscovery" of old principles, such as isotachophoresis. Moreover, in addition to spreading on hearth (in at least one of its many facets it is now a standard technique in most life-science laboratories) electrophoresis has been recently taken into the outer space in Apollo and Apollo-Soyuz flights[3], thus pointing out the enormous interest in this methodology, second perhaps only to chromatographic techniques.

We will try here to summarize the evolution of electrophoresis from its onset to present-day applications. As early as in 1809 Reuss[4] described the migration of clay particles under an electric field and the unwanted phenomenon of electro-osmosis, an ever present hazard of all electrophoretic applications, not completely eliminated as yet[5]. It was only at the turning of the century, though, with the pioneering work of Picton and Linder[6], Hardy[7,8], Ellis[9] and Powis[10], that the potential of electrophoresis for the separation and characterization of biomacromolecules was realized. At that time, also, the possibility of using electrophoresis for the physico-chemical characterization of polyelectrolytes, such as the determination of isoelectric points by migration experiments at different pH's, was demonstrated by Michaelis[11]. During the twenties and thirties, though, most of the efforts were devoted to the development of preparative electrophoresis apparatuses in all-liquid media, of the multi-membrane type (for reviews, see Svensson[12], Bier[13] and Polson and Largier[14]). In the late thirties, some new developments were made which dramatically changed the course of electrophoretic techniques. One could in fact draw an historical sketch just by following these new trends, *i.e.* "moving boundary electrophoresis" and "zone electrophoresis" which was partly developed to counteract the drawbaks of the former.

MOVING BOUNDARY ELECTROPHORESIS

In 1937, the description by Tiselius[15] of an innovative U-tube apparatus opened a new epoch both in analytical and preparative electrophoresis. A drawing of this

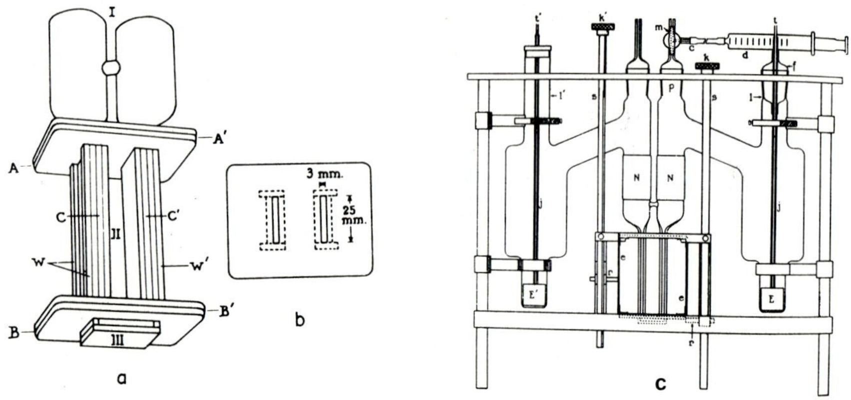

Fig. 1. Tiselius cell (a and b) and apparatus (c). (a): the U-shaped channel consists of three sections (I, II and III) that may be slid over each other along the planes AA' and BB'. The height of center section II is 92 mm, leaving 86 mm clear height for boundary observation. (b): view from above of a horizontal plate. The channel width is 3 mm and its depth 25 mm. (c): electrophoresis cell, electrode vessels and support. *k*, *k'*: knobs for shifting the horizontal plate of bottom section III; *t* and *t'*: glass-enclosed silver tubes for metallic contact with electrodes E and E'; *l* and *l'*: levels for buffer filling (any excess buffer overflows through *c* via the hollow stopper *p*); *f*: glass stopper; *d*: syringe for countercurrent flow (from Longsworth[71]; source: Academic Press).

apparatus and of the separation cell is shown in Fig. 1. By controlling the electric field and the pH in the neighborhood of the moving protein boundary, by dissipating the Joule heat generated by the current, by providing for constant electrode volumes and excluding foreign ions from penetrating in the region of the moving boundary and by a built-in system of Schlieren optics and astigmatic camera for the observation and photography of the boundaries, Tiselius made possible for the first time the accurate measurement of absolute mobilities of proteins. Moreover, since the movement of the macroions was in a free-liquid phase, not even stabilized by a sucrose density gradient, the parameter measured was also the free-mobility, unhindered by the viscous drag of density gradients or by the tortuous channels of support media in "zone electrophoresis". Since both, the ascending and descending boundaries in the two arms of the cell U-tube could be monitored, it was soon realized that departure from enantiography (mirror images of each other) afforded clues to interactions among the species under investigation. Thus, the very limit of "boundary electrophoresis", *i.e.* the failure to permit complete separation of the components of a mixture, became an advantage for the study of many protein-protein, protein small ion and protein-small molecule interactions (for a review see Brown and Timasheff[16]). The Tiselius apparatus could also be used for preparative work, since the Schlieren optics made possible the isolation of components under optical control. Moreover, since the separation in carried out at the density maximum of water, by simply cooling the apparatus to 4°C it was possible to increase its maximum possible load. The U-tube could also be subdivided into seven compartments, allowing for the isolation of four to six fractions and its full lenght could be exploited for the separation by the principle of countercurrent flow[17]. Unfortunately, the usefulness of "boundary electrophoresis" for protein purification is marginal since at best only parts of the fastest and slowest components may be obtained pure from the ascending and descending limbs, respectively. Moreover, the method was born at the onset of World War II and, as Svensson[12] put it: "the comparatively high costs of cuvettes and optical parts in the Tiselius instruments have rendered them unavailable for many laboratoires, especially in countries that have suffered much under the war and post was conditions".

ZONE ELECTROPHORESIS

The overlapping "boundary separation" was soon substituted by a "zone separation" in which each component was allowed to form a zone separated from the others by empty buffer regions. Zones of this kind are gravitationally unstable and would give rise to convection in ordinary electrophoresis tubes. To avoid this, all sorts of stabilizing media were described.

Papers and fabrics

The development of paper strip electrophoresis received a great stimulus from the marked success of paper chromatography. The use of filter paper made of very pure cotton fibers (Whatman, Schleicher & Schüll, Macherey, Nagel & Co., Munktell, etc.)[18] has been described, as well as of polyvinyl chloride (Rhovyl paper of Schleicher and Schüll)[19], of glass-fibers[20] and of silk[21].

Powders and granulated gels

Effective materials employed include potato starch and "soluble" starch[22], cellulose powder[23] and ethanolyzed cellulose[24], glass powder[25], prepared by grinding glass wool under water and fractional sedimentation, plastics, such as polyvinyl chloride (Geon 426)[26,27] and pevikon C-870 (a copolymer of polyvinyl chloride and polyvinyl acetate)[28], minerals (Hyflo-Super Cel)[29] and even asbestos[30]. In the early sixties, with the advent of molecular sieving techniques,a new wave of supporting media was introduced, such as Sephadex (dextrans cross-linked with hepichlorohydrine)[31,32] and granulated agarose[33] and polyacrylamide[34]. The powders are made into a thick, homogeneous paste with the buffer, poured as a layer into the trough and then pressed thoroughly dry with filter paper. In columns, a thin suspension is poured evenly, as in the filling of chromatographic columns, and the solid material is then allowed to settle firmly.

Gels

In the first zone electrophoretic separation, that of amino acids, carried out by Consden *et al.*[35], a silica gel was employed as supporting layer. However, especially for analytical purposes, the advantages of continuous matrices over powders or granulated gels was soon realized.

Agar gel. Even older is the use of agar gel, first reported in 1923 by Kendall and Crittenden[36] for inorganic ions, rediscovered then for the separation of organic substances[37] and now, in its purified form, agarose, of fundamental importance in all immuno-electrophoretic techniques[38-42].

Starch gel. Its use was first described in the classical "horizontal" tray[43], then converted into a "vertical" chamber[44] (in order to avoid electrodecantation of the sample). This method, coupled to zymogram techniques, was of fundamental importance in the study of genetic polymorphism. It is still one of the methods most widely used by genetists in the analysis of phenotypes, notwithstanding the superior resolution of polyacrylamide gels.

Various gels. Other media suggested have been finely porous rubber sponges[45], gelatin hardened with formaldehyde[46] and gels of polyvinyl alcohol[47]. This last gel has properties similar to polyacrylamide, but it is less transparent. Of special interest, particularly in clinical chemistry, was the introduction of cellulose acetate (acetylated cellulose), developed by Kohn[48,49], and used either as such or gelatinized[50,51].

Polyacrylamide gel. The use of polyacrylamide as a medium for electrophoresis was first suggested by Raymond and Weintraub[52] in 1959, at which time starch was the major supporting medium. While Raymond and co-workers at the University of Pennsylvania, introduced the technique of polyacrylamide slab electrophoresis[53], Ornstein[54] and Davis[55] at Mount Sinai Hospital, N.Y., introduced disc-gel electrophoresis (the use of discontinuous buffers, though, had been already tried by Poulik[56]). Discontinuous electrophoresis was a major step forward and paved the way to modern, high resolution techniques, such as isoelectric focusing and isotachophoresis.

Special gels. A host of modified polyacrylamide gels, for special separation purposes, have been described: highly cross-linked, *i.e.* highly porous, gels[57]; polyacrylamide cross-linked with ethylene diacrylate[58], or with diallyl tartardiamide [59] or with amido methylol bisacrylamide[60] or with N,N'-bis-acrylyl cistamine[61]. Several composite gels have also been reported: agarose-polyacrylamide[62,63], cellulose acetate impregnated with polyacrylamide[64], polymerizable acrylamide-sephadex gels[65] and agar-agar gels containing liquid linear polyacrylamide[66].

Density gradients

This method, as compared to zone electrophoresis in any other inert media, has the advantage of being entirely free from adsorption effects. The liquid phase, in order to counteract the gravitational field, is usually a density gradient (mostly sucrose, or glycerol or sorbitol). The first attempt at using this principle is due to Philpot[67]. An extensive survey has been written by Svensson[68].

CONCLUDING REMARKS

Another was of trying to make an historical survey of electrophoretic techniques would be to follow the evolution of the different apparatuses developed over the years, but this would take up as much space as we have used so far, therefore we have to refer to the extensive reviews of Hanning[69], Hochstrasser *et al.*[70], Longsworth[71], Pucar[72], Kolin[73,74], Rilbe[75] and Righetti[76]. Several books have also been published in the field of electrophoresis, starting from the early days of paper techniques[77-81], to agar gels[82], blocks and columns[83], starch and polyacrylamide[84] to high resolution techniques, such as disc-electrophoresis[85], isoelectric focusing [86-88] and isotachophoresis[89]. Books covering more general subjects have also appeared[90-96].

From an analytical point of view, with the advent of high resolution techniques such as disc-electrophoresis, isoelectric focusing, isotachophoresis, and with the more extended use of two-dimensional separations, such as the O'Farrell technique[97], we feel that present-day electrophoretic methods are working very close to their resolving limits. Perhaps, as stressed by Chrambach and Nguyen[98], more work should be spent in developing preparative electrophoretic techniques, which, notwithstanding the tremendous deal of work done in the past, are today largely unused. There are hints that we are withnessing a revival of preparative methods, as documented by the proceedings of a recent meeting in Uppsala[99-101].

We realize it is foolish to try to squeeze 80 years of history of electrophoresis in a handful of pages. So much has been left out, due to severe space limitations and to our ignorance in the field, that we feel ashamed of it. Therefore, we have asked our printer to set this chapter in the book in special, ready-to-tear pages for scientists who will feel too enraged by our disastrous historical attempt.

REFERENCES

1. Morris, C.J.O.R. (1978) Nature 173,326-326
2. Bier, M. (ed.) (1959 and 1967) Electrophoresis: Theory, Methods and Applications, Vols. I and II, Academic Press, New York
3. Vanderhoff, J.W. and van Oss, C.J. (1979) this book, pp. 257-273
4. Reuss, F.F. (1809) Mémoires de la Société Imperiale des Naturalistes de Moscou 2, 324-347
5. Vanderhoff, J.W. and Micale, J. (1979) this book, pp. 81-93
6. Picton, H. and Linder, S.E. (1897) J. Chem. Soc. 71, 568-573
7. Hardy, W.B. (1899) J. Physiol. 24, 200 293
8. Hardy, W.B. (1900) Proc. Royal Soc. (London) 66, 110-125
9. Ellis, R. (1912) Z. Physik. Chem. (Leipzig) 80, 597-603
10. Powis, F. (1915) Z. Physik. Chem. (Leipzig) 89, 186-195

11. Michaelis, L. (1909) Biochem. Z. 16, 81-92

12. Svensson, H. (1948) in Advances in Protein Chemistry (Anson, M.L. and Edsall, J. T., eds.) vol. IV, Academic Press, New York, pp. 251-295

13. Bier, M. (1959) in ref. 2, pp. 263-316

14. Polson, A. and Largier, J.F. (1960) in A Laboratory Manual of Analytical Methods of Protein Chemistry (Alexander, P. and Block, R.J., eds.) Pergamon Press, London, pp. 161-191

15. Tiselius, A. (1937) Trans. Faraday Soc. 33, 524-535

16. Brown, R.A. and Timasheff, S.N. (1959) in ref. 2, pp. 317-367

17. Tiselius, A. (1938) Kolloid Z. 85, 129-140

18. Kunkel, H.G. and Tiselius, A. (1952) J. Gen. Physiol. 35, 89-99

19. Stegemann, H. (1956) Naturwissenschaften 43, 518-522

20. Bourne, E.J., Foster, A.B. and Grant, P.M. (1956) J. Chem. Soc. IV, 4311-4314

21. Geldmacher-Mallinckrodt, M. and Weinland, H. (1953) Z. Physiol. Chem. Hoppe Seyler's 292, 65-70

22. Kunkel, H.G. and Slater, R.J. (1952) Proc. Soc. Exptl. Biol. Med. 80, 42-51

23. Haglund, H. and Tiselius, A. (1950) Acta Chem. Scand. 4, 957-965

24. Flodin, P. and Kupke, P.W. (1956) Biochim. Biophys. Acta 21, 368-375

25. Hocevar, B.J. and Northcote, D.H. (1957) Nature 179, 488-490

26. Müller-Eberhard, H.J. and Kunkel, H.G. (1956) J. Exptl. Med. 104, 253-260

27. Stein, W.H., Kunkel, H.G., Cole, R.D., Spackmann, D.H. and Moore, S. (1957) Biochim. Biophys. Acta 24, 640-650

28. Müller-Eberhardt, H.J. (1960) Scand. J. Clin. Lab. Invest. 12, 33-40

29. Strain, H.H. (1939) J. Amer. Chem. Soc. 61, 1291-1295

30. Butler, J.A.V. and Stephen, J.M.L. (1947) Nature 160, 469-471

31. Porath, J.K. and Flodin, P. (1959) Nature 183, 1657-1658

32. Radola, B.J. (1973) Ann. N. Y. Acad. Sci. 209, 127-143

33. Hjertén, S. (1961) Biochim. Biophys. Acta 53, 514-523

34. Hjertén, S. (1962) Arch. Biochem. Biophys. Suppl. 1, 147-156

35. Consden, R., Gordon, A.H. and Martin, A.J.P. (1946) Biochem. J. 40, 33-40

36. Kendall, J. and Crittenden, E.D. (1923) Proc. Natl. Acad. Sci. U.S. 9, 75-82

37. Gordon, A.H. and Reichardt, D. (1951) Biochem. J. 48, 1229-1235

38. Axelsen, N.H., Kroll, J. and Weeke, B. (eds.) (1973) A Manual of Quantitative Immunoelectrophoresis, Scand. J. Immunol., vol. 2, Suppl. 1, Universitetsforlaget, Oslo

39. Axelsen, N.H. (ed.) (1975) Quantitative Immunoelectrophoresis, Scand. J. Immunol. Suppl. 2, Universitetsforlaget, Oslo

40. Laurell, C.B. (ed.) (1972) Electrophoretic and Electroimmunochemical Analysis of Proteins, Scand. J. Clin. Lab. Invest. 29, Suppl. 124

41. Grabar, P. and Burtin, P. (1964) Immunoelectrophoretic Analysis, Elsevier, Amsterdam

42. Clausen, J. (1969) Immunochemical Techniques for the Identification and Estimation of Macromolecules, North Holland/American Elsevier, Amsterdam and New York

43. Smithies, O. (1955) Biochem. J. 61, 629-641

44. Smithies, O. (1959) Biochem. J. 71, 585-587

45. Mitchell, H.K. and Herrenberg, L.A. (1957) Anal. Chem. 29, 1229-1233

46. Van Orden, D.E. (1971) Immunochemistry 8, 869-875

47. Reich, G. and Sieber, H. (1966) Z. Chem. 6, 351-356

48. Kohn, J. (1957) Clin. Chim. Acta 2, 297-303

49. Kohn, J. (1957) Nature 180, 986-988

50. Le Bouton, A.V. (1967) Anal. Biochem. 20, 551-558

51. Del Campo, G.B. (1968) Clin. Chim. Acta 22, 475-481

52. Raymond, S. and Weintraub, L. (1959) Science 130, 711-713

53. Raymond, S. (1964) Ann. N. Y. Acad. Sci. 121, 350-360

54. Ornstein, L. (1964) Ann. N. Y. Acad. Sci. 121, 321-349

55. Davis, B.J. (1964) Ann. N. Y. Acad. Sci. 121, 404-427

56. Poulik, M.D. (1957) Nature 180, 1477-1487

57. Baumann, G. and Chrambach, A. (1976) Anal. Biochem. 70, 32-43

58. Paus, N. (1971) Anal. Biochem. 42, 372-376

59. Anker, H.S. (1970) FEBS Letters 7, 293-295

60. O'Connel, P.H.B. and Brady, C.J. (1976) Anal. Biochem. 76, 63-73

61. Hansen, J.N. (1976) Anal. Biochem. 76, 37-44

62. Uriel, J. (1966) Bull. Soc. Chim. Biol. 48, 969-975

63. Peacock, A.C. and Dingman, C.W. (1967) Biochemistry 6, 1818-1825

64. Frenkel, M.J. and Blagrove, R.J. (1978) Anal. Biochem. 84, 583-588

65. Ziegler, A. and Kohler, G. (1976) FEBS Letters 64, 48-51

66. Bode, H.J. (1977) Anal. Biochem. 83, 204-210

67. Philpot, J.S.L. (1940) Trans. Faraday Soc. 36, 39-46

68. Svensson, H. (1960) in A Laboratory Manual of Analytical Methods of Protein Chemistry (Alexander, P. and Block, R.J., eds.) Pergamon Press, London, pp. 193-244

69. Hanning, K. (1967) in ref. 2, pp. 423-471

70. Hochstrasser, H., Lerner, H. and Skeggs Jr., L.T. (1967) in ref. 2, pp. 473-491

71. Longsworth, L.G. (1959) in ref. 2, pp. 137-177

72. Pucar, Z. (1960) J. Chromatogr. 4, 261-318

73. Kolin, A. (1970) in Methods in Medical Research (Olson, R.E., ed.), Year Book Med. Publ. Inc., Chicago, pp. 326-358

74. Kolin, A. (1958) in Methods of Biochemical Analysis (Glick, D., ed.) Interscience Publ., New York, pp. 259-288

75. Rilbe, H. (1969) in Protides of the Biological Fluids (Peeters, H., ed.) vol.17, Pergamon Press, London, pp. 369-382

76. Righetti, P.G. (1975) Separ. Purif. Methods 4, 23-72

77. Lederer, M. (1957) An Introduction to Paper Electrophoresis and Related Methods, Van Nostrand, Princeton, New Jersey

78. Wunderly, C. (1961) Principles and Applications of Paper Electrophoresis, Elsevier, Amsterdam

79. Antweiler, H.J. (1952) Die Quantitative Electrophorese in der Medizin, Springer Verlag, Berlin

80. Block, R.J., Durrum, E.L. and Zweig, G. (1958) A Manual of Paper Chromatography

and Paper Electrophoresis, Academic Press, New York

81. Ribeiro, L.P., Mitidieri, E. and Affonso, O.R. (1961) Paper Electrophoresis, Elsevier, Amsterdam

82. Wieme, R.J. (1965) Agar Gel Electrophoresis, Elsevier, Amsterdam

83. Bloemendal, H. (1963) Zone Electrophoresis in Blocks and Columns, Elsevier, Amsterdam

84. Gordon, A.H. (1975) Electrophoresis of Proteins in Polyacrylamide and Starch Gels, North Holland/American Elsevier, Amsterdam and New York

85. Maurer, H.R. (1971) Disc Electrophoresis, de Gruyter, Berlin

86. Allen, R.C. and Maurer, H.R. (eds.) (1974) Electrophoresis and Isoelectric Focusing in Polyacrylamide Gels, de Gruyter, Berlin

87. Righetti, P.G. and Drysdale, J.W. (1976) Isoelectric Focusing, North Holland/American Elsevier, Amsterdam and New York

88. Catsimpoolas, N. (ed.) (1976) Isoelectric Focusing, Academic Press, New York

89. Everaerts, F.M., Beckers, J.L. and Verheggen, Th. P.M. (1976) Isotachophoresis: Theory, Instrumentation and Applications, Elsevier, Amsterdam

90. Sargent, J.R. and George, S.G. (1975) Methods in Zone Electrophoresis, BDH Chemicals Ltd., Poole

91. Smith, I. and Seakins, J.W.T. (1976) Zone Electrophoresis (vol. 2), Heineman Mecal Books, London and Year Book Med. Publ., New York

92. Nakamura (1965) Cross Electrophoresis, Elsevier, Amsterdam

93. Abramson, H.A., Moyer, L.S. and Corin, M.H. (1942) Electrophoresis of Proteins and Chemistry of Cell Surfaces, Van Nostrand-Reinhold Publ. Co., Princeton, New Jersey

94. Moore, D.H. (1968) Electrophoresis; Physical Techniques in Biological Research, vol. II, Part A, 2nd ed., Academic Press, New York

95. Shaw, D.J. (1969) Electrophoresis, Academic Press, New York

96. Ambrose, E.J. (1969) Electrophoresis, Academic Press, New York

97. O'Farrell, P. (1975) J. Biol. Chem. 250, 4007-4021

98. Chrambach, A. and Nguyen, N. (1979) this book, pp. 337-368

99. Hjertén, S. and Porath, J. (eds.) (1978) 1st Tiselius Symposium on Modern Biochemical Separation Techniques, J.Chromatogr. 159, pp. 1-205

100. Martin, A.J.P. and Hampson, F. (1978) J. Chromatogr. 159, 101-110

101. Rilbe, H. (1978) J. Chromatogr. 159, 193-205

Electrokinetic Separation Methods,
P.G. Righetti, C.J. van Oss and J.W. Vanderhoff

THE ELECTRICAL DOUBLE LAYER AND THE THEORY OF ELECTROPHORESIS

J.TH.G. OVERBEEK AND B.H. BIJSTERBOSCH
Van 't Hoff Laboratory for Physical and Colloid Chemistry, State University Utrecht and Laboratory for Physical and Colloid Chemistry, Agricultural University, Wageningen (The Netherlands)

I. THE ELECTRICAL DOUBLE LAYER

I.1 INTRODUCTION

As early as in 1809 Reuss[1] observed that in an aqueous dispersion of clay particles the latter were migrating under the influence of an applied electric field. Since that time, numerous investigations have shown this migration to be quite a general phenomenon that is displayed by colloidal particles of very diverse chemical composition and of a wide variety of sizes and shapes. It has been given the name electrophoresis. Although electrophoresis is qualitatively identical to the migration of ions in electrolyte solutions, quantitatively its description is more involved and continues to be the subject of theoretical investigations.

For practical purposes a sound theoretical foundation of electrophoresis is of considerable importance. It enables information to be obtained on the state of charge of the particles and on the hydrodynamic resistance they encounter during migration. Moreover, it predicts how electrophoretic mobilities can purposely be varied, an essential topic in separation processes.

The similarity of electrophoresis and ionic migration indicates that colloidal particles undergoing electrophoresis are electrically charged relative to the continuous medium. As to the manner in which the particles acquire their charge, four different mechanisms can be distinguished. For the above mentioned clay particles, to begin with, Si^{4+} ions in the crystal lattice of these compounds can be isomorphically substituted by Al^{3+} ions. This process provides them with a negative charge relative to the solution. The crystal structure of dispersed particles can also be operative in another way in the acquisition of surface charge. Several inorganic materials (e.g. the silver halides) adsorb the cations and anions that constitute their crystal lattice preferentially and usually to a different extent. Along with the ensuing surface charge, the potential difference between solid and liquid also changes in this process and therefore these ions are called *potential determining ions*. (Between a metal and a redox system, electrons play the role of potential determining ions.) In passing it may be remarked that this change in potential actually is a variation in the difference in Galvani potential between the inner core of the particles and the bulk of the solution and is not necessarily just a surface effect. The Galvani potential difference can also vary due to other causes, such as changes in the amount and type of adsorbed dipoles. In the context of this chapter, however, only those variations that involve a redistribution of free charges (as contrasted to dipoles) will be dealt with. The third mechanism is found when surfactant ions are present in the solution. When these accumulate at the interface, as they usually do, they create a surface charge. Other ions may also be preferentially adsorbed, although usually to a lesser degree than surfactant ions. Last but not least particles may carry surface groups that can dissociate, e.g. carboxyl groups. The degree of dissociation of these groups usually depends on the pH and the composition of the solution. This fourth mechanism is particularly important for oxides, polyelectrolytes and proteins.

When the dispersed particles carry a net charge, electroneutrality of the system as a whole requires the solution surrounding them to carry a charge of the same magnitude, but with the opposite sign. The charge on the particle and that in the solution form an *electrical double layer*. The charges in solution are mobile, contrary to those on the surface. This has important consequences for the charge- (and potential-) distribution relative to the surface of the central particle. The tendency to minimize the (Coulomb) energy results in attraction of ions with a sign opposite to that of the surface (called counterions) and repulsion of ions with equal sign (co-ions). On the other hand, in order to maximize the entropy both ion types tend to reach a homogeneous distribution. When the two tendencies balance, thus when the free energy is at a minimum, the situation is very much like that in electrolyte solutions as described in the Debye-Hückel theory. An important difference with simple electrolyte solutions, however, is the circumstance that in colloidal systems the potentials can be much higher and the particles are larger, which requires a more complicated theoretical treatment.

Because all theories of electrophoresis are related to the structure of the electrical double layer around the moving entities, we start with a discussion of the latter. As a first approach only situations without an applied electrical field will be considered.

I.2 THEORY OF THE DIFFUSE DOUBLE LAYER

The theory of the diffuse electrical double layer has been developed independently by Gouy[2] and Chapman[3]. Their basic approach is very much like that of the Debye-Hückel theory, which was developed about 10 years later. Assuming all ions to be pointlike charges both theories express the combined result of electrostatic interaction and Brownian movement in terms of a Boltzmann-distribution for all ion types. This distribution can be expressed as

$$n_i = n_{i\infty} \exp(-\, z_i e\psi/kT) \qquad \text{(I-1)}$$

n_i is the number of ions of type i per unit volume, at a distance x from the particle surface. At this position the potential has a value ψ relative to the solution at very large x. Thus n_i and ψ are both functions of the coordinate x. $n_{i\infty}$ is the value of n_i in the bulk of the solution, where $\psi = 0$; e is the elementary charge, k the Boltzmann constant, T the absolute temperature and z_i the charge number of the ions, including their sign. Eq. (I-1) is in agreement with the statement that counterions are attracted and that co-ions are repelled. This situation is schematically depicted in fig. 1.

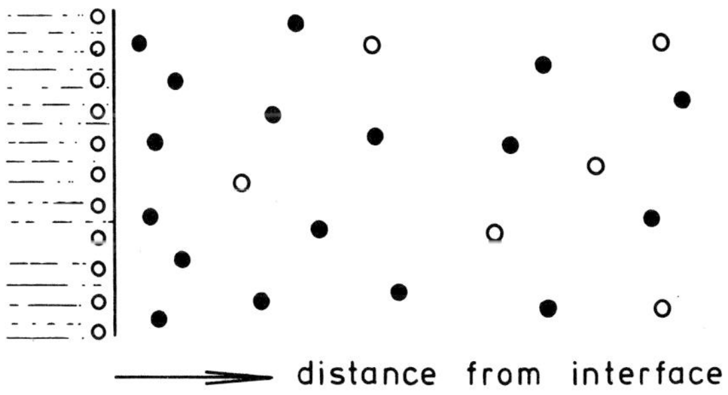

Fig. 1. Schematic distribution of negative counterions (full circles) and positive co-ions (open circles) in solution, relative to the surface of a positively charged particle (hatched area).

The net charge per unit volume of solution, the space charge density ρ, is also a function of x. It amounts to

$$\rho = \sum_i n_i z_i e \qquad \text{(I-2)}$$

When further elaboration of this equation is restricted to the presence of just one symmetric electrolyte ($z_+ = - z_- = z$; $n_{+\infty} = n_{-\infty} = n$), (I-2) can be written

$$\rho = ze(n_+ - n_-) \qquad \text{(I-3)}$$

Substitution of (I-1) in (I-3) leads to

$$\rho = - 2zen \sinh(ze\psi/kT) \qquad \text{(I-4)}$$

which applies for every value of x. An additional relation between x, ψ and ρ is given by the Poisson equation. Assuming the relative dielectric constant of the solution, ε_r, to be independent of the field strength and applying rationalized four quantities equations this equation reads

$$\mathrm{div}(\mathrm{grad}\ \psi) = \nabla^2\psi = -\ (\rho/\varepsilon_r\varepsilon_o) \tag{I-5}$$

where ε_o is the permittivity of free space (= $8.854 \times 10^{-12}\ C^{+2}.N^{-1}.m^{-2}$). Combination of (I-4) and (I-5) gives the Poisson-Boltzmann equation

$$\nabla^2\psi = (2zen/\varepsilon_r\varepsilon_o)\ \sinh(ze\psi/kT) \tag{I-6}$$

which forms the basis of the Gouy-Chapman theory.

Before considering the general merits of this equation, it is worthwhile to present the approximate form of (I-6) for low potentials ($\psi \leqslant 25/z$ mV). In that case $\sinh(ze\psi/kT) \simeq ze\psi/kT$ and (I-6) simplifies to

$$\nabla^2\psi = (2ne^2z^2/\varepsilon_o\varepsilon_r kT)\psi = \kappa^2\psi \tag{I-7}$$

In passing it can be mentioned that this step is essentially the linearization used in the Debye-Hückel theory.

In the r.h.s. of (I-7) the abbreviation $\kappa^2 = (2ne^2z^2/\varepsilon_o\varepsilon_r kT)$ has been introduced. κ has the dimension of a reciprocal length. $1/\kappa$ is an important quantity that is called the Debye-length or "the thickness of the double layer". Whereas the physical justification of the latter term has to await the discussion of eq. (I-11), we mention here that for a mixture of electrolytes κ is defined by

$$\kappa^2 = (e^2/\varepsilon_o\varepsilon_r kT)\sum_i n_i z_i^2 = \frac{2F^2\omega}{\varepsilon_o\varepsilon_r RT} \tag{I-8}$$

κ^2 is thus closely related to the ionic strength ω, which is defined by $\omega = \frac{1}{2}\sum_i c_i z_i^2$; the concentrations c_i are expressed in mol/unit volume. Table 1 gives a few typical values for $1/\kappa$ and ω for aqueous solutions of 25°C.

TABLE 1

Debye-length at 25°C and ionic strength for a number of electrolyte solutions

1-1 electrolyte c in mol/l	$1/\kappa$ m	ω mol/l	2-2 electrolyte c in mol/l	$1/\kappa$ m	ω mol/l
10^{-5}	1.0×10^{-7}	1.0×10^{-5}	10^{-5}	5.0×10^{-8}	4.0×10^{-5}
10^{-3}	1.0×10^{-8}	1.0×10^{-3}	10^{-3}	5.0×10^{-9}	4.0×10^{-3}
10^{-1}	1.0×10^{-9}	1.0×10^{-1}	10^{-1}	5.0×10^{-10}	4.0×10^{-1}

In non-polar or slightly polar organic solvents the solubility and the degree of dissociation of electrolytes is usually much lower than in water. ε_r is also lower for these solutions, but this does not compensate for the reduced solubility. Consequently for electrolyte solutions in such organic solvents one generally has to reckon with relatively thick double layers.

2.1 Poisson-Boltzmann equation for flat plates

If one deals with flat plates and if the potential does not vary in directions parallel to the plate the Poisson-Boltzmann (from hereon abbreviated as P.B.) equation takes the form

$$d^2\psi/dx^2 = (2zen/\varepsilon_o\varepsilon_r)\ \sinh(ze\psi/kT) \tag{I-9}$$

where x is the coordinate perpendicular to the plate.

2.1.1 Potential-distance relation.

Obviously (I-9) has to be integrated twice in order to express ψ in terms of x, which has its origin on the particle surface. The first integration is performed using the boundary conditions $\psi = 0$ and $d\psi/dx = 0$ at $x = \infty$. The result is

$$d\psi/dx = -\sqrt{8nkT/\varepsilon_o\varepsilon_r}\ \sinh(ze\psi/2kT) \tag{I-10}$$

In the second integration the boundary condition $\psi = \psi_o$ at $x = 0$ is introduced. ψ_o is called the surface potential. The result can be given as

$$\tanh(ze\psi/4kT) = \tanh(ze\psi_o/4kT)\ \exp(-\ \kappa x) \tag{I-11}$$

Thus (I-11) is the general solution of the P.B. equation for flat plates. For small potentials, that is in case $ze\psi_o/4kT \ll 1$, the approximation $\tanh x \simeq x$ can be used and then (I-11) reduces to

$$\psi = \psi_o\ \exp(-\ \kappa x) \tag{I-12}$$

This equation shows that $\psi = \psi_o\ \exp(-1)$, when $x = 1/\kappa$, illustrating why $1/\kappa$ is often called the thickness of the double layer. It is also easily proved that in this approximation the center of gravity of the double layer is exactly at a distance $1/\kappa$ from the surface. An increase in κ makes the double layer thinner ("compresses the double layer") and causes the potential to change more rapidly with the distance not only in the approximation (I-12) but also for the complete eq. (I-11).

In the derivation of (I-11) it has been assumed that the double layer is fully developed, i.e. that the double layers of individual particles do not interfere. However, in relatively concentrated dispersions interference can occur, a phenomenon that is known as double layer overlap. This situation, which is particularly relevant for colliding colloidal particles, requires integration of (I-9) with adapted boundary conditions. It has extensively been dealt with by Verwey and Overbeek[4], to which monograph we refer for further details.

2.1.2 Surface charge density and capacitance. The result obtained in the preceding section also enables us to derive a relation between the surface charge density σ_o and the surface potential ψ_o. Because σ_o is compensated by the space charge in the solution

$$\sigma_o = -\int_o^\infty \rho \mathrm{d}x \tag{I-13}$$

Substitution of (I-5) into (I-13), taking into account that potential and charge variations along the y- and z-coordinates are assumed to be absent, leads to

$$\sigma_o = \varepsilon_o\varepsilon_r \int_o^\infty \mathrm{d}^2\psi/\mathrm{d}x^2\mathrm{d}x = -\varepsilon_o\varepsilon_r(\mathrm{d}\psi/\mathrm{d}x)_{x=0} \tag{I-14}$$

and insertion of (I-10) into (I-14) results in

$$\sigma_o = \sqrt{8n\varepsilon_o\varepsilon_r kT}\,\sinh(ze\psi_o/2kT) \tag{I-15}$$

For small values of ψ_o (I-15) simplifies to

$$\sigma_o = \varepsilon_o\varepsilon_r\kappa\psi_o \tag{I-16}$$

This actually is the equation for the capacitance per unit area (σ_o/ψ_o) of a plane condenser with plate distance $1/\kappa$. So again a physical meaning of $1/\kappa$ emerges.

Differentiation of (I-15) with respect to ψ_o provides the general equation for the (differential) capacitance of the diffuse double layer

$$\begin{aligned} C_d = \mathrm{d}\sigma_o/\mathrm{d}\psi_o &= \sqrt{2n\varepsilon_o\varepsilon_r z^2e^2/kT}\,\cosh(ze\psi_o/2kT) \\ &= \kappa\varepsilon_o\varepsilon_r\,\cosh(ze\psi_o/2kT) \end{aligned} \tag{I-17}$$

Equations (I-16) and (I-17) demonstrate that for a constant surface potential the capacitance, and thus the surface charge increases with increasing κ or with increasing electrolyte concentration.

2.2 Poisson-Boltzmann equation for spheres

For spherical symmetry the P.B. equation takes the form

$$(1/r^2)\,\frac{d}{dr}\,(r^2\,\frac{d\psi}{dr}) = (2zen/\varepsilon_o\varepsilon_r)\,\sinh(ze\psi/kT) \qquad \text{(I-18)}$$

In contrast with the situation for flat plates this equation cannot be solved analytically, except when the linearization for low potentials is applied. In that case the result is

$$\psi = \psi_a(a/r)\,\exp\,[-\kappa(r-a)] \qquad \text{(I-19)}$$

where ψ_a is the surface potential of a spherical particle with radius a, and r is the distance to its center. From a comparison of (I-12) and (I-19) it can be concluded that for high values of κa double layers around spherical particles can be treated as if they have flat geometry.

From (I-19) it can rather easily be derived that for low surface potentials the total surface charge Q of a spherical particle amounts to

$$Q = 4\pi\varepsilon_o\varepsilon_r a(1 + \kappa a)\psi_a \qquad \text{(I-20)}$$

Because $Q = 4\pi a^2\sigma_o$ it is obvious that for $\kappa a \gg 1$, eq. (I-20) also reduces to the corresponding equation, (I-16), for flat plates.

When the potentials are too high to allow linearization, numerical solutions of (I-18) have to take the place of analytical ones. Extensive numerical data, for a wide range of κa-values and also including asymmetrical electrolytes, have been tabulated by Loeb et al.[5]. Space does not allow to cover them extensively in this text. However, in fig. 2 some results for spheres are compared with those for flat plates. These graphs visualize that for spherical symmetry the potential falls off more rapidly than for flat plates, and that is more pronounced the smaller the radius of the particle. It is also seen that for spheres as well as for flat plates, bivalent ions are much more effective in compressing the double layer.

Stigter[6] formulated interpolation equations for the tables of Loeb et al., whereas White[7] recently published approximate analytical solutions. For finite concentrations of the dispersed particles, again the possibility of double layer overlap has to be taken into account. This has recently been done in numerical calculations of Mille and Vanderkooi[8].

Examples of practical systems with spherical dispersed particles are emulsions, (monodisperse) latices, small surfactant micelles and some proteins.

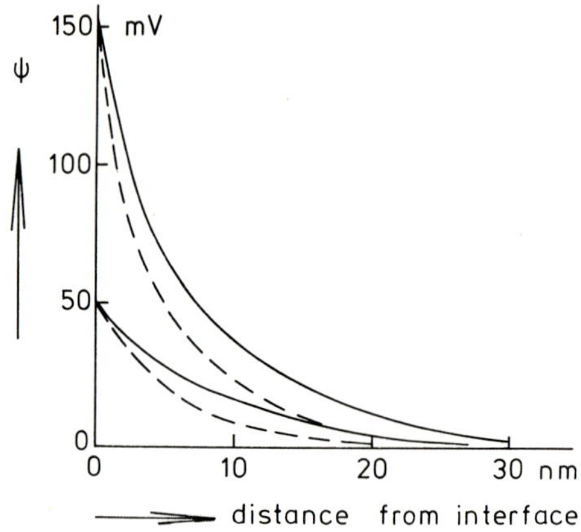
150
mV
ψ
100
50
0
0
10
20
30 nm
distance from interface

a

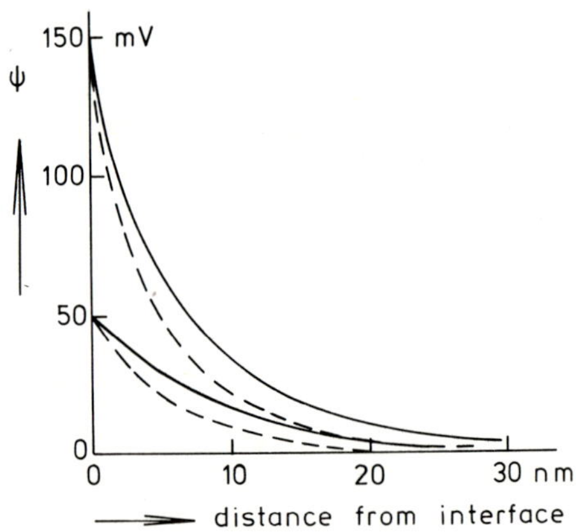
150
mV
ψ
100
50
0
0
10
20
30 nm
distance from interface

b

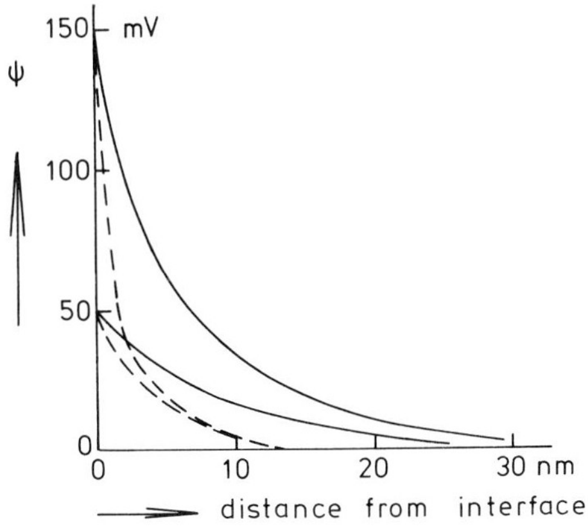

Fig. 2. Potential ψ as a function of the distance from the particle surface. Upper curves refer to ψ_o = 154 mV, lower curves to ψ_o = 51 mV.

a) 10^{-3} M 1-1 electrolyte; drawn lines: flat plates; dashed lines: spheres with a radius of 10 nm.

b) 10^{-3} M 1-1 electrolyte; drawn lines: spheres with a radius of 100 nm; dashed lines: spheres with a radius of 10 nm.

c) spheres with a radius of 100 nm in 10^{-3} M 1-1 electrolyte (drawn curves) and in 10^{-3} M 2-2 electrolyte (dashed curves).

2.3 Poisson-Boltzmann equation for cylinders

Polyelectrolytes are long chain polymer molecules containing a large number of ionizable groups along the chain. Although in electrophoresis some polyelectrolytes can occasionally be considered as impermeable or partially permeable spheres, most polyelectrolytes, being linear molecules, can better be treated as (kinked) cylindrical rods. Thus, for these cases it is essential that the P.B. equation is solved for cylindrical geometry. The Laplace operator then reads $\nabla^2 = (1/r) \times d[r(d\psi/dr)]/dr$. Just like in case of spherical particles the ensuing equation cannot be solved analytically, so that one has to recur again to numerical procedures or to approximate solutions. The first approximation was introduced by Dube[9] and Gorin[10]. However, it suffers from the drawback that it is only valid for small potentials. Of much wider applicability is the approximation introduced by Philip and Wooding[11]. These authors proceed by dividing the double layer into two parts, the boundary between them being characterized by the value of κr that makes $ze\psi/kT$

equal to one. For the inner region it is assumed that all co-ions are absent, so that the corresponding exponential term in the r.h.s. of the P.B. equation can be neglected. In passing, it can be mentioned that exclusion of co-ions from regions close to the central particle is quite a general phenomenon for any type of double layer; it is known as negative adsorption. As to the outer region the complete linearized equation is used. The system of equations thus arising can be solved analytically. Both Philip and Wooding[11] and v.d. Drift[12] have compared the results of this approximation with exact numerical solutions and arrived at the important conclusion that the available analytical solutions, although not being exact, have such a degree of accuracy that they can safely be used for subsequent purposes, e.g. in a theory of electrophoresis.

A discussion on polyelectrolytes as charged linear particles must also include the concepts of Manning[13]. Manning considers the particles as rods without a volume, thus reducing the charge to a line charge. It then appears that the linear charge density cannot exceed a certain value and any dissociation above that limit is offset by a "condensation" of counterions onto the line charge. The extent of this condensation is dependent on the charge number of the counterions and on the ionic strength[14]. The remaining cylindrical double layer may then be treated using the linearized Debye-Hückel theory.

Mannings theory aims in particular at the explanation of thermodynamic properties of polyelectrolytes and of the mobility of the counterions. So far it has not been used to estimate the electrophoretic mobility of the polyion.

2.4 Modifications of the Poisson-Boltzmann equation

In the preceding derivation and integrations of the P.B. equation a substantial number of assumptions has been made, e.g. that the ions are point charges and that the dielectric constant is independent of the concentration and the (often high) electrical field strength. These assumptions have been the subject of many critical objections, a review of which has been given by Bell and Levine[15]. From extensive theoretical studies these authors[15,16] conclude that many of the effects are mutually compensating and that for flat plates the corrections to the P.B. equation are not excessive for 1-1 electrolyte concentrations below 0.1 M and surface potentials not exceeding 75 mV. Another aspect deserving attention is the assumed homogeneous distribution of the surface charges. However, despite the actually discrete character of these charges and their often relatively large mutual separations, it can be shown by statistical mechanical arguments that the potential distribution in the double layer may be described by a continuum theory[17].

I.3 STERN-CORRECTION

Contrary to the assumption in diffuse double layer theory, ions do have a finite size and treating them as point charges is a simplification. This finite size shows up in the first place in preventing the center of the ions, so the center of the charge, to approach the interface to distances smaller than the ion radius. Therefore, in an actual situation a charge-free region is present between the surface charge and the diffuse charge in the solution. Taking this aspect into account eliminates the impossible consequence, implicit in simple diffuse double layer theory, that concentrations at the surface can grow as high as 200 M. This improvement was first introduced by Stern[18], who also pointed out that counterions not only interact with the surface by electrical forces, but frequently by chemical forces too, in which case specific adsorption is said to occur. The surface charge in then compensated by the sum of the charges in the diffuse and in the Stern-layer:

$$\sigma_o = -(\sigma_s + \sigma_d) \qquad \text{(I-21)}$$

Following Stern, one usually describes σ_s by a Langmuir isotherm. In its simplest form, when only specific adsorption of counterions is considered, it reads in the case of a negatively charged surface

$$\sigma_s = \frac{ze\Gamma_o}{1 + (1/Kc)\exp(ze\psi_d + \phi)/kT} \qquad \text{(I-22)}$$

Γ_o is the total number of available adsorption sites per unit area, K a proportionality constant and c the bulk concentration of the ions. ϕ is called the specific adsorption potential, although it is an energy-, rather than a potential term. ϕ may discriminate between ions of identical charge number, thus explaining such specific effects as the lyotropic order. Moreover, ϕ may be a function of the surface charge.

Stern's approach has since been refined, particularly by D.C. Grahame, who introduced the conceptions Inner and Outer Helmholtz Plane. The specifically adsorbed ions are in the I.H.P., the diffuse layer ends at the O.H.P. and these two planes are in general not identical. For a detailed treatment of these notions and of refined electrical double layer theories in general we refer to the monograph by Sparnaay[19].

Of course the presence of the Stern layer (frequently also called molecular condenser) has important consequences for the potential distribution in the solution. In fig. 3 the distribution has been sketched that goes along with the presence of merely a charge-free layer. It can be seen that across the molecular condenser the potential decays linearly from ψ_o to the value ψ_d at the boundary between Stern and Gouy layer. This implies that actually in all diffuse double layer equations ψ_o has to be replaced by ψ_d. Farther away from the surface than

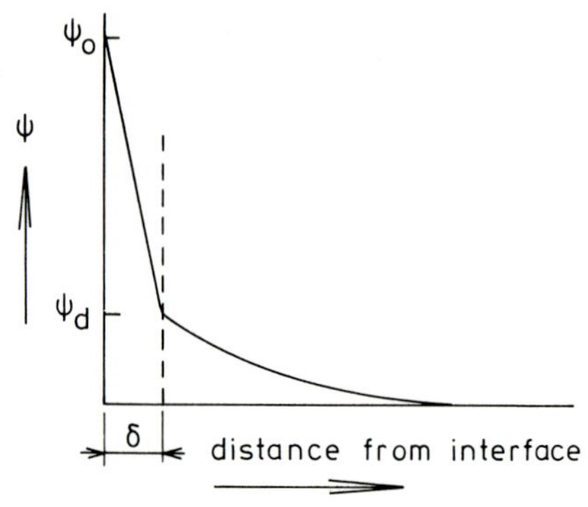

Fig. 3. Potential as a function of distance in the presence of a charge-free Stern layer of thickness δ and a diffuse layer for $x > \delta$.

ψ_d the character of the Gouy layer is not altered e.g. with regard to the compressing effect of κ.

The overall value of the capacitance of the double layer, C, in the presence of a molecular condenser results from the combination of the capacitances of Stern and Gouy layer placed in series:

$$(1/C) = (1/C_s) + (1/C_d) \qquad \text{(I-23)}$$

It can be inferred from (I-23) that C is dominated by the smallest of its two contributions. Because C_d is an increasing function of the concentration and of ψ_d (cf. eq. (I-17)), C will mainly reflect C_d at low values of ψ_d and c, and C_s at higher values of these parameters. In the latter case any dependence of C_s on σ_o will also become manifest.

I.4 EXPERIMENTAL VERIFICATION OF DOUBLE LAYER THEORIES

Double layer theories have been the subject of numerous investigations. The system most extensively studied in this respect has been the interface between mercury and aqueous electrolyte solutions. Experimentally either electrocapillary curves or differential capacitances were measured. In some studies both approaches were used and the results could be compared. For a review of all the conclusions to be drawn from these data we refer to Sparnaay[19] again. Suffice it to say that in general the Stern-Gouy-Chapman picture has been confirmed.

In the mercury-aqueous electrolyte system the surface charge and the surface potential are modified by applying an external voltage. In most colloidal systems, however, the surface charge is obtained by adsorption of ions or - in particular with proteins and other polyelectrolytes - by dissociation of surface groups. In favorable cases the charge may be determined by chemical analysis e.g. by titration techniques. Various oxides and the silver halides have been used for such studies on the double layer. Among the latter silver iodide is most outstanding, because it combines a relatively wide range of accessible potentials with the possibility to relate direct double layer data to data on colloidal stability and electrokinetics. In a recent review[20] it has been outlined that for silver iodide the basic ideas of current double layer theory are also applicable. At the same time it is observed that quantitative differences with mercury exist, that emphasize the importance of the specific nature of the Stern layer. In fig. 4 some data representative for silver iodide are reproduced. They illustrate, e.g., that at

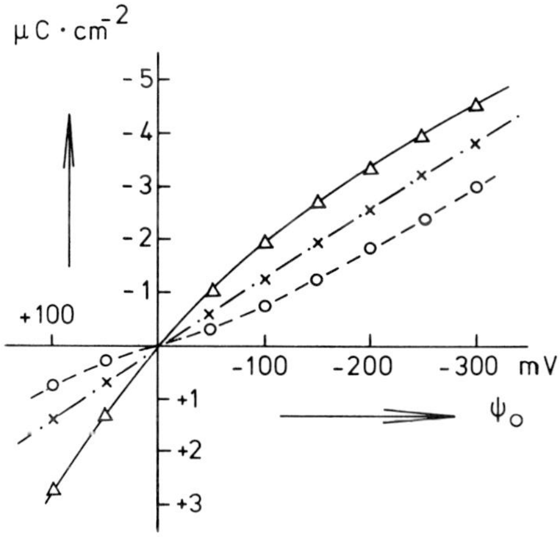

Fig. 4. Surface charge σ_o as a function of surface potential ψ_o for silver iodide in aqueous KNO_3 solutions.
Δ——Δ 0.1 M X—·—X 0.01 M O – – O 0.001 M

low ionic strength (10^{-3} M) and in the neighborhood of the point of zero charge σ_o progressively increases with ψ_o, in accordance with diffuse double layer theory (e.g. eq. (I-15)). At high ionic strength (10^{-1} M) the differential capacitance, being the slope of the curves, is usually much higher than at low ionic strength. Because C is now dominated by the Stern layer its variation reflects the dependence of C_s on σ_o as was anticipated in the discussion of section I.3.

In practical systems compression of the double layer results in an increase of the charge and/or a decrease of the potential. The precise conditions (e.g. the amount of potential determining electrolyte and the total surface area) dictate whether the surface charge or the surface potential remains constant, or that both quantities are adapted, upon changing the ionic strength of the solution.

I.5 THE POSITION OF THE SLIPPING PLANE IN ELECTROKINETICS

All electrokinetic phenomena deal with the tangential movement of particle and surrounding liquid. In discussing these processes it is important to know the exact position of the slipping plane relative to the two phases. Unfortunately, no unambiguous statements can be made as to this position because a complete theory on the slipping process is not available. However, from a vast amount of experimental data it may be concluded that the slipping plane does not coincide with the particle surface, but is situated at some distance in the solution. Apparently a stagnant layer prevents displacement of the liquid close to the surface. However, this qualitative notion still leaves us with the question about the *exact* position of the slipping plane and thus about the value of the potential (the ζ-potential) at this position. Considering double layer-, stability- and electrokinetic data on silver iodide, Lyklema[21] recently concluded that, although exact answers still cannot be given, it remains a reasonable assumption to set ζ equal to ψ_d.

The supposed equality of ζ and ψ_d of course implies that ζ is equally sensitive to changes in surface charge, surface potential and ionic strength as ψ_d is. So, here we have the justification for the ample attention given to diffuse double layer theory in the preceding sections. The relation between ζ and electrophoretic mobility will be discussed in the following sections.

II. THE THEORY OF ELECTROPHORESIS

II.1 INTRODUCTION

The most simple theoretical approach to electrophoresis would lead us to expect that the electrophoretic velocity, v, is equal to the force on the particle (charge, Q, times field strength, E) divided by the friction constant, f ($f = 6\pi\eta a$ for a sphere of radius a in a medium with viscosity η). Or, introducing the mobility, u, we would have

$$u \equiv \frac{v}{E} = \frac{Q}{f} \qquad \text{(II-1)}$$

This approach, however, neglects the fact that the countercharge, $-Q$, is often quite close to the particle surface and is subjected to a force, $-Q\times E$. By hydrodynamic interaction a substantial fraction of this counterforce is transmitted to the particle, thus reducing its electrophoretic mobility, often by one or more orders of magnitude. This *hydrodynamic* effect is called *electrophoretic retardation*. Moreover, since particle and countercharge move in opposite directions, and a certain time, the relaxation time τ, is required to replace the counterions that have moved away by new ones, the double layer is distorted. This distortion creates a field opposite to the applied field, E, and this *relaxation field* slows down the particle by *electric interaction*. The *relaxation effect* is usually not as large as the electrophoretic retardation, but a reduction of the mobility by 10-50% is not exceptional.

The main difficulty in the theory of electrophoresis is the correct calculation of these two retarding effects.

Theoretically, the electrophoretic mobility is expected to be independent of the field strength, as long as the latter remains below 10000 V cm^{-1}, and this is confirmed empirically. We shall limit our discussion to low field strengths.

II.2 ELECTROPHORESIS OF SPHERICAL PARTICLES

We shall deal with the electrophoresis of spheres fairly extensively, not because most particles are spherical (although emulsions, latices, micelles and some proteins offer good examples), but because this case has been studied so thoroughly. To begin with, we omit the relaxation effect (i.e. we treat the double layer as undistorted), but it will be included later in this section.

2.1 Method of Hückel and Onsager

Closely connected with his work with Debye[22] on the conductivity of strong electrolytes, Hückel[23] published a paper on the electrophoresis of the sphere. In this paper he solved the Navier-Stokes equations for the stationary motion of a sphere surrounded by an electrical double layer in an applied field, E, and found for the mobility, u,

$$u = \frac{2\varepsilon_o \varepsilon_r \zeta}{3\eta} \tag{II-2a}$$

or rather, since he used non-rationalized three quantity equations, without a separate electrical dimension,

$$u = \frac{\varepsilon_r \zeta}{6\pi\eta} \tag{II-2b}$$

In these equations ζ is the surface potential (more precisely, the potential at the slipping plane) and η is the viscosity of the liquid.

A few years later, Onsager[24] presented a simple and illuminating derivation of eq.(II-2) that we shall repeat here. The ionic atmosphere around the particle can be divided in concentric spherical shells of radius, r, and thickness, dr, as shown in figure 5. The field, E, exerts a force $4\pi r^2 \rho E \mathrm{d}r$ on the shell and this results in the shell moving with a velocity, v, with respect to its surroundings, but drawing

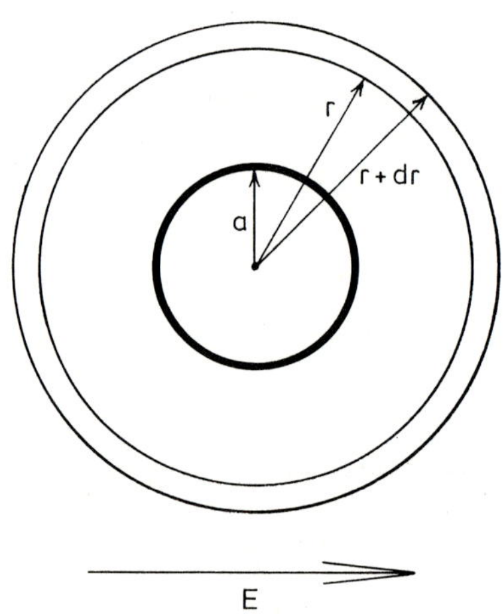

Fig. 5. Particle with radius a, surrounded by an ionic atmosphere, consisting of spherical shells, one of which is shown, in an electric field, E.

everything (liquid and particle) within it along. v is found by the application of Stokes' law.

$$v = \frac{4\pi r^2 \rho E \mathrm{d}r}{6\pi\eta r} \tag{II-3}$$

The total ionic atmosphere gives its innermost layer and thus the particle a velocity

$$v_{\mathrm{atm}} = \int_a^\infty \frac{4\pi r^2 \rho E \mathrm{d}r}{6\pi\eta r} \tag{II-4}$$

The field, acting on the particle charge, Q, gives the particle a velocity v_{part} with respect to its immediate surroundings

$$v_{\mathrm{part}} = \frac{QE}{6\pi\eta a} \tag{II-5}$$

The net velocity of the particle with respect to the liquid at infinity, v, is the sum of (II-4) and (II-5), which counteract one another, since ρ and Q have opposite signs. The atmosphere decreases the velocity of the particle.

$$v = \frac{QE}{6\pi\eta a} + \int_a^\infty \frac{4\pi r^2 \rho E \mathrm{d}r}{6\pi\eta r} \tag{II-6}$$

This expression is easily transformed into (II-2) by using

$$Q = -\int_a^\infty 4\pi r^2 \rho \mathrm{d}r \tag{II-7}$$

which expresses the electroneutrality of particle plus atmosphere and by applying Poisson's law for spherical symmetry:

$$\rho = -\varepsilon_r \varepsilon_o \nabla^2 \psi = -\frac{\varepsilon_r \varepsilon_o}{r} \frac{\partial^2}{\partial r^2}(r\psi) = -\frac{\varepsilon_r \varepsilon_o}{r^2} \frac{\partial}{\partial r}\left(r^2 \frac{\partial \psi}{\partial r}\right) . \tag{II-8}$$

The transformation runs as follows:

$$v = +\frac{2\varepsilon_r \varepsilon_o E}{3\eta a} \int_a^\infty \frac{\partial}{\partial r}\left(r^2 \frac{\partial \psi}{\partial r}\right) \mathrm{d}r - \frac{2\varepsilon_r \varepsilon_o E}{3\eta} \int_a^\infty \frac{\partial^2}{\partial r^2}(r\psi) \mathrm{d}r =$$

$$= \frac{2\varepsilon_r \varepsilon_o E}{3\eta}\left[-\frac{1}{a} a^2 \left(\frac{\partial \psi}{\partial r}\right)_a + \left(\frac{\partial}{\partial r}(r\psi)\right)_a\right] = +\frac{2\varepsilon_r \varepsilon_o \zeta E}{3\eta} \tag{II-9}$$

which proves eq.(II-2a).

Remarkably enough, neither the size of the particle nor the thickness of the double layer shows explicitly, if the mobility is expressed as a function of the surface potential, ζ.

2.2 Smoluchowski's equation

Already in 1903 and especially in 1914, von Smoluchowski[25] derived an equation very similar to Hückel's eq.(II-2b), but differing from it by having 4π instead of 6π in the denominator.

$$u\ (\text{Smol.}) = \frac{\varepsilon_r \zeta}{4\pi\eta} = \frac{\varepsilon_o \varepsilon_r \zeta}{\eta} \tag{II-10}$$

In his very elegant derivation, which applies to particles of arbitrary shape, Smoluchowski assumes that the double layer is thin compared to any dimension of the particles and that the applied field is distorted so as to run parallel to the particle-liquid interface at that interface and close to it (see fig.6).

As a consequence of the distortion of the field, the atmosphere is on the average in a field smaller than E and thus it retards the particle less than in Hückel's and Onsager's calculation.

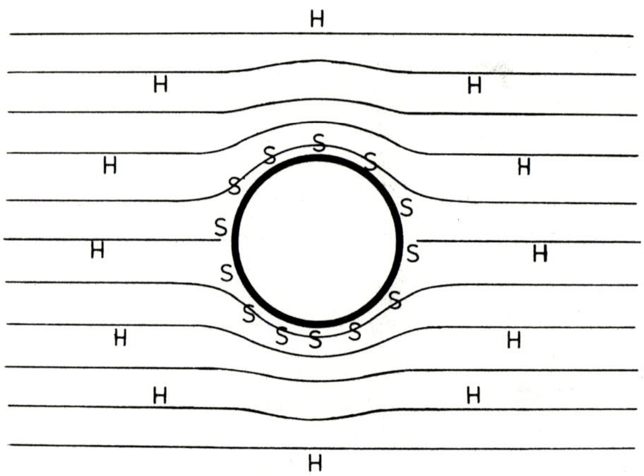

Fig. 6. Illustrating that a thin double layer (S for Smoluchowski, $1/\kappa << a$) is in the distortion region, whereas a thick double layer (H for Hückel, $1/\kappa >> a$) is mostly in the undisturbed field.

Eq.(II-10) with the sign reversed is also valid for electro-osmosis, as Smoluchowski demonstrated as follows. An electric field, E, is applied parallel to the interface between a solid and a liquid. The solid carries a surface charge compensated by a diffuse double layer in the liquid. We consider a thin layer of liquid of area, A, and thickness, $\mathrm{d}x$, parallel to the surface at a distance, x. The electric field E exerts a force, $A\rho E\mathrm{d}x$, on this layer. The viscous friction tends to retard the layer with a force, $-A\eta \mathrm{d}v/\mathrm{d}x$ at the plane x, whereas it tends to accelerate it with a force $+A\eta(\mathrm{d}v/\mathrm{d}x)_{x+\mathrm{d}x}$ at the plane $x+\mathrm{d}x$, where v is the velocity of the liquid parallel to the interface. In the stationary situation, the net force on the layer is zero, or

$$A\rho E\mathrm{d}x - A\eta(\mathrm{d}v/\mathrm{d}x)_x + A\eta(\mathrm{d}v/\mathrm{d}x)_{x+\mathrm{d}x} = 0 \quad \text{(II-11)}$$

$$\text{or: } A(\rho E\mathrm{d}x + \eta \frac{\mathrm{d}^2 v}{\mathrm{d}x^2}\mathrm{d}x) = 0 \quad \text{(II-12)}$$

With $\rho = -\varepsilon_r\varepsilon_o \mathrm{d}^2\psi/\mathrm{d}x^2$ we find:

$$\varepsilon_r\varepsilon_o E\mathrm{d}^2\psi/\mathrm{d}x^2 = \eta \mathrm{d}^2 v/\mathrm{d}x^2 \quad \text{(II-13)}$$

A first integration with $\mathrm{d}\psi/\mathrm{d}x = \mathrm{d}v/\mathrm{d}x = 0$ for $x \to \infty$ leads to

$$\varepsilon_r\varepsilon_o E\mathrm{d}\psi/\mathrm{d}x = \eta \mathrm{d}v/\mathrm{d}x \quad \text{(II-14)}$$

The second integration with $\psi = \zeta$ and $v = 0$ for $x = 0$ and $\psi = 0$ and $v = v$ (electro-osmosis) for $x \to \infty$ gives

$$\frac{v(\text{electro-osmosis})}{E} = -\frac{\varepsilon_r\varepsilon_o\zeta}{\eta} \quad \text{(II-15)}$$

2.3 Henry's contribution

A few years after the appearance of Hückel's paper[23], Henry[26] showed conclusively that the Smoluchowski approach is correct for thin double layers ($\kappa a >> 1$), and the Hückel-Onsager equation is correct for thick double layers ($\kappa a << 1$). Henry bridged the gap and gave equations for intermediate values of κa both for spheres and for cylinders, using the linearized Debye-Hückel approximation (eqs.I-7,I-19) for the structure of the double layer. For that reason, and because he neglected the relaxation effect, his results are only valid for small values of ζ. Figure 6 illustrates the different influences of the field distortion for thin and for thick double layers, and figure 7 gives a graphic representation of Henry's results.

Table II gives a few numerical values of $3u\eta/2\varepsilon_r\varepsilon_o\zeta$ for the non-conducting sphere.

TABLE II

$f(\kappa a) = 3u(\text{sphere})\ \eta/2\varepsilon_r\varepsilon_o\zeta$ as a function of κa

κa	0.1	0.3	1	3	5	10	20	50	100	1000
$f(\kappa a)$	1.000545	1.00398	1.0267	1.1005	1.163	1.25	1.34	1.424	1.458	1.495

The cases

a) no field distortion or conductivity of sphere and medium are equal, and

b) the perfectly conducting sphere

have hardly any real practical significance, since the smallest amount of polarization is sufficient to prevent the current to cross the particle medium interface.

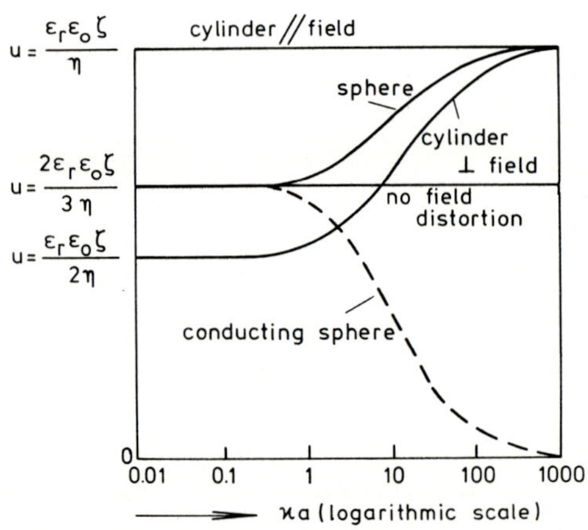

Fig. 7. Electrophoretic mobility for spheres and cylinders as a function of log(κa). Debye-Hückel linearization (eq. I-7) used.

2.4 The relaxation effect

So far, we have neglected the distortion of the ionic atmosphere, which leads to the relaxation effect. Figure 8 illustrates how an originally symmetric atmosphere, which is forced to move with respect to the particle, is dispersed at one side and rebuilt at the other, but with a time lag, which causes the atmosphere to become asymmetric, creating a field, ΔE, opposed in direction to the applied field E and slowing down the particle.

The actual calculation of the relaxation field is difficult. It already formed one of the more difficult aspects in the Debye-Hückel theory, where it was possible to consider the ions as point charges.

The great difference in size between the particle and the atmospheric ions and the necessity to go beyond the linear approximation of the P.B. equation make the problem all the more difficult. In the nineteen forties, Booth[27] and Overbeek[28] formulated solutions in the form of expansions in a power series of increasing powers of $ze\zeta/kT$, which showed that the relaxation effect is particularly important in the region $1 < \kappa a < 10$ and that it vanishes at low and at high values of κa. Later, Wiersema et al.[29] applied a computerized numerical procedure to solve eight simultaneous differential equations (three components of the liquid velocity, pressure in the liquid, concentrations of cations and anions and local potential

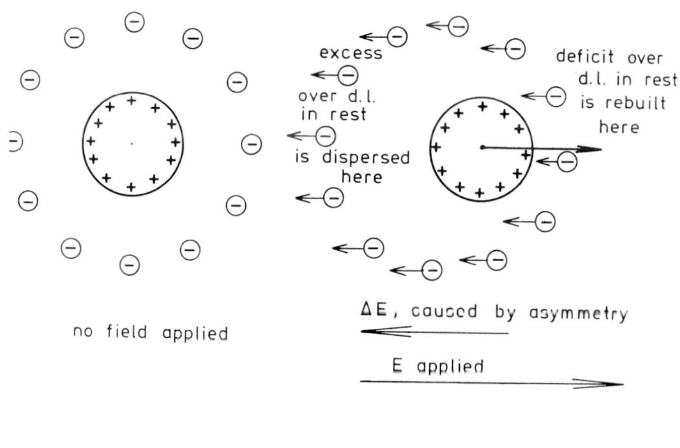

Fig. 8. Illustrating schematically the formation of the relaxation field, ΔE.

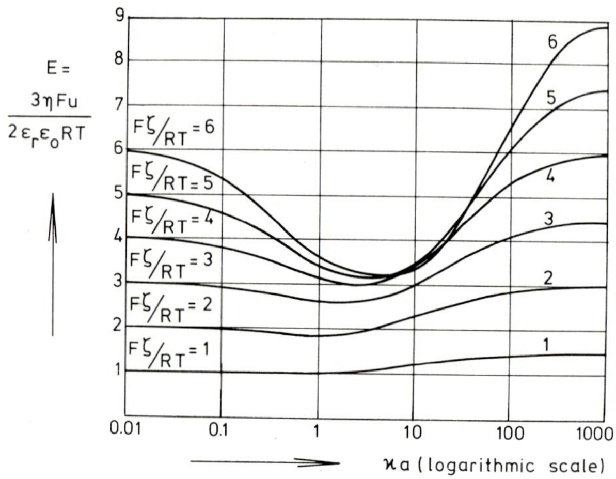

Fig. 9. Dimensionless electrophoretic mobility, E, as a function of κa for a series of values of the ζ-potential. The electrophoretic mobility $u = \frac{2\varepsilon_r\varepsilon_o RT}{3\eta F} E$. The curves are given for $\zeta/(RT/F) = \zeta/(kT/e) = 1, 2, 3, 4, 5$ and 6.

inside and outside the sphere, using the complete P.B. equation (I-6)) in successive approximations. They confirmed the results of Booth and Overbeek for low values of ζ and extended the calculations to $z\zeta \sim 150$ mV. Figure 9 gives their results for particles embedded in 1 - 1 electrolytes of conductivity equal to that of KCl. The influence of the ionic mobility proved to be of only secondary importance. For other charge numbers of the electrolyte and for other ionic mobilities, the reader is referred to the original paper[29] or to a more condensed version[30]. It is obvious from figure 9 that at high and at low κa the mobility and ζ-potential are nearly proportional and that this is also true at intermediate values of κa for low mobilities and low ζ (say $\zeta < 1.5 \frac{kT}{e} \sim 40$ mV). But for high ζ-potentials and intermediate values of κa (say $2 < \kappa a < 50$) the mobilities become nearly independent of ζ. In that region electrophoresis is not a good method to obtain information on surface charge or surface potential.

In most practical situations, it is essential to take both electrophoretic retardation and relaxation effect into account, although it is often possible to use analytical approximations instead of the computer values of figure 9, see [29].

2.5 Stern layer and slipping plane

It should be emphasized that all models treated above describe the particle as a rigid sphere, with a double layer containing point ions which may approach the surface infinitely closely. A more realistic model with finite size ions leaving a "Stern layer" of thickness, δ, free of charge, is still covered by the calculations, if it is assumed that the slipping plane (surface of shear) coincides with the O.H.P. at a distance δ from the surface and that the layer between the O.H.P. and the surface does not contain any mobile ions. In that case, ζ-potential and Stern-potential are identical. For locations of the slipping plane further outside than the O.H.P., electrophoresis leads to a ζ-potential which is lower than ψ_δ, but if the slipping plane is closer to the surface than the O.H.P., the ζ-potential cannot be found from Wiersema's theory since it does not take a charge free layer that is still mobile into account.

II.3 ELECTROPHORESIS OF CYLINDRICAL PARTICLES

When considering the electrophoresis of cylindrical particles, two facts should be kept in mind. In the first place, orientation is important, and secondly, most considerations about cylinders neglect end-effects, the results being obtained by considering infinitely long cylinders. The cylinder may be an acceptable model for rod-like particles such as tobacco mosaic virus, and for stiff or flexible long chain polyelectrolytes for which DNA and polyacrylic or polymethacrylic acid are examples.

3.1 Neglecting the relaxation effect

Henry[26], again neglecting the relaxation effect, solved the hydrodynamic equations for two orientations of the cylinder, viz. parallel or perpendicular to the applied field. His results are given in figure 7 for cylinders of radius a. For a cylinder parallel to the field, the field strength has the value E and is parallel to the surface everywhere in the solution. Therefore, Smoluchowski's equation (II-10) applies, irrespective of the value of κa. The field around a cylinder in perpendicular orientation is distorted in a similar way as sketched in figure 6, and this results in an electrophoretic mobility that depends on κa, with the Smoluchowski value (II-10) at high κa and only half that value at low κa.

$$u\ (\text{cyl.} \perp E)_{\kappa a \to 0} = \frac{\varepsilon_r \zeta}{8\pi\eta} = \frac{\varepsilon_0 \varepsilon_r \zeta}{2\eta} \qquad \text{(II-16)}$$

It should further be realized, that at the extreme values of κa eqs. (II-2), (II-10) and (II-16) are valid, even if ζ is large ($|e\zeta/kT| > 1$) and that this is also true for the cylinder parallel to the field at any κa. But the transition for intermediate values of κa, as given in fig. 7, has been found by Henry using the linear Debye-Hückel approximation and is, therefore, not strictly valid for high ζ. However, the necessary corrections are smaller than those for the relaxation effect.

3.2 Averaging over different orientations

For a cylinder oriented at an arbitrary angle, θ, between its axis and the field, the field may be split into two components; one, $E\cos\theta$, parallel to the cylinder axis and one, $E\sin\theta$, perpendicular to it. Each of these components results in a velocity of the cylinder, one parallel and the other perpendicular to the axis. Vectorial addition of these two velocity components gives the net velocity of the cylinder and when this velocity is averaged over a random distribution of orientations, a simple result is obtained, viz.,

$$u \text{ (average over all orientations)} = \frac{1}{3}u_{//} + \frac{2}{3}u_{\perp} \qquad \text{(II-17)}$$

where $u_{//}$ and $u_{\perp}$ are the mobilities for parallel and perpendicular orientation resp.. De Keizer, Van der Drift and Overbeek[31] have shown this relation to be valid in the absence of relaxation. Stigter[32] has extended the proof so as to include the relaxation effect.

At low field strengths, as long as Ohm's law is obeyed, the field does not orient the cylinder against the randomizing effect of Brownian motion. Similarly, a flexible polyelectrolyte coil may be considered as a collection of cylindrical pieces with on the average random orientation.

3.3 Relaxation effect for cylinders

Stigter[32], using an approach very similar to that of Wiersema et al.[29], computed the mobility of the cylinder perpendicular to the field, including the relaxation effect. He further concluded that for parallel orientation the relaxation effect may be neglected, and then using eq.(II-17) applied his results to polyelectrolytes such as DNA, PAA and PMA. In the middle range of κa he obtained relaxation effects as high as 20%.

Shortly before, Van der Drift[12] had used Philips' and Wooding's[11] description of the cylindrical double layer at high potentials ($e\zeta/kT > 1$) to improve Henry's mobility values, but he applied a semi-empirical method (Möller et al.[33]) based on the essential equality of the relaxation force on the particle and on the charge in its atmosphere, to correct for the relaxation effect. For half-ionized PMA, Van der Drift found relaxation corrections of 40-60%.

We conclude that until Stigter's results[32] are completely published, the theoretical interpretation of the electrophoretic mobility of cylindrical particles must be limited to low ζ-potentials, using eq.(II-17) for averaging over orientations, and using the numerical values obtained by Abramson, Gorin and Moyer[34] for Henry's equation[26] for the cylinder perpendicular to the field.

II.4 ELECTROPHORESIS OF THE POROUS SPHERE

The electrophoresis of polyelectrolytes has been interpreted in terms of the randomly oriented cylinder as was mentioned in the previous section. An older model is that of the porous sphere. The porous sphere, with the solvent flowing more or less freely through it, was quite a successful model for hydrodynamic properties (viscosity, diffusion, sedimentation) of solutions of uncharged polymers. The model has been extended to polyelectrolytes, the assumptions being made of a uniform or Gaussian distribution of the fixed charges, a silimar distribution of the chain elements, each element having the same friction factor with respect to liquid flowing past it and a Debye-Hückel type interaction between the free, low molecular weight ions and the fixed charges.

If the relaxation effect and the deformation of the applied field by the difference in conductivity inside and outside the porous sphere are neglected, the further mathematical treatment is simple. Overbeek and Stigter[35] obtained the result that for small κR (R is the radius of the sphere) the countercharge has no influence on the mobility and the sphere moves with the friction factor of a partially drained coil (Debye and Bueche[36], Brinkman[37])). For large κR, the sphere behaves as freely drained; its friction factor is the sum of the friction factors, f, of the chain elements and the counterions move through the porous sphere with the same velocity that they would have in free solution. Figure 10 shows the transition between the two cases.

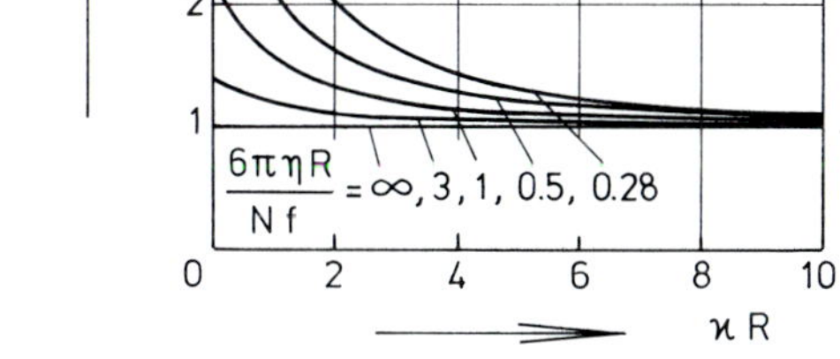

Fig. 10. Electrophoretic mobility of a porous sphere with radius R, charge Ze, containing N elements, each with a friction constant, f.

Although the model is in qualitative agreement with experiments, close consideration[12] shows that it is only realistic for compact, highly crosslinked polymers. The usual polyelectrolyte coils are so open, that a uniform distribution of chain elements, fixed charges and countercharges is quite unrealistic. A much better model is that of a loosely coiled cylinder surrounded by a fairly compact double layer as shown schematically in figure 11.

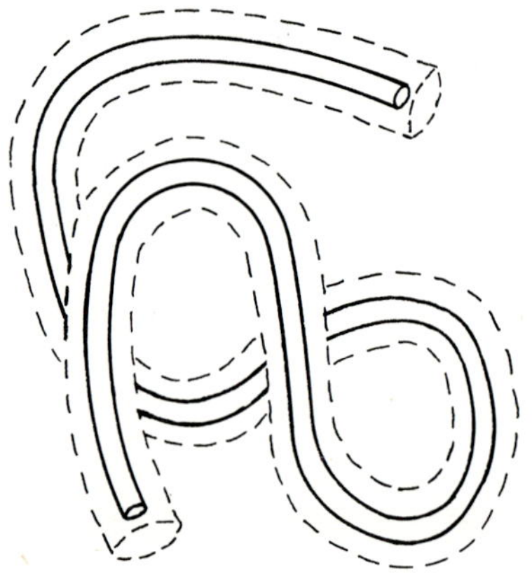

Fig. 11. Coiled cylinder surrounded by a double layer.

II.5 PARTICLES OF ARBITRARY SHAPE

After the discussion of the difficulties arising in the formulation of an electrophoresis theory for three rather simple models, one might wonder whether any useful interpretation can be given for particles of arbitrary shape. In fact, however, relatively simple interpretations are possible, at least for the limits of very high and very low values of κ (i.e. of the electrolyte concentration).

5.1 Small κ

If the largest linear dimension, b, of a particle is much smaller than $1/\kappa$, or in other words $\kappa b \ll 1$, ionic atmosphere effects, both electrophoretic and relaxation retardation, are bound to be minor, and the mobility is simply equal to the charge, Ze, divided by the friction factor, f,

$$u = \frac{Ze}{f} \tag{II-18}$$

The same friction factor can be found from rates of diffusion or sedimentation, and since the charge may be determined by analysis, eq.(II-18) can be tested or be used for the calculation of one of its three elements from the two others. In aqueous media, however, it is difficult to decrease κ below $(100\ \text{nm})^{-1}$, and thus the case $\kappa b \ll 1$ only applies to small particles (say < 10 nm or even < 1 nm). In non-polar media, where $1/\kappa$ may easily be several µm, eq.(II-18) is the obvious one for the interpretation of electrophoresis.

5.2 Large κ

In the context of irregularly shaped particles, large κ means $\kappa a \gg 1$, where a is the smallest local radius of curvature of the particle surface. In that case, the Smoluchowski equation (II-10) applies, even at large charge and ζ-potential. Overbeek[38] has shown that under these circumstances the relaxation effect becomes negligible (see also our figures 7 and 9). The electrophoretic retardation, however, is then quite large, slowing down the mobility as compared to eq.(II-18) in the ratio of about $1/\kappa b$. It is in particular under these conditions of large particle size and/or thin double layers, that the ζ-potential has such a unique position in the interpretation of electrophoresis.

II.6 ELECTROPHORESIS AS A FUNCTION OF THE CHARGE OF THE PARTICLES AND OF THE POTENTIAL DIFFERENCE BETWEEN THE PHASES

So far, we have interpreted the electrophoretic mobility nearly exclusively in terms of the ζ-potential, with some correction for relaxation. Unfortunately, there is no independent way to measure ζ and no way either to determine the exact location of the slipping plane. So it should be asked whether an interpretation in terms of the surface charge, rather than the ζ-potential, might not have advantages. One obvious advantage is the fact that the particle charge is carried by ions, and thus can be determined in principle by chemical analysis. Moreover, the charge can be manipulated by addition of easily adsorbable ions, by a change in pH, or by chemical reaction. On the other hand, the charge as determined analytically, q_1, and the electrokinetic charge, q_2, need not be identical, and in general are not identical, the difference being formed by that part of the atmospheric charge that is situated between the surface and the slipping plane. Nevertheless, a comparison between q_1 and q_2 remains of interest, since it gives at least some information on the location of the slipping plane. At low ionic strength, the difference between q_1 and q_2 may be expected to be small, but at high ionic strength it may be large due to the compression of the double layer and the steep decay of potential with distance at high ionic strength. Figure 12 illustrates this.

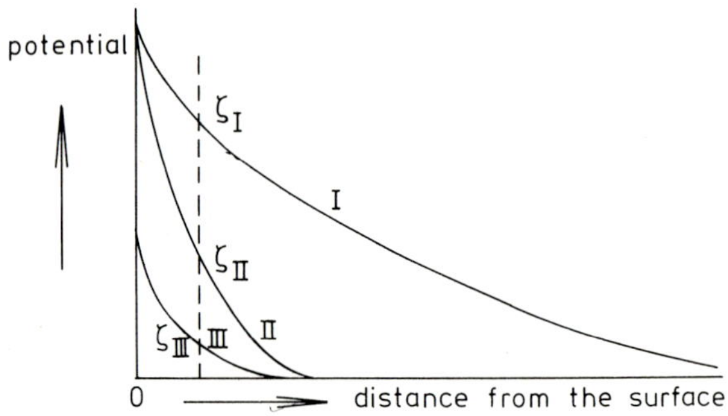

Fig. 12. Illustrating the influence of compression of the double layer on ζ. The dotted line represents the slipping plane.
I. Low ionic strength
II. High ionic strength. Surface potential, ψ_o, as in I.
III. High ionic strength. Surface charge, q_1, as in I.
The surface charge density is proportional to the slope of the potential — distance line at $x = 0$ (eq.I-14). It can be seen that $q_2(\mathrm{I}) \sim q_1(\mathrm{I})$ but $q_2(\mathrm{II,III}) < q_1(\mathrm{II,III})$.

The ζ-potential may also be compared to the surface potential, ψ_o, which in many cases is equal to the potential difference between the particle phase and the solution phase, as it contributes to the e.m.f. of a galvanic cell. Since the absolute value of such a potential difference is not accessible to measurement, the point of zero charge should be taken as the point where ψ_o and ζ are both zero. Such a comparison, if made for a number of ionic strengths, may show whether the assumption that $\zeta \sim \psi_d$ (the Stern-potential) is justified (cf. Lyklema[21]).

Not enough work has been aimed in this direction, and what has been done has not always been too successfully interpreted. One conclusion might be that the slipping plane shifts away from the surface with decreasing ionic strength.

The electrophoretic mobility and the ζ-potential reflect changes in ψ_o, q_1, κ, and also in the detailed composition of the solution, since adsorption of ions or neutral substances may affect ζ.

On the whole, changes in the surface potential ψ_o are reflected somewhat more weakly (say as a fraction between 0.5 and 1) in ζ. Increases in the ionic strength lead to a pronounced decrease in ζ, the more so the higher the charge number of the counterions, this on account of the term $\exp(-z_i e\psi/kT)$ in the P.B. equation (eq.I-1,I-6).

Changes in the ζ-potential may further be used to determine the presence and the extent of adsorption. Adsorption of ions changes the surface charge directly, adsorption of neutral substances may lead to a change in the charge as a secondary effect and may also shift the location of the slipping plane.

II-7 APPLICATIONS

At least two types of applications of the theoretical knowledge on electrophoresis may be envisaged, viz. a. the comparison of calculated and experimental values of the mobility, with the aim of obtaining a better understanding of the size and shape of the particle and especially of the quantitative structure of its double layer, and b. the improvement of electrophoretic separations by manipulating the mobilities.

7.1 The interpretation of electrophoretic mobility in terms of properties of particles or molecules and of their double layers

The conversion of the electrophoretic mobility into a ζ-potential or an electrophoretic charge can be used to check the validity of certain models of particles or molecules, e.g. to check whether the rheology of DNA is sufficiently well described by a cylinder model with a suitably chosen radius. Adsorption of ions changes the charge of the particle and adsorption isotherms can be obtained from electrophoresis. It has been observed that the adsorption of proteins on a variety of particles often changes the electrophoretic mobility from its original value for the bare particles into one very similar to that of the free protein, indicating a shift of the slipping plane to a position comparable to that of the free protein.

Such applications require absolute values of the mobility. Consequently, they cannot be based on paper- or gel-electrophoresis, where electro-osmosis, adsorption of the particles and the incluence of tortuosity are hard to quantify. Moreover, a series of values of u at different compositions of the solution is required to disentangle the influence of the Stern-layer, of adsorption of ions, of the location of the slipping plane, etc.

Although this type of application of electrophoresis has led to interesting and useful results, even in the most favorable cases doubts remain, connected with the uncertainty in the location of the slipping plane (cf. Lyklema[39]).

7.2 Improvement of electrophoretic separations

In electrophoretic separation processes, we use the difference in mobilities between different types of particles. For better separations we are interested in increasing the absolute value of the differences between mobilities, although sometimes the relative differences are also important.

To begin with, we review the essential relations. The mobility is nearly proportional to the ζ-potential with a relatively modest correction for relaxation, largely determined by κ, at least when ζ is not too large. For high ζ and intermediate values of κa, the mobility becomes quite insensitive to ζ. ζ itself increases less than proportionally with the charge, with a very strong influence of κ. Size and shape of the particles have a relatively minor influence on the mobility for a given surface charge density. The temperature affects the mobility mainly through its influence on ε and η, but the effect will not be different for different particles.

The particle charge depends strongly on the activity of potential determining ions, in many cases the pH, and occasionally on the specific adsorption of ions, e.g. detergent ions.

Electrophoretic separations are based upon a natural difference in surface charge density and ζ-potential between different types of particles. We want to enhance these differences by a suitable choice of the solution.

Separations are carried out preferably at moderate to low ionic strength, since high ionic strengths combine the drawbacks of low mobilities, high heat production, evaporation and convections.

At *moderate ionic strength*, say 0.1 M, $1/\kappa$ is about 1 nm, in many cases $\kappa a > 1$, ζ-potentials are fairly low, the mobilities and the differences between them are not very large, but for proteins and other biopolymers a moderate ionic strength corresponds to the natural environment and may be essential for preventing denaturation.

On changing the pH or the activity of other potential determining ions, B, usually one or more fairly narrow regions of pH or pB can be found, where surface charge and potential change rapidly. We mention the region around the pK of acidic or basic surface groups or the region around the p.z.c. of metal oxides, hydroxides, or silver halides. Good conditions for separation are often found near one of the ends of such a region of steep change.

At *low ionic strength* ζ-potentials and mobilities are higher. In general this leads to larger differences in mobility, but it should be clear from figure 9 that the relaxation effect has a kind of buffering action on the mobilities at high ζ-potentials in a fairly broad range of κa-values. Nevertheless, if other conditions (such as avoiding denaturation) permit, electrophoresis at low ionic strength has the additional advantage of low ohmic losses and thus the possibility of applying high voltages.

In choosing the best circumstances for separation, the qualitative rather than the precise quantitative aspects of the theory are important, but to use these at greatest advantage, the general features and the orders of magnitude involved in the theory should be well understood.

REFERENCES

1. Reuss, F.F. (1809), Mémoires de la Société Impériale des Naturalistes de Moscou 2, 327.
2. Gouy, G. (1910), J. Phys. 9, 457.
3. Chapman, D.L. (1913), Phil. Mag. 25, 475.
4. Verwey, E.J.W. and Overbeek, J.Th.G. (1948), Theory of the Stability of Lyophobic Colloids, Elsevier, Amsterdam.
5. Loeb, A.L., Overbeek, J.Th.G. and Wiersema, P.H. (1961), The Electrical Double Layer around a Spherical Colloid Particle, M.I.T. Press, Cambridge, Mass.
6. Stigter, D. (1972), J. Electroanal. Chem. 37, 61-64.
7. White, L.R. (1977), J. Chem. Soc. Faraday II 73, 577-596.
8. Mille, M. and Vanderkooi, C. (1977), J. Colloid Interface Sci. 59, 211-224.
9. Dube, G.P. (1943), Indian J. Phys. 17, 189.
10. Gorin, M.H. (1942) in Electrophoresis of Proteins (Abramson, H.A., Moyer, L.S. and Gorin, M.H., eds.), Reinhold, New York, p. 126.
11. Philip, J.R. and Wooding, R.A. (1970), J. Chem. Phys. 52, 953.
12. Van der Drift, W.P.J.T. (1975), Thesis, State University of Utrecht.
13. Manning, G.S. (1974) in Polyelectrolytes (Sélègny, E., ed.), Reidel, Dordrecht, pp. 9-37, and other references given there.
14. Manning, G.S. (1977), Biophys. Chem. 7, 95-102.
15. Bell, G.M. and Levine, S. (1966) in Chemical Physics of Ionic Solutions (Conway, B.E. and Barradas, R.G., eds.), Wiley, New York, pp. 409-454.
16. Levine, S. and Bell, G.M. (1966), Disc. Faraday Soc. 42, 69-80.
17. Levine, S. (1965), Kolloid-Z.u.Z.Polymere 205, 100.
18. Stern, O. (1924), Z. Elektrochemie 30, 508.
19. Sparnaay, M.J. (1972), The Electrical Double Layer, Pergamon Press, Oxford.
20. Bijsterbosch, B.H. and Lyklema, J. (1978), Advances Colloid and Interface Science 9 no.3.
21. Lyklema, J. (1977), J. Colloid Interface Sci. 58, 242-250.
22. Debye, P. and Hückel, E. (1923), Physik. Z. 24, 305.
23. Hückel, E. (1924), Physik. Z. 25, 204.
24. Onsager, L. (1926), Physik. Z. 27, 388.
25. Smoluchowski, M. von (1903), Bull.Intern.Acad.Sci. Cracovic, p.184; (1914) in Graetz, Handbuch der Elektrizität und des Magnetismus Vol.II, p.366.
26. Henry, D.C. (1931), Proc. Roy. Soc. London A133, 106.
27. Booth, F. (1948), Nature 161, 83; (1950), Proc. Roy. Soc. London A203, 514.
28. Overbeek, J.Th.G. (1941), Thesis, State University of Utrecht; (1943), Kolloid Beihefte 54, 287.
29. Wiersema, P.H. (1964), Thesis, State University of Utrecht.
Wiersema, P.H., Loeb, A.L. and Overbeek, J.Th.G. (1966), J. Colloid Interface Sci. 22, 78-99.

30. Overbeek, J.Th.G. and Wiersema, P.H. (1966), in Electrophoresis. Theory, Methods and Applications, Vol.2, Bier, M., ed., Ac.Press, New York, London, pp. 1-52.

31. Keizer, A. de, Van der Drift, W.P.J.T. and Overbeek, J.Th.G. (1975), Biophys. Chemistry 3, 107-108.

32. Stigter, D., Private communication, to be published in J. Phys. Chem.

33. Möller, W.J.H.M., Os, G.A.J. van and Overbeek, J.Th.G. (1961), Trans. Faraday Soc. 57, 325.

34. Abramson, H.A., Gorin, M.H. and Moyer, L.S. (1939), Chem. Revs. 24, 345.

35. Overbeek, J.Th.G. and Stigter, D. (1956), Rec. Trav. Chim. 75, 543.

36. Debye, P. and Bueche, A.M. (1948), J. Chem. Phys. 16, 573.

37. Brinkman, H.C. (1947), Proc. Koninkl.Nederl.Akad.Wetensch. 50, 618, 821.

38. Overbeek, J.Th.G. (1946), Philips Res. Reports 1, 315-319.

39. Lyklema, J. (1976), Pure and Appl. Chem. 46, 149.

© 1979 Elsevier/North-Holland Biomedical Press
Electrokinetic Separation Methods,
P.G. Righetti, C.J. van Oss and J.W. Vanderhoff

ZONE ELECTROPHORESIS ON PAPER, THIN LAYERS AND PEVIKON BLOCK

IVOR SMITH
Courtauld Institute of Biochemistry, Middlesex Hospital Medical School, London W1P 7PN, England

INTRODUCTION

Any ion in solution, be it as small as a hydrogen ion or as large as a protein or nucleic acid ion, will migrate when placed in an electric field. In order to make use of this property to separate the components of a mixture, two practical points must be observed. First, the mixture of ions must be applied in as narrow a band or zone as possible to allow the maximum distance between the electrodes to be available for separation of the component ions. Second, a method must be available to prevent rediffusion i.e. remixing, once the experiment is complete. The solution of these two problems made available a technique which was simple to perform, so inexpensive that it was available to all and allowed the investigation of a host of problems that were previously either extremely difficult research problems or quite impossible for routine investigation.

In principle, as shown in figures 1 and 2, a strip of paper was moistened with buffer and suspended either in the form of an inverted V or horizontally with its two ends dipping into separate solutions of buffer, and a solution of the mixture to be separated was applied as a narrow zone at an appropriate place on the paper. Current was applied for a period of time, which was determined empirically, until a suitable separation had occurred. The strip was immediately removed and dried thereby fixing the separated components on the paper. Subsequently some chemical reagent which reacted with the components of interest was applied and their positions were indicated by the appearance of colored or stained bands. This technique was named paper electrophoresis and was first applied to the separation of serum proteins. Although it was, and still is, an extremely powerful technique, it had some disadvantages and so other stabilising media which would replace the paper were sought. Cellulose acetate, a much less polar medium, and cellulose, silica gel and polyamide thin layers have all been found of great value for electrophoresis and have displaced paper for much of this type of separation.

These media were used mainly analytically and applied to the separation of a whole host of different families of ionisable compounds.

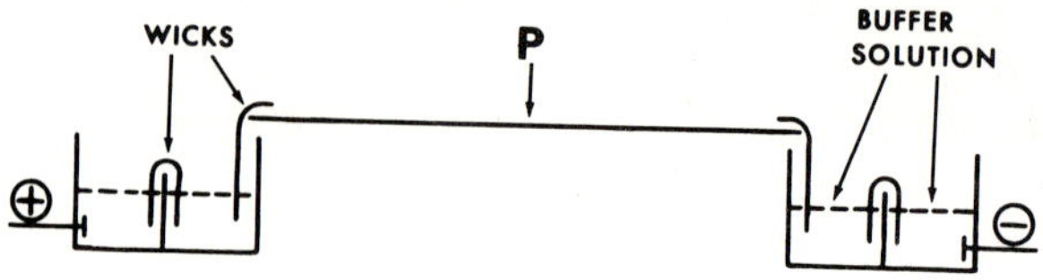

Fig. 1. Current flows through the buffer, paper wicks connecting buffer to the electrophoresis medium, P, and across the medium. It can be seen that the voltage drop across P may be much less than across the whole apparatus. Each electrode vessel is shown as a two compartment unit but many modern electrode vessels are now single compartment units.

However methods were also sought which would enable the technique to be used preparatively on the milligram and, later, on the gram scale. It was found that paper sheets and thin layers could be used to separate milligram quantities but another inert medium, namely Pevikon, could be suspended in buffer and formed into thick, wide blocks which were capable of allowing the separation of up to gram amounts.

Conditions affecting separation[1,2,3]

The stabilising medium. In principle, the paper, cellulose acetate, thin layer or pevikon block can each be considered as an inert medium whose sole functions are to hold the buffer during the electrophoresis, and then to allow the buffer to be removed and quickly dried so that the separated components become fixed on the medium. Thus separations on any of these media will indicate the type of separation to be expected on any other of the media and this is particularly useful when going from an analytical to a preparative procedure. Of course, due to the different particle and pore sizes, the rates of migration will be different and there will be minor changes in relative mobility but these are of little consequence.

pH, buffer composition and related phenomena. Because electrolysis occurs at the electrodes it is essential to use a buffer adequate to negate the reverse flow of ions which would change the pH and so alter the rate of migration as an ion will only migrate reproducibly under controlled conditions in which its charge remains constant. Buffer ions will also migrate and therefore their concentration must be adjusted to a reasonably low level in order that the rate of migration of the ions of interest is not reduced too much. Buffer ions may also play a more positive role and, for example, phosphate buffers are well known to affect protein mobility whereas borate buffers will affect the migration of glycoproteins and also cause neutral sugars to form sugar borate ions which will then migrate freely. As these techniques are now so well known, excellent buffers of defined composition are available for all the common families of compounds such as proteins, nucleic acids, peptides, nucleotides, aminoacids, dyes, simple and complex metal ions, etc.[2]

Voltage and heating effects. As the potential drop (voltage) between the electrodes increases, the rate of migration will increase but, unfortunately, so will the heat generated in the medium. Distillation will then occur from the medium, causing replacement buffer to flow in with consequent increase in buffer salt concentration. With low voltages such as 5-10 volts per cm drop along the medium this is of little consequence but with increasing drops of from 10 to 100 volts per cm precautions must be taken to overcome this. Such precautions include the use of organic rather than inorganic buffers because of their much lower conductivity and, at the higher ranges, the use of a cooling platen or sandwich which also prevents evaporation. As higher voltages are lethal all modern apparatus designed for this type of work includes a power cut-out which operates automatically should the apparatus be inadvertantly opened when in use. Although in general the rate of migration of an ion increases with the applied voltage this is not so for proteins which appear to denature for reasons which are not understood and so these types of compound are always separated at voltages not exceeding 10-20 volts per cm.[5] It must be remembered that the voltage shown on the power supply voltmeter is by no means the potential drop across the actual medium. Voltage loss occurs through the liquid buffer between each electrode and the medium and also along any paper wicks which may be used, so the true voltage across the medium must be measured, if desired, by means of a separate voltmeter whose probes are applied to each edge of the medium itself.

Other factors. Electro-endosmosis, syphoning and diffusion phenomena are discussed elsewhere in this book and so will not be considered here.

Two way separations. The above discussion has dealt with the separation of the components of a mixture in one way or dimension, i.e. between the path connecting the electrodes. This is perfectly adequate for the separation of serum proteins on paper or acetate as five clear, interpretable bands are observed and variations of concentration within these five bands are diagnostically significant of pathological changes. However, for the vast majority of separations the number of components of interest may vary from seven or eight up to thirty to fifty in such studies as fingerprinting of proteins and t-RNA's and even up to a hundred or more in other studies. As it is impossible to separate such complex mixtures by any one way procedure, recourse is made to a two way procedure. Here the electrophoresis separation is carried out along one edge of a large sheet which is then dried and chromatography is run at right angles to the original separation thus providing a two way or two dimensional separation, variations on this procedure will be described later. As chromatographic techniques are not within the scope of this chapter, readers unfamiliar with these are referred to reference 2.

PAPER ELECTROPHORESIS AT LOW VOLTAGE[1, 2, 3, 4]

This was the first of the electrophoresis techniques to be described independently and simultaneously by Flynn & de Mayo, and by Durrum. Apparatus similar to that which they described is shown in figure 2 and is still in common use. Strips or sheets of paper, such as Whatman No 1 or 3MM, of appropriate dimensions are taken and one or more origins are marked in pencil. Origins can be placed at the center of the paper if a mixture of anions and cations is to be separated or near one end if, say, only anions are to be separated. A suitable buffer is applied to the paper either by dipping the paper, if it is not too large and unwieldy, through the buffer or by laying larger sheets flat on a pad of absorbent paper and applying the buffer as evenly as possible with the aid of a shaving brush. Subsequently the paper is well blotted. It may then be placed in the apparatus and the sample applied. Alternatively the origin line is raised by placing two glass rods underneath and on either side of it, a small volume of sample is applied and the paper is now placed in the apparatus. If it is necessary to apply large volumes of samples of stable compounds, such as amino acids, they can be applied to the dry paper with intermittent drying and the main area of the paper can then be wetted as described above but leaving about 0.5cm on either side of the origin in a dry state. The paper is then placed in the tank and buffer diffuses towards the origin from each side until the origin area is completely wetted; this has the added advantage that it compacts the origin into a very narrow band. Separation is allowed to proceed for a suitable time.

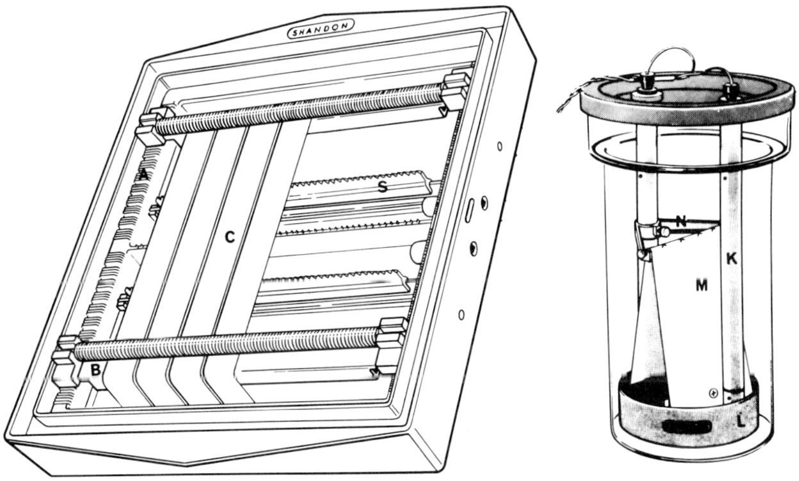

Fig. 2. Apparatus for low voltage electrophoresis[1,2]; many similar types of apparatus are available.[14] The Vertical 'Unikit' apparatus[1]: K, vertical polythene hollow pillars with shock-proof sockets for electrical connection to power supply and platinised nickel-iron electrodes making contact with buffer in each of the outer compartments of buffer-containing vessel L; N, movable support with glass rod for supporting paper, M, of length varying from 8-40 cm. The whole can be assembled before placing in the tank, placing the lid in position and connecting the electrodes. The reverse set of steps enables the paper, still on the glass rod to be removed easily, by holding one lower edge of the paper, blotting excess buffer from the edges and then hanging vertically to dry. The horizontal apparatus also has adjustable supports B, for different lengths of paper, C, and for longer strips intermediate combs, S, act as further supports. The main supports, B, consist of a lower unit which is often adequate alone and an upper, removable piece which is used to sandwich the paper and keep it taut, with the far ends of paper dipping into the buffer solution and acting as wicks also. See also Fig. 11.

The paper is removed, rapidly dried and treated with a suitable reagent which will react with the compounds of interest; for example, ninhydrin for aminoacids or bromphenol blue for proteins. If it is possible to dissolve the reagent in a solvent in which the compounds are insoluble, e.g. ninhydrin in acetone, then the paper can be dipped through the reagent as a much more even distribution is possible this way. Alternatively the reagent solution must be applied by spraying.

Examples of one and two way separations are shown in fig. 3.

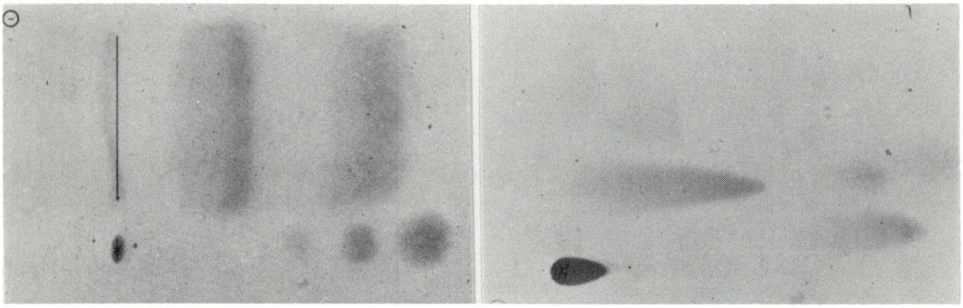

Fig. 3. A, one way analytical and preparative separation of a mixture of indicator dyes.[1] An analytical separation of a mixture of congo red, bromphenol blue and phenol red is shown together with a preparative separation of a 'universal Indicator' mixture. B, a two way separation of the component pigments of a writing ink, using electrophoresis and chromatography. In both cases the experiments were run on a sheet 30 x 10 cm in the Unikit tank shown in fig. 2. with the origins at the apex using ammonia-ammonium acetate buffer for the electrophoresis and butanol-ethanol-ammonia for the chromatography. Most of the spots in the figure are of different colours and shades.

To obtain absolutely identical separations on different occasions, colored markers can be used and run to a fixed position previously marked on the paper. For serum or other proteins, albumin stained with a drop of bromphenol blue is frequently used but other colored proteins are equally valuable. More frequently, mixtures of stable dyes are used but the exact mixture depends on the medium used for the electrophoresis and the rate of migration of particular compounds of interest; dyes which have been used include bromphenol blue, phenol red, methyl orange, xylene cyanol FF, acridine orange, etc. Another standard procedure is to run a mixture of synthetic compounds in parallel with the unknown e.g. a mixture of known dyes when examining a food extract for permitted colour additives, or a mixture of amino acids when examining blood or urine for inborn errors of amino acid metabolism.

Preparative paper electrophoresis. The identical apparatus and techniques described above are eminently suitable for the separation of milligram amounts of material. Obviously the mixture is applied to the origin drawn across most of the width of the dry paper with intermittant drying and re-application if necessary. Assuming it is intended to isolate one compound, it is best to investigate a number of different pH buffers on the analytical scale and to choose that which gives the best separation from neighbouring compounds. One run may be inadequate for complete separation, in which case the strip can be cut out and re-run in a second buffer after sewing the strip into a second sheet as described later. The high voltage methods can also be used preparatively on this scale. To locate colorless substances, a thin strip is cut from the paper and identified with a suitable reagent especially as its approximate position will be known already. The separated substance can be eluted from the strip by setting it up as a descending chromatogram using a solvent in which it is so soluble that it moves with or near the front and drips off into a small test tube.

PAPER ELECTROPHORESIS AT HIGH VOLTAGE (HVPE)[2,6.7]

The original low voltage techniques described usually required many hours and even overnight for adequate separation to occur. Unfortunately, these prolonged periods allowed diffusion of the separating substances to occur as well and so partly negated the effect of the separation. It became obvious that if separation could be speeded up, the diffusion could be minimised and so increasingly larger voltages were applied. For reasons previously stated, techniques had to be modified to prevent evaporation and two procedures have now become standard as illustrated in figure 4. In both cases the paper is wetted and the sample applied as before. The paper is then either sandwiched under pressure between non-conducting, water-cooled platens with only the wicks protruding into buffer solution, or it is entirely immersed in a cooled, non-conducting, water-immiscible oil, such as Varsol, in which the substances to be separated are insoluble. Up to 250 volts/cm can be applied with the platen technique whereas 50 volts/cm is a more usual maximum with the oil immersion method. In principle, there seems little to choose between the two methods but, in practice, the latter has been used most frequently for finger-printing and sequencing of proteins and nucleic

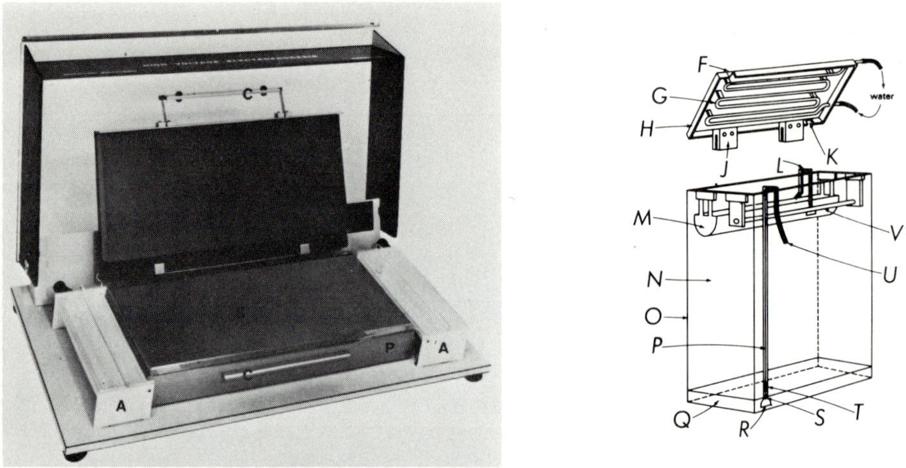

Fig. 4. The cooled metal plate, platen or flat bed apparatus (left)[14]. Apparatus is available to take sheets up to 25 cm square or 60 cm by 28 cm although narrow strips can be run[6]. The apparatus is built on a levelling board. Buffer troughs, A, are placed at each end, the paper sandwiched between plastic sheeting is placed on the lower part B which is a soft, compressible plastic sheet fully enclosed in a polythene envelope; B rests on a metal plate P with a clip C at the front and rear inlet and outlet pipes for fast-flowing cooling water; the upper plate is then closed and the handles locked. A Perspex (lucite) lid with an automatic cut-out micro-switch encloses the whole. A detailed description of the interleaving of paper and insulation sheeting is shown in figure 5. The water-immiscible, organic-liquid method (right). Any glass or Perspex trough may be used and the size of the paper cut accordingly. F and K, recesses for electrodes; G, cooling coil; H, hinged lid; J, tufnol clamp to fit side of tank; L, short electrode; M, electrolyte trough; N, toluene or Varsol filling tank to above M; O, tank; P, glass tubing containing insulating wire; Q, lower electrolyte buffer; R, platinum foil electrode; S, tungsten wire; T, mercury connection; U, connection to power pack; V, paper support rod. For narrow strips, 11 and 21 cylinders have been used as well as the Unikit tank with its convenient upper, internal ledge well suited to support the upper trough M. Alternatively a set-up such as that shown in Fig. 2 is used with a Perspex set of fitments.

acids although this may well be because it is cheap and simple to set up using any large tank available and predated satisfactory and more expensive platen types of apparatus. Larger tanks may contain anything from 30-100 litres of toxic, highly inflammable solvent and should be set up in a special fume cupboard or room after consultation with a fire officer.

Buffers. Very dilute organic buffers are always used in order to minimise heating effects. Table 1 lists a set of buffers in common use which covers the whole pH range; these can be made up in 10 l amounts and stored. With the flat-bed apparatus, the buffer is often used only a few times as it is inexpensive and simple to replace. However, if the electrode vessels are filled fairly full the buffer may be used a number of times especially if the current flow is occasionally reversed between experiments. The organic liquid method usually employs a much larger volume of buffer in the bottom of the tank and this is usually used for up to a week depending on the number of experiments performed although if obvious discoloration of buffer occurs this is a good sign that change is necessary; the Varsol needs to be changed about every three months at most. Provided that the relative proportions of organic solvents remain the same, the dilution with water can be varied over a fair range; this explains the apparently different compositions of pyridine-acetate and other buffers of the same pH. Nevertheless these different dil-

TABLE 1

BUFFERS FOR HIGH VOLTAGE ELECTROPHORESIS

pH	Components	Parts by volume[a]	Comments
1.6	Formic acid:water	80:920	flat bed only
1.6-2.2	Formic acid-Acetic acid-Water	45: 60:895	flat bed only
3.5	Pyridine:Acetic acid:Water	10:100:890	both methods
4.4	Pyridine:Acetic acid:Water	4: 8:988	both methods
5.3	Pyridine:Acetic acid:Water	10:3.2:987	both methods
6.1	Pyridine:Acetic acid:Water	40: 3:957	both methods
6.5	Pyridine:Acetic acid:Water	40: 2:958	both methods
9.0	Ammonium carbonate 1%:Water		both methods

[a] Appropriate volumes are pipetted into a cylinder and made up to the mark with water; the pH is always checked and adjusted if necessary.

utions are compatible with very similar results as the large volumes of buffer used can cope with the electrolytic effects especially as the potential drop is also varied to deal with the different heating effects and the time allowed for the separation.

Setting up the experiment

Flat bed method, Figure 5. As has been mentioned, soft plastics, polythene-enclosed foam pads cover the metal plates. Nevertheless, further polythene or melinex sheets, at least 0.01" thick, are also used both for extra insulation protection and to ensure that the paper is completely enclosed so that evaporation is reduced to a minimum and these sheets should be checked periodically for wear. The lower sheet should be wider than the platen and run from one electrode vessel to the other. A thick paper wick, e.g. Whatman 3MM, is wetted, blotted and placed so that it just squarely overlaps the platen and runs back into the buffer. The prepared electrophoresis sheet or strip is now placed in position so that it just overlaps the wicks at each end by 1-2 cm. A second filter paper wick and insulating sheet are placed in position and the upper plate of the apparatus closed as described above. The apparatus usually includes a set of insulating sheets and many manufacturers supply special wicks which can be used many times for the same pH provided that they are not allowed to dry out. Thick wicks are essential to reduce voltage loss between the buffer and electrophoresis paper.

Organic liquid method. The prepared paper is lowered through the Varsol until it dips squarely into the buffer. It is then brought over the support rod and into the upper electrolyte trough where it is held by a long horizontal rod. This method is quite satisfactory but various methods have been used to simplify the operation. Thus, short glass rods can be slipped through two small holes punched at each side of the lower edge so that the paper will pass easily downwards, or two holes can be punched along each vertical side and a long glass rod with its upper end bent over into a U shape can be threaded through the holes such that the U fits the dimensions over the support rod and into the upper trough. However, many workers prefer the inverted V method mentioned in the legend to Fig. 2, because of its ease.

Marking the paper. Precise descriptions cannot be given because the actual measurements will depend on the dimensions of the apparatus. In general the origin should be at least 4 cm from the wick or buffer solution and, with long strips, up to 10 cm is used. For a set of one way separations on the same sheet, origins should be 2-4 cm apart depending on the subsequent procedure. One way separations are adequate for simple mixtures such as a set of dyes, the mono-, di- and tri-phosphates of a single purine or pyrimidine base or a single PTH- or dansyl-amino acid obtained during sequencing.

However, the vast majority of separations require a further chromatographic run in the second direction to effect a complete - or, at least, a more complete - separation and the most common examples are fingerprinting of protein and nucleic acid partial digests. Such separations can be run on a single sheet by placing the origin in one corner but this is not usually done as large numbers of separations are required and the necessary time can be reduced by using the 'sewing in' technique. Origins are placed about 5 cm apart and electrophoresis carried out as described. The individual strips are

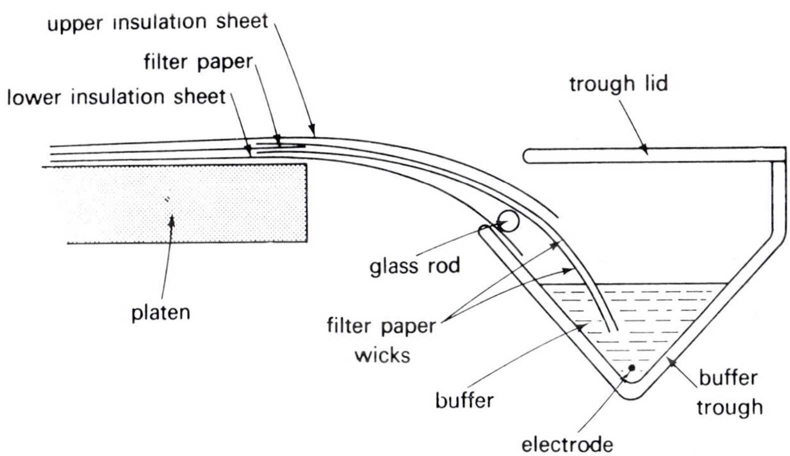

Fig.5. Detail of the procedure used to sandwich the electrophoresis (filter) paper between wicks and polythene sheets in the flat bed method (see text).

cut out and sewn onto one or two clean pieces of paper, a narrower one which will act as the wick and dip into the chromatography solvent (for descending chromatography) and the longer one will be used for the actual separation as shown in Fig. 6.

This method has the added advantage that the separations are exactly comparable, an essential prerequisite for many experiments such as the fingerprinting of normal and abnormal hemoglobins.

The type of paper used is a matter of personal choice as both thin and thick papers are used frequently. Thick sheets are often chosen solely because they are easier to handle when wet.

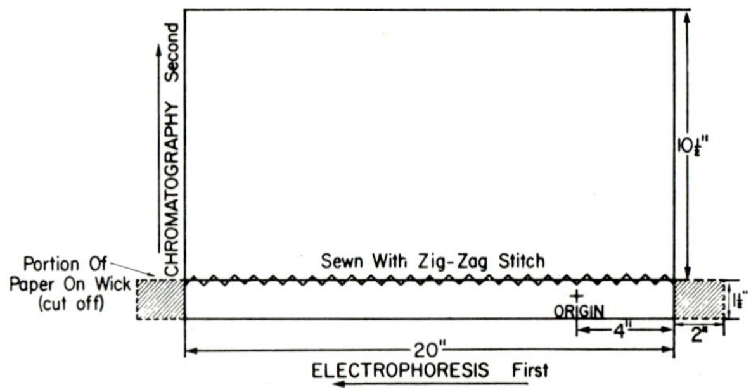

Fig. 6. The sewing in technique, see text. Electrophoresis is carried out and the strip trimmed so that the origin is near to the part to be sewn in to the chromatography sheet and this used for ascending chromatography using a frame. For descending chromatography, a second strip is sewn in on the far side from the sheet for placing in the solvent trough. The dimensions shown must be adjusted for each particular apparatus.

A number of other variations in technique are in use for specific purposes and one example will be mentioned. Radioactive nucleotide mixtures from partial nucleic acid digests have been separated first by electrophoresis on cellulose acetate and then by electrophoresis on DEAE paper. The acetate is brittle when dry but tough when wet whereas the DEAE paper is strong when dry but tears easily when wet. The acetate is wetted as follows. A trough or container, somewhat longer and a little wider than the strip is nearly filled with buffer and the strip is gently lain on the surface. It takes up buffer and when wet it is immersed until thoroughly soaked - it cannot be dipped as described for cellulose paper. It is then sandwiched between two pads very wet with buffer to prevent it drying out but with the origin area just exposed and suspended over two glass rods. The mixture is applied to the origin, the strip blotted to remove liquid drops of buffer and placed as rapidly as possible in an organic liquid type of apparatus. Because the acetate dries out rapidly, single strip separations with only one origin are used. The strip is removed rapidly and hung vertically whilst the Varsol runs down off. Whilst still wet with buffer, the acetate strip is lain on the dry, horizontal, marked origin of the DEAE paper resting on a glass plate and a thick water-wet pad placed on top with a glass weight pressing down evenly. As the DEAE paper sucks up water, this passes through the acetate, thereby causing a transfer of the partly separated compounds as well, where the acidic nucleotides are held by ion exchange forces and without diffusion. After a suitable time the DEAE paper is dried and then set up for electrophoresis at pH 2, although in the general technique chromatography could be used as a second separation; transfer of material to DEAE can be as much as 95%.

Results

Some examples of two-way separations are shown in figures 7, 8 and 9. Amino acids and peptides are now separated by standard procedures and located with ninhydrin or other reagents. Studies of metabolism using physiological doses of compounds can be effected with radio-isotopically labelled compounds such as 2-^{14}C tyramine. The separated compounds are located by autoradiography or the spark chamber method[2,15].

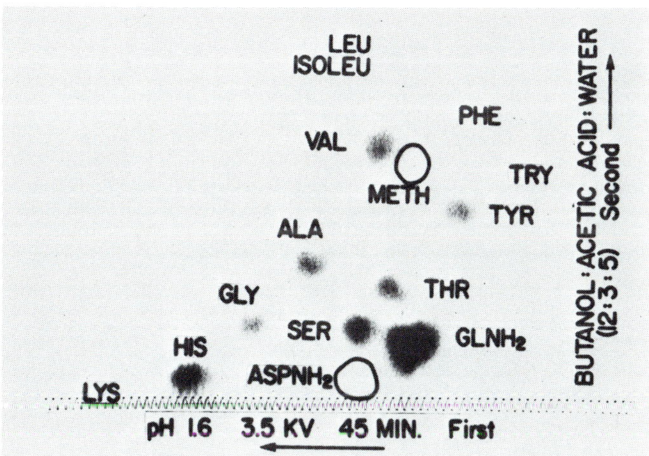

Fig. 7. Amino acids in 1 ml normal plasma run as shown. Note the stitching. For descending chromatography a second strip of paper can be sewn to the lower edge of the paper, (See reference 2, p. 46).

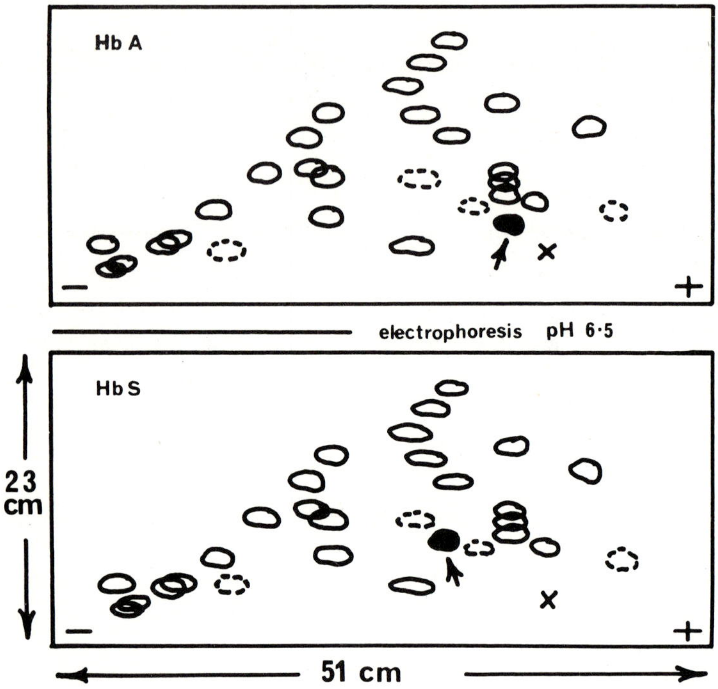

Fig. 8. Comparison of fingerprints of human hemoglobins A and S, run as shown. The black spot marked with an arrow indicates the peptide which differs in the two fingerprints. (See reference 2, p. 56). Separation of metabolites of tyramine found in different species. 2Kv, pH 5.3, 10 mins. 1,12 N-Acetyl tyramine; 2,11, Tyrosol; 3,10, tyramine-O-sulphate; 4,9, p-hydroxyphenylacetic acid sulphate; 5,8, Tyramine, octopamine, dopamine mixture; 6,7, Phenolic acid mixture including p-hydroxyphenyl acetic acid.

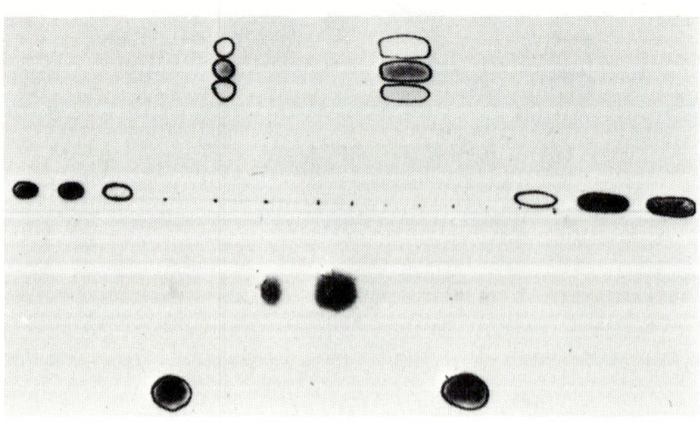

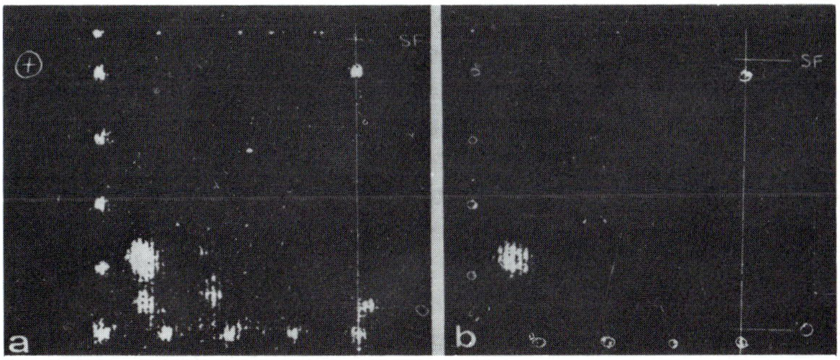

Fig. 9. Comparison of the metabolism of oral radioactive tyramine in rat and rabbit on 10 x 10" 3MM paper. The ether extract was chromatographed in iPrOH-880 NH_3-H_2O (8-1-1) first followed by electrophoresis in pyridine-acetate pH 5.3 buffer at 2 Kv and 80 mA for 40 mins. Cold standards were placed on the same origins. Radioactivity was located with a 10-20 min exposure in the spark chamber[15] and cold standards with the diazo reagent for phenolic acids. The 5 aligned horizontal and vertical spots are a 'marker grid' for comparison purposes and are placed outside of the separation area.

THIN LAYER ELECTROPHORESIS (TLE)[2,3,8,9,10,11,14]

Thin layers are composed of microcrystalline cellulose, various types of silica gel, polyamide or other materials bonded to inert supports such as glass, aluminium foil or plastic. They are usually supplied as 20 cm squares and the foil and plastic sheets have the great advantage that they can be readily cut to any desired size. They were originally designed as alternatives to paper for chromatography but are now commonly used for electrophoresis and practically all types of ionic compounds, except proteins, have been extensively studied. Indeed many workers prefer TLE to paper electrophoresis because as good or better results can be obtained more quickly and on smaller areas which makes for easier handling. Much of the apparatus already available for paper can be used directly or modified simply to take the thin layer sheets and, as voltages of not more than 500-1000 V are more than adequate, the procedure is much safer than HVPE. Again, better results are obtained if the plates are cooled but adequate results can be observed without cooling if a somewhat lower voltage is used.

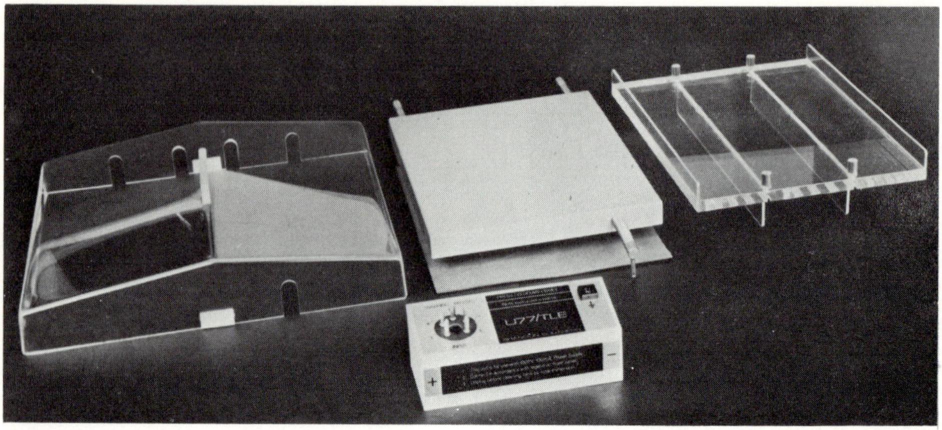

Fig. 10. Plastic base plate (on right) to support cooling plate and modified cover to permit passage of cooling water inlet and outlet tubes. Wicks and polythene sheeting, if used, can be placed as shown in fig. 5.

Buffers. Buffers described in Table 1 are equally suitable here.

Applying the sample and buffer. The sample is applied to the dry plate either as a small spot or streak. Buffer is then sprayed onto the plate until it just glistens - usually 20-40 ml is required for a 20 x 20 cm plate - and it is placed immediately into the tank.

Apparatus. The horizontal tank shown in fig. 2. is suitable. The supports are removed and replaced by a plastic support which takes the cooling plate and a modified lid, fig. 10. The plate is laid face upwards on the cooling plate, thick paper wicks are applied along the length of the layer, a glass plate is placed above this to hold the wicks down onto the layer to ensure good contact and the lid placed in position. Some workers also use polythene sheets as shown for HVPE in fig. 5. An alternative procedure is to use the original supports with thick paper wicks covered with buffer-impregnated, un-coated cellophane to reduce endosmotic effects. The thin layer plate is then laid, face down, on the wicks and covered with an aluminium sheet about 0.5 cm thick sealed in a polythene envelope; this sheet acts as a heat sink and prevents the layer heating up; fig. 11.

Size of plate. Although most work has been done on the full size

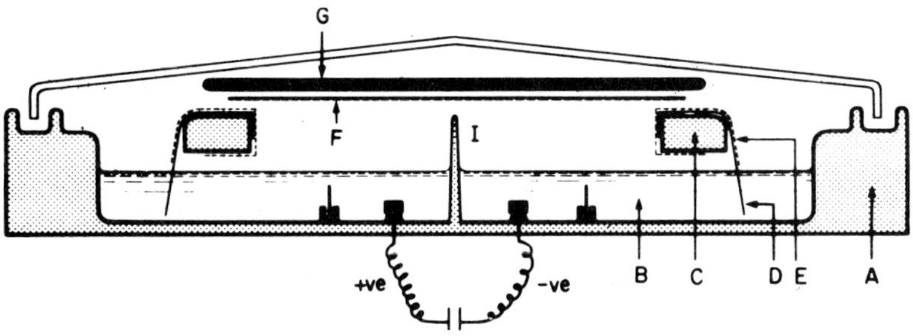

Fig. 11. TLE - alternative set up. Tank A, buffer B, shoulder pieces C with wicks D and cellophane sheet cover E. Thin layer plate F is covered with aluminium cooling plate G which also helps to ensure good contact between plate and wicks.

plates, it is quite possible to use both 10 x 10 cm and 5 x 5 cm size plates for adequate one and two way separations. However, it is not necessary to cut the plates to size before the experiment. For example, if it is desired to run four 5 x 5 cm plates, the electrophoresis can be run on a 7 x 20 cm sheet, as shown in fig. 11, and the sheet can then be cut into squares for the chromatographic run in the second direction.

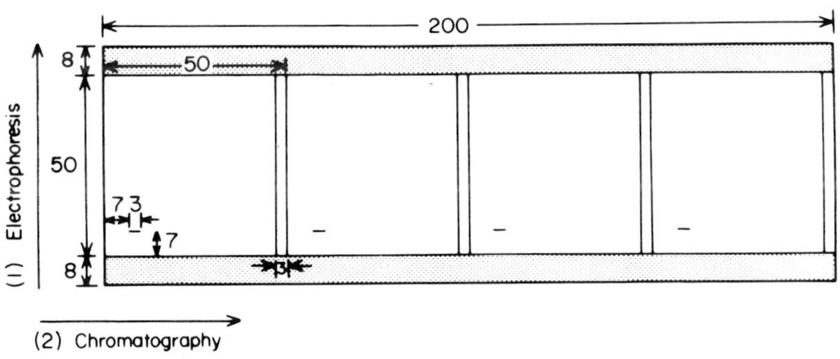

Fig. 12. TL sheet marked out for four 5 x 5 cm two way runs on a sheet 7 x 20 cm. The shaded areas are cut off after electrophoresis.

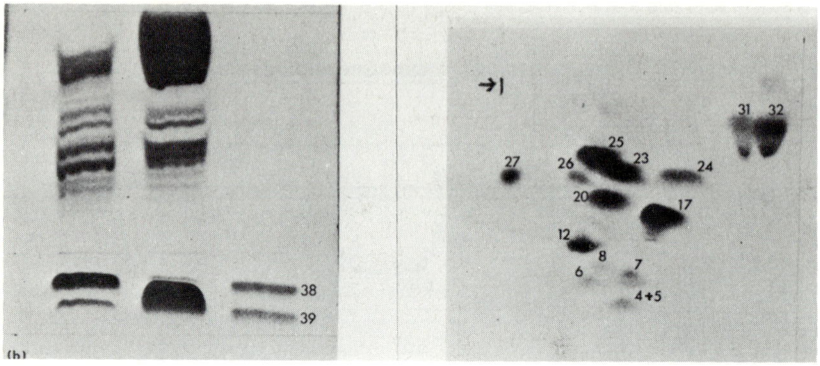

Fig. 13. One way and two way runs on 5 cm squares. Samples left to right: oxidised homocystinuric urine, cystinuric urine and mixture of homocysteic acid (38) and cysteic acid (39). Electrochromatography of generalised amino aciduria. Figures 11, 12 and 13 taken from Ersser and Krykwawych, with permission. Electrophoresis was at pH 1.9 for 15 minutes using 50 V/cm and chromatography was a double development with n-butanol-acetone-acetic acid-water = 7-7-2-4.[10]

PREPARATIVE PEVIKON BLOCK ELECTROPHORESIS[3,12,13,14]

Pevikon C-870 is a copolymer of polyvinyl chloride and polyvinyl acetate which is non-swelling and has a negligible electro-endosmotic flow; it is consequently far superior to starch and other materials previously used. Unfortunately it is not manufactured for this purpose and must be extensively washed by first shaking with ten volumes of 4M HCl, filtered on a sintered glass funnel, washing with a further 2 x 5 volumes of acid and copiously with water. It is now resuspended in ten volumes of 2M acetic acid, shaken for 15 minutes, refiltered and washed with water to neutral pH. Finally it is resuspended in water and the fines decanted. For use, the filtered Pevikon is suspended in a volume of buffer just fluid enough to pour.

Apparatus and technique. The apparatus consists of a plate glass support, up to 60 cm long x 25 cm wide, resting on and overlapping the electrode vessels and with two plastic bars running the length of the plate; the bars are of 1.5 cm cross section and graduated at 1 cm intervals. A polythene sheet covers the bars and plate and at each end is a dry 'labshammie' or thick paper pad to contain the Pevikon. The Pevikon in an appropriate buffer is poured into the apparatus to fill it completely. Excess liquid is absorbed into the pads and, when it appears sufficiently set, the pads are replaced by a dry set. The block can be gently 'rolled' to produce a good surface. A slit is cut in the block, leaving a 2 cm width at each side and the Pevikon removed. A strip of moist thick paper is placed along each long side of the slit which is widened by gently inserting a 2 mm wide strip of plastic. The solution to be run is mixed to a thin slurry with Pevikon, the plastic strip removed and the slurry inserted, with a Pasteur pipette, to fill the gap completely; up to 10 ml can be added. The moist paper strips are removed, the slit is rolled to close it up, the pads at each end removed, fresh wicks of knitted nylon or orlon are applied and the whole covered with another sheet of polythene. The apparatus is placed in the cold and a voltage of not more than 5 V/cm applied for an appropriate time. The apparatus and technique can be modified for high voltage work in a manner similar to the flat bed method described previously but the original paper should be consulted first; up to 150 V/cm can be used. Solutions containing up to 1 g of dissolved solids have been separated successfully.

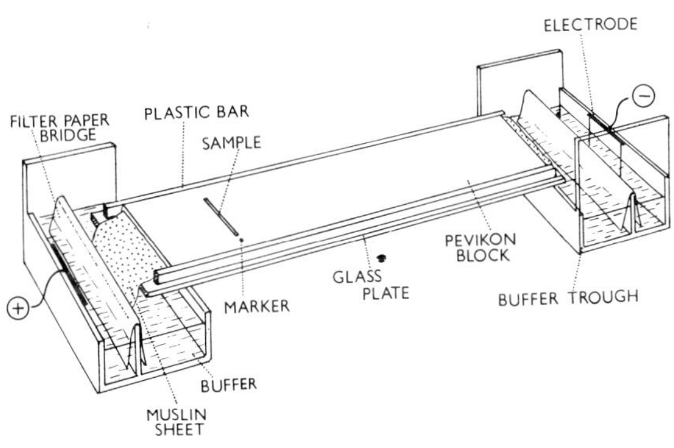

Fig. 14. Pevikon block preparative electrophoresis; for description see text. For simplicity the polythene sheets are not shown.

Location of substances. If the substances have no visible color or fluorescence then the marker dye procedure can be used having first ascertained the best dye mixture using analytical paper electrophoresis. Alternatively a strip of paper can be laid along the length of the Pevikon to absorb solution, removed, dried and located with a suitable chemical reagent.

Isolation of separated material. The Pevikon is gently cut on either side of the substance(s) required with a spatula, gently removed into a long sintered filter tube and the liquid sucked through under vacuum. Two washes with 1 ml of water or buffer yield almost total recovery even of proteins. Of course, if many substances are to be recovered, 1 cm width cuts can be removed, one at a time, and separately eluted.

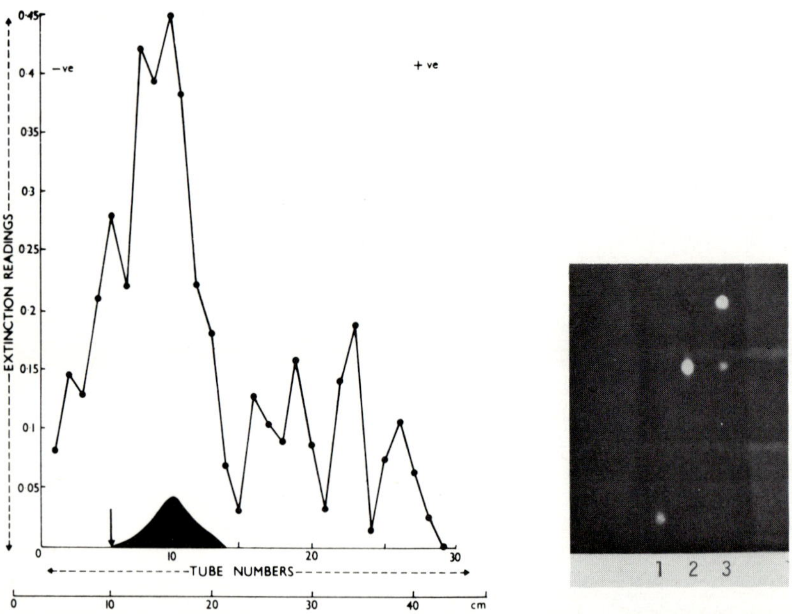

Fig. 15. Preparative separations on Pevikon block. Left: separation of urinary glycopeptides using a pyridine-acetate buffer pH 6.5, potential difference of 5 V/cm for twelve hours, with the origin 10 cm from the cathode end. 1 cm widths were removed, extracted as described and the extinction of the orcinol-sulphuric acid colour reaction of each fraction plotted against tube number. Right: a mixture of 450 mg of thiamine and its mono- and di-phosphates was run on a block 40 x 23.5 x 0.5 cm. The block buffer was 5% pyridine in 2% acetic acid, diluted 1 in 20, and the electrode buffer was 0.2 M sodium dihydrogen phosphate diluted 1 in 17 at pH 6. Buffer characteristics were: conductance G = 0.001Ω^{-1}, conductivity K = 0.71$\Omega^{-1}m^{-1}$. Electrophoresis was using 5 Kv at 135 mA. The separated fractions were eluted and rerun on analytical electrophoresis: fraction 1 = thiamine, fraction 2 = monophosphate and fraction 3 = diphosphate. The minor spot in fraction 3 probably represents a small amount of hydrolysis.

REFERENCES

Only key references are listed here; each contains a further full discussion of technique together with a large number of references to the original literature.

1. Smith, I. and Feinberg, J.G. (1968) Paper and Thin Layer Chromatography and Electrophoresis - A Teaching Level Manual, pp. 1-241, Shandon Southern Instruments, Runcorn, Cheshire, England.
2. Smith, I. and Seakins, J.W.T. (1976) Chromatographic and Electrophoretic Techniques, 4th Edn., Vol. 1, Paper and Thin Layer Chromatography, pp. 1-464; Vol. 2, Zone Electrophoresis, pp. 1-485. Heinemann Medical Books, London and Year Book Med. Publ., New York.
3. Sargent, J.R. and George, S.G. (1975) Methods in Zone Electrophoresis, 2nd Edn., pp. 1-154, British Drug House Chemical Ltd., Poole, England.
4. Deyl, Z., Kopecky, J., Davidek, J., Juricova, M. and Helm, R. (1975) Bibliography of Electrophoresis 1968-72 (Supplementary Vol. No. 4, 1975, to J. Chromatog.) Elsevier Sci. Publ. Co. Amsterdam, Oxford and New York.
5. Kohn, J. (1976) in ref. 2, Vol. 2, pp. 90-137.
6. Sanger, F. (1965) J. Mol. Biol., 13, 373-398.
7. Brownlee, G.G. (1975) in Laboratory Techniques in Biochemistry and Molecular Biology (Work, T.S. and Work, E., eds) Vol. 3(1) pp. 1-265, North Holland Publ. Co. Amsterdam and New York.
8. Ritschard, W.J. (1976) in ref. 2, Vol. 2, pp. 66-89.
9. Stahl, E. (1967) Thin Layer Chromatography, pp. 1-822, Springer-Verlag, Berlin and New York.
10. Ersser, R.S. and Krywawych, S. (1974) Med. Lab. Technol., 31, 235-244.
11. Stoker, B.M., Roux, S.J. and Brown, W.E. (1978) Nature, 271, 180-2.
12. Weiss, J.B. (1976) in ref. 2, Vol. 2, pp. 367-377.
13. Lote, C.J. and Weiss, J.B. (1971) FEBS Lett., 16, 81-84.
14. All the commercial apparatus described here is available from Shandon Southern Instruments, Runcorn, England, and USA. Other manufacturers include: Camag, Switzerland and USA; Savant Instruments, USA; Chandos Products, Stockport, England.
15. Birchover Instruments, Spirella Building, Letchworth, England.

Electrokinetic Separation Methods,
P.G. Righetti, C.J. van Oss and J.W. Vanderhoff

IMMUNOELECTROPHORESIS

JØRGEN CLAUSEN
University of Roskilde, Institute of Life Sciences, Laboratory of Biochemistry and Toxicology, 4000 Roskilde and The Neurochemical Institute, Rådmandsgade 58, 2200 Copenhagen N, Denmark.

INTRODUCTION

Two different principles for immuno-electrophoresis, i.e. simple and crossed immuno-electrophoresis exist. They will be treated separately below.

1.1. General principles of simple immuno-electrophoresis (electro-immunodiffusion)

Simple immunoelectrophoresis consists of a combination of electrophoresis and immunodiffusion in gels. It is based on the fact that in agar gel, the movement of molecules in an electric field is similar to that in a liquid medium, with the advantage that free diffusion during and after electrophoresis is lessened. The constituents of a mixture are then defined both by their electrophoretic mobilities and by their antigenic specificities.

This technique may be performed in a macro scale, as originally demonstrated by Grabar and Williams[1] and Poulik[2,3] or as a micromethod[4]. The antigen sample is applied (Fig. 4) to a hole placed in the middle of a glass plate coated with buffered gel. Electrophoresis separates the antigen mixture into various zones. A longitudinal trough parallel to the long edge of the plate is then made in the gel and is filled with antiserum against the antigen mixture. A passive, double diffusion takes place; the antiserum diffuses into the gel and the antigens diffuse radially in all directions from the electrophoretic zones. The antigen-antibody complexes are formed at equivalence points; the number of precipitates formed corresponds to the number of independent antigens present (Fig. 1 & 2). The precipitates are dispersed in a characteristic system of arcs with precipitation lines located alongside the electrophoretic zones.

Two- and three-dimensional immunoelectrophoresis or one-dimensional electrophoresis combined with two-dimensional chromato-immunodiffusion have been suggested[6,7,8].

Constituents in the mixture may be identified serologically in a complex spectrum of arcs by matching each individual arc by means of a marker arc produced by a monospecific reactant (antigen or antibody) added to the plate in a different location from the polyvalent serum. Alternatively the polyvalent antiserum may be absorbed by a pure antigen and then used for comparison with the unabsorbed serum.

For simple immuno-electrophoresis different types of supporting media can be used. Agar (agarose and agaro-pectin)[9] is often used. Agarose shows the lowest electro-endoosmotic flow. The use of an agarose gel instead of an ordinary agar gel has advantages as the gel is free from ionized groups and does not cause pronounced electroendomosis

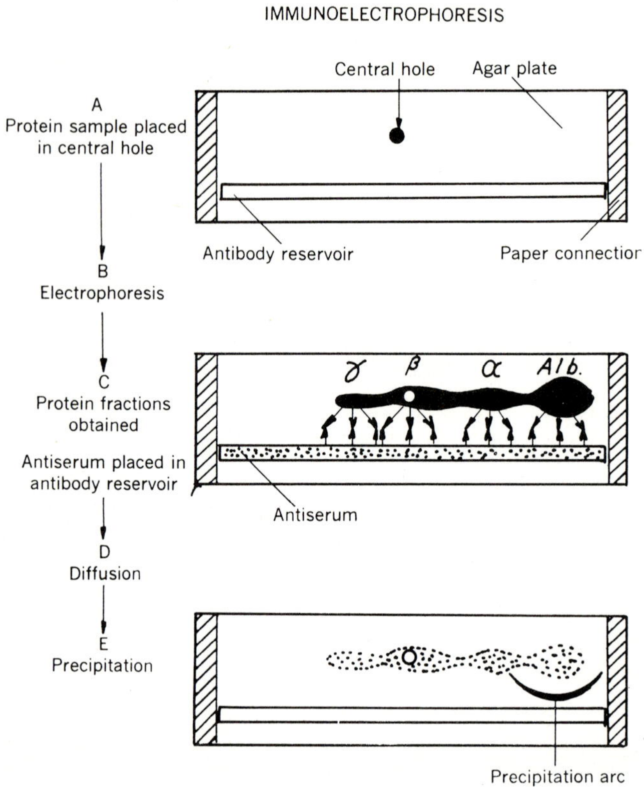

Fig. 1. Method for micro-immunoelectrophoresis. A: The microscope slide is covered with buffered agar gel. The central hole (for insertion of the antigen) is dug out of the agar gel. B: Electrophoresis is performed for 1½ h (10 V/cm). C: The electrophoretic pattern is obtained; antiserum solution is placed in the antibody reservoir which is dug out of gel after electrophoresis. D: Diffusion is allowed to proceed in a moist chamber. E: As the result of diffusion, antigen-antibody reactions occur between corresponding fractions, and precipitation arcs or lines appear; these are specific for each antigen. The precipitation arc for albumin is illustrated here (Clausen[5])

or interaction with certain basic proteins such as lysozyme[10]. Agarose is more resistant to acids than agaro-pectin and therefore electrophoresis may be carried out at low pH. On the other hand, the complicated preparation of agarose may limit its use ; it is available commercially.

Also cellulose acetate can be used[11,12,13,14]. The technique suffers from difficulties involved in applying the antiserum evenly on the paper. It is therefore not recommended for the beginner in the field of immunochemistry.

More rarely starch gel may be used. Horizontal or vertical starch gel electrophoresis[15,16] of protein antigens may be followed by diffusion in gel by cutting the electrophoretic starch gel vertically in two halves. One half is stained for proteins and the other half is placed in a melted agar buffer solution. After gelling, an antibodytrough is made in the gel in the anode-cathode direction, and an antiserum is applied in this trough. The antigens separated in the starch gel will now diffuse into the agar gel and meet the corresponding antibody and thus create multiple precipitation lines. This method is laborious, because the electrophoresis and immunodiffusion cannot be performed in one step. However, starch gel electrophoresis has a higher resolving power for the antigen mixture than agar has. Horizontal starch gel electrophoresis may produce 18 electrophoretic zones in separation of serum proteins, whereas agar gel electrophoresis results in only seven.

During recent years polyacrylamide gel has been used[17,18] (for details see later).

For special purposes fibrin-agar may be used; thus in studies of proteolytic enzymes, their activators and inhibitors. Such studies make it possible to characterize the number of individual steps in the catalytic processes, for example blood coagulation, by examination of the reaction products[19,20]. The method is based on the combination of immunoelectrophoresis and the classical test of fibrinolysis performed by means of the fibrin-plate technique[21]. By mixing agar with buffer, fibrinogen and calcium ions followed by appropriate heating, a medium is obtained which possesses the electrophoretic properties of the agar and the substrate properties required for proteolytic digestion. After electrophoretic separation and incubation, every electrophoretic zone containing a proteinase will appear as a clear zone of lysis. By cutting out troughs for antisera, precipitation lines corresponding to the antigenic proteinases may appear. The troughs may also be used for application of an activator or inhibitor of proteinase function. An electrophoretic zone containing a proenzyme of a proteinase may, during diffusion towards the trough containing the activator, cause the formation of a clear zone of digested fibrin as a sign of the activation of the proenzyme. Combined with immunodiffusion on the second half of the slide, this method offers possibilities for localization of enzymes and proenzymes belonging to the group of proteinases.

1.2. General principles of crossed immuno-electrophoresis and electrophoresis in antibody-containing media (Laurell technique)

In this technique, the antigen-antibody reaction occurs during the electrophoresis of an antigen mixture in an antibody containing medium[22,23]. Both antigen and antibody move according to their electrophoretic mobilities, and they also react together, resulting in flame-shaped precipitin zones of antigen antibody complexes (fig. 2). Under the influence of the electric field, unbound antigen within the peak of the flame-shaped precipitate migrates into the precipitate which redissolves in the excess antigen. Thus, the leading edge of the flame is gradually displaced in the direction the antigen is migrating. The amount of antigen within the leading boundary edge is successively diminished because of the formation of soluble antigen-antibody complexes. When the antigen is diminished to equivalence with the antibody, the antibody-antigen complex can no longer be dissolved and a stable precipitate will form at the leading edge, which is thereafter stationary. The distance finally travelled

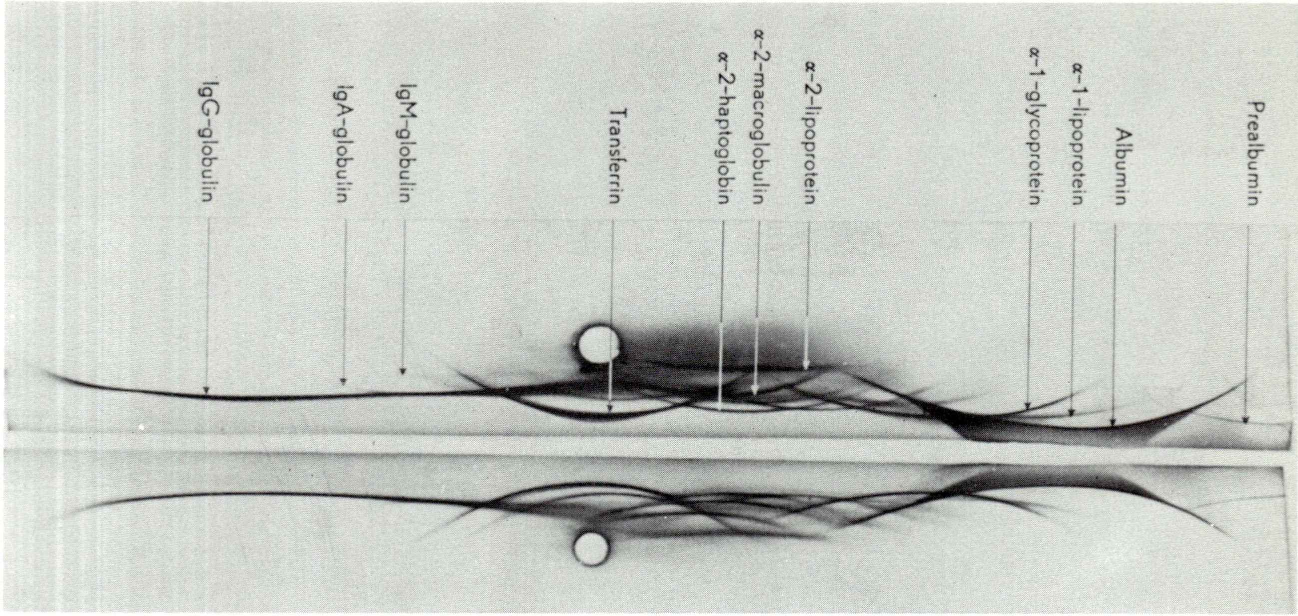

Fig. 2. Micro-immunoelectrophoresis of normal human serum. Serum was placed in the two holes, and after electrophoresis precipitation pattern was developed with a horse antiserum against pooled normal human serum (placed in central trough). The main precipitation arcs are marked. Experimental conditions - Antigen: 1.5 ul normal human serum (6.5 g protein/100 ml) have been applied in the two antigen wells. Electrophoresis: 10 V/cm for 70 min, buffer: 0.05 M Na-veronal buffer pH 8.6. Antiserum: 80 ul polyvalent antiserum from a horse against normal pooled human serum is applied in central trough. Immunodiffusion: 16 hr at 24°C. Staining: After washing by standing for two days in 0.9 % (w/v) aqueous NaCl and for 1 hr in distilled water, the slide is dried in open air and stained for protein with Amido Black.

by the peak depends on the relative excess of antigen over antibody and can be used as a measure of the amount of antigen present. Every precipitin band featured as a flame represents an individual antigen. By absorbing the antibody with pure antigens prior to the migration it is possible to identify the pattern[24]. Greater resolution can be obtained if, prior to the electrophoresis in the agar gel containing antiserum the proteins are separated by starch gel electrophoresis; a number of precipitates of albumin and ceruloplasmin have thus been obtained[24]. Electrophoresis in an antibody-containing medium may thus be applied to studies of antigenic sites.

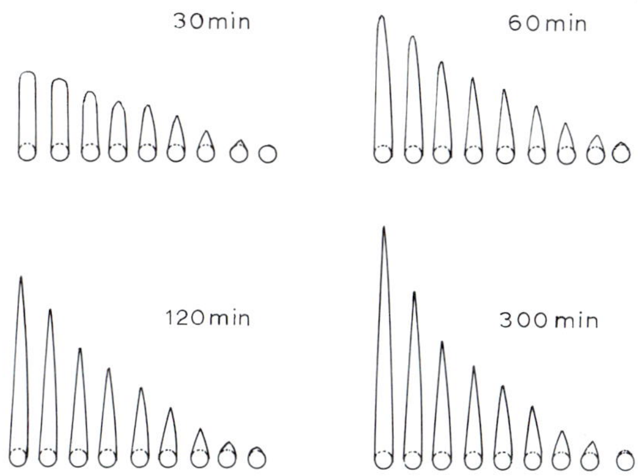

Fig. 3. Electrophoresis of serum albumin in antibody-containing agarose gel. Serum albumin dilutions run for 30, 60, 120 and 300 min in agarose gel containing 2.5 % rabbit anti-albumin serum. Amounts of albumin in each series from left to right: 5, 3.75, 2.5, 2.0, 1.5, 1.0, 0.5, 0.25 and 0.125 ug[25].

The electrophoresis of an antigen mixture in an antibody-containing agar gel has been used one-dimensionally or two-dimensionally, the socalled crossed immunoelectrophoresis. The techniques (see Fig. 3, Laurell´s method) have been used for quantitative estimation of antigens in solutions containing from 500 to 12.5 ug/ml[26,25] provided the isoelectric point of the antigen differs from that of the antibody. Systematic studies of the migration rate in an agarose medium containing the antibody [25] revealed empirically that within a certain time (2 to 10 hr) a rectilinear relationship between final migration distance of the leading edge and the amount of antigen may be established. This method is appropriate for estimation of serum proteins or other antigens in different biological fluids (vide infra). The migration rate of the antigens must be faster than that of the IgG globulin.

Like simple immunoelectrophoresis crossed immunoelectrophoresis may be used for tracing and identification of protein antigens in a solution as well as for quantitative assay of individual antigens. The electrophoresis in the first dimension characterizes the antigens by their mobilities. The electrophoresis in the second dimension i.e. their migration into antibody containing gel, however, also characterize the antigens because the shape of the individual immunoprecipitate is determined by the electrophoretic distribution of the protein antigen in the gel of the first dimension, by its antigenic properties and by its migration rate in the antibody containing gel. The identification of the individual immunoprecipitate by the two methods can be performed by different methods. Thus for instance the follwoing identifications may be used:

a. The morphological properties of the immunoprecipitate
b. Immunoabsorption of an antibody by its purified antigen[27]
c. Characterization of the immunoprecipitate by means of its enzymic properties
d. Absorption experiments and mixing of an unknown antigen solution with a known antigen solution and comparison of the immunoelectrophoretic pattern prior to and after mixing[27]
e. Tandem crossed immunoelectrophoresis[27,28]
f. The migration length in the second dimension as in rocket immunoelectrophoresis[25]
g. Fusion of immunoprecipitates as in fused rocket immunoelectrophoresis[28,29]
h. Line immunoelectrophoresis[27,28]

Below are mentioned some technical, qualitative and quantitative aspects of immunoelectrophoretic techniques:

The morphological properties of the immunoprecipitate may be used for the identification of the individual antigens. Thus in electrophoresis of biological fluids of mammals the albumin immunoprecipitates is the major anodically migrating protein. On the other hand the γ-globulins are cathodically situated precipitates covering a broard area of mobility intervals due to their different electrophoretic mobilities, but with identical immunological properties. Also the high molecular proteins such as serum α-2-lipoprotein are slowly migrating into the antibody containing gel. Some proteins may furthermore be characterized by their biological properties such as α-1-antitrypsin and α-2-macroglobulin by their antiproteinase activities and lipoprotein by their content of lipid respectively.

Studies on proteins only partially soluble in buffers or of particulate fractions may also be made by crossed immunoelectrophoresis. Thus Hejgaard & Bøg-Hansen[30] have used the characteristic major precipitation lines of Barley and Malt extract to compare the similarities and differences between precipitates of these two sources and Jørgensen et al.[31] have used crossed immunoelectrophoresis of particulate brain fractions such as the synaptosomes for an elucidation of synaptosomal membrane characteristics. In the clinical chemistry crossed immunoelectrophoresis has been used for comparison of proteins in biological fluids, i.e. serum, urine or cerebrospinal fluid, especially during pathological states[32,33,34]. Some serum proteins, however, need special technical solutions in order to be quantitated. Thus γ-globulin and β-lipoprotein migrate so slowly that quantitation may be difficult. Carbomylation of Igs and delipidation of β-lipoprotein may solve these questions[32,35]

Identification of precipitation lines on basis of their enzymic properties: In simple as well as in crossed immunoelectrophoresis antigens having enzymic properties often possess these properties intact after complex formation with their corresponding antibodies[36]. Thus esterases, peptidases, oxido-reductases may be traced in immunoelectrophoretic tracings. Thus as an example it can be mentioned that Hejgaard and Bech Sørensen[37] were able to follow the changes in the antigenic pattern and the development of esterase positive antigens during germination of Barley. This made comparison possible between different strains of Barley.

As in classical immunoelectrophoresis (electro-immuno-diffusion) the mere absorption of the antibody with a purified antigen makes identification possible. Furthermore mixing of two unknown solutions makes comparison possible by tracing the immunoprecipitates prior to and after mixing.

The tandem crossed immunoelectrophoresis is a modification of the crossed immunoelectrophoresis. It is an immunoelectrophoretic method in which two protein mixtures are compared in one schedule. The two samples are applied to the first dimension gel in two holes placed in front of each other in such a way that the immuno-chemically identical precipitate peaks from the two different samples fuse together to double peaks during the immunoelectrophoresis in the second dimension. Immunochemical identical proteins from the two different samples fuse their immunoelectrophoretic peaks. The origin of the two peak maxima in the double arcs is determined by the position of the two application wells and furthermore a semi-quantitative comparison can be based upon peak heights[30].

The Rocket immunoelectrophoresis has already been mentioned. It is a simple and quick method for the estimation of a single protein in a series of samples of protein mixtures. The individual samples are applied in a row of circular wells in the antibody containing gel and electrophoresis is started. The proteins in each sample migrate and form rocket shaped precipitates with their corresponding antibodies. The method needs specific antibodies to the individual antigens. The identification of the antigens must be based upon specific antibody, the staining for enzymic activities or chemical composition of the antigen. The peak height is a measure of the amount of antigen applied.

The fused rocket immunoelectrophoresis can be applied to studies on collected protein fractions, for instance collected during column chromatography during a protein separation experiment. A small volume of each fraction is transferred to a series of holes in antibody free agarose gel. The proteins in the samples are allowed to diffuse into the gel and electrophoresis into an antibody containing gel is performed perpendicular to the line of holes. In the resulting pattern each protein will form a precipitation line which may fuse with neighbouring precipitates. Thus the elution profile of each individual protein may be obtained in one experiment. The fused rocked method was used to follow the separation by DEAE-sephadec ion exchange chromatography of proteins from Barley extracts and make it possible to trace individual antigens[30].

Line immunoelectrophoresis: When a protein mixture is uniformly distributed in one direction of the gel and subjected to immunoelectrophoresis per pendicularily to this direction, parallel precipitation lines are obtained and the difference of migration of the individual proteins is proportional to the protein - antibody ratio. It could be used for comparison of different protein samples.

The crossed immunoelectrophoresis has mainly been performed in agarose since this medium permitsa free electrophoretic migration. In order to increase the electrophoretic separation, however, the crossed immunoelectrophoresis may be combined with other methods, such as isoelectric focusing[37,38,39]

The isoelectric focusing - crossed immunoelectrophoresis may be performed in polyacrylamide gel containing ampholines of a selected pH interval. After the electrophoresis has been made, the polyacrylamide gel containing the separated proteins may be placed directly on the surface of or ajacent to an antibody containing agarose gel and immunoelectrophoresis is performed perpendicular to the electrofocusing direction[37]. The main problem in using polyacrylamide gel in the first direction and agarose in the second direction in crossed immunoelectrophoresis is the different electroendomotic properties of the two gels. Most earlier work has been based on "moulding in" techniques, i.e. polyacrylamide gels containing separated antigens moulded in an antibody-containing agarose. To overcome the problems caused by differences in electroendosmotic properties most earlier attemps were directed towards reducing the electroendosmosis of agarose gels.

Purified agarose mixed with methyl-cellulose[40], polyacrylamide-agarose gels[41], chemically modified agarose[42] and purified agarose containing linear acrylamide[43] have been recommended to reduce the accumulation of water at the contact zone between polyacrylamide and agarose gels.

A simple washing step in distilled water combined with a "laying-on" method seems to be the most convenient and rapid way to obtain reproducible results using commercially available agarose. It is possible that the use of highly purified or of chemically modified agarose will further improve the immuno-precipitation patterns.

The loss of proteins in the washing step will depend upon the pore size of the polyacrylamide gel, the molecular size of the proteins and the washing conditions.

Since isoelectric focusing in polyacrylamide gel may allow separation of even poorly soluble antigens such as those found in particulate fractions and plant material, this method may be the tool of choice for studying detergent soluble antigens, i.e. membrane components.

Berzins et al.[39] have thus used this method for tracing multiple form of esterase active antigens of deoxycholate-Lubrol W solubilized rat liver microsomes using α-naphthyl-proprionate as a substrate and Fast Blue B-salt as a coupling chromophore. Eight esterase active antigens were traced in liver microsomes. Finally polyacrylamide gel electrophoresis combined with crossed immuno-electrophoresis permits allotypic tracing of serum proteins[44].

Thin layer gel filtration - crossed immunoelectrophoresis. The thin layer gel filtration may be performed on Sephadex G-150 superfine applied on a 200 x 100 mm glass plate according to Pharmacias directions. The Sephadex gel is applied upon a gel plate containing an agarose layer containing the immunoelectrophoretic buffer. After the one-dimensional separation of the protein antigen mixture in the thin layer gel, the one side of the Sephadex gel is removed and exchanged with an antibody containing agarose gel. Immunoelectrophoresis performed perpendicularly to the gel filtration may now permit the tracing of the antigens in the solution.

The methods mentioned above may all be used for quantitative as well as qualitative purposes since for most proteins the rate of migration of the indivudual antigens in the antibody containing gel is proportional with the amount of antigen in the agarose.

Also by chemical modification of the antigens the electrophoretic mobilities may be changed to a certain extent making the electrophoretic separation more appropriate for immunoelectrophoretic studies. Thus Weeke[32] showed that by cabamylation of human immunoglobulins the migration rate of immunoglobulins in an antibody containing agarose is promoted to such an extent that they can be quantitatively assayed.

2. Material and Methods 4.

2.1. Techniques of immunoelectrophoresis 4.

2.1.1. Choice of buffers. The buffers used in the gel have an influence on electrophoretic mobility and on the position of the immunoprecipitate. The pH of the buffer must be chosen with the isoelectric point of the antigen in mind, it being desirable usually to work at a pH slightly above the isoelectric point[36], so that the proteins have a negative charge and do not react with the agar. A pH of 8.2-8.4 is usually used for serum proteins.

Immunological reactions take place only between pH of about 4.5 and 9.0. In studies of antigens with different isoelectric points, it is important to remember this fact. Thus, glycoproteins with low isoelectric points should be separated at pH between 2.5 and 4.5; however, the subsequent antigen-antibody reaction cannot be performed at this acid pH. Therefore, another buffer must be introduced into the gel prior to immunodiffusion. To do this, an appropriate buffer, with a pH of about 6.5 is placed in the antibody trough prior to the application of antiserum. In this way, a zone of diffusing buffer neutralizes the acid electrophoretic medium prior to immunodiffusion and

Fig. 4. The pattern of crossed immunoelectrophoresis of normal human serum. The application hole is localized below to the left. 5 ul serum is applied. The electrophoretic separation was performed in 0.05 M sodium barbital buffer (pH 8.6) in 1% (w/v) agarose (Agarose A 45 L´Industrie Biologique Francaise, Grennevilliers, France). The cathode-anode direction is from left to the right. Perpendicularily to the primary electrophoretic migration route the electrophoretic fractions are electrophoresed upwardly into an antibody containing agarose (1 ml rabbit anti-human antiserum per 10 ml buffered 1 % agarose mixed at 45°C). Several immunoprecipitates appear, the height of which are proportional with the protein concentration in the serum sample.

antigen-antibody reaction. The ionic strength is also of great importance in gel electrophoresis; it if is too high, heating is liable to occur, while too low a buffer concentration does not keep the pH constant during the passage of current. Unless cooling is used, the ionic strength should not exceed 0.05 M.

2.1.2. Choice of method for simple immunoelectrophoresis. Two basic methods are available, i.e. the original macro technique[1] and the micro technique is time-consuming, taking up to 14 days for immunodiffusion; furthermore, it requires a large amount of antiserum and is therefore not used as a routine method. However, for some purposes the macro technique is more convenient; it is especially useful for revealing weak precipitation lines. In studies of the binding capacities of proteins for drugs and hormones, weakly binding proteins giving rise to weak precipitation lines may be revealed by radioactive labelling. The macro method may also be used for the demonstration of enzyme activities of individual antigens, using conventional agar or fibrin agar. For other purposes, particularly in routine work, micro techniques are preferable. Here the times of electrophoresis (1-1.5 hr) and diffusion (16 hr) are relatively short and only 100 ul of antiserum and 1-3 ul of antigen are used.

The separating power of microimmunoelectrophoresis may be enhanced by carrying out the electrophoresis below light petroleum ether[45] (Fig. 7). This enables a constant

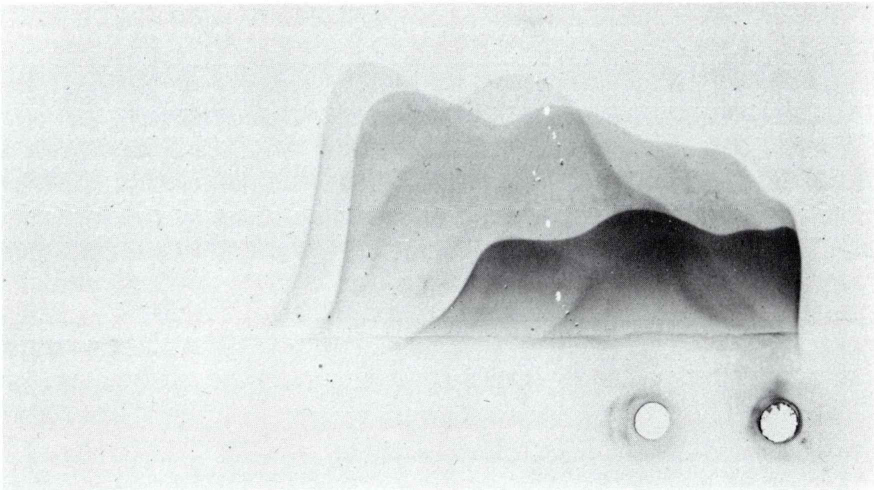

Fig. 5. The pattern of tandem crossed immunoelectrophoresis of two different serum samples. The application holes are below to the right. The primary electrophoretic migration is from right to the left, and the migration into the antibody-containng gel is from below to the top. The protein antigens with antigenic similarities are fused together into precipitates with double arcs (experimental details cf. Fig. 4).

temperature to be maintained, since heating of the electrophoretic slide during the run will cause evaporation of the petroleum ether, with consequent cooling. A field strength up to 20 V/cm can be used, because of the cooling process; this is higher than is possible in ordinary gel electrophoresis (5 to 7 V/cm). The higher field strength results in enhanced separation and faster movement of fractions.

2.1.3. Choises of type of pattern for simple immunoelectrophoresis are exemplified in Figs. 8 and 9. As a general rule, the antigen samples are applied in circular holes. However, when using the Wieme technique[45] with its higher resolving power, the antigen is placed in rectangular holes in order to avoid overlapping of adjacent fractions.

The choice of pattern is governed by the purpose of the immunoelectrophoresis and by the fact that the location of the precipitate in gel diffusion is influenced by the concentration of the antibody as well as of the antigen. For routine analysis therefore, a standard reference antigen solution of known concentration and electrophoretic mobility[36] should always be included. For instance, in order to cover the biological variations in studies of the antigen composition of different samples of a biological fluid, a "normal" sample obtained by pooling 10 or more samples, should be used as a standard. The standard is run on the same slide and against the same

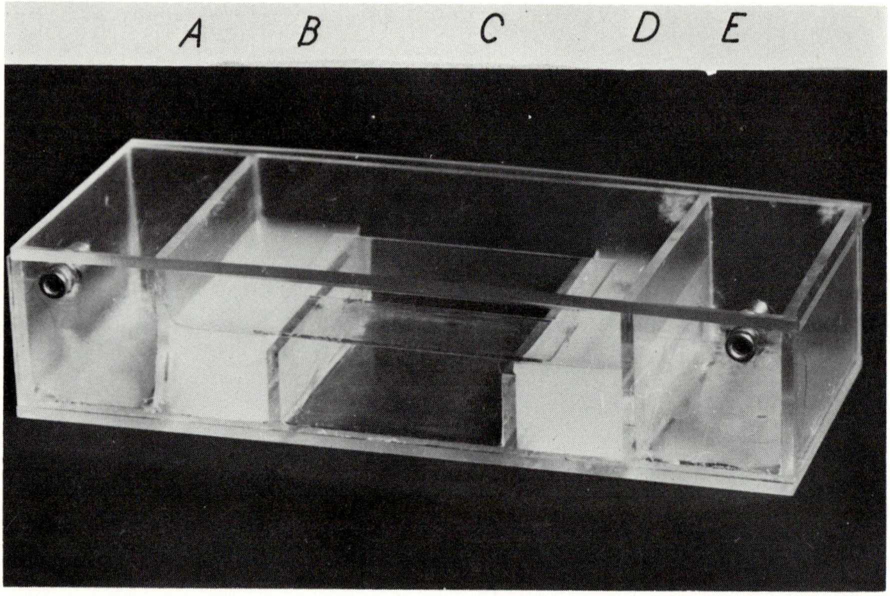

Fig. 6. A chamber for agar gel micro electrophoresis[45]. A: Anode vessel for buffer (3.2x4.0x7.5 cm); B: Anode vessel for agar buffer (2.5x2.2x7.5 cm); C: Central reservoir for petroleum ether (6.8x2.2x 7.5 cm); D: Cathode vessel for agar buffer (dimensions as B); E: Cathode vessel for buffer (dimensions as A). The walls between A and B and D and E have holes (4x1 cm) to provide electrical connections.

antiserum used for the unknown sample. The standard immunoelectrophoretic pattern in these cases may consist of two antigen wells cut equidistant from the antibody trough (Fig. 8). In this set-up the amount of an antigen may be roughly estimated from the location of the apex of the corresponding immunoprecipitate and the standards precipitates relative to the common antibody trough. As mentioned previously, an increased amount of antigen in the sample will cause the point of optimal precipitation to be placed closer to the antibody trough.

In conclusion, in order to evaluate an immunoelectrophoretic slide the individual arcs of the precipitation pattern produced by the sample and by the reference standard arcs are compared as follows:

1. the electrophoretic position of the precipitate;
2. the distance from the apex of the precipitin arc to the antiserum trough;
3. the length of the precipitation line;

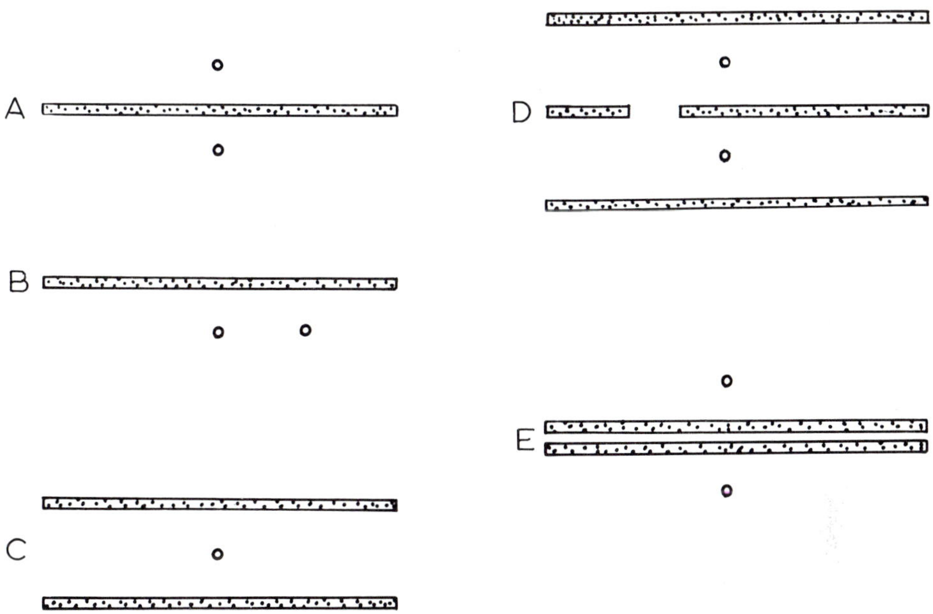

Fig.7. Different arrangement of wells and antibody troughs for immunoelectrophoresis. A: Conventional set-up for comparison to two different antigen mixtures; B: Another set-up for comparison of two different antigens; C: Set-up for the comparison of two different antisera, or for absorption experiments; D: Set-up for the combined diffusion method (cf. Fig. 9); E: Set-up for comparison of two different antigens or antibodies, when the antibodies originate from two different animal species.

4. the shape of the precipitate;
5. the sharpness and thickness of the precipitate; and
6. the interaction with other precipitates or the presence of double humped or multi-humped precipitate arcs (eventually fusion with other lines)

2.1.4. Crossed immunoelectrophoresis. The crossed immunoelectrophoresis which consists of an electrophoretic separation of protein antigens in an agarose gel in the first dimension followed by migration of the protein antigens into an antibody containing gel in the second dimension can be used for (semi)-quantitative assays of protein antigens.

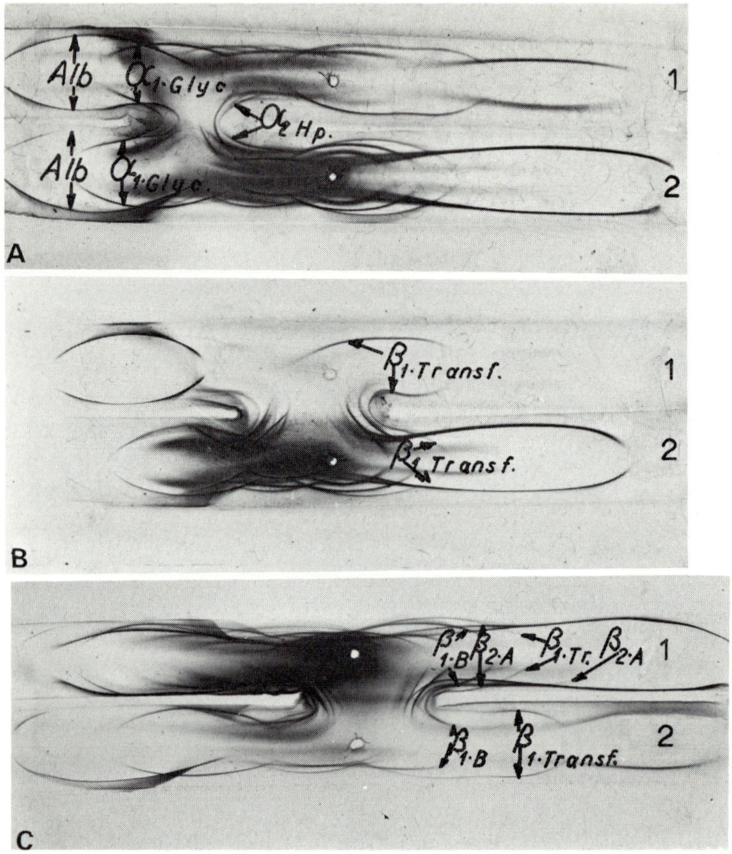

Fig. 8. The combined diffusion method of immunoelectrophoresis. Identification of the serum proteins found in concentrated normal cerebrospinal fluid by means of the immunoelectrophoretic tracings of normal human serum.
A: Identification of the serum proteins in the albumin and α-area. An intermediate agar bridge remains in the α-1-areas.(1) Immunoelectrophoresis of normal cerebrospinal fluid, concentrated 100 times.(2) Immunoelectrophoresis of normal human serum.
B: Identification of the serum proteins in the α-2- and β-1-area. An intermediate agar bridge remains in the α-2-area.(1) Immunoelectrophoresis of normal concentrated cerebrospinal fluid. (2) Immunoelectrophoresis of normal human serum.
C: Identification of the serum proteins in the β-2-area. (1) Immunoelectrophoresis of normal human serum. (2) Immunoelectrophoresis of normal cerebrospinal fluid concentrated 100 times.
Corresponding precipitation lines are similarly identified. In all experiments the three antibody reservoirs have been filled with horse-anti-human antiserum [36].

The choice of pattern has been mentioned above i.e. normal technique, tandem techniques and rocket technique.

It is important, especially when the student is working with an unknown protein solution in which the solubility properties of the indivudual protein antigens are unknown or if he is working with high molecular components or with membrane or plant proteins to evaluate results in different agarose gels with different electro-endo-osmotic properties. Thus certain firms e.g. Miles Labs. Inc. USA or Litex Company Ltd. Glostrup, Denmark, offer agarose batches with socalled normal, low and especially low electro-endo-osmotic properties (the batches of Litex are entitled HSA Litex, HBS Litex and HSC Litex respectively). Furthermore in such experiments often it is wise to add a thiol containing reagent such as cysteinamine to the antigen solution in order to split polymeric complexes. This is of importance since the quantitative assay is based upon the assumption that all identical protein antigens are precipitating the same amount of antibody. This may not be the case for dimers or trimers. Therefore the de-polymerisation is necessary in order to avoid irregular precipitation of antibodies. The electrophoretic separation in an agarose gel may be increased by exchanging the agarose electrophoresis in the first dimension by either polyacrylamide gel electrophoresis or thin layer gel filtration or even thin layer electrofocusing. Technical difficulties may however arise by the use of Sephadex gel in the first dimension either in thin layer chromatographic or thin layer electrophoretic procedures due to the increased electro-endo-osmosis which is cause by the sephadex gel. Secondly this causes distension of the precipitate formed by migration of the individual antigens into the antibody containing agarose. The precipitates are hereby blurred but the difficulties may partially be compensated by the use of agarose in the antibody containing gel with a specially low endo-osmosis such as Litex type HSC.

2.2. The immunochemical reactants. The immunoelectrophoretic methods enables qualitative and quantitative tracing of both antigens and antibodies.

2.2.1. Tracing of antigens: The tracing of antigens (protein- (lipo)-polysaccharide-virus- and tissue antigens[36]) is based upon the manufacture of corresponding antibodies (antiserum).

Preparation of antiserum. It is well known that injection of "pure" protein antigens into a number of animals of the same species will not cause the same immune response in all of them. Some animals may form only one specific antibody; others may form antibodies against other antigens. This "polyvalent" response to a "pure" antigen may be explained by adsorption of a few foreign antigen molecules onto it. It is therefore largely a question of luck as to whether a particular animal will produce only one antibody. This is the reason for immunizing several animals at the same time, hoping that at least one of them may respond by producing a specific antiserum.

Many different species of animals may be used for antibody production, for instance horse, goat, sheep, duck, hen, rabbit and guinea pig. Horses are expensive, but the horse type of antibody is characterized by forming well-defined precipitation arcs because its antigen-antibody complex is soluble in excess of both antigen and antibody. However, horse antibody may easily give rise to multiple immunoprecipitates under unbalanced or unstable conditions. In order to avoid these effects, rabbit antisera may be preferred. For the purpose of making specific antisera, rabbits are also convenient because they are inexpensive and so it is possible to use larger groups of animals from which those containing only the specific antibody with optimal titre may be selected. Rabbit-type antibodies are also present in goats and guinea pigs and most other species mentioned above.

Birds, especially hens, produce also good antibodies, but their antisera are not directly applicable in immuno-chemical reactions, because the sera are hyperlipaemic. Avian antibodies are to some extent different in their properties from mammalian antibodies, thus, in free buffer solutions their precipitating power seems maximal at salt concentrations about 10 times that normally used for mammalian antisera[46]. In cases of avian antisera where the lipid content disturbs the immunochemical reactions, the serum may be cleared by flotation or by a specific complex formation combined with flotation in solutions of the organic sulphate derivatives,i.e. heparin or dextran sulphate.

Several schedules for antigen injection have been suggested by different authors. Many of these are intended to decrease the immunization time by daily intravenous injections of antigens. These methods, however, seem inappropriate because the daily injections consume large amounts of antigen, which is often undesirable, especially when making specific antibodies, e.g. against certain protein hormones. Also daily injections of protein antigens may saturate the reticuloendothelial system[47].

In the authors opinion, it is wizest not to be too impatient in provoking antibody response by abandoning the classical extended immunization schedules.

Use of adjuvants. Not all macromolecules are equally effective antigens and it is often difficult to elicit any significant antibody response by injection of a pure antigen. Various local irritants and other adjuvants have been introduced, which overcome this difficulty to some extent. Thus, tapioca, nucleic acids, potassium aluminium hydroxide and various mineral oils emulsified with mycobacteria as for instance *Mycobacterium butyricum* and *M. smegmatis* or *haemophilus pertussis* and *corynebacterium parvum* have been used. Adjuvants containing the last two bacterias are available commercially.

The antigen is generally mixed with the adjuvant and injected subcutaneously. The exact mechanism of operation of these adjuvants is unknown[48], but in all probability the adjuvants induce local inflammation with proliferation of macrophages and bring macrophages into immediate contact with the antigen.

Absorption of antisera. In order to avoid the complicated and time-consuming procedures for isolation of a pure protein, mono- or oligospecific antisera against certain antigens may be prepared by certain "tricks". It is not possible in this chapter to cover all variables but some advice may be given. Newborn sera from man and certain animals lack some of the proteins present in adult sera, for instance haptoglobin and IgA globulin [47]. Thus, by absorbing a polyvalent antiserum against adult human serum with foetal serum in a dilution series (from 1:1 to 1:200) it is possible by means of immunoelectrophoresis to chose the dilution giving an antiserum specific against haptoglobin and IgA. Sera from patients genetically lacking albumin, α- or β-lipoprotein, IgG, IgA, β-1-AC, haptoglobin, etc., may be used similarly to prepare the corresponding antisera. Early in the lactation period, human milk contains no IgG globulin, but an immunoglobulin closely related to IgA[36]. Therefore, an anti-human milk antiserum does not contain antibodies against the gamma-chain of IgG. This antiserum may be used specifically against IgA after absorption with newborn serum.

Normal cerebrospinal fluid and urine contain only traces of α-2-macroglobulin and α-2-lipoprotein. These biological fluids added in a series of experiments in increasing amount to polyvalent antiserum against human serum may give a specific antiserum against these two macroglobulins.

Inulin added to serum causes precipitation of β-1-C (third part of complement). This precipitate can be used for preparing a specific antiserum[49].

The immunoelectrophoretical methods finally also permit quantitative and qualitative assay of antibodies provided their corresponding antigens are available.

2.4. Practical performance of immunoelectrophoretic assays

2.4.1. Simple immunoelectrophoresis:

2.4.1.1. Preparation of agar slides for simple immunoelectrophoresis

The buffered agar (2 % agar heated (95^0) in sodium veronalbuffer, 0.05 M pH 8.6) is stored at 4^0C in a stoppered flask or bottle. The flask is placed in a boiling water bath until the agar is completely melted. A 10 ml pipette is placed in the flask so that the pipette is at the same temperature as the agar. Microscope slides are degreased successively in 90 % (v/v) ethanol, ethanol:ethyl ether (1:1v/v) and ether. They are dried and numbered consecutively in agreement with protocol. The slides are placed on horizontal wooden blocks with their ends free. (A commercially available rack can be used). With the pipette, each slide is covered., free hand, with 1 ml of fluid agar, avoiding air bubble formation in the agar. If bubbles form, these can be removed by tilting the pipette slightly, at the same time touching the bubbles. After the agar has set on the slides they are placed in pairs in Petrie dishes containing a few drops of water or some damp cotton wool to avoid drying out of the agar. The dishes are stored at 4^0C until used.

It is especially important that agar should be of equal depth throughout the slide; otherwise, during the electrophoretic separation or the immunodiffusion, irregularities will cause asymmetrical migrations and diffusions. Furthermore, the thickness of the agar gel must not be over 1 mm, otherwise there may be difficulties in washing out the unreacted antigen and antibody and also peeling of the agar gel surface from the slide may occur after drying. It is wise to use special casting racks made of plastic which are commercially available. Equipment for cutting wells and troughs is also available commercially.

Cutting the pattern in gels of micro-immunoelectrophoresis and immunodiffusion: The cutting for immunodiffusion or immunoelectrophoresis may be made by free hand or by means of templates. The template technique is based upon the use of plastic or metal cutting and sucking devices for making appropriate wells and troughs[50,51]. Templates for immunoelectrophoretic analysis are commercially available. They are, however, expensive and the necessary equipment may well be home-made with advantage. Working with only a few slides it may be sufficient to place the agar plate on white paper on which wells and troughs are drawn. It is then possible to trace the pattern (vide infra).

The holes on the slides can be made by a syringe needle, 1 mm in diameter, square cut and ground at the end. By forcing the needle through the gel and applying continous suction by mouth through a rubber tube it is possible to make a hole with a capacity of about 1.5 ul. In the macro procedure the holes may be made by means of forcing metal cylinders (e.g. cork borer) of an appropriate diameter into the gel.

2.4.1.2. Techniques for immunoelectrophoresis

Manufacture of antibodies: As already mentioned antibodies are made by weekly subcutaneous (0.2 to 0.5 ml) or intradermal (0.010 to 0.050 ml) injection of an antigen solution in animals - appropriately rabbits. After immunization for four to eight weeks, the animal is bled exactly eight days after the last injection. The animal is bled by heart puncture or by cutting an ear vein after cutaneous irritation with to toluene[36].

Electrophoretic step:

In ordinary microscale immunoelectrophoresis the slide prepared by the micro procedure[4] allow the electrophoretic separation of proteins with isoelectric point below 6.8 within 1½ hr. at 5-10 V/cm, at pH 8.6. The electrophoretic stage may be performed in a custom-built or a commercially available apparatus, allowing from one to 18 slides to electrophorese simultaneously. A home-made box for electrophoresis may be prepared from a plastic box used for storing material in refrigerators, e.g. 20 x 20 x 5 cm. The box is equipped with platinum electrodes and with two transverse plastic boards in the middle of the box 5 cm from each other. These boards support the slides, and divide the box into anode and cathode compartments. The plastic boards are fixed to bottom and side of the box by dissolving a part of the corresponding edges of the boards in chloroform; slight pressure and heating will waterproof the board and fix it to the box.

Filter paper soaked in buffer and laid on both ends of the slides or frame connects the buffer solution in the anode and cathode compartments with the agar surface.

2.4.1.3. Immunodiffusion:

The application of antiserum for immunoelectrophoresis can best be performed by means of ordinary constriction pipettes. It is wise to apply a defined volume of antiserum (80-100 ul in the micro method). As already mentioned, the trough for antiserum is not made until after electrophoretic separation. The immunodiffusion must be performed in a moist chamber, i.e. the electrophoretic box, at constant temperature for 16 hr. Thereafter the precipitates may be inspected directly or the pattern may be washed, dried and stained.

2.4.2. Immunoelectrophoresis in an antibody-containing medium for quantitative estimation of antigens[25]

1 % (w/v) agarose in Na-barbital buffer (ionic strength 0.07, 2 mM calcium lactate, pH 8.6) is melted in boiling water and cooled to 45^0C.1 ml of antiserum (45^0) is added to 10 ml melted agarose and mixed.A mould is prepared from plate glass rectangles 1 cm wide and 1.5 mm thick, two placed along the short edges and one along the long edge of a plate (150 x 90 x 3 mm). A similar plate is then placed on top and the whole held together by spring paper clips. The mould 50^0 warm is placed with its open edge uppermost and filled with the agarose-antiserum mixture at a temperature of 40-45^0C through a pipette moved along the opening. No leakage will occur if the corners of the mould have been properly fitted together.The mould is left for at least 30 min to cool to room temperature. The paper clips are removed. The mould is placed flat on a bench and the top glass plate is carefully pushed off. This is facilitated by placing the long edge against a 4 mm thick stop plate to prevent the intermediate strips from following the sliding cover plate.

Holes (3-3.5 mm diameter) are cut in the agarose gel along a line 2 cm from and parallel with, one of the long edges. The distance between the centers of adjacent holes must be at least 8 mm. The holes are filled with 10 ul of antigen solution. 24 samples can be run on each gel. Dilutions of a standard antigen are used for calibration.

A cooling system, for instance that used in high voltage electrophoresis, may be used to cool the gels during electrophoresis.

The glass plate with the strips still in position is placed on the cooled surface (4^0C) of the cooling system and connected with the electrode vessel by agarose sheets (3 mm thick) or linen bridges. The agarose sheets (2 cm x 3 mm) are cut out of an agarose plate by a double-bladed knife and used as conducting bridges. Electrophoresis is run at 10 V/cm for 2 to 10 hr., depending on the charge and amount of antigen applied in relation to antibody concentration and on desired exactness of the estimation of the peak height.

The gels on the glass plates are placed in 0.2 M NaCl until the following day, then dried and stained in the usual way. The height of the various peaks e.g. the maximal migration distance of the boundary of antigen excess can be measure directly under darkfield illumination, with an error not exceeding 0.2 mm, if the peak heights are between 2 and 5 cm. A standard curve is constructed, plotting concentration of standard antigen against peak height. If the electrophoresis has been allowed to continue until all the antigen excess has been consumed and the precipitin arcs are stationary, the relationship is linear and only two antigen concentrations are needed as standard for calibration. In shorter experiments, a series of concentrations of the standard of the standard must be used covering the desired concentration range.

2.4.3. Crossed immunoelectrophoresis

Electrophoretic buffers: A buffer containing calcium lactate is especially recommended when the electrophoresis shall be used for separation of high molecular components veronal buffer, pH 8.6 (24.5 mM) containing tris (73 mM) and calcium lactate (0.36 mM) may be used). This solution is prepared by dilution 1:4 of the stem-solution mentioned below:
Dissolve 22.4 g diethylbarbituric acid (Veronal Merck), 44.3 g tris-(hydroxymethyl-aminomethane)(Fluka) and 0.53 g calcium lactate (May & Baker 97 %) mix in 1 l of distilled water.

Agarose: Three types of agarose may be used. Thus agarose with normal electro-endo-osmotic properties (such as type HSA Litex, Glostrup, Denmark), a batch with a low electro-endo-osmotic property (type HSB Litex, Glostrup, Denmark) or especially low electro-endo-osmotic properties (type HSC, Litex, Glostrup, Denmark).

Antisera: Initial trials must be performed in order to select the antisera with the highest number of precipitates and those which gives rise to the strongest immunoprecipitates.

Coating of the electrophoretic slides: Electrophoretic slides (8 x 8 cm or 10 x 10 cm) are defatted in ether and ethanol, dried and covered with a 1 mm thick layer of 0.5 g agarose dissolved in 100 ml distilled water. The coating agarose solution is heated to 100°C for about a 1/4 of an hour (until the suspension is homogenous) and by means of a preheated pipette the solution is spread in a 1 mm thick layer first on the upper part of the glass slide. The glass slide is hereafter dried below hot air, for instance from a hair drier. After drying the slides are covered on the other side in a similar manner and dried.

Preparation of the immunoelectrophoretic slides: The pre-coated slides are covered with an agarose solution containing 1 g of the appropriate agarose dissolved in 100 ml of the veronal buffer mentioned above. After heating in a water bath to the boiling point the solution is continously heated until it is homogeneous, the upper part of the glass slide is covered with a 1 mm thick layer of agarose buffer by means of a preheated pipette filled with the agarose solution (the slide being placed on two matches on a horizontal table). After solidification the slide is placed on a millimeter paper and an appropriate hole (for instance 1 or 2 mm in diameter) is punched in the gel, 1.5 cm from the cathodic edge and 1 cm from the edge which is parallel to the electrophoretic migration direction (first dimension). The hole is made by an appropriate plan-grinded needle or capillary metal tube or by means of a commercially available punch apparatus. . The hole is filled with the antigen solution. The amount should be defined for instance 1 or 5 ul. The slide is now placed in the electrophoretic apparatus (for instance the water cooled apparatus of Laurell[25]. The orientation of the hole in relationship to the anode and cathode depends on the isoelectric point (IP) of the protein antigens and the endo-osmotic flow in the gel and on the buffer selected. For separation of serum proteins, all having IP below 6.8, the veronal buffer will give rise to a negative change of the proteins. The electrophoretic strips e.g.filter papers sucked into the buffer are placed on the cathodic and anodic edges such that they connect these with the anodic and cathodic buffers solutions. After starting the cooling by circulation of tap-water (10-18^{0}) below the electrophoretic surface the electrophoretic chamber is now closed and sealed with vaseline. The electrophoresis is now performed in the first dimension at 3 volt pr. cm for about 1.5 to 5 hours depending on the electrophoretic migration velocity of the protein antigens under study. During the electrophoretic run the slides may be taken out and inspected below an ordinary office lamp. Hereby the migration lengths of proteins mowing into the agarose can be traced as dark shadows in the gel. After an appropriate migration length has been obtained the electrophoretic slide is now placed on the millimeter paper again and the gel gel being outside the electrophoretic zone is now cut away parallel to the anode-cathode direction 1 cm from the electrophoretic hole. This cutting may be made by a long knife with a planar edge. The unused gel is now replaced by an antibody containing gel.

Preparation of the antibody containing gel: The buffered agarose mentioned above is preheated to about 80^{0}C in a water bath. Hereafter it is cooled down and placed in a thermostated water bath at 45-48^{0}C and 10 ml of the buffered agarose solution of that temperature is now mixed in a preheated glass tube with 1 ml preheated (45-48^{0}C) antiserum. After thorough mixing the antibody containing agarose is now applied with a preheated pipette (45-48^{0}C) on the electrophoretic slide covering the part of the slide from which the agarose has been removed. Thus the two agarose gels are now in close contact. After solidification the electrophoretic slide is now placed again in the electrophoretic chamber in such a way that now the agarose strip containing the electrophoretic pattern is forming the cathodic part of the slide. The ½ cm of the anodic and the cathodic edge of the electrophoretic slide is now covered with a dialysis membrane which is bent down the edge of the electrophoretic chamber. The dialysis membrane is now covered with the electrophoretic strips connecting the slide with the anode and the cathode buffer solutions (the dialysis membrane may inhibit the proteins of the antiserum to migrate from the slide down in the buffer and hereby contaminating this solution. This procedure is not necessary if the buffers are renewed after every electrophoretic run or if the antibodies of the antiserum have been isolated by salting-out procedure). The electrophoretic chamber is now sealed again, cooled and the electrophoresis is performed with 1 volt pr. cm for about 18 hours.

Washing of the electrophoretic slide: The electrophoretic slide is placed in a horizontal glass container with 0.9 % saline solution for 2 to 24 hours. In order to avoid the gel phase to flote out from the glass surface, it is wise to place the slide in one of the corners of the glass container and along the two free edges of the slide two liminating brass blocks. After washing out non-reacted antigen and antibody the saline solution may be sucked off with a water pump and exchanged with distilled water for one hour. Hereafter the slide is dried in open air or below filter paper.

Protein staining: The slides are stained in an appropriate chamber in a solution of 0.5 % Coomassie Brilliant Blue R-250 in the destaining solution (450 ml ethanol + 100 ml glacial acedic acid + 450 ml distilled water). After 15 min of staining the slide is destained for 10 min in the destaining soluzion and dried in open air.

2.4.4. Crossed immunoelectrophoresis in a system combining thin layer electrofocussing in Polyacryl-amide gel with immuno-electrophoresis in an agarose gel:

Electrofocusing in a thin-layer polyacrylamide gel

Solution A: 29.1 % (w/v) Acrylamide in distilled water.
Solution B: 0.9 % (w/v) N,N´-methylenebisacrylamide (Canalco) in distilled water.
Solution C: 20.5 % (w/v) sucrose in distilled water.
Solutions A and B were stored at 4^{0}C and filtered before use.
The polyacrylamide gel was made from a mixture of the following composition:

Solution A	13.0 ml
Solution B	13.0 ml
Solution C	30.6 ml
Ampholine (LKB Ltd., Stockholm, Sweden) carrier ampholytes (pH 4-6(	3.0 ml
TEMED	100 ul
Ammonium persulphate 1 % (w/v)	2.0 ml

Polymerization occurs within 30 min at room temperature and the gel is ready for use after 10 min. at 4^{0}C. This gives a total gel concentration (T) of 6.3 % and a degree of crosslinkage (C) of 3 %.

The serum samples to be analyzed are applied on the cathode-side of the gel using 10 x 4 mm filter paper strips (Schleicher & Schüll 2043 a Mg 1). Up to 34 samples can be applied on one gel plate (e.g. 8.4 x 9.4 cm glass plate). Electrofocusing can be made for instance with the LKB 2117-010 Multiphor

The voltage is gradually increased with the current set at 50 mA. Between 600 V and 1080 V the current should be set at 25 mA.

The efficient cooling system of the LKB 2117 Multiphor makes it possible to keep the running time to less than 2 hour.

Fixing and staining can be performed by immersing the gel in a solution of 0.75 g Coomassie brilliant blue R-250 (Merck), 22.5 g sulfosalicylic acid and 225 ml methanol in 465 ml distilled water, for 60 min at 80°C. For destaining, a solution of water-ethanol-acetic acid (8:3:1) can be used.

3. Evaluation of results

3.1. Simple immunoelectrophoresis

The shape of the precipitates in immunoelectrophoresis is influenced both by the electrophoretic mobility of the antigens[52] and by the fact that double diffusion in two dimensions has taken place. The morphology of precipitates can be used for identification. Fusion and line deviation and spur phenomena all occur under appropriate conditions.[36] Electrophoretically homogenous antigens will migrate in the gel as a zone in which the distribution of the individual molecules follows a Gaussian distribution. In that case, a more or less circular precipitation zone is formed. On the other hand, substances with identical antigenic properties but possessing different electrophoretic mobilities will produce immunoprecipitates located in accordance with the electrophoretic mobilities. Thus, immunoglobulins precipitated by the corresponding antibodies develop as protracted lines extending from the slow γ-area (M_r = 0.00) towards the α-2-area (M_r = 0.50). This indicates a family of globulins with identical immunological (antigenic) properties, but different electrophoretic mobilities[36].

Double-humped arcs indicate the presence of an antigen with two main electrophoretic mobilities, but with identical or with partially identical immunological properties. Immunoelectrophoretic studies of biological fluids have revealed double-humped arcs for the transferrin of cerebropsinal fluid and for the β-1-AC globulin (the third part of the complement) present in serum stored for more than a few days[36].

Uriel and Grabar[53] demonstrated two fast-moving lipoproteins in immunoelectrophoresis, one, present in fresh serum, possessing an electrophoretic mobility of an α-1-globulin, and one present in stored serum with pre-albumin mobility. This change during storage was later shown to be caused by a lipase[54].

The electrophoretic mobility may thus be used to indicate successive changes in the composition of stored antigens. In studies comparing two or more immunoelectrophoretic patterns, the electrophoretic mobility of an antigen cannot be used solely as a criterion of identity of different antigens precipitated in different patterns, since, besides that mentioned above, in groups of individuals the electrophoretic mobility of the so-called group-specific proteins may differ from one group to another (allotypia). The differences in migration rate of antigens with identical immunological properties, seems to be genetically determined[55]

The shape of the immunoprecipitate can be an indicator of the conformational state of the protein antigen and can show whether the protein is native, denatured or split into subfractions by proteolytic digestion. Endopeptidases may split proteins into subfractions which may have different electrophoretic mobilities, but possess the antigenic sites of the original protein antigens. After immunodiffusion every subfraction will therefore give rise to an immunoprecipitate. However, because of the immunological similarities, the immunoprecipitates may fuse, causing double-humped or multiple-humped precipitates. Similar changes are encountered during denaturation of proteins by chemical treatment, for example, during iodination, if more than two molecules of iodine have been introduced into one molecule of protein in which case the corresponding immunoprecipitate is double-humped, shows asymmetric arcs or a striated pattern in the gel. Also hydrazine treatment is known to change the mobility of a β-1-C in human serum[56].

Immunoelectrophoresis in an antibody-containing gel (crossed immunoelectrophoresis and related techniques) is based on the electrophoretic migration of the zone of excess of antigen through the antibody containing gel. The migration of the arcs through the gel ahead of the zone of antigen excess depends on the properties of the antigen. The migration of certain proteins can indicate additional antigenic sites especially in proteins where allotypic forms exist[36]. Double precipitation arcs develop, when the antigens possess different electrophoretic mobilities. The enzymic and chemical properties of the antigens may be used for their tracing[36].

As already mentioned in the present chapter it is possible to trace and identify protein antigens in the solution by crossed immunoelectrophoresis. Thus protein antigens with different electrophoretic mobilities but with partial immunochemical identity may in the crossed immunoelectrophoretic pattern show spur-phenomena where precipitation lines interlink the major precipitates corresponding to each of the two electrophoretic areas. Another way of identification of protein antigens may be by specific staining of the immunoprecipitate for certain chemical groups or by enzymic staining for the enzymatic properties of the corresponding antigens. Also prior absorption of an antiserum with a corresponding purified antigen may identify the antigen by comparison of the crossed immunoelectrophoretic pattern prior to an after absorption of the antiserum. As already mentioned, furthermore the tandem-crossed immunoelectrophoresis, the fused rocket immunoelectrophoresis and the line immuno-electrophoresis may be used for identification purposes because of the spurs linking of two identical precipitates of the two protein antigens in the electrophoretic pattern. The line immuno-electrophoretic method may also be used for comparison purposes of different antisera by electrophoresis in the second dimension with a common protein antigen mixture.

As already mentioned above the immunoelectrophoretic methods involving electrophoresis in an antibody containing medium may be used for quantitative estimation of antigens[36].

REFERENCES

1. Grabar, P. and Williams, Jr. (1953) Biochim. Biophys. Acta 10, 193-194.
2. Poulik, M.D. (1952) Can. J. Med. Sci. 30, 417-419.
3. Poulik, M.D. (1958) Fed. Proc. 17, 530.
4. Scheidegger, J.J. (1955) Intern. Arch. Allergy Appl. Immunol. 7, 103-110.
5. Clausen, J. (1960) World Neurol. 1, 479-490.

6. Blanc, B. (1959) Bull. Soc. Chim. Biol. 41, 891-899.

7. Blanc, B. (1961) in Protides of the Biological Fluids (Peeters, H., ed.) p. 159-161, Elsevier, Amsterdam.

8. Peeters, A. and Vuylsteke, P. (1960) in Protides of the Biological Fluids (Peeters, H., ed.) p. 31-35, Elsevier, Amsterdam.

9. Araki, C. (1937) J. Chem. Soc. Japan, 58, 1338-1340.

10. Uriel, J., Avrameas, S. and Grabar, P. (1964) in Protides of the Biological Fluids (Peeters, H., ed.) p. 355-359, Elsevier, Amsterdam.

11. Kohn, J. (1958) Clin. Chim. Acta 3, 450-454.

12. Kohn, J. (1962) in Protides of the Biological Fluids (Peeters, H., ed.) p. 120-122, Elsevier, Amsterdam.

13. Kohn, J. (1968) Nature (Lond.) 217, 1261-1262.

14. Lomanto, B. and Vergani, C. (1967) Clin. Chim. Acta 15, 169-171.

15. Smithies, O. (1955) Biochem. J. 61, 629-641.

16. Smithies, O. (1959) Biochem. J. 71, 585-587.

17. Antoine, B. (1962) Science 138, 977-978.

18. Keutel, H.J. (1964) Ann. N.Y. Acad. Sci. 121, 484-489.

19. Heimburger, N. and Schwick, G. (1962) Thromb. Diath. Haemorrhag. 7, 432-443.

20. Schwick, G. (1967) in Biochemie der Fibrinolyse (Künzer, W., Winckelmann, G. and Walther, C. eds.), p. 27, Verh. Deut. Arbeitsgemeinschaft, Butgerinnungsf. Schattauer-Verlag, Stutgart.

21. Astrup, T. and Müllertz, S. (1952) Arch. Biochem. Biophys. 40, 346-351.

22. Bussard, A. (1959) Biochim. Biophys. Acta 34, 258-260.

23. Ressler, N. (1960) Clin. Chim. Acta 5, 795-800.

24. Ressler, N. (1960) Clin. Chim. Acta 5, 359-365.

25. Laurell, C.B. (1966) Analyt. Biochem. 15, 45-52.

26. Clarke, H.M.G. and Freeman, T. (1967) in Protides of the Biological Fluids (Peeters, H., ed.) p. 503-509, Elsevier, Amsterdam.

27. Krøll, J. (1968) Scand. J. Clin. Lab. Invest. 22, 79-81.

28. Axelsen, N.H., Krøll, J. and Weeke, B. (1973) Scand. J. Immun. Suppl. 1, 15-169.

29. Svendsen, P.J. and Rose, . (1970) Science Tools, 17, 13-20.

30. Hejgaard, J. and Bøg-Hansen, T.C. (1974) J. Inst. Brew. 80, 436-442.

31. Jørgensen, A.O., Subrahmanyan, L., Turnbull, C. and V.J. Kalnins (1976) Proc. Natl. Acad. Sci. 73, 3192-3196.

32. Weeke, B. (1968) Scand. J. Clin. Lab. Invest. 21, 351-354.

33. Weeke, B. (1970) Scand. J. Clin. Lab. Invest. 25, 269-275.

34. Krøll, J. (1969) Scand. J. Clin. Lab. Invest. 24, 55-60.

35. Curry, M.O., McConathy, W.J., Alaupovic, P., Ledford, H.J. and Propovic, M. (1976) Biochim. Biophys. Acta 439, 413-425.

36. Clausen, J. (1969) in Laboratory Techniques in Biochemistry and Molecular Biology: Immunochemical techniques for the identification and estimation of macromolecules (Work, T.S. and Work, E., edts) North-Holland, Amsterdam.

37. Hejgaard, J. and Sørensen, S.B. (1975) Comp. Rend. Lab. Carlsberg, 40, 187-203.

38. Söderholm, J., Smyth, C.J. and Wadström, T. (1975) Scand. J. Immunol. 4, 107-113.

39. Berzins, K. Blomberg, F., Kjellgren, Smyth, C. and Wadström, T. (1976) FEBS Letters, 61, 77-80.

40. Johansson, B.G. and Stenflo, J. (1971) Anal. Biochem. 40, 232-236.

41. Skude, G. and Jeppsson, J.-O. (1972) Scand. J. Clin. Lab. Invest. 29, Suppl. 124, 55-58.

42. Grubb, A. (1973) Anal. Biochem. 55, 582-592.

43. Johansson, B.G. and Hjerten, S. (1974) Anal. Biochem. 59, 200-213.

44. Arnaud, Ph. and Cryssel, R. (1975) Applic. Note LKB., LKB. Ltd., Stockholm. Sweden, 185,.

45. Wieme, R.J. (1969) Studies on agar gel electrophoresis, Arcia Uitgaven, N.V., Brussels and Elsevier, Amsterdam, 1965.

46. Crowle, A.J. (1961) Immunodiffusion, Acad. Press, Inc., N.Y.

47. Sweet, L.C., Abrams, G.D. and Johnson, A.G. (1965) J. Immunol. 94, 105-110.

48. Schultze, H.E. (1961) Immunitätslehre (Cobet, R., Gutzeit, K. and Bock, H.E.,eds.) Klin. der Gegenwart. Handbuch der praktischen Med., Verlag von Urban & Schwarsenberg, München-Berlin.

49. Brummerstedt-Hansen, E. (1967) The serum proteins of the pig (Munksgaard, ed.) Copenhagen.

50. Heremans, J. (1960) Les globulins sēriques du systeme gamma (Arcia S.A., ed.) Brussels.

51. Crowle, A.J. and Leuker, D.C. (1962) J. Lab. Clin. Med. 59, 697-704.

52. Hirschfeld, J. (1960) Nature (Lond.) 185, 164-165.

53. Uriel, J. and Grabar, P. (1965) Bull. Soc. Chim. Biol. 38, 1253-1269.

54. Schultze, H.E. and Schwick, G. (1957) Behringwerk-Mitt. 33, 11-38.

55. Hirschfeld, J. (1962) The Ge system. Immunoelectrophoretic studies of normal human sera with special reference to a new genetically determined system (Gc). Thesis. Progr. Allergy 6, 155-186.

56. Bammer, H. (1964) J. Neurol. Sci. 1, 129-151.

Electrokinetic Separation Methods,
P.G. Righetti, C.J. van Oss and J.W. Vanderhoff

INFLUENCE OF ELECTROOSMOSIS

J. W. VANDERHOFF AND F. J. MICALE

Center for Surface and Coatings Research, Lehigh University, Bethlehem, Pennsylvania 18015 (USA)

INTRODUCTION

An electric potential applied to a dispersion of particles in a fluid medium induces two different types of electrokinetic phenomena - electrophoresis and electroosmosis. Electrophoresis is the movement of the dispersed particles toward the electrode of opposite charge, and is the result of the electrical double-layer at the particle/liquid interface. Electroosmosis is the backward flow of the liquid medium along the walls of the electrophoresis chamber. The driving force for electroosmosis is the same as for electrophoresis, except that, in this case, the solid is fixed and the liquid moves with respect to the solid. Since electrophoresis and electroosmosis was discovered by Reuss[1] in 1808, these phenomena have been used for the characterization of solid/liquid interfaces and the separation of biological cells. Although an analytical investigation or preparative separation of particles or cells by definition involves the electrophoretic movement of particles in an electric field, electroosmosis is always involved to a greater or less degree because the experimental conditions require the presence of a wall in contact with the liquid phase, which invariably has an electrical double-layer associated with it. The actual movement of particles under these experimental conditions, therefore, is the summation of the electrophoretic velocity and the solvent velocity due to electroosmosis. The subject of this chapter is the electroosmotic flow of solvent in free-fluid (static) systems, where it is important in predicting and interpreting the behavior of particles in both analytical and preparative electrophoresis experiments.

THEORY OF ELECTROOSMOSIS

An electrophoresis chamber may be of any geometry, but is usually constructed with a circular or "rectangular" cross-section. The walls of such chambers may be made of any non-conducting material, such as glass or various plastics. The type of material used for the walls and the ionic composition of the liquid medium in contact with the walls invariably results in the formation of an electrical double-layer at the wall/liquid interface, which is the driving force for electroosmosis. When the electrophoresis chamber is a closed system, which is the case for free-fluid electrophoresis, the liquid flows along the wall of the chamber and there is a return flow in the center of the channel, resulting in levels of zero solvent flow. The solvent flow profile is governed by two principles of hydrodynamic theory: 1) the net volume flow of solvent across any section perpendicular to the direction of flow is zero in the steady state; and 2) the solvent velocity profile as a function of position in the channel is parabolic. A schematic representation of these principles is shown in Figure 1, where the channel wall is negatively charged with respect to the liquid and the shaded area marked A is equal to the sum of the two shaded areas marked B.

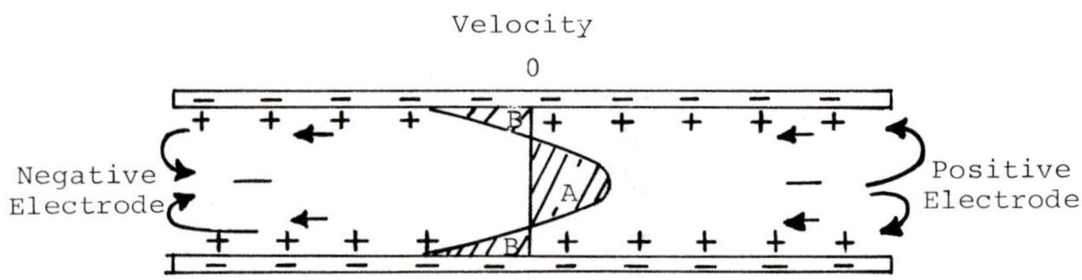

Figure 1: Schematic Representation of Electroosmotic Flow in Electrophoresis Cell.

When colloidal particles dispersed in a fluid medium are subjected to an electric field, they move with a velocity which is a function of their electrical double-layer properties. If there is no electroosmosis, the particles move with a constant velocity, independent of their position in the cell. When there is an electroosmotic flow, the total particle velocity v_t will be the sum of the electrophoretic velocity v_e and the solvent velocity v_s according to the equation:

$$v_t = v_e + v_s \qquad (1)$$

where v_e is constant for a given applied potential and v_s is a function of position in the channel. When the channel is cylindrical, the solvent velocity is defined by the Equation:

$$v_s = v_o \{(2r^2/a^2) - 1\} \qquad (2)$$

where a is the channel radius, r the radial distance from the center of the channel, and v_o the solvent velocity at the channel wall/liquid interface, i.e., at $r = a$. The condition for zero solvent flow, i.e., $v_s = 0$, which is important for analytical electrophoresis, is defined by the equation:

$$r = (a/2)^{1/2} \qquad (3)$$

The driving force for the solvent velocity is the potential of the electrical double-layer as expressed in the following equation derived by Helmholtz[2]:

$$v_o = E\varepsilon\zeta/4\pi\eta \qquad (4)$$

where $\underline{E}$ is the strength of the applied electric field, ε the dielectric constant of the fluid, η the fluid viscosity, and ζ the potential of the liquid layer which does not move relative to the channel wall, i.e., the potential at the slipping plane or the zeta potential. The solvent velocity for a given liquid medium and applied potential, therefore, can be decreased by a surface treatment which results in a decreased zeta potential or an increased viscosity at the slipping plane.

A novel electrophoresis apparatus has been proposed by Smith and Lisse[3] in which the electrode compartments are connected by two cylindrical channels with identical electrical double-layer properties. The length $\underline{l}$ and radius $\underline{a}$ of the two channels, defined by the subscripts 1 and 2, must satisfy the relationship:

$$l_2/l_1 = (a_2/a_1)^2[(a_2/a_1)^2 - 2] \qquad (5)$$

When the above conditions are satisfied, there is no return flow of solvent in the smaller channel and the solvent velocity for the small channel will be defined by the equation:

$$v_s = v_o \frac{r^2}{a^2} \qquad (6)$$

where the solvent velocity is equal to zero in the center of the channel, i.e., at $\underline{r} = 0$. This cell has been proposed for analytical electrophoresis because both the solvent velocity and the solvent velocity gradient are equal to zero at the same position as opposed to a single-channel cell where the solvent velocity is equal to zero at a position where the solvent velocity gradient is large. The potential advantage of a double-channel cell for preparative electrophoresis is that there is no return flow of solvent and the magnitude of the resulting parabolic flow profile is decreased by a factor of two. Figure 2 shows the parabolic flow profiles for single-channel (Equation 2) and double-channel cells (Equation 6) when the electroosmotic velocities at the channel walls are identical for both cases.

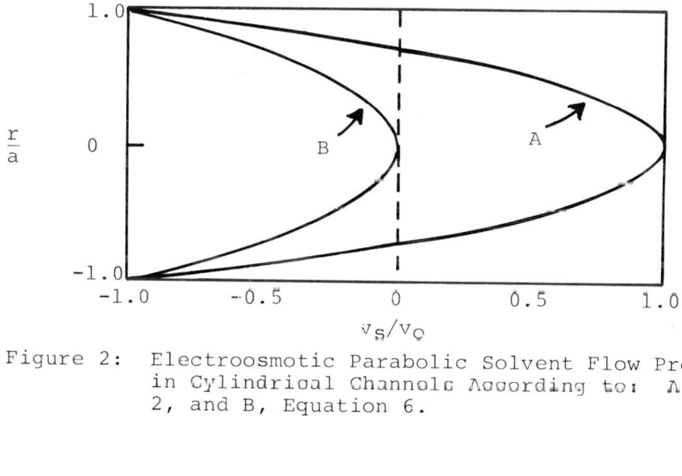

Figure 2: Electroosmotic Parabolic Solvent Flow Profile in Cylindrical Channels According to: A, Equation 2, and B, Equation 6.

The double-channel cell has not been used extensively, apparently because the advantages of the solvent flow profile do not offset the disadvantages of the two-channel system, especially for analytical electorphoresis. It has a potential advantage, however, for preparative electrophoresis, where the resolution of separation is improved by a 50% reduction in the magnitude of the parabolic solvent flow profile.

Another type of cell consists of a channel with a rectangular cross section. The electroosmotic solvent profile for a cell with this geometry has been developed by Komagata[4], according to the following equation, which uses the nomenclature of Figure 3:

$$\frac{v_s(x = 0)}{v_o} = 1 - \frac{3}{2}\left[1 - \frac{y^2}{b^2}\right]\left[1 - \frac{192}{5K}\right]^{-1} \tag{7}$$

where $v_s(x = 0)$ is the solvent velocity measured as a function of height y from the center of the channel and at the mid-width v_o the solvent velocity at the channel wall/liquid interface, i.e., at $y = b$, b one-half the channel heights, and K the ratio

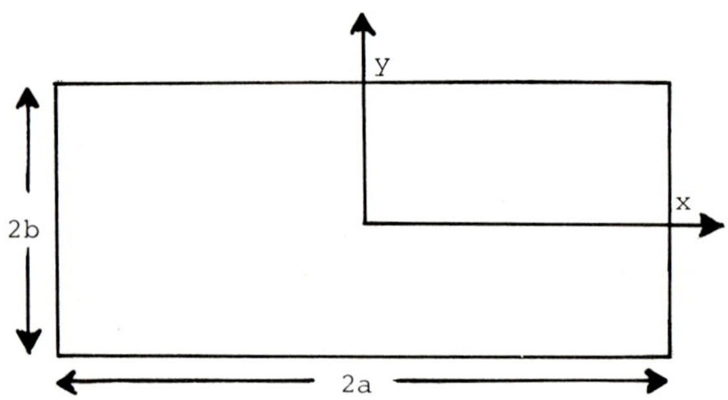

Figure 3: Cross Section of Rectangular Electrophoresis Channel.

of channel width to channel height.

Figure 4 shows the linear form of the calculated parabolic flow profiles, normalized with respect to the electroosmotic flow at the channel wall, for rectangular channels with different values of K, for cylindrical channels, and for double cylindrical channels. The minimum electroosmotic flow is observed for the double cylindrical channels, followed by rectangular channels with high K values, and finally, by the cylindrical channels. Although the geometry of the system affects the magnitude of the electroosmotic flow, the degree of change for the systems described here is not great, with the greatest decrease being observed for the compelx double cylindrical channel cell. Therefore, a more promising approach to decrease the electroosmotic mobility is to decrease the magnitude of the electricle double-layer at the channel wall/liquid interface by molecular adsorption or coating techniques.

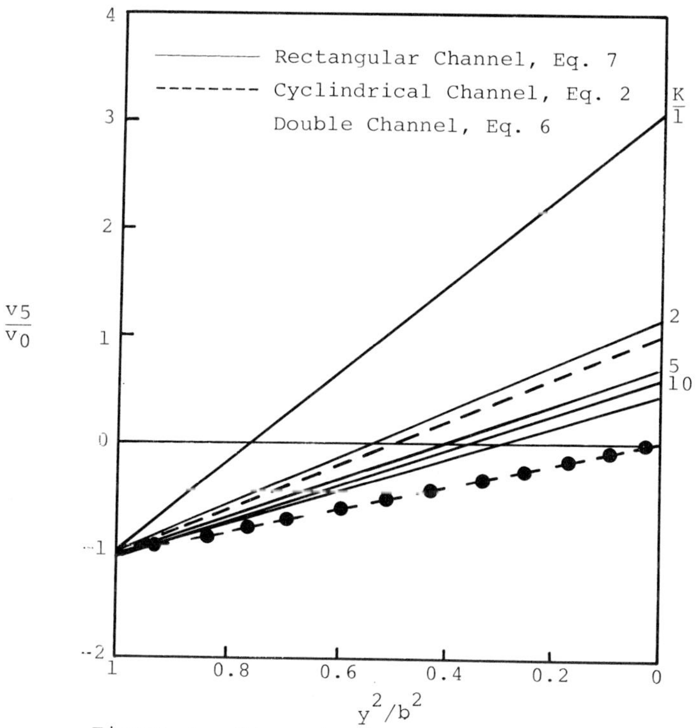

Figure 4: Linearized Parabolic Flow Profiles for Rectangular Electrophoresis Channels with Different Values of K, Cylindrical Channel, and Double-Cylindrical Channels.

LOW-ELECTROOSMOTIC-MOBILITY COATINGS

NASA has initiated a program on the electrophoretic separation of biological cells in space. The absence of thermal convection and gravitational settling promises a resolution of separation that cannot be obtained on earth. On the Apollo 16 mission, the free-fluid (static) electrophoretic separation of mixed 0.23μm- and 0.80μm-diameter monodisperse polystyrene latex particles gave elongated, bullet-shaped particle profiles because of the strong electroosmotic flow. Analysis of the flight photographs and ground-based experiments showed that the two sizes separated to form two nested parabolic cones, in accord with a mathematical model based on theory[5]. Figure 5 shows the results of this experiment after 3 minutes and demonstrates that a practical separation was obviated by the strong electroosmotic flow of the medium, which gave the parabolic particle profiles. These results demonstrated the need for a reliable low-electroosmotic-mobility coating which could be used for future electrophoretic separations in space. The following section describes the development of such a coating and an experimental method for the evaluation of such coatings[6].

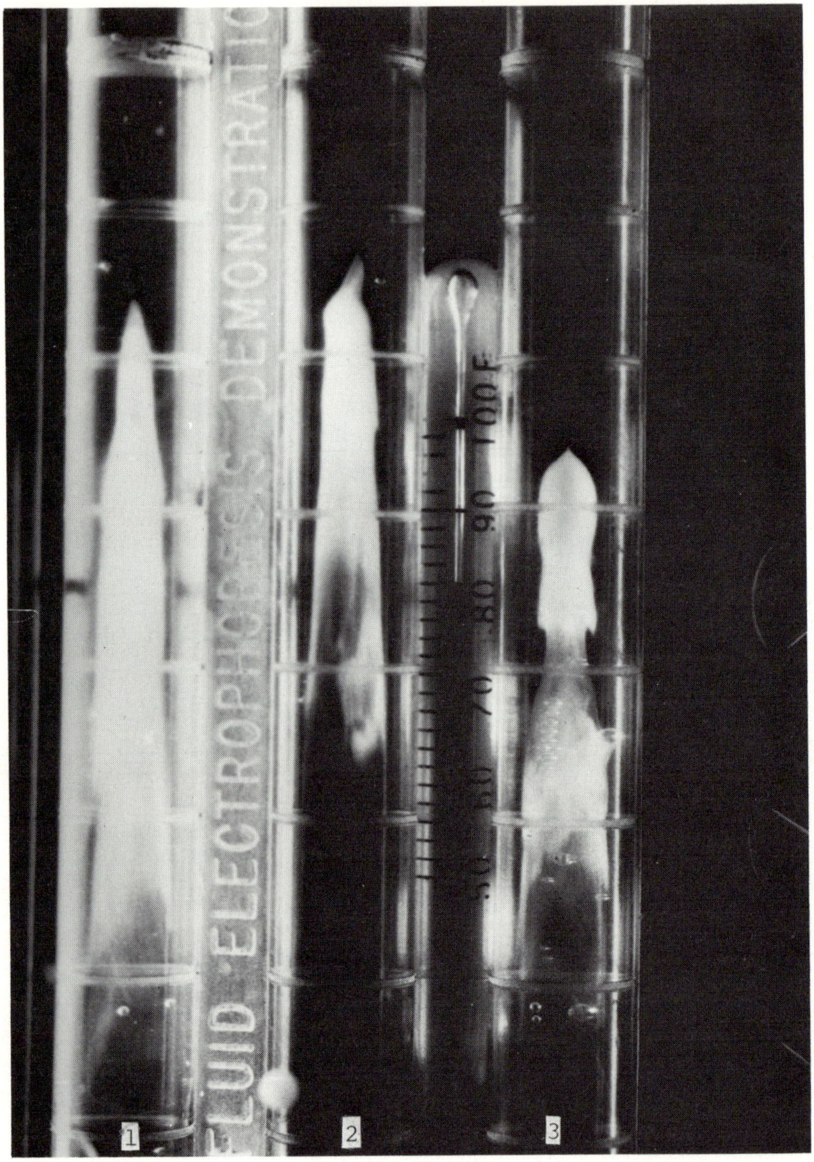

Figure 5: Electrophoretic Separation of Polystyrene Latexes on Apollo 16 after Three Minutes. Tube No. 1 - Combination of 0.80 and 0.2μm latexes, 2 - 0.8μm latex, and 3 - 0.36μm latex.

Experimental. The method developed in this study was to construct both cylindrical and rectangular microelectrophoresis cells from the materials under investigation and to measure directly the velocity of individual particles of known electrophoretic mobility in an applied electric field as a function of position in the cell. The experimental design required that the channel be easily removed from the cell. Figure 6 shows a diagram of an electrophoresis cell designed for cylindrical channels, constructed with threaded nylon caps and O rings, which seal the channel in the cell and allow for easy replacement of the channel. A metal holder was constructed to clamp the electrode compartments in a fixed position. The cell was designed to fit into the constant-temperature bath of the Rank Brothers electrophoresis

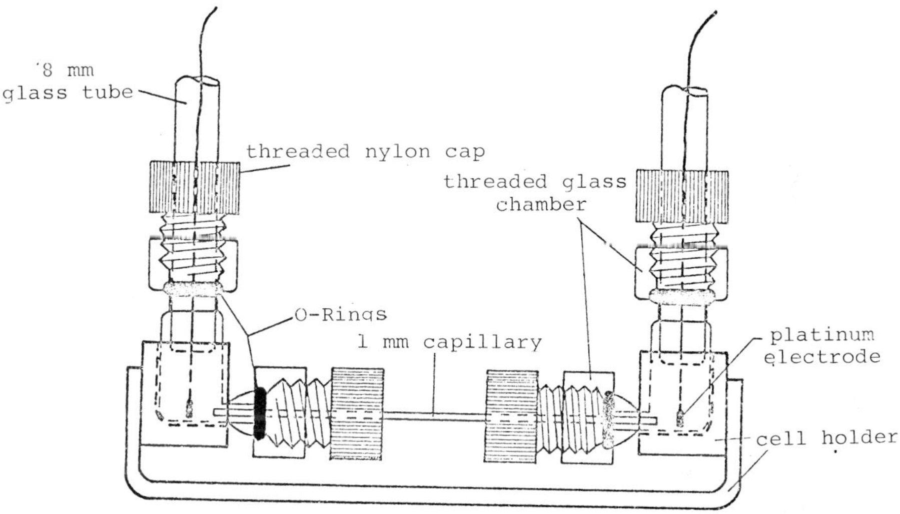

Figure 6: Electrophoresis Cell with Removable Capillary Channel.

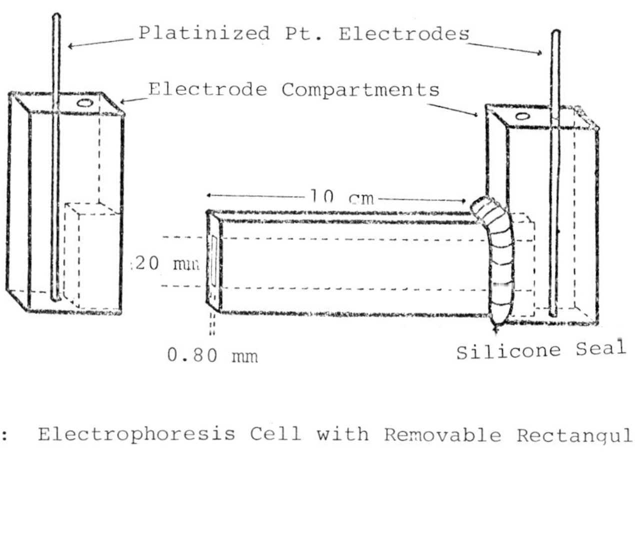

Figure 7: Electrophoresis Cell with Removable Rectangular Channel.

apparatus, Figure 7 shows a diagram of an analogous cell with a replaceable rectangular channel. This cell was used when the materials under investigation could only be obtained in the form of sheets.

Figures 8 and 9 shows the linearized form of the variation of observed electrophoretic mobility of 0.80μm polystyrene latex with position in the channel for glass capillaries and Lexan rectangular channels, respectively. The electroosmotic mobility values in the different media were calculated from the electrophoretic mobility in that medium, using Equations 2 and 7. This experimental technique was used to calculate the electroosmotic mobility values for the different surfaces coated in various ways.

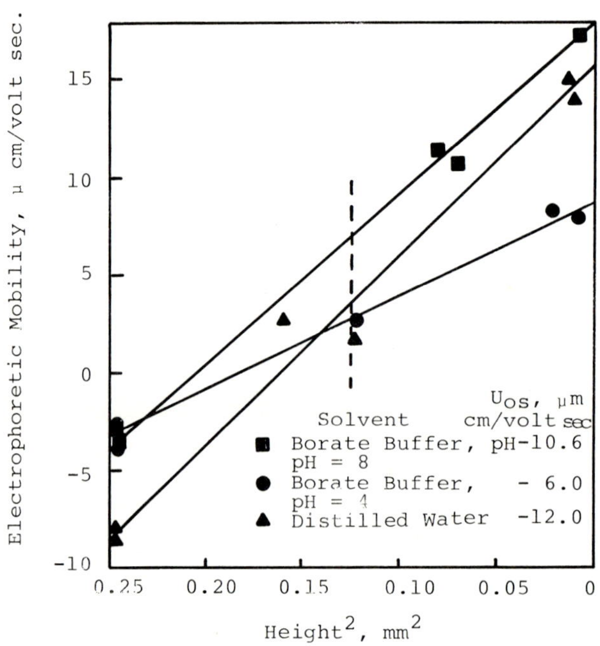

Figure 8: Observed Electrophoretic Mobility Parabolic Profile in Uncoated Glass Capillary.

<u>Results</u>. The most effective material for reducing electroosmosis was found to be methylcellulose, which was shown, however, to desorb from the surface in less than an hour in agreement with the work of Hjerten[7]. Therefore, the methylcellulose was more strongly adsorbed to the glass surface by treating the surface with γ-glycidoxypropyltrimethoxysilane (Dow Corning Corp. Z-6040) which binds to the glass, then using the epoxide group to link with the hydroxyl groups of the methylcellulose molecules. The proposed mechanism[8] for the binding of this and other trialkoxysilanes to glass surfaces is as follows:

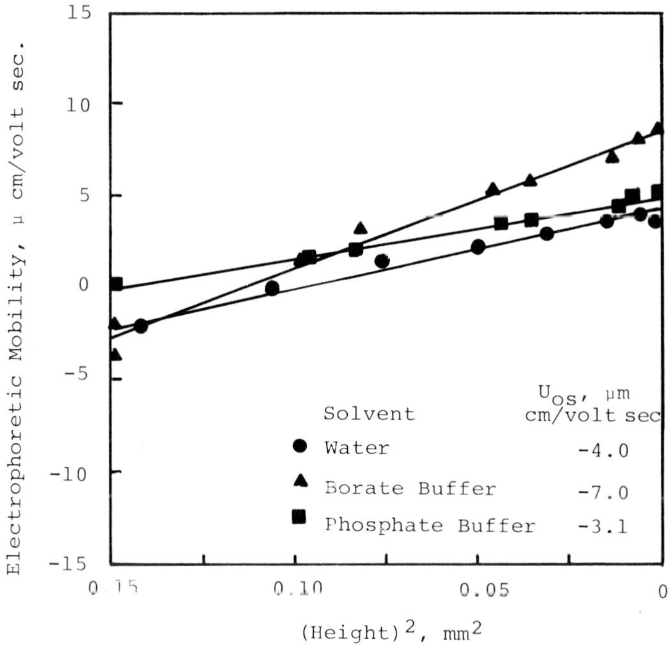

Figure 9: Observed Electrophoretic Mobility Parabolic Profile in Uncoated Rectangular Lexan Channel.

1. hydrolysis of the trialkoxysilane to the corresponding silanetriol;
2. chemisorption of the silanetriol on the glass surface (interaction between a hydroxyl group of the glass with one of the hydroxyl groups of the silanetriol);
3. formation of hydrogen bonds between hydroxyl groups of adjacent chemisorbed silanetriol molecules;
4. polymerization of adjacent silanetriol molecules during drying by splitting out water from hydrogen-bonded hydroxyl groups.

For the Dow Corning Z-6040, such a process would give a polysilanetriol chemisorbed to the glass, with an exposed surface of epoxide groups, which are free to form bonds with compounds containing active hydrogen, such as methylcellulose. The methylcellulose (Dow Methocel MOC) is applied over the Z6040-treated surface and heated to increase the interaction between the epoxide groups of the Z6040 and the hydroxyl groups of the methylcellulose.

Part of this methylcellulose coating can be removed by rigorous washing -- this is referred to as "strongly bound". This distinction is important because the "physically adsorbed" methylcellulose apparently is governed by an adsorption-desorption equilibrium and thus, when a colloidal sol with partially covered

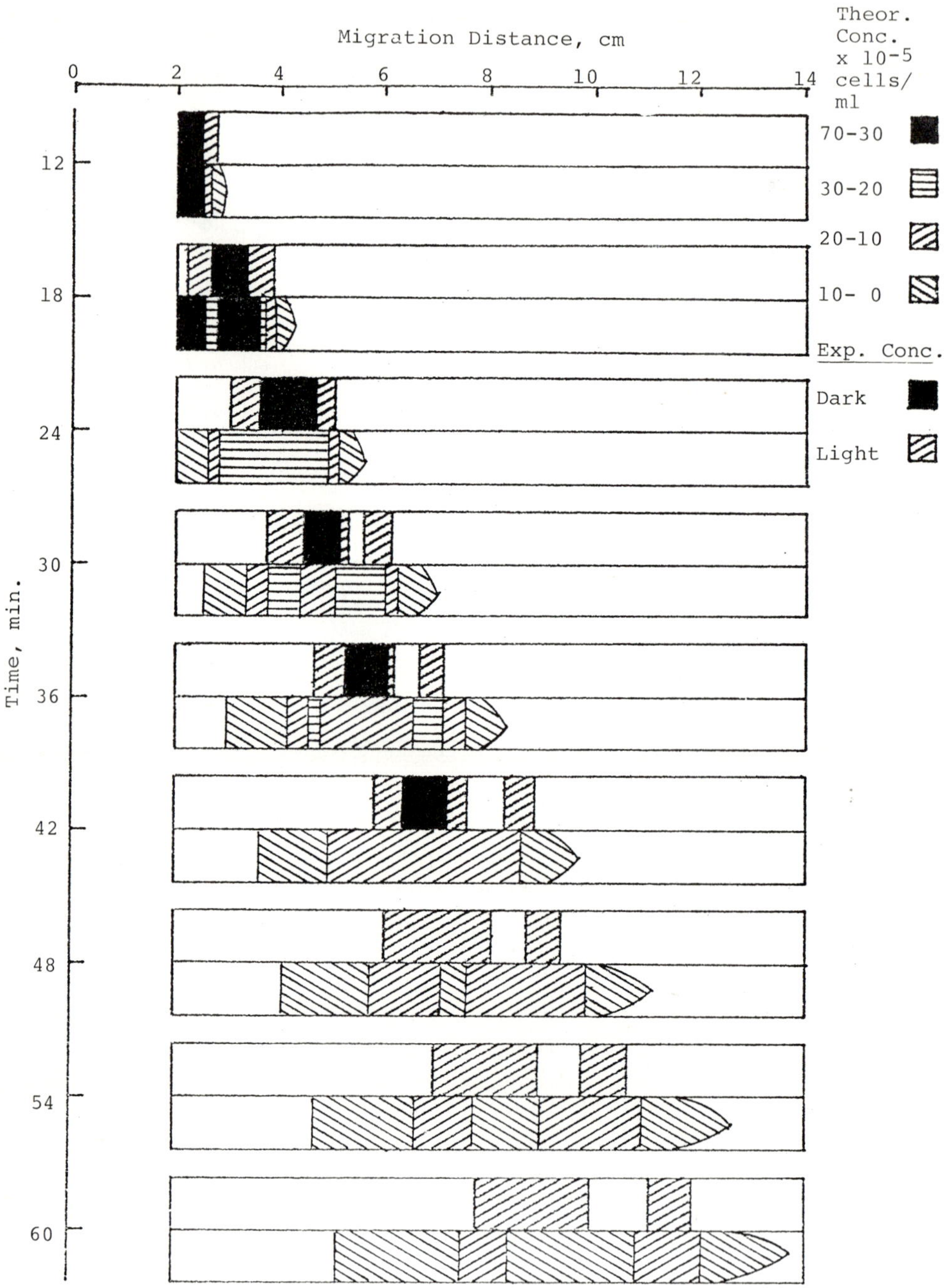

Figure 10: Position of Bands as Determined from Visual Observation of Flight Film and Theoretical Calculations.

surfaces is added to the cell, the methylcellulose may desorb from the cell wall and readsorb on the colloidal particles, reducing their electrophoretic mobility to that of methylcellulose (zero). Such desorption and readsorption would obviate the electrophoretic separation. Fortunately, the "physically adsorbed" methylcellulose can be removed by rigorous washing, leaving the "strongly bound" methylcellulose, which does not desorb under these conditions.

The methylcellulose forms a layer of uncharged hydrated polymer on the surfaces of both the colloidal particles and the cell walls, and thus effectively reduces the zeta potential to zero. The mechanism proposed for this reduction in zeta potential comprises an increase in viscosity at the distance δ from the surface (usually the distance of the slipping plane) such that the slipping plane is moved outward from the surface to a distance where the potential is close to zero.

The glass capillary tubes coated with different variations of the foregoing procedure were inserted in the electrophoresis cell shown in Figure 6 and their electroosmotic mobilities were determined from the parabolic electrophoretic velocity profiles. Table I shows the results for glass capillary tubes coated with different variations of the Z6040-MOC combination and tested using 0.80μm-diameter monodisperse polystyrene latex particles (LS-1117-B; the Dow Chemical Co.). Although many more measurements were made than are listed in Table I, the results given are representative and reproducible to within + 0.2μm cm/volt sec. Further details concerning the experimental results on different plastic surfaces can be found in an earlier publication[6].

This low-electroosmotic-mobility coating was used in the Apollo-Soyuz Test Project (ASTP) carried out in July 1975. The experiments comprised a series of free-fluid electrophoretic and isotachophoretic separations of mixtures of fresh and aldehyde-fixed red blood cells, human B and T lymphocytes, and human kidney cells. The free-fluid electrophoretic separations included human B and T lymphocytes, human kidney cells, and a human-rabbit-horse mixture of fixed red blood cells. The purpose of the first two types of separations was to return the separated fractions to earth in viable form; the purpose of the third type was to demonstrate the efficiency of electrophoretic separation of biological cells in microgravity. All electrophoretic separations were carried out using a cylindrical glass electrophoresis channel which could be inserted in the separation apparatus. The glass channels were coated as described in the foregoing section, and their measured electroosmotic mobilities were in the range -0.10 to -0.25μm cm/volt sec.

The analysis was carried out from the flight photographs taken at 3-minute intervals. The migration distances of these separated particle bands varied linearly with time as expected, and the rate of migration corresponded to the electrophoretic mobility values for the individual cell species at a voltage gradient of 10.6 volts/cm. Moreover, the sharp flat boundaries confirmed the very low value of the electroosmotic mobility.

To compare these photographic results with theory, the expected particle profiles were calculated using the best known values for the parameters. Figure 10 compares the calculated particle concentrations divided into four concentration regions with the experimental particle profiles at six minute intervals. The results are in good agreement with theory and confirm the low-electroosmotic-

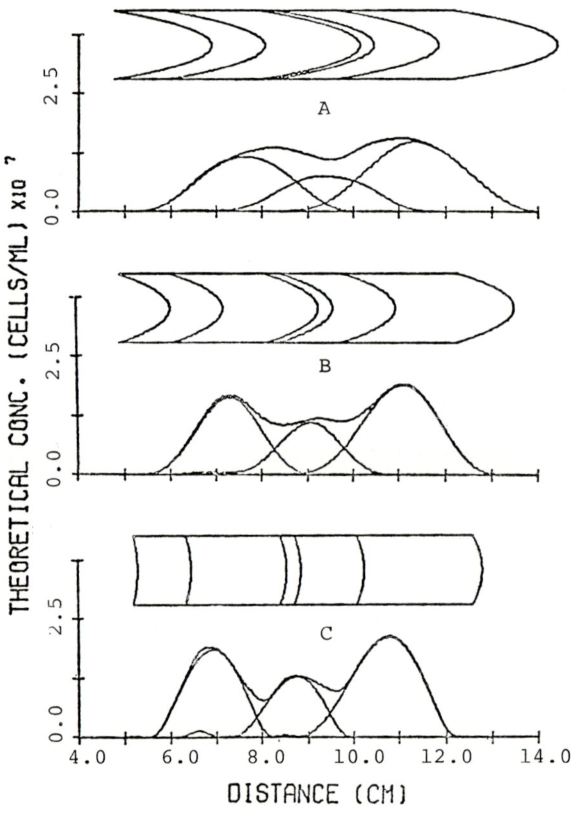

Figure 11: Computed Cell Displacement Showing the Effect of Electroosmosis on Cell Separation: A. U_{OS} = -0.4μm cm/volt sec., B. U_{OS} = -0.2μm cm/volt sec., and C. U_{OS} = 0.0μm cm/volt sec.

mobility of the coatings, as well as the importance of the electroosmotic flow in free-fluid electrophoretic separations.

The effect of electroosmosis on the resolution of separation for a band of cells in a free fluid electrophoresis channel of circular geometry can conveniently be demonstrated by a computer simulation of the separation[9]. The simulated conditions assume a potential gradient of 10 volts/cm, the sample band ratio, R, or sample diameter to channel diameter of 0.75, and a sample band thickness, θ, of 0.3 cm. The three sample cells are assumed to have electrophoretic mobility values of 1.4, 1.8 and 2.3μm cm/volt sec, and the mobility distributions are assumed to be parabolic with a percent standard deviation of 10. Figure 11 shows a concentration profile and a cross-section of the bands after 60 minutes of separation for electroosmotic mobility values of 0.4, 0.2 and 0.0μm cm/volt sec for A through C, respectively. The effect of electroosmotic flow, even at these low values, is readily apparent from these computations. Although the resolution increases as the electroosmotic flow decreases in a static system of this type, different types of electrophoresis apparatus can dictate different requirements as demonstrated in Chapter 9 for Continuous flow Electrophoresis.

REFERENCES

1. Reuss, F. F. (1809) Memoires de la Société Imperiale des Naturalistes de Mskou, 2, 324-347.
2. Abramson, H. A. (1934) Electrokinetic Phenomena, The Chemical Catalog Co. Inc., New York, 43-51.
3. Smith, M. E. and Lissie, M. W. (1936) J. Phys. Chem., 40, 399-412.
4. Komagata, S. (1933) Researches of the Electro-Technical Lab. No. 348, 97-110.
5. Micale, F. J., Vanderhoff, J. W., and Snyder, R. S. (1976) Sep. and Purification Methods, 5, 361-383.
6. Vanderhoff, J. W., Micale, F. J., and Krumrine, P. H. (1977) Sep. and Purification Methods, 6, 61-87.
7. Hjerten, S. (1967) Chromatog. Rev., 9, 122-219.
8. Lee, L. H. (1968) J. Colloid and Interface Sci., 27, 751-760.
9. Krumrine, P. H. (1978), Modeling of Two Electrophoretic Separation Techniques, Ph.D. Theses, Lehigh University, 131-142.

© 1979 Elsevier/North-Holland Biomedical Press
Electrokinetic Separation Methods,
P.G. Righetti, C.J. van Oss and J.W. Vanderhoff

ANALYTICAL CELL ELECTROPHORESIS

G.V.F. SEAMAN and D.E. BROOKS
University of Oregon Health Sciences Center, Portland, Oregon 97201, USA and Department of Pathology, University of British Columbia, Vancouver, Canada V6T 1W5

INTRODUCTION

Analytical particle electrophoresis has been variously termed microcataphoresis[1], microelectrophoresis[2], cell electrophoresis[3], cellular electrophoresis[4], microscopic electrophoresis[5] or often simply electrophoresis. The technique may be used purely for analytical purposes, for example to help elucidate the nature, number and distribution of charge groups in the peripheral zone of biological cells, or to survey the distribution of electrophoretic mobilities in a mixture of cells or particles with the intent of devising appropriate means to separate subpopulations from such systems by preparative electrophoresis or other procedures. Analytical and preparative electrophoresis are distinguished by the facts that in the former technique no spatial separation between populations of different mobilities is achieved while in the latter the absolute mobilities of the separated populations cannot generally be deduced.

Electrophoretic methods have provided information on the surface structure of a diverse range of particles. One of the strengths of the method is the facility with which absolute data on particles may be acquired without measurably altering their surface structure. The surface characteristics of biological cells have been examined without producing significant destruction of the cellular organization or loss of cell viability. In addition, the course of any chemical, enzymatic, immunological or viral reaction which changes the number or nature of surface charge groups on particles or cells may be followed by such procedures.

The aims of this review are to discuss briefly the applicability of electrokinetic theory to cells, to describe currently available electrophoretic equipment and to compare the manual and automatic measurement techniques presently in use. Because of the authors' backgrounds this review will focus on the application of electrophoresis to studies on biological cells, but virtually all the following discussion applies equally well to electrophoretic examination of non-biological particles.

Earlier work on the electrophoresis of particles has been covered by Abramson[5] and Abramson *et al*[6]. More recent reviews include Brinton and Lauffer[7], Ambrose[3], Moore[8], Shaw[9], Mehrishi[10] and Seaman[11].

ELECTROKINETIC THEORY

The basic aspects of electrokinetic theory will not be discussed here as they are treated in detail in other articles in this volume. Under most circumstances the theory applicable to electrophoresis of large insulating bodies for which the double layer thickness is much less than any radius of curvature on the particle can be applied to work on cells. If these conditions are satisfied, and assuming the cell surface is smooth and uniform with respect to charge properties, the Helmholtz-Smoluchowski equation provides a valid relationship between cellular electrophoretic mobility, u, and zeta potential, ζ (the electrostatic potential at the hydrodynamic surface of the particle):

$$u = \frac{\varepsilon\zeta}{4\pi\eta} \qquad (1)$$

where v is the particle steady state velocity, E the applied electric field strength and ε and η the dielectric constant and viscosity, respectively, within the electrical double layer which are assumed to be the same as the bulk values for the suspending medium. Since equation (1) has been shown to apply for an insulating body of arbitrary shape[12] zeta potentials may be calculated for cells of various geometries or cells which exhibit different degrees of aggregation directly from the mobilities without making any adjustment for particle dimensions.

The parameter of interest in analytical cell electrophoresis is usually the surface charge density, σ. This parameter is calculated from the zeta potential using an expression derived by solving the Poisson-Boltzmann equation with appropriate approximations and boundary conditions. If a flat surface with a uniform charge density is assumed, and if ζ < 25mV at 25°C the Gouy-Chapman equation may be used which leads to:

$$\sigma = \kappa\eta u \qquad (2)$$

where κ is the Debye-Huckel parameter. In this expression σ refers to the apparent charge density at the shear plane of the cell, the (flat) surface over which the hydrodynamic boundary condition of zero relative fluid velocity is satisfied. For higher zeta potentials more complex expressions similar to (2) are available but none provide a particularly satisfactory estimate of the cell surface charge density, largely because the assumptions made in deriving (2) and its analogues are particularly unrealistic insofar as cells are concerned.

There is good evidence from biochemical[13] and electrokinetic[14] studies that the dissociable chemical groups which make up the cell surface charge are distributed through a significant depth normal to the plane of the membrane. The charges are born by protein, glycoprotein and glycolipid molecules whose two dimensional distribution in the plane of the membrane is not known. The charges certainly do not lie in a plane, nor are they likely to be distributed uniformly parallel with the surface as assumed in deriving equation (2). As yet no satisfactory theoretical treatments are avilable which take into account either of these effects. Attempts have been made to correct for counterion penetration behind the shear plane using versions of the Poisson-Boltzmann equation modified to allow for excluded volume effects[15,16]. The models used don't allow for the polyelectrolyte nature of the macromolecules which are probably found in this region, however. Aside from predicting that the charge density calculated from (2) is underestimated they are of little practical assistance.

Independent measurements of σ are required to check the applicability of theoretical expressions analogous to (2). Such measurements are in a sense made possible for cells like erythrocytes through the use of the enzyme neuraminidase (E.C. 3.2.1.18). This enzyme specifically releases virtually all of the sialic acid from the external surface of red cells. Since sialic acid is known to provide about 60% of the net negative surface charge of human erythrocytes[17], the action of neuraminidase can be followed both by assaying for free sialic acid in the suspending medium and by measuring the decrease in electrophoretic mobility of the cells. The results of such experiments show that the rate at which the mobility decreases closely parallels the rate of appearance of free sialic acid in the medium if both quantities are expressed as a percentage of the maximum effect found[18]. The total amount of charge lost, as calculated from the mobility change, is underestimated by a factor of 2 to 3, however, when compared to the amount of sialic acid released[19]. It seems somewhat unlikely that the major portion of the membrane sialic acid would be buried behind the shear plane yet be as readily accessible to the enzyme as that at the hydrodynamic surface. An alternative explanation for the results obtained is that an inadequate theory was applied in analyzing the electrophoretic results. This would appear to be a fruitful area for theoretical investigation.

ANALYTICAL PARTICLE ELECTROPHORESIS EQUIPMENT

The roles of analytical and preparative electrophoresis. Analytical and preparative electrophoretic procedures usually serve complementary roles. In the analytical methods normally no concentration gradients are formed and no product is isolated. One is concerned instead with the surface properties of particles or cells in very dilute suspension where particle-particle interactions are negligible. In preparative electrophoresis the aim is to isolate samples of particles or cells which differ in electrophoretic properties, often for further nonelectrophoretic use. Efforts aimed at eliminating the principal limitations of the technique have usually been concerned with methods of reducing the disturbing influences of thermal convection and the gravitational settling of particles during separation. The problems have been tackled in various ways including an electromagnetic method of stabilization in which the medium rotates in an annular gap[20], the use of density gradients[21,22], the mechanical rotation of a cylindrical electrophoresis tube[23] and by continuous flow particle electrophoresis[24]. In principle one can obtain a mean electrophoretic mobility of a particle population from any of these methods and in the case of continuous flow particle electrophoresis an accuracy close to that of analytical particle electrophoresis is claimed[24]. In general they seem to be better suited to relative than absolute mobility determinations, however.

MANUALLY OPERATED EQUIPMENT

The electrophoretic mobility of microscopically or ultramicroscopically visible particles or cells can be measured directly by classical analytical particle electrophoresis carried out using manually timed individual particle displacements for velocity data. In some instances the electrophoretic properties of soluble materials can also be examined by adsorption onto a suitable carrier particle. However, incomplete coverage of the surface of the carrier particle or change in conformation of the soluble molecules upon adsorption may render such measurements unreliable.

Essential components of a suitable apparatus. The essentials of the classical particle electrophoresis apparatus comprise a transparent chamber, usually of cylindrical or rectangular cross section, equipped with electrodes; a means of measuring the rate of migration (μm sec^{-1}) of the particles (usually by manually timing migration over a known distance) and a means of determining the location of the particle whose mobility is being measured relative to the walls of the chamber. After determination of the applied electrical field strength (V cm^{-1}), the electrophoretic mobility (μm sec^{-1} V^{-1} cm) can be calculated.

Evaluation of chamber designs. Two main types of measuring chamber geometry are in general use, a circular cross section (cylindrical chamber) and a rectangular cross section chamber. Numerous more complex variations on these primary designs have been described, generally aimed at eliminating or modifying the electroosmotic flow in the chamber[5,6,11]. The literature on many of these designs is referenced in a review by Seaman[11] and the earlier literature is covered in Abramson[5] and Abramson *et al.*[6]. When correctly set up and operated there is good agreement between electrophoretic data obtained from cylindrical and rectangular equipment. Along with simplicity and ease of filling the following factors should be taken into consideration in the choice of chamber design:

(a) choice of materials which are likely to have reproducible and uniform surface properties and/or are capable of being coated with a stable low zeta potential substance;

(b) dimensions such that:

  (i) there is not too marked a change in electroosmotic velocity with depth at the stationary level,

  (ii) the chamber crosssectional area will not permit too high a current (and hence too much heating and convection) for the given applied electric field at the suspending media ionic strengths routinely used in the studies,

  (iii) the chamber wall is thin enough to facilitate rapid dissipation of joule heating,

  (iv) the chamber length is sufficient to avoid contamination of the suspending medium by electrode products in the volume under observation;

(c) non-polarizable electrodes which will yield a known uniform electrical field within the observation volume.

The principle advantage enjoyed by the rectangular cross section chamber is the possibility of using phase contrast optics, which is more difficult with a circular cross section. Cylindrical chambers on the other hand are generally of smaller volume and have a more shallow electroosmotic flow profile which produces smaller errors associated with finite depth of focus at the stationary level.

Electroosmosis and its elimination. In a closed cylindrical chamber bearing a uniform wall charge electroosmotic flow near the wall returns down the center of the tube, setting up a parabolic velocity profile across the tube. The fluid velocity thus adds to or subtracts from the velocity of all particles undergoing electrophoresis except those located at the radial position where the electroosmotic and return flows balance (the stationary level). It can be shown[25] that the particle velocity v_o observed at any distance x from the axis is given by:

$$v_o = v_e + v_s \left(2 \frac{x^2}{a^2} - 1\right) \qquad (3)$$

where v_s is the electroosmotic fluid velocity at the wall, v_e is the particle velocity due to electrophoresis and a is the tube radius. If the observations are made at $x = a/2^{\frac{1}{2}}$, which defines the stationary level, $v_o = v_e$ as required. Marked heterogeneity in the wall charge distribution will lead to non-uniform electroosmosis with a resultant shift in the location of the stationary level. Velocities measured at $x = a/2^{\frac{1}{2}}$ would then contain a non-electrophoretic component and be in error.

Since one of the principle sources of uncertainty in mobility measurements is inaccuracy of focussing at the stationary level, chamber coatings have been devised to eliminate the zeta potential at the wall and hence eliminate electroosmotic flow in the chamber. The wall potential may be reduced or eliminated by the use of adhesive films[24], covalently bonded materials[25] or physically adsorbed substances[27,28]. Coatings used are designed to either neutralize the net wall charge or move the shear plane away from the charged wall with neutral material sufficiently far that the zeta potential approaches zero. Testing of the stability and completeness of the chamber coating as well as criteria for selection of suitable chamber coatings are discussed by Nordt *et al.*[28].

Accuracy and variability in electrophoretic measurements. The electrophoretic mobility values of particles or cells are influenced by intrinsic variations in surface charge density in the case of the former and biological variation within the sample in the case of the latter. No matter what the degree of sophistication or automation of the equipment there will be an inherent dispersity in the particle or cell population due to these factors. In principle broadening of a recorded electrophoretic mobility distribution will occur due to particle Brownian motion but for typical biological cells of diameter 5-15 x 10^{-4} cm diffusional displacements are much less than 1% of electrophoretic displacements and can be ignored. In chambers with significant electroosmosis a more serious cell size-dependent uncertainty is introduced by the tendency of large cells to be affected by non-zero fluid flow on either side of the stationary level. Cells with larger cross sections which are not precisely centered in the stationary level interact more strongly with the flow field than smaller cells. Due to the finite depth of focus of the optics such cells may be accepted for timing but contributions of up to several percent can be made to the observed velocity by fluid flow in unfavourable cases. Other sources of experimental error in the manual technique include imprecise timing and the degree of objectivity involved in the selection of particles for measurement. Temperature effects (particularly convective drift), errors associated with locating the stationary level, assymmetric electroosmotic flow in the chamber, inaccuracies in estimating the magnitude of the electric field and contamination of the sample (particularly by the growth of organisms in suspensions maintained at room temperature for a few days) are all common sources of difficulties encountered with the classical procedure. A more detailed discussion, including tips on trouble shooting a cylindrical chamber apparatus, is provided in the review by Seaman[11].

Electrophoretic calibration particles. There is a clear need for the development of a stable calibration particle whose suspensions have known electrophoretic characteristics which can be used to monitor the performance of electrophoresis equipment. The human red blood cell has been widely used as a standard particle because of the consistency of its electrophoretic behavior and its relative insensitivity to the presence of moderate amounts of a variety of macromolecules and ions[10,11]. This latter characteristic is not shared by most polymer latex bead preparations[29] which would in many other respects be ideal. As a primary standard the erythrocyte is also subject to a number of defects, however. A major shortcoming is the limited range of conditions of buffer composition, ionic strength, pH and temperature over which these cells are stable[30], and the very limited storage life. The mobility of the human erythrocyte is rather low in standard saline and phosphate buffer which limits the accuracy of some calibration procedures. The surface of the red cell is also very complex and liable to change in a variety of circumstances. A more useful standard particle may prove to be the aldehyde-fixed red cell[31,32] which has a much greater range of electrokinetic stability than the normal red cell and, under appropriate conditions, indefinite storage life. At present a completely satisfactory standard particle with a fairly simple constitution, long storage life, and relative insensitivity to trace contaminants has yet to be found. Such a standard calibration particle could be used throughout the world for monitoring the performance specifications of both analytical and preparative electrophoretic equipment.

AUTOMATED EQUIPMENT

There are a number of reasons for wishing to automate analytical particle electrophoresis, most of which are associated with the need to measure the mobilities of a large number of cells (several hundred) rapidly and accurately in order to characterize most naturally occurring, metabolically active cell populations. A variety of modifications can be made to the classical procedure which will speed up or make more accurate the data collection and analysis. These include the use of photography or video taping to collect the maximum amount of data on a sample, on-line analysis of data fed from an electronic timer into a calculator or computer or automatic interpretation of the rate at which a prism must be made to rotate to freeze the image of a suspension of particles undergoing electrophoresis (Laser Zee^{tm} Meter, Pen Kem Inc., Croton-on-Hudson, New York). All of these techniques involve manual velocity measurements, however, which are subject to the problems discussed above. More direct approaches have therefore been developed which eliminate manual intervention in the measurement process. As well as providing speed and convenience these techniques permit the application of some novel methods for eliminating electroosmotic interference.

Electroosmosis in automated analytical particle electrophoresis (AAPE). It has been shown by Vorob'eva *et al*[33] that the electroosmotic flow profile in a chamber with rectangular cross-section is significantly blunted if an alternating electric field is applied. This blunting provides a core region in the center of the chamber throughout which the electrophoretic mobility of a particle will be independent of position. Mobilities measured in this region will therefore not be subject to errors associated with a particle location other than in the stationary level. If rapid velocity measurements are possible particles throughout this volume can be measured, providing much more data than can be obtained by observing only the stationary level. The effect cannot be taken advantage of with manual techniques because the frequencies of the applied field required are such that particle oscillation amplitudes are too small to be conveniently measured.

The theoretical frequency dependence of the fluid velocity amplitude profile for a long cylindrical chamber of radius a = 0.1 cm in which an axial sinusoidal electric field exists is shown in Fig. 1[34]. The results are expressed as the absolute value of the velocity amplitude, $|u|$, relative to the wall velocity u_o (the velocity changes sign where it meets the r/a axis) as a function of distance from the center of the tube (chamber wall at r/a = 1). It is seen that at frequencies of 50-100 Hz the velocity amplitude is small and independent of position over most of the cross section of the tube.

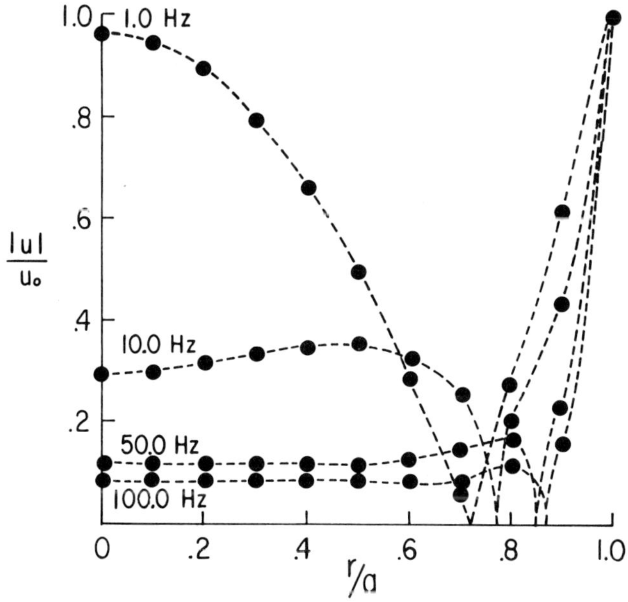

Fig. 1. Fluid velocity profile produced by electroosmosis in a sinusoidal electric field; for explanation see text.

Another approach to eliminating electroosmosis has been taken by Uzgiris and his collaborators[35] who utilize a pair of closely spaced, albumin-coated Pt electrodes located so far from the walls of the container that no electroosmotic flow develops. Alternating fields are used to avoid problems associated with electrode polarization and mobilities are determined by laser Doppler velocimetry (vide infra). While no electroosmosis is present, other difficulties appear associated with fringing fields and harmonic distortions in the apparent mobility distributions due to the a.c. field used[36].

Automation via laser Doppler velocimetry. The application of laser Doppler velocimetry (LDV) to the problem of measuring electrophoretic mobility distributions of cells has been developed independently by Dr. Ben Ware of Harvard University and Dr. E.E. Uzgiris of General Electric Research and Development Center. In this technique a laser beam is directed through a suspension of particles undergoing electrophoresis. Light scattered from a moving particle is shifted in frequency by an amount proportional to its velocity. The spectrum of frequency differences between the entrance and scattered beams is detected by optically mixing the two and deriving the beat frequency spectrum from a photocell with a real time spectrum analyzer. This spectrum represents the velocity distribution of particles intersecting the beam and is therefore proportional to the electrophoretic mobility distribution of the population. For a detailed description of the technique the original literature should be consulted[35,37,38].

The LDV approach offers a number of advantages over manually operated measurement procedures. The mobility distribution is obtained rapidly; an acceptable spectrum from particles present in the scattering volume can be collected in less than one minute under ideal conditions and in under 10 minutes for most samples. Sample volumes can be small (under 1 ml) depending on chamber design and a distribution can be obtained on less than 10^5 particles. Moreover, the technique is very versatile with respect to the size and shape of particles which can be examined, spectra having been obtained on everything from macromolecules (for which diffusion constants can also be derived by this means) to large biological cells.

The accuracy and reproducibility of mean mobilities obtained by LDV on cell populations appear to be comparable to those obtained by classical methods, at least at low ionic strength ($\pm$ 3-5%). In physiological ionic strength somewhat more difficulty is experienced but if proper attention is paid to heat dissipation and electroosmosis suppression satisfactory results may be obtained. An illustration is provided in Figure 2, taken from Ware's work[38], in which the mobility distribution of human lymphoblasts suspended in a medium of physiological ionic strength before and after treatment with neuraminidase is presented as determined by laser Doppler velocimetry. For comparison a mobility distribution obtained under similar conditions with aldehyde-fixed human erythrocytes using a cylindrical chamber microelectrophoresis apparatus (Rank Bros., Bottisham, Cambs., England) is given in Figure 3. Interpretation of the details of the peak shape in Fig. 2 is not clear (vide infra) but it is apparent that the LDV and manually obtained results are comparable and the modal lymphoblast mobilities reasonable.

Although LDV offers the advantages of automation and speed of data collection there are some difficulties with this approach. The most serious of these was referred to earlier with regard to interpretation of the shape of the apparent mobility distribution. The output of the spectrum analyzer is a plot of scattered light intensity as a function of beat frequency. While the beat frequency is directly proportional to cell velocity, the scattered light intensity is not uniquely defined in terms of the number of cells moving at a certain velocity in the scattering volume. The intensity of light scattered at a given angle will depend on particle or cell size, shape, orientation and

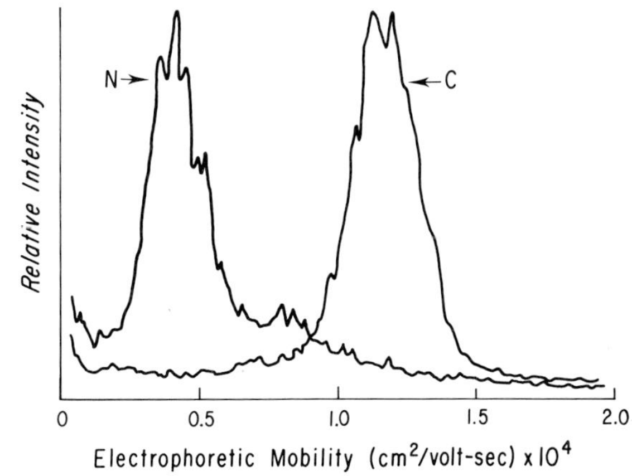

Fig. 2. The electrophoretic mobility distribution measured by LDV of human lymphoblasts suspended in physiological ionic strength buffer. C represents native sample, N denotes treatment with neuraminidase before measurement; T=20°C, E=25V cm^{-1} (data of Smith and Ware[38]).

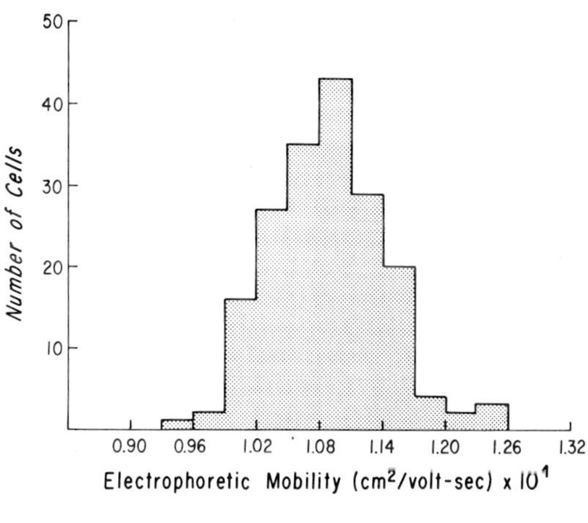

Fig. 3. The electrophoretic mobility distribution measured by the manual method of formaldehyde-fixed human erythrocytes suspended in 0.15M NaCl, pH 7.2, buffered with 0.15M $NaHCO_3$; T=25°C, E=4V cm^{-1}; based on timings of 182 cells in cylindrical chamber apparatus.

refractive index. Unless all these properties are known as a function of mobility, the true distribution of cell number vs mobility cannot in principle be obtained. In an unknown cell population these parameters could vary widely, producing apparent peaks in the distribution that could lead to misinterpretation of the spectrum. Another problem with the LDV method involves sample chamber design. There does not yet seem to be a simple, versatile design suitable for routine use. This is primarily due to the problem of electroosmosis which is made more acute by the fact that LDV utilizes a sampling volume so that the stationary level cannot be utilized. Electroosmotic disturbances therefore have to be eliminated from the region traversed by the laser beam. This has either been approached using chamber wall coatings with methylcellulose to reduce the wall zeta potential[38] or utilizing closely spaced Pt electrodes well away from the chamber boundaries[35]. The methylcellulose coating procedure is fairly involved and time consuming, however, and has to be repeated after about 10 hr. of use. The difficulties with the Pt system have already been mentioned. A final limitation on the use of LDV for electrophoresis automation lies in the general availability of the technique. There is no LDV electrophoresis apparatus based on this principle commercially available and unfortunately only a few laboratories possess the interest and resources required for its continued development.

Pen Kem Laser Zee^{tm} System 3000[33]. The only AAPE currently available commercially is that produced by Pen Kem, Incorporated. The System 3000 was initially designed to operate as a process control monitor for the pulp and paper industry, but its potential applications extend to biological cells. In the Pen Kem apparatus a vertical curtain of laser illumination approximately 10μ thick traverses the stationary level of a horizontally mounted cylindrical electrophoresis chamber. Light scattered by particles undergoing electrophoresis at the stationary level is collected at 90° from the beam and focussed through an objective onto the surface of a rotating radial grating. Light transmitted through the grating impinges in pulses on a photomultiplier tube whose output is analyzed. Because the image of each particle on the grating is moving at a velocity proportional to its electrophoretic mobility the frequency of light pulses entering the photomultiplier is somewhat different from that produced when the particles are stationary. This difference in frequency is proportional to the particle mobility so when a population of particles undergoing electrophoresis is examined the frequency difference spectrum, which is provided as the output of the System 3000 in its most general form, is proportional to the electrophoretic mobility distribution of the population.

A principal strength of the System 3000 is that electroosmosis does not affect the output since measurements are carried out at the stationary level. Moreover, this approach would seem to be compatible with a.c. field application if larger sample sizes are desirable. Mobility distributions can be generated fairly rapidly with an accuracy similar to that of LDV. It has the practical advantage that it is less sensitive to vibration than LDV and does not require elaborate mechanical isolation. Finally, although expensive, the System 3000 is commercially available with a variety of options for continuous sampling and data manipulation.

Since the output of the System 3000 is generated from scattered light, it is subject to the same uncertainty as that of LDV, namely that the scattered light intensity depends on other parameters in addition to cell number. These parameters will not in general be known as a function of mobility for a given sample population resulting in uncertainties in the interpretation of apparent mobility distributions. Nonetheless, much valuable information can potentially be obtained utilizing this instrument and it is to be hoped that its availability will generate sufficient laboratory experience to establish operating limitations, particularly in biological systems.

AAPE via computer analysis of video images. This approach is currently being pursued by a group under the direction of Dr. Peter Bartels at the University of Arizona who are supported by funds from the National Aeronautics and Space Administration of the USA (NASA). NASA has for many years been interested in the potential of preparative electrophoresis for materials processing in space and is supporting development of an AAPE system of advanced design to provide a facility for the determination of absolute mobility distributions of candidate systems for separation. The Bartels apparatus is based on the classical microelectrophoresis technique with the operator replaced by a vidicom camera interfaced to a video compressor and computer. The apparatus is projected to be operational by 1979, at which time a detailed description will be published. It features a cylindrical chamber, phase contrast optics and the ability to track the trajectories of several particles simultaneously while making a sequence of velocity measurements on each one. Data accumulation and analysis will be carried out by a software package which will include the ability to discriminate between particles with different optical characteristics and to analyze the mobility data with a number of sophisticated statistical routines.

The automated video image analysis technique offers the outstanding advantage that the mobility distributions obtained are built up from absolute velocity determinations on single cells and are hence amenable to complete statistical analysis. This is not the case for data derived from scattered light intensities, as discussed above. Since the microscope is focussed at the stationary level, no significant interference from electroosmosis occurs. Because of the multiple estimates of velocity made per particle the individual mobility measurements are all quite accurate ($\pm$ 1%) so small differences in mobility distributions should be detectable. This is a particularly important feature for possible applications in clinical medicine. In principle, since a visual image is analyzed, the system could correlate mobility values with other cell parameters such as size, shape or presence of stain or fluorescent markers. The rate of data collection will be similar to that of the System 3000, at least for biological cells, since both techniques accept data only from cells in the stationary level, but will be somewhat slower than LDV. Aside from the high optical quality of the cylindrical tube, the chamber is of standard design and is very straightforward to use. The apparatus is designed to be readily operably by research technicians after a few hours of instruction with the aim of examining its applicability in as wide a variety of experimental and clinical paradigms as possible.

The major limitation on the applicability of computerized video analysis is, of course, the availability of the apparatus. Its development has been reasonably expensive but, once the prototype machine is fully operational and the software written,it will be considerably cheaper to reproduce. It is to be hoped that funding agencies will support the dissemination of machines of this type to make them available to the electrophoresis user community as a whole.

TABLE 1

COMPARISON OF AUTOMATION TECHNIQUES WITH MANUAL MICROELECTROPHORESIS

	Speed[a]	Fidelity of dist'n[b]	Cost; ease of access	# cells in dist'n	Correlate with cell markers
Manual	0	0	0	0	0
LDV	+3	-2	-2	+3	-2
System 3000	+2	-1	-1	+2	-1
Video analysis	+1	+3	-3	+2	0

[a]scale compares AAPE techniques with manual microelectrophoresis procedure. Increasingly positive rating implies relatively greater usefulness, 0 implies equivalence with manual technique; increasingly negative rating implies relatively less desirable feature.
[b]For cells in medium of physiological ionic strength.

It appears from the ratings in Table 1 that the Bartels apparatus comes closest to fitting the description of an ideal AAPE.

Comparison of automation approaches. A subjective assessment of the relative strengths of the automation techniques described above is given in Table 1.

ACKNOWLEDGEMENTS

Support is acknowledged from Contract No. NAS8-32609 from NASA, USPHS Grant HL 18284 from the National Heart, Lung and Blood Institute of the USA and Grant MT-5759 from the Medical Research Council of Canada. We thank Dr. Ben Ware, Mr. John Penniman and Dr. Peter Bartels for providing pre-publication reference material.

REFERENCES

1. Brown, H.C. and Broom, J.C. (1936) Proc. Roy. Soc. B., 119, 231-244.
2. Seaman, G.V.F. (1958) Microelectrophoresis of Red Blood Cells, Ph.D. Dissertation, Univ. of Cambridge.
3. Ambrose, E.J. (1965) Cell Electrophoresis, J. and A. Churchill, Ltd., London.
4. Weiss, L. (1966) J. Natl. Cancer Inst., 36, 837-847.
5. Abramson, H.A. (1934) Electrokinetic Phenomena and Their Application to Biology of Medicine, The Chemical Catalog Co. Inc., New York.
6. Abramson, H.A., Moyer, L.S and Gorin, M.H. (1942) Electrophoresis of Proteins and the Chemistry of Cell Surfaces, Van Nostrand-Reinhold, Princeton, New Jersey.
7. Brinton, C.C. and Lauffer, M.A. (1959) in Electrophoresis; Theory, Methods and Applications, Bier, M., ed., Academic Press, New York, pp. 427-492.
8. Moore, D.H. (1968) Electrophoresis; Physical Techniques in Biological Research, V. II, Part A, 2nd Edition, Academic Press, New York.
9. Shaw, D.J. (1969) Electrophoresis, Academic Press, New York.
10. Mehrishi, J. (1972) Progr. Biophys. Mol. Biol. 25, 1-70.
11. Seaman, G.V.F. (1975) in: The Red Blood Cell, MacSurgenor, D., ed., Academic Press, New York, pp. 1135-1229.
12. Morrison, F.A. (1970) J. Coll. Int. Sci., 34, 210-214.
13. Tomita, M. and Marchesi, V.T. (1975) Proc. Nat. Acad. Sci. USA, 72, 2964-2968.
14. Cook, G.M.W., Heard, D. and Seaman, G.V.F. (1961) Nature, 191, 44-47.
15. Haydon, D.A. (1961) Biochim. Biophys. Acta, 50, 450-457.
16. Brooks, D.E. (1973) J. Colloid Interface Sci., 43, 687-699.
17. Haydon, D.A. and Seaman, G.V.F. (1967) Arch. Biochem. Biophys., 122, 126-136.
18. Luner, S.J., Sturgeon, P., Szklarek, D. and McQuiston, D.T. (1975) Vox Sanguinis, 28, 184-199.
19. Eylar, E.H., Madoff, M.A., Brody, O.V. and Oncley, J.L (1962) J. Biol. Chem., 237, 1992-2000.
20. Kolin, A. (1960) Proc. Natl. Acad. Sci. U.S., 46, 509-523.
21. Mel, H.C. (1964) J. Theoret. Biol. 6, 307-324.
22. Boltz, R.C., Todd, P., Streibel and Louie, M.K. (1973) Prep. Biochem., 3, 383-401.
23. Hjerten, S. (1976) Meth. Protein Sep., 2, 219-231.
24. Strickler, A. and Sacks, T. (1973) Ann. N.Y. Acad. Sci., 209, 497-514.
25. Bangham, A.D., Flemans, R., Heard, D.H. and Seaman, G.V.F. (1958) Nature 182, 642-644.
26. Lee, L.H. (1968) J. Coll. Inter. Sci., 27, 751-760.
27. Van Oss, C.J., Fike, R.M., Good, R.J. and Reinig, J.M. (1974) Anal. Biochem., 60, 242-251.

28. Nordt, F.J., Knox, R.J. and Seaman, G.V.F. (1976) in Hydrogels for Medical and Related Applications, Andrade, J.D., ed., American Chemical Society, ACS Symposium Series 31, pp. 225-240.
29. Saunders, F.L. (1968) J. Coll. Inter. Sci. 28, 475-480.
30. Heard, D.H. and Seaman, G.V.F. (1960) J. Gen. Physiol. 43, 635-654.
31. Heard, D.H. and Seaman, G.V.F. (1961) Biochim. Biophys. Acta, 53, 366-374.
32. Vassar, P.S., Hards, J.M., Brooks, D.E., Hagenberger, B. and Seaman, G.V.F. (1972) J. Cell Biol. 53, 809-818.
33. Vorob'eva, T.A., Vlodavets, I.N. and Dukhin, S.S. (1970) Colloid J. (USSR), 32, 152-156.
34. Brooks, D.E. (1976) Final report on NASA Contract No. NAS8-31386.
35. Uzgiris, E.E. (1974) Rev. Sci. Instrum., 45, 74-80.
36. Bennet, A.J. and Uzgiris, E.E. (1973) Phys. Rev. A8, 2662-2669.
37. Ware, B.R. (1974) Advan. Colloid Interface Sci., 4, 1-44.
38. Smith, B.A. and Ware, B.R. Current Topics in Analytical and Clinical Chemistry, in press.
39. Goetz, P.J. and Penniman, J.G. (1977) International Seminar on Paper Mill Chemistry, Amsterday, Sept. 11-13 (1977). Available from Pen Kem, Inc., Box 364, Croton-on-Hudson, New York 10520.

Electrokinetic Separation Methods,
P.G. Righetti, C.J. van Oss and J.W. Vanderhoff

SIMPLIFIED CELL MICROELECTROPHORESIS WITH UNIFORM ELECTROOSMOTIC BACKFLOW

CAREL J. van OSS and RICHARD M. FIKE
Department of Microbiology, State University of New York at Buffalo, Buffalo, New York 14214 (U.S.A.)

INTRODUCTION

Microelectrophoresis is becoming an increasingly important method in research as well as in clinical laboratories, for the determination of the electrophoretic mobilities of a variety of blood cells [1-3]. However, particularly for routine use, the classical microelectrophoresis method is rendered difficult and tedious on account of the effects of the electroosmotic backflow in the vicinity of the inner wall of the electrophoresis chamber. This causes the observed mobilities of cells to vary considerably, according to their position in the electrophoresis chamber. Thus, in order to obtain true electrophoretic mobilities, one either must plot many observed cell mobilities versus their individual depths of focusing and obtain their true mobility by interpolation on the parabolic section of the paraboloid of mobilities [4,5], or one must do all observations while focusing precisely on the "stationary level" [4,5]. In both cases these critical mobility points lie on those parts of the paraboloid that make a sharp angle with the wall of the capillary or chamber. This enhances the likelihood of error in the determination of the true mobility; a small deviation (caused by, e.g., only slightly up or down focusing) in a direction perpendicular to the capillary or chamber axis, tends to produce a large shift in the value of the true mobility. In addition, rather large cells, such as macrophages, have a strong tendency to sediment even during short time intervals, which allows too brief a presence of the cells at the stationary level to clock them before they disappear from the stationary level.

All these complications can be obviated by rendering the electroosmotic backflow uniform throughout the chamber's cross section, i.e., by reducing the electroosmotic backflow to plug-flow [6,7].

MATERIALS AND METHODS

Preparation of capillary chambers. Good results are obtained with "Drummond Microcaps" (50 μl, Drummond Scientific Co., Broomall, PA 19008), which have an internal diameter of 0.80 mm and are 100 mm long. The capillaries are coated on the inside with several layers of an agarose solution (Indubiose A-45, Fisher Scientific Co., Rochester, NY), using a 3 ml syringe with an 18 gauge needle to which a length of plastic tubing is attached for connection to one end of the capillary. The 2.5% agarose solution (with the electrophoresis buffer as solvent) should be maintained at 55° C. By drawing warm agarose solution through the capillary and into the syringe, and by immediately lifting the free capillary end out of the agarose while still maintaining gentle suction with the syringe, a layer of agarose is deposited along the inner walls of the capillary. Gentle suction should be maintained until agarose solidification occurs, to prevent plugging of the lumen of the capillary by the gel. This procedure is repeated a few times, until a layer of ≈ 0.1 mm is deposited over the entire inner wall of the capillary. Once filled with buffer, the capillary should be inspected under the microscope for completeness of coating; the agarose coating need not be completely smooth but it should cover the entire inner surface. The ultimate test for completeness of coating can only be done by inspecting the electrophoretic movement of particles or cells in it: an agarose coated and plugged capillary is only acceptable when particles near the capillary wall move at the same rate in an electric field as those near the axis (see below).

Loading and plugging of the capillary. Once a capillary is coated on the inside with a layer of agarose, it can be filled with a dilute suspension of cells ($\approx 10^6$/ml, amounting to $\approx 3 \times 10^4$/capillary), the mobility of which is to be determined. This is done by drawing the cell suspension up into the capillary, by capillary action. The capillary is then sealed at both ends with a gel-plug made from the same agarose solution with which the inner capillary wall has been coated. This is done by allowing the warm agarose solution to exude slowly from a Pasteur pipette over the ends of the capillary, until two bulb-like structures have been formed (see Figure 1).

Microscope. Horizontal placement of the optical axis of the microscope (Microstar, American Optical Co., Buffalo, NY) is preferred, as it allows cell mobility determinations for a much more extended period of time than is possible with the more commonly used vertical axis (see Figure 2). As the suspended cells or particles start to settle to the bottom of the capillary, which is easily observed with the horizontal axis arrangement, occasionally rotating the capillary $\approx 180°$ by hand will keep the cells or particles indefinitely in suspension. And as such electroosmotic backflow as persists has been reduced to a horizontal plugflow, movement of particles in a plane perpendicular to the electric field is no longer connected with their apparent electrophoretic mobility. An ocular micrometer scale in one of the microscope's eyepieces provides the calibration of length units, which, together with measurements of time intervals with a stopwatch, allows the determination of mobilities in an electric field. A 100 x enlargement is used most generally.

Electrodes and electrode vessels. Ag/AgCl electrodes are best used. These electrodes are placed in compartments filled with 1.0 M NaCl. The NaCl compartments are connected, via inverted glass U tubes filled with a 2.5% agarose gel with the electrophoresis buffer as the fluid phase, to electrolyte vessels filled with electrophoresis buffer. The capillary is placed across the two buffer vessels, and is held in place by shallow notches in the walls of these vessels, in such a

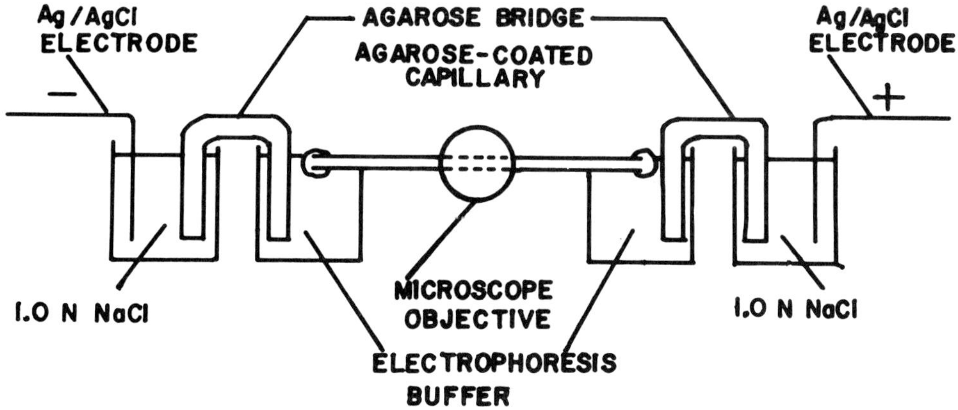

Fig. 1. Schematic representation of the simplified cell microelectrophoresis device. The capillary (coated on the inside with a thin layer of a 2.5% agarose gel) is visible in the center of the drawing; the two bulb-like agarose structures that serve to enclose it are visible at each end of the capillary. The microscope is left out of the drawing, but the emplacement of the objective is indicated by a circle in the middle of the capillary. See Fig. 2 for a photographic view.

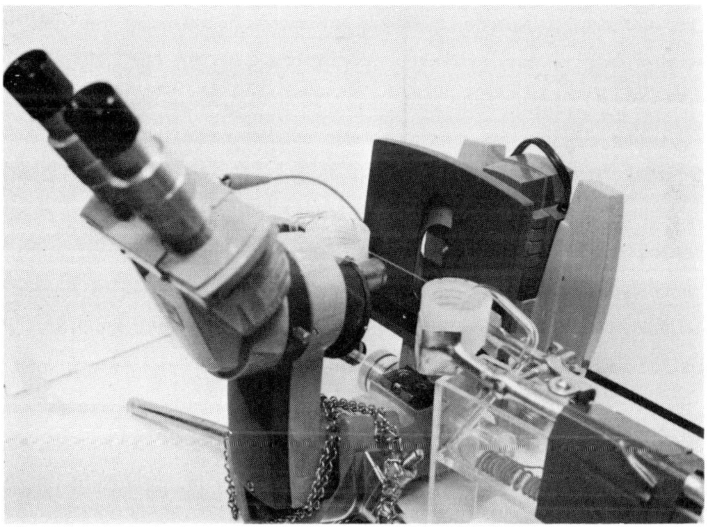

Fig. 2. Photograph of the simplified cell microelectrophoresis device, with the microscope placed horizontally. The capillary is clearly visible between the objective and the microscope stage as a sharp white line. Its right extremity is seen to rest in a notch of the white cylindrical buffer cup which is in its turn connected to a glass inverted U tube, dipping with its other branch into a rectangular plastic electrode vessel, with a spiral electrode at the bottom.

manner that the agarose bulbs at the ends of the closed capillaries are immersed in the buffer (see Figure 1).

Buffers. For different purposes it can be practical to have several buffers at one's disposal. The first one should be a buffer of physiological ionic strength ($\mu = 0.15$), consisting of, e.g., 0.145 M NaCl buffered to pH 7.3 with 0.075 g/l Na_2HCO_3 [6,7]. Phosphate buffered saline (PBS), of pH 7.2 is also quite useful; it consists of 6.79 g NaCl, 1.48 g Na_2HPO_4, 0.43 g KH_2PO_4, and distilled water, to 1 liter. Not only the high conductivity of these buffers, but also the relatively low electrophoretic mobility that particles and cells attain in them, and thus the longer observation times needed, make it imperative to cool the electrophoresis capillary by dripping cooling water through a trough in which the capillary rests [6,7], when measurements at physiological ionic strength are required.

With a low ionic strenth (LIS) buffer, the lower conductivity, coupled with higher electrophoretic mobilities, render water cooling of the capillary superfluous. A useful LIS consists of 96.65 ml of 0.2 M Na_2HPO_4, 6.35 ml of 0.1 M citric acid, 81.8 g glucose and 1600 ml distilled water. It has an ionic strength $\mu = 0.015$, pH 7.4, conductivity $\approx 10^{-3}$ mho/cm, and is isotonic for cells, due to its glucose content [6,7].

An intermediate ionic strength (IIS) buffer is useful when microelectrophoretic characterizations have to be made of fractions of cells separated on a preparative scale (for which an IIS buffer is particularly well suited [8]). The IIS buffer used in this laboratory consists of 1.272 g anhydrous Na_2HPO_4, 0.15 g KH_2PO_4, 1.272 g NaCl, 0.25 g Na_2EDTA, 40.0 g sucrose, 0.05 g glucose and 1 liter distilled water. It has an ionic strength $\mu = 0.04$, pH 7.35 and is isotonic for cells. Although the electrophoretic mobilities of cells in this buffer are lower than in the LIS buffer, its cell-separating power is somewhat better [8]. The Na_2EDTA is needed in a buffer that is also used in preparative cell electrophoresis, as it counteracts clumping of cells when they are present in high concentrations [8].

Electrophoresis. The mobility measurements are taken by using both a "normal" and a reversed polarity for each cell measured, so that with the reversible Ag/AgCl electrodes the amperage at the "normal" remains equal to that of the reversed polarity. The electric field strength is determined by direct measurement of the voltage by touching the bulbs at the ends of the capillary with the probes of the voltmeter, which yields the voltage over a distance of ≈ 102 mm. The mobilities found are then expressed in $\mu m\ sec^{-1}\ V^{-1}\ cm$ (or, in cgs units in $10^{-4}\ cm^2\ V^{-1}\ sec^{-1}$, or in $10^{-8}\ m^2\ V^{-1}\ sec^{-1}$ SI units).

Determination of the electroosmotic backflow. The value of the small electroosmotic backflow caused by the slightly negatively charged agarose coating and plugs is best determined by the electrophoresis of dextran in a slab of the same agarose gel used in the coating and plugging of the capillary, in the same buffer, and at the same gel concentration. The displacement of dextran in the agarose can be visualized by precipitation with a 5 : 50 : 25 (V : V : V) mixture of acetic acid, ethanol and water [6,7,9,10]. The electroosmotic backflow characteristics should be determined anew for each new batch of agarose [9,10]. The extremely slight electrophoretic mobility of the dextran used (Dextran T-40, molecular weight ≈ 40,000, lot No. To 5402, Pharmacia, Piscataway, NJ) was determined in a Tiselius electrophoresis apparatus (Spinco Model H, Beckman Instruments, Palo Alto, CA), to be -0.022 μm sec^{-1} V^{-1} cm in the LIS buffer (pH 7.4), and -0.002 μm sec^{-1} V^{-1} cm at $\mu = 0.15$ (pH 7.3) [6,7]. The total electroosmotic flow in the agarose used in our laboratory (see above) was thus determined to be -0.13 μm sec^{-1} V^{-1} cm in the LIS buffer, and -0.08 μm sec^{-1} V^{-1} cm at $\mu = 0.15$ [6,7]. The appropriate electroosmotic flow value due to the agarose (being negative, as is the electrophoretic mobility of cells at the pH used) must be <u>added</u> to the electrophoretic mobilities found for cells migrating to the positive electrode.

RESULTS AND DISCUSSION

With the LIS buffer the electrophoretic mobilities of human and chicken erythrocytes was determined with the apparatus described above, and the mobilities found were compared with those obtained in the classical manner, with a Rank Particle Microelectrophoresis Appartus, Mark II, (Rank Bros., Brottisham, Cambridge, England) [6,7]. The mobilities obtained with the two methods were essentially identical; they were, for respectively human and chicken erythrocytes -1.9 and -1.4 μm sec^{-1} V^{-1} cm [6,7].

With agarose-coated and -plugged capillaries mobilities of human and chicken erythrocytes were also determined in a buffer of physiological ionic strength (with a flow-by cooling system [6,7]); these mobilities were, respectively, -1.2 and -0.9 μm sec^{-1} V^{-1} cm; they were essentially the same as those quoted in the literature [1,6,7,11,12].

With the LIS buffer the electrophoretic mobilities of a number of tissue cells, monodispersed in various ways (by means of treatment with trypsin, or with EDTA, or by scraping) were measured [7,13], and compared with electroosmotic liquid mobilities obtained with the same buffer, through capillaries coated on the inside with monolayers of the cells [7,13]. It was found that of all methods studied for dispersing attached cells with the aim of subjecting them to microelectrophoresis, treatment with EDTA is the least damaging to the cells and also has the smallest influence on the cells' ζ-potential [7,13]. The values for the electrophoretic mobilities of a number of representative cells, in LIS buffer, are given in Table 1 [7,13].

Human peripheral lymphocytes were analyzed by microelectrophoresis in the LIS buffer; a pronounced bimodal mobility distribution was observed, see the histogram in Figure 3.

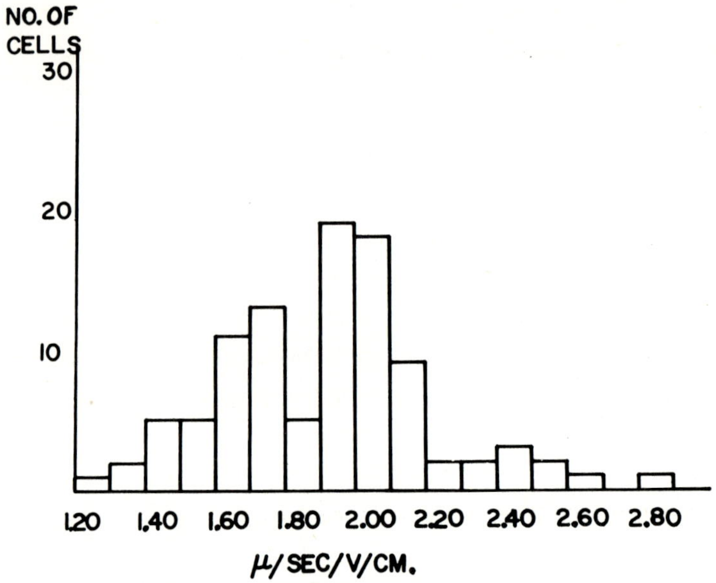

Fig. 3. Histogram of normal human peripheral lymphocytes in LIS buffer (μ = 0.015). The bimodal mobility distribution is quite evident.

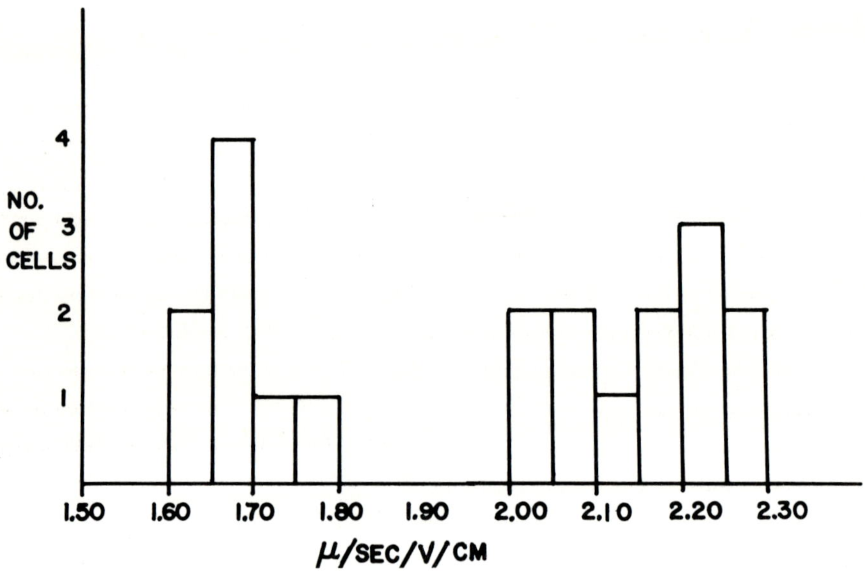

Fig. 4. Histogram of normal human peripheral granulocytes in LIS buffer (μ = 0.015). The bimodal mobility distribution is striking.

TABLE 1

ELECTROPHORETIC MOBILITIES OF DISPERSED MONOLAYER CELLS IN LIS BUFFER [7,13]

Cells	Mobility in $\mu m\ sec^{-1}\ V^{-1}\ cm$
Hamster melanoma cells (RPMI 1846)	-2.0
Transformed epithelial cells (HEP 2)	-1.5
Buffalo green monkey kidney cells	-1.5

Guinea pig alveolar and peritoneal macrophages have been compared in the LIS buffer, as well as in phosphate buffered saline (PBS). The mobilities are given in Table 2.

TABLE 2

AVERAGE ELECTROPHORETIC MOBILITIES OF GUINEA PIG MACROPHAGES

Cells	Mobility (in $\mu m\ sec^{-1}\ V^{-1}\ cm$)	pH	Ionic strength (μ)
Alveolar macrophages	-1.82	7.4	0.015 (LIS)
	-0.79	7.2	0.15 (PBS)
Peritoneal macrophages	-1.65	7.4	0.015 (LIS)
	-0.75	7.2	0.15 (PBS)

In all cases a bimodal mobility distribution was observed; the average mobility of alveolar macrophages was (in LIS) noticeably higher than that of peritoneal macrophages. The average phagocytic activities of both types of macrophages were identical.

Human peripheral polymorphonuclear granulocytes (PMN's) have (in LIS) an average electrophoretic mobility of -1.7 $\mu m\ sec^{-1}\ V^{-1} cm$; their mobilities also show a pronounced bimodal distribution (see Figure 4). The influence of the length of treatment with neuraminidase (at 37° C) on their average electrophoretic mobility (in LIS) is shown in Table 3. Remarkable enough, the phagocytic activities as well as the interfacial free energies (as measured by contact angle determinations of sessile drops of saline water [14,15]) of these PMN's remained unaltered, and their bimodal electrophoretic mobility distribution persisted after treatment with neuraminidase. Thus, the hydrophilicity of the cell surface of PMN's appears unrelated to their ζ-potential, contrary to what occurs with erythrocytes that are subjected to the same treatment [16]. Neuraminidase not only reduces the electrophoretic mobility of human erythrocytes by $\approx$ 70% but it also makes them significantly more hydrophobic (as judged by an increase in contact angle from $\approx$ 15° to 19°-21°) [15]. Treatment of human erythrocytes with papain reduces their electrophoretic mobility $\approx$ 30% and induces an increase in contact angle to $\approx$ 23° [15].

TABLE 3

AVERAGE ELECTROPHORETIC MOBILITIES IN LIS BUFFER OF HUMAN PMN'S AS A FUNCTION OF THE LENGTH OF TREATMENT WITH NEURAMINIDASE

Duration of treatment (in minutes)	Average electrophoretic mobility (in μm sec^{-1} V^{-1} cm)	Cell viability [a]
0	-1.7	80
30	-1.3	82
60	-1.1	82
50	-1.0	50

[a] Expressed as % of cells excluding Trypan blue.

Guinea pig pancreas cells, dispersed with collagenase and trypsin, showed an extensive range of electrophoretic mobilities (from -1.3 to -2.8 μm sec^{-1} V^{-1} cm, in IIS buffer, μ=0.04, pH 7.3 (see Figure 5).

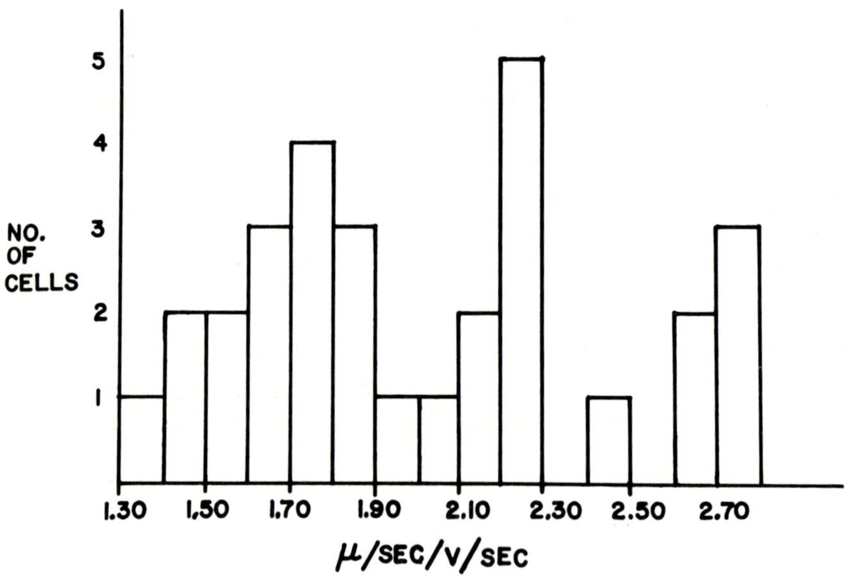

Fig. 5. Histogram of trypsinized guinea pig pancreas cells in I IS buffer (μ = 0.04). The range of electrophoretic mobilities is quite wide, particularly considering that at medium high ionic strength the range, as well as the absolute mobility, tends to be less than at very low ionic strength (compare with Figs. 3 and 4).

The macrophage electrophoretic mobility (MEM) test [2,3] may well become available to most clinical laboratories, thanks to the simplified microelectrophoresis method described above, which essentially only requires a microscope and a DC power supply. The intrinsic value of the MEM test as a tool in cancer diagnosis has been confirmed by this method [17], and results were obtained that indicate that electrophoretic mobility tests on the lymphocytes of patients possibly may directly yield the results that hitherto could only be obtained with electrophoretic analysis of guinea pig macrophages after treatment with the supernatants of patients' lymphocytes [17].

CONCLUSION

The paraboloid distribution of mobilities resulting from electroosmotic backflow that forms one of the principal drawbacks of the classical method of cell microelectrophoresis, can be entirely obviated by reducing it to a uniform plugflow. This is easiest done by coating the inner wall of a capillary with a gel layer of 2.5% agarose, and, once cells are introduced into that capillary, by closing both ends with the same gel. Under these conditions the observed electrophoretic mobility of cells remains the same at all depths of focusing. Undue sedimentation of especially large cells can now easily be circumvented by occasional rotation of the capillary around its cylindrical axis.

ACKNOWLEDGEMENTS

This work was partly supported through NASA contracts Nos. NAS G-15365; R.M.F. was supported through postdoctoral fellowship No. F32 - CA 05453.

REFERENCES

1. Ruhenstroth-Bauer, G. (1965) in Cell Electrophoresis, Ambrose, E.J., ed., Little, Brown, Boston, pp. 66-72.

2. Field, E.J. and Caspary, E.A. (1972) Brit. J. Cancer, 26, 164-173.

3. Lubran, M.M. (1964) Ann. Clin. Lab. Sci., 4, 121-124.

4. Abramson, H.A., Moyer, L.S. and Gorin, M.H. (1964) Electrophoresis of Proteins, Hafner, New York, pp. 48-52.

5. Seaman, G.V.F. (1965) in Cell Electrophoresis, Ambrose, E.J., ed., Little, Brown, Boston, pp. 4-21.

6. van Oss, C.J., Fike, R.M., Good, R.J. and Reinig, J.M. (1974) Anal. Biochem., 60, 242-251.

7. Fike, R.M. (1975) Zeta-Potentials of Cells, Measured by Microelectrophoresis and Microelectroosmosis, Dissertation, State University of New York at Buffalo.

8. Gillman, C.F., Bigazzi, P.F., Bronson, P.M. and van Oss, C.J. (1974) Prep. Biochem., 4, 457-472.

9. Duckworth, M. and Yaphe, W. (1971) Carbohydr. Res., 16, 189-197.

10. Young, K., Duckworth, M. and Yaphe, W. (1971) Carbohydr. Res., 16, 446-448.

11. Seaman, G.V.F. and Uhlenbruck, G. (1963) Arch. Biochem. Biophys., 100, 493-502.

12. Straub, E. (1965) in Cell Electrophoresis, Ambrose, E.J., ed., Little, Brown, Boston, pp. 125-140.

13. Fike, R.M. and van Oss, C.J. (1977) In Vitro, 12, 428-435.

14. van Oss, C.J. and Gillman, C.F. (1972) J. Reticuloendothel. Soc., 12, 497-502.

15. van Oss, C.J., Gillman, C.F. and Neumann, A.W. (1975) Phagocytic Engulfment and Cell Adhesiveness, Marcel Dekker, New York, pp. 7-19.

16. van Oss, C.J., Gillman, C.F. and Neumann, A.W. (1975) Phagocytic Engulfment and Cell Adhesiveness, Marcel Dekker, New York, pp. 129-130.

17. Fike, R.M. and van Oss, C.J. (1978) Fed. Proc., in press.

© 1979 Elsevier/North-Holland Biomedical Press
Electrokinetic Separation Methods,
P.G. Righetti, C.J. van Oss and J.W. Vanderhoff

CONTINUOUS FLOW ELECTROPHORESIS

J. W. Vanderhoff, F. J. Micale, and P. H. Krumrine
Center for Surface and Coatings Research, Building #7, Lehigh University,
Bethlehem, Pennsylvania 18015 (U.S.A.)

INTRODUCTION

Continuous flow electrophoresis (CFE), sometimes referred to as continuous particle electrophoresis (CPE), has been developed over the past 15 years primarily by two independent investigators, Hannig[1-3] and Strickler[4-6]. During this time, the CPE has been used primarily as a preparative method for the separation of biological materials[7,8]. The commercial Beckman CPE has also veen investigated as a method for the separation of latex particles according to size[3,9]. This instrument developed by Strickler, and similar in design to the instrument developed by Hannig, was available from Beckman Instruments for a limited period of time. The commercial instrument, which was designed primarily as a preparative tool, has also been modified and used as an analytical tool for the absolute measurement of electrophoretic mobilities[10]. This analytical capability has been utilized to investigate the separation capability of the CFE and, along with a theoretical development, has allowed a direct comparison between experiment and theory.

EXPERIMENTAL

Figure 1 shows a schematic representation of the Beckman CPE instrument. The 0.15 x 4.5 x 50-cm electrophoresis channel is positioned vertically with the 30.5-cm long electrodes on opposite sides. Electrolyte solution is pumped into the top of the channel to form the electrolyte curtain, which flows downward through the channel and empties at the bottom through 48 1-mm O.D. stainless steel tubes set edge-to-edge, so that the electrolyte curtain is separated into 48 fractions. The electrodes are separated from the electrophoresis channel by semi-permeable membranes, and a separate pumping system circulates another supply of the same electrolyte solution through the electrode compartments to remove any products of electrolysis. The sample to be separated is pumped into the center of the electrolyte curtain near the left side of the cell. This sample stream descends vertically in the absence of a voltage gradient, but when the voltage gradient is applied, it is displaced to the right according to the electrophoretic mobility of the particles. When the sample stream passes beyond the electrodes, it again descends vertically, but now displaced to the right. A viewing port with a millimeter scale allows the measurement of the migration distance, i.e., the lateral displacement of the particle stream. The electrolyte curtain and electrode rinse pumping systems use peristaltic pumps or gravity feed. The sample pumping system uses gas generated

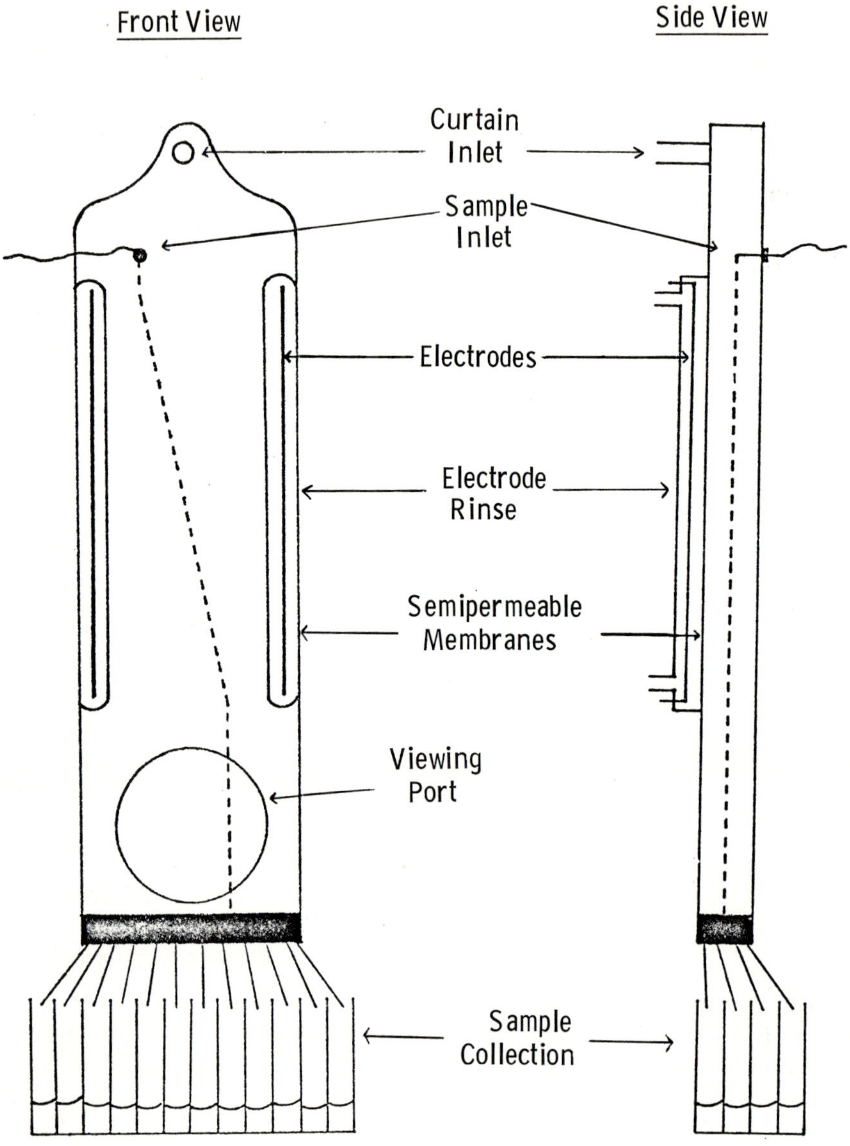

Figure I. Schematic Representation of the Beckman CPE.

by the electrolysis of water, which in turn is governed by a rheostat.

The Beckman CPE Instrument was modified in two ways. The sample delivery system, which was originally controlled by an electrochemically-induced pressure head, was replaced by a micrometer syringe attached to a variable-speed motor. This modification resulted in a more reproducible sample delivery rate, which could be controlled over a wider range of flow rates and which results in a more accurate definition of the sample stream diameter. The second modification concerns the determination of the lateral displacement of the particle stream due to electrophoresis. The Plexiglas viewing port at the bottom of the separation cell, which is used for visual observation of the sample stream by means of reflected light, was replaced by a quartz window which is transparent to ultraviolet light. A small optical bench with an ultraviolet light source and slit for control of beam dimensions was placed in front of the window. The optical bench was attached to a motor, which moves it horizontally across the window to scan the particle stream after the separation, and a resistance box, to establish the precise position of the ultraviolet light beam electronically. A UV-sensitive phototube was located in a fixed position on the other side of the window. A Houston Instruments Model 2000 X-Y recorder was used to record the intensity of ultraviolet radiation on the Y-axis and the horizontal position of the optical bench on the X-axis. The horizontal position on the recorder is magnified by a factor of 18.2X.

The principles of flow in the CPE system have been described by Strickler[6] and Hannig[3]. A schematic representation of the electroosmotic and induced flow in the curtain is presented in Figure 2. The parabolic flow profile due to electroosmosis is in the X-Z plane and is either positive or negative in the X direction, while the induced parabolic flow profile is in the Y-Z plane and is always positive in the Y direction. The sample is injected into the center of the stream in a cylindrical configuration and moves with a constant electrophoretic velocity in the X direction in the presence of an applied potential. The migration of the particles is affected by two factors, both of which are a function of their position in the Z direction: the electroosmosis of the electrolyte medium which affects the net particle velocity; and the induced parabolic flow profile which affects the velocity of the particle in the Y direction and hence the time of exposure of the particle to the electric field. Both factors ultimately affect the migration distance of the particles and the configuration of the sample stream. Both the electroosmotic and induced parabolic flow profiles of the electrolyte medium have compensating effects, i.e., as the particle position moves away from the center of the channel, the net particle velocity decreases in both the Y and X directions. This means that the slower-moving particles will be subjected to electrophoretic migration for a longer period of time than the faster-moving particles in the center of the curtain flow. In principle, therefore, it is possible to match the electroosmotic flow to the induced flow so that the sample streams will remain undistorted after separation. Under

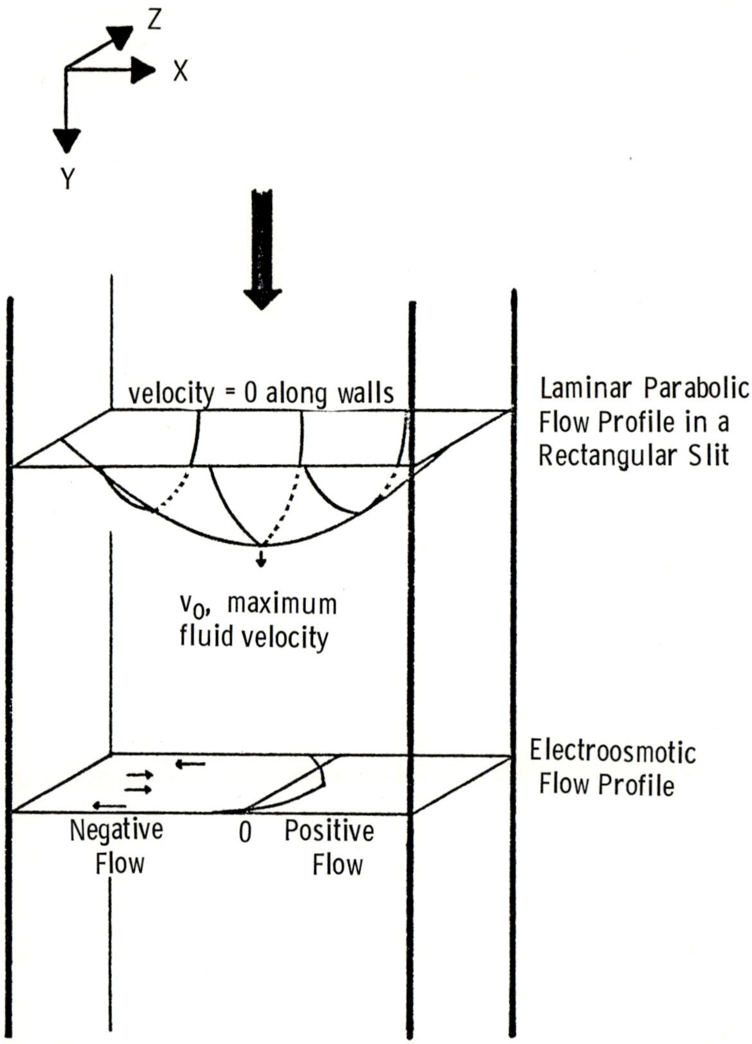

Figure 2. Schematic Representation of Flow Profiles Present in Electrolyte Curtain of CPE.

these conditions, the displacement of the sample stream will be controlled by the particles in the center of the curtain where the velocity of buffer in the _Y_ direction and the positive velocity of buffer in the _X_ direction are both at a maximum.

COMPUTER MODELING OF CPE

To determine the effect of the variables affecting resolution in continuous particle electrophoresis, a computer program was developed to simulate the separation of particles of different electrophoretic mobilities under a wide variety of conditions. This program[10] is capable of predicting the cross-sections of the separated particle streams or bands at any stage or separation as well as the variation of particle concentration with separation distance as would be detected by a scanning device mounted in front of the cell. These predictions are useful in determining exactly what effect each parameter has on the resolution. In this way, the variation of experimental parameters can be simulated to predict performance or to compare theoretical computations with experimental performance.

The principle equation which describes the displacement of a particle in the flow fields has been derived by Strickler[6].

$$x = \frac{HE}{v_o}\left(\frac{U_{os} + U_e}{1 - z^2/d^2} - \frac{3}{2}U_{os}\right) \qquad (1)$$

where x is the horizontal displacement distance (μm)

H is the electric field length through which the particle travels vertically (cm)

E is the applied potential (volt/cm)

v_o is the maximum fluid velocity at the midplane (cm/sec)

U_{os} is the electroosmotic mobility of the walls (μm cm/volt sec)

U_e is the electrophoretic mobility of the particles (μm cm/volt sec)

d is the curtain half-thickness (cm)

z is the particle depth from the curtain midplane (cm)

This equation describes the displacement of a particle as a function of the electrode length, applied voltage, curtain flow rate, electroosmotic and electrophoretic mobilities, and the particle depth to curtain half-thickness ratio. The migration distance is directly proportional to the electric field length and applied voltage, and inversely proportional to the curtain flow rate. Changing any of these three parameters will affect the particle migration and therefore the attainable separation. All three parameters, however, affect the Joule heating of the curtain fluid thereby creating an upper limit to this ratio of HE/v_o above which mixing from convection will occur and result in unstable flow.

Equation 1 assumes that the electroosmosis of both walls is the same, which may not be the case if two different materials are used. Equation 1 also derived by

Strickler[6], takes this effect into account:

$$x = \frac{HE}{v_o}\left[\frac{1}{2}\left(\frac{U_{osr}}{1 - z/d} + \frac{U_{osf}}{1 + z/d}\right) - \frac{3}{4}(U_{osr} + U_{osf}) + \frac{U_e}{1 - z^2/d^2}\right] \quad (2)$$

where U_{osr} and U_{osf} are the electroosmotic mobilities of the rear and front walls respectively. Equation 2 reduces to Equation 1 if U_{osr} is equal to U_{osf}.

An important consideraton in the CPE, as in any form of electrophoresis, is the effect of temperature on electrophoretic mobility. Because of the high potentials applied, the temperature of the curtain fluid increases, giving rise to a temperature gradient between the center of the channel and the walls. This temperature gradient will be small in a narrow-channel CPE, but, in a thicker channel, it will be much greater. Both the viscosity and dielectric constant of the fluid medium are temperature-dependent and affect the mobility values. Therefore, if the mobility is measured at a standard temperature of 25°C, and a CPE separation is carried out at a different temperature or with cooling in the face plate, a correction should be made to take into account the effect of temperature. This may be done by modifying Equation 2 to:

$$x = \frac{HE}{v_o}\left[\frac{1}{2}R_w\left(\frac{U_{osr}}{1 - z/d} + \frac{U_{osf}}{1 + z/d}\right) - \frac{3}{4}R_w(U_{osr} + U_{osf}) + \frac{U_e R_z}{1 - z^2/d^2}\right] \quad (3)$$

where,

$$R_w = \frac{\varepsilon_w \eta_{298}}{\eta_w \varepsilon_{293}} \quad , \quad R_z = \frac{\varepsilon_z \eta_{298}}{\eta_z \varepsilon_{298}} \quad (4)$$

where ε and η are the dielectric constants and viscosities, respectively, at the wall position, w, and at the particle position, z. Thus, these ratios R_w and R_z correct the electroosmotic mobility values to the temperature at the walls where the driving force is located and the electrophoretic mobility values to the temperature at the particle position. This procedure allows the effect of a temperature gradient to be superimposed on the particle migration. This temperature gradient is not constant as the particles pass through the electric field, but gradually increases to a maximum as the particles leave the electric field. This effect may be taken into account in a computer simulation.

This treatment accounts for changes in electrophoretic mobility, but not for changes in the flow profiles due to the temperature-dependent fluid viscosity. This latter effect, however, is considered to be small for the Beckman CPE in which the temperature gradient is relatively small.

Another factor which the theory does not cover is the laminar flow profile in the direction of migration, X, as shown in Figure 2. Although the width of the channel is large in comparison to the thickness (about thirty times larger in the Beckman CPE, 4.5 cm. versus 0.15 cm.) and the sample stream is confined to the middle third of the channel (1.5 cm. from both sides), the velocity variation from this source can effect the exact separation distance of each particle band. This effect has been accounted for in the computer simulation by a trial-and-error procedure to determine an average vertical velocity as a function of the migration distance.

These equations and modifications describe the movement of a particle anywhere within the CPE curtain; however, actual separations are carried with particle samples that may have a distribution of electrophoretic mobilities. To model this effect the initial band cross-section is divided into a number of points. Then, by calculating the displacement as a function of position, the total cross-sectional image and particle distribution may be obtained by adding the contribution from each segment. Most samples, however, exhibit a range of electrophoretic mobilities; therefore, it is necessary to characterize this mobility distribution in order to determine the breadth of the sample band and to add the contribution from each segment to gain the overall concentration profile. The computer program provides four standard mobility distribution shapes --- Gaussian, triangular, rectangular, and parabolic --- plus a provision for a nth-order polynomial computer-fitted to any random shape such as a bimodal distribution. Once a distribution shape can be mathematically defined, it can be applied to each point into which the initial band cross-section has been divided. This summing process can be done easily and quickly by the computer thereby allowing the band to be divided into many points for a high degree of accuracy. In this manner, a concentration profile can be calculated as a function of all the parameters that affect the migration distance.

RESULTS

This section presents theoretical predictions for electrophoretic separations in the Beckman CPE, although the computer program can be applied to electrophoretic channels of any dimensions. The electrophoretic mobility distributions of the sample particles used in these calculations are shown in Figure 3 and Table I. Also, the relative concentrations of the four species are varied to show the concentration effect on the peak height. The electroosmosis value has been matched to the mean particle mobility of -4.0μm cm/volt sec, which gives the best resolution under these conditions. Figure 4 shows an example of hypothetical separation of four particle species of different electrophoretic mobility. Above the particle concentration profile is shown the band cross-sections to further demonstrate how the bands appear and are separated.

All theoretical predictions which will be presented are for the CPE originally manufactured by Beckman Instruments, namely 1.5 mm. thick by 4.5 cm. wide by 30 cm. long for the electrode region. The computer program, however, can be used to

Table I

Assumed Particle Mobility Distribution Data

Particle	Mean Mobility*	Range	Relative Conc.	Distribution Shape
1	-3.0	± 0.1	60.0	Parabolic
2	-3.7	± 0.1	80.0	Parabolic
3	-4.0	± 0.2	80.0	Parabolic
4	-4.5	± 0.2	100.0	Parabolic

Electroosmotic Mobility Front and Rear Walls = 4.00 um cm/volt sec

* - denotes negatively charged particles

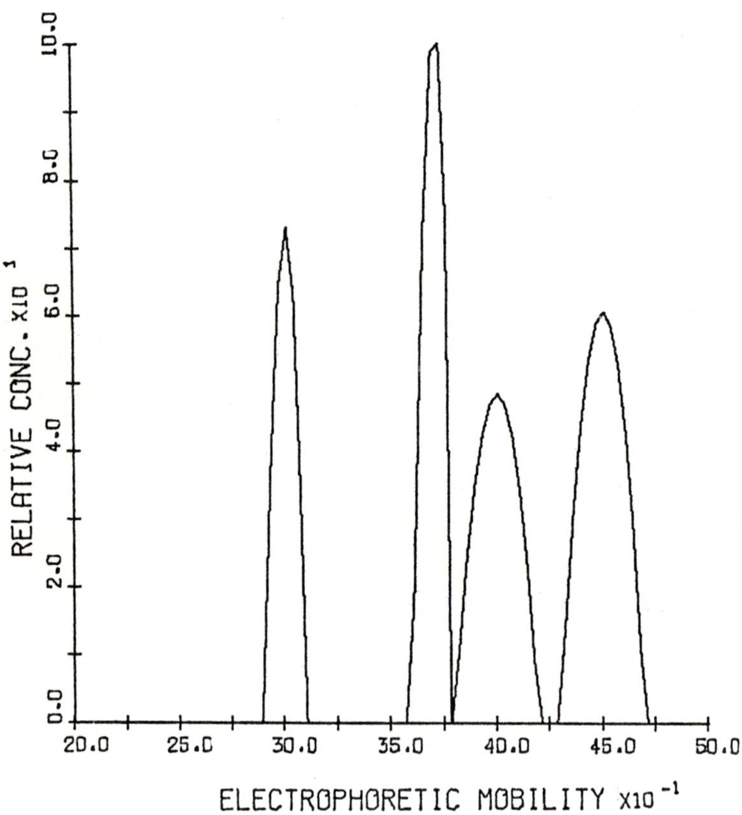

Figure 3. Mobility Distributions of Four Particles Which Were Used for Comparison of Various Parameters.

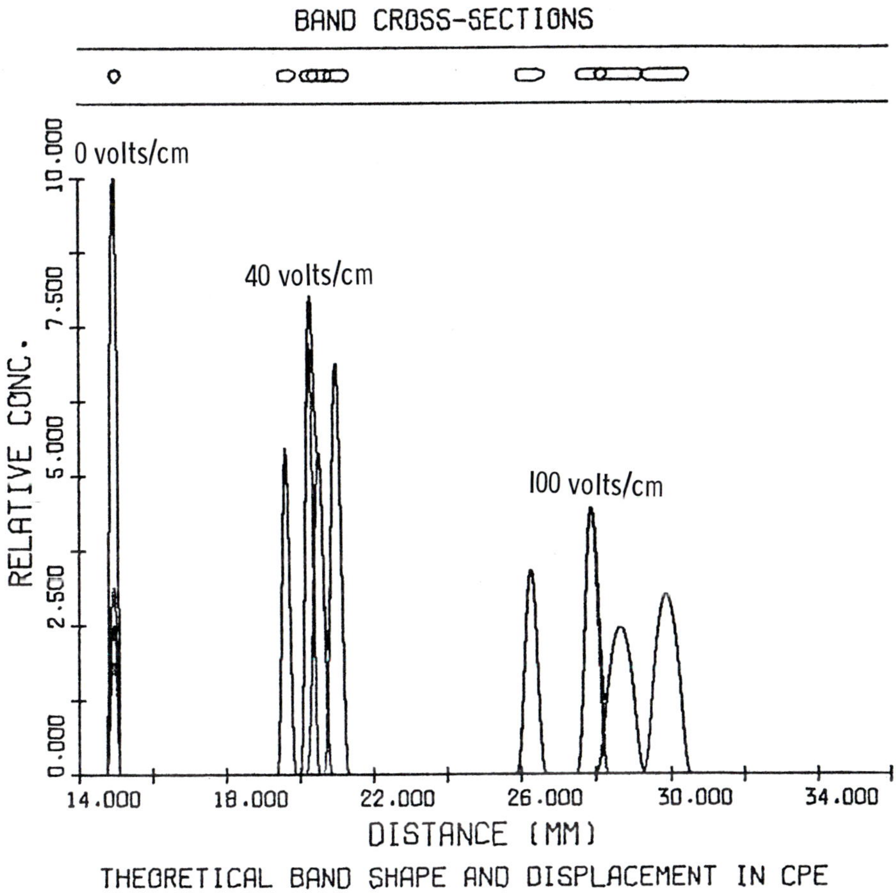

Figure 4. Hypothetical Separation at Two Stages of a Mixture of Four Species of Differing Mobility Distributions Under Ideal Conditions.

compute results for channels of any dimensions.

The separation is shown at two stages, one corresponds to a potential of 40 and 100 volts/cm. corresponding to HE/v_o values of 0.0864 and 0.216 volt sec/μm respectively, whre H is 30 cm. and the volumetric flow rate is 25 cm^3/minute. As Figure 4 shows, the migration distance is proportional to the value of HE/v_o. In this case the particles are shown injected in a cylindrical band where the ratio of band diameter to channel thickness, R, is 0.2, and the band is in the center of the channel. The migration distance scale runs from 14.0 to 36.0 mm. in absolute distance from the curtain edge, which roughly covers the middle third of the column, with the sample being injected at 15.0 mm. Also, a 2°C temperature gradient is assumed to develop as the particles progress through the electrode region.

When the foregoing particle mixture is taken as a standard model, it is possible to change the different experimental parameters to determine the specific effect of each parameter. Figure 5 shows the effect of injecting the particles off-center in the curtain stream. The further off-center the bands are injected, the poorer the resolution. This poor resolution results because the difference between the electrophoretic mobility of the fast or slow particles and the electroosmotic mobility is large; this distorts the bands as they are positioned closer to the wall. The resolution would be decreased even further if the electroosmotic mobility was higher or lower than the mean mobility value of -4.0μm cm/volt sec.

Figure 6 shows the effect when the electroosmotic mobilities of the front and back walls are different. The greater the difference in electroosmotic mobility between the two walls, the more distorted will be the sample bands. The migration distance remains about the same in all cases because the average electroosmotic mobility for both walls remains the same. If the average mobility increased or decreased, then the migration distance would have increased or decreased proportionately.

Figure 7 shows the effect of increasing the injected band radius. This is similar to the effect shown in Figure 5 and results from a mismatch of the electroosmotic mobility and the electrophoretic mobility of the leading and trailing bands. In many cases, the amount of processed sample or throughput rate is important and Figure 7 demonstrates the trade-off in resolution by selecting higher band volume. Again, the resolution would appear much worse if the electroosmotic mobility were not matched to the mean particle mobility.

When the sample throughput volume is increased, it is interesting to determine which dimension, i.e., in the X or Z direction as defined in Figure 2, is the most detrimental to the resolution. Figure 8 shows this effect when the sample is injected in a rectangular cross-section rather than a circular cross-section. For circular geometry, if the radius is doubled, the area and sample throughput increases by a factor of four. However, for rectangular geometry, one dimension

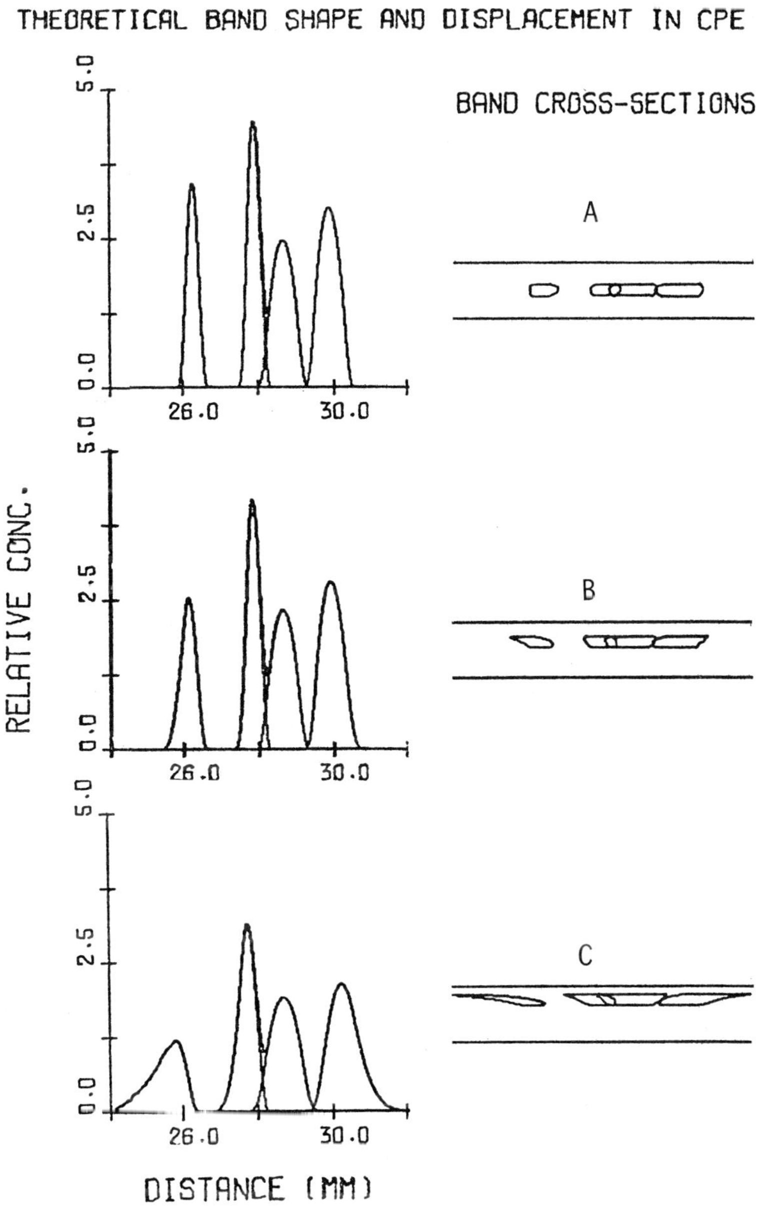

Figure 5. Computed Particle Displacement Showing the Effect of Off-Center Injection on Band Separation with U_{OS} = -4.00 μm cm/volt sec., and R = 0.2, A. Centered, B. 0.25 Off-Center, and C. 0.5 Off-Center.

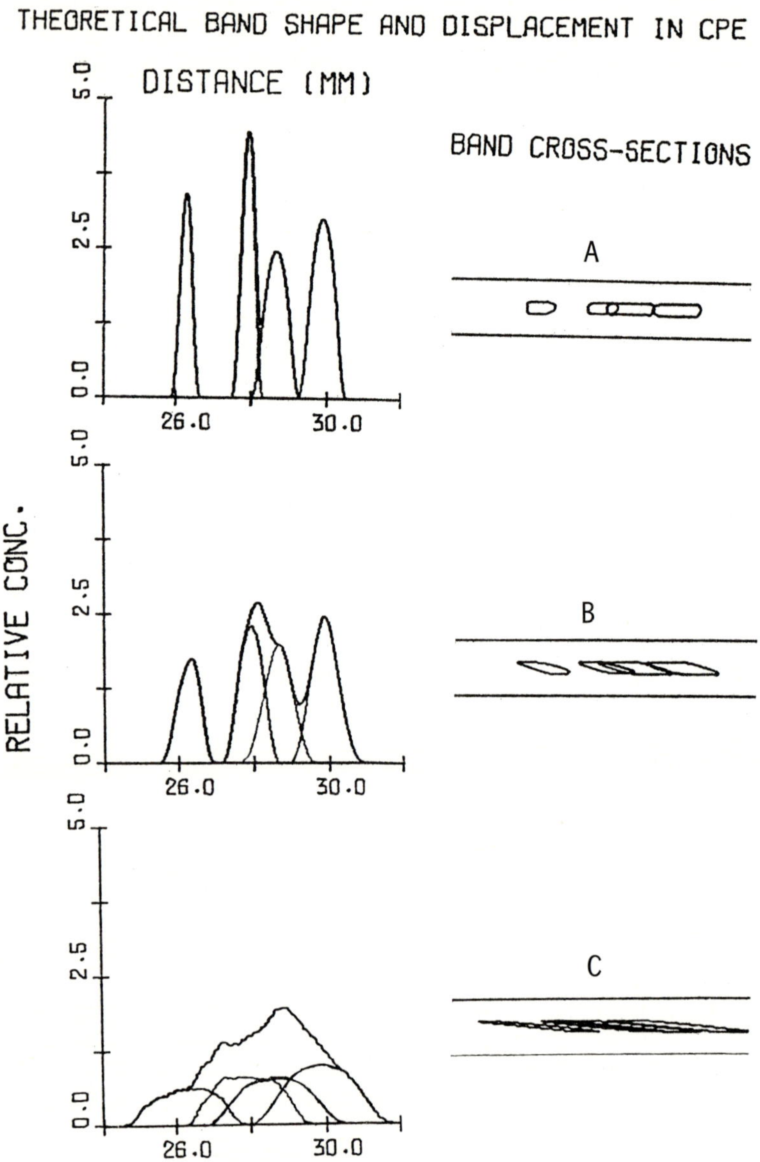

Figure 6. Computed Particle Displacement Showing the Effect of Different Values of U_{os} for the Front and Rear Walls with R = 0.2, and Average U_{os} = -4.00 μm cm/volt sec.; A. U_{osf} = U_{osr} = -4.00 μm cm/volt sec., B. U_{osf} = -5.00 μm cm/volt sec, U_{osr} = -3.00 μm cm/volt sec., C. U_{osf} = -7.00 μm cm/volt sec, U_{osr} = -1.00 μm cm/volt sec.

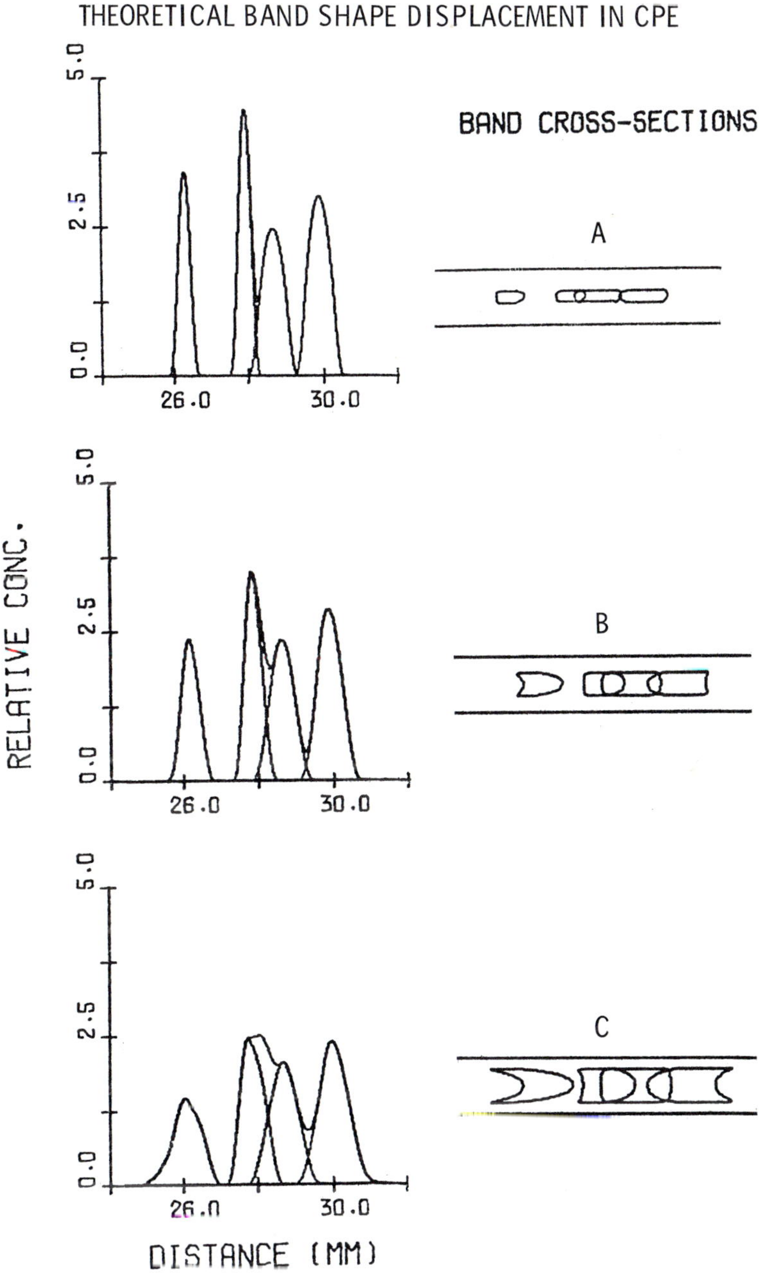

Figure 7. Computed Particle Displacement Showing the Effect of Increasing the Injection Radius with U_{os} = -4.00 μm cm/volt sec.; A. R = 0.2, B. R = 0.4, and C. R = 0.6.

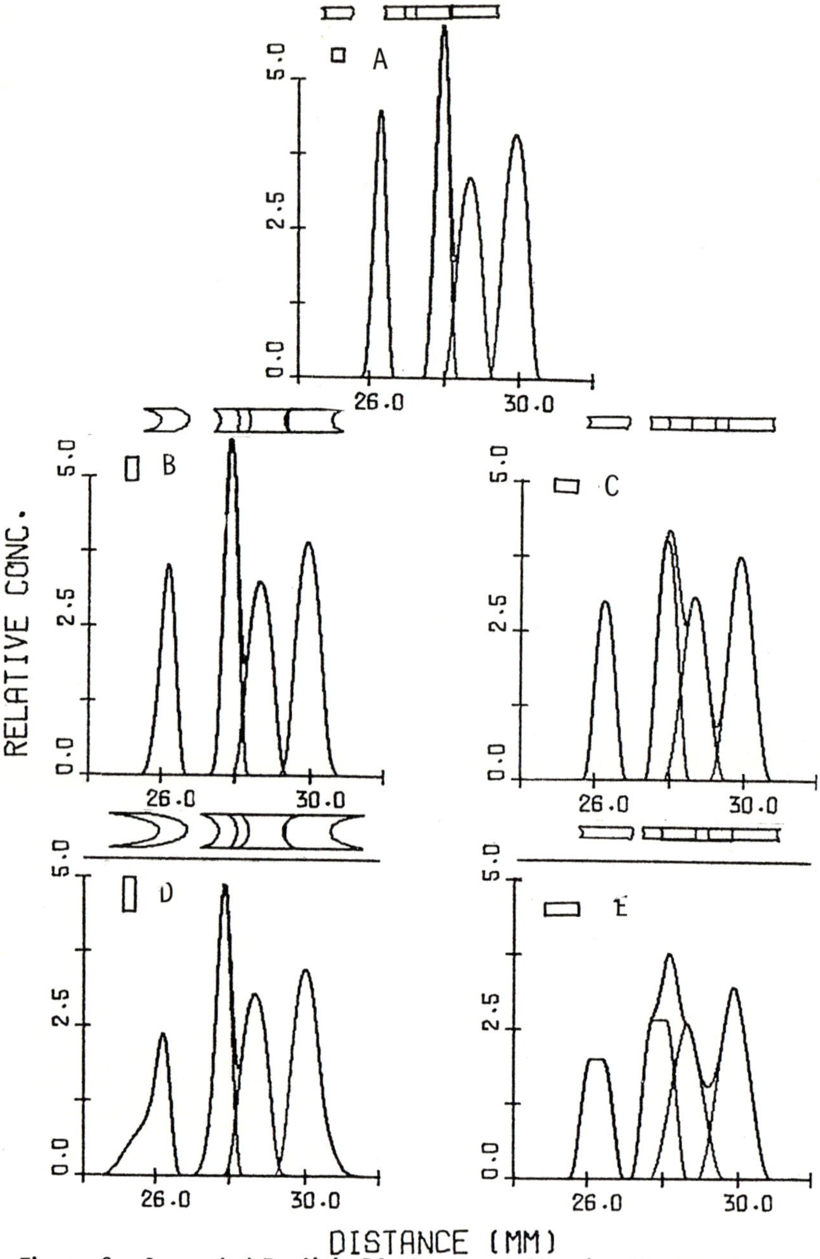

Figure 8. Computed Particle Displacement Showing the Effect of Increasing Band Volume in the R or θ Direction with U_{OS} = -4.00 μm cm/volt sec.; A. R = 0.2, θ = 0.2, B. R = 0.4, θ = 0.2, C. R = 0.2, θ = 0.4, D. R = 0.6, θ = 0.2, and E. R = 0.2, θ = 0.6.

can be doubled or tripled, resulting in a corresponding doubling or tripling of sample throughput to determine the optimum direction of sample increase. In the upper left hand corner of each diagram, the original size of the sample stream relative to the cell thickness is shown. From these results, it may be concluded that it is better to increase the sample width, or increase the R value, rather than the sample thickness, θ, in the direction of migration. This conclusion is not necessarily always true, as shown in Figure 9 where the electroosmotic mobility has been decreased from the mean particle mobility of -4.0 to -1.0μm cm/volt sec. Increasing the band width, i.e., the R value, in this case results in highly elongated cones, while increasing the thickness only causes minor distrotions. In this case, it would be better to increase the stream thickness to gain more throughput. These results show that the relative effects of the sample width and thickness on resolution are a sensitive function of the electroosmotic mobility relative to the particle mobilities. In general, the value of the electroosmotic mobility relative to the particle mobilities is the most important consideration in obtaining high resolution and sample throughput.

Finally, it is important to look at the effect of electroosmosis alone as presented in Figure 10. Electroosmosis is a horizontal motion of the fluid in the curtain, as shown in Figure 2, which is additive to the particle velocity. Therefore, at the same applied potential but different electroosmotic mobilities, the particle displacement distances will be different, as seen in Figures 10A, 10B, and 10C for electroosmotic mobilities of -4.00, -1.00, and -7.00μm cm/volt second, respectively. Figures 10D and 10E show the results when the applied voltages were adjusted to the appropriate values to obtain the same migration distance as in 10A. When this is done the interpeak distances is maximized for low electroosmotic values, Figure 10D, while degree of sample purity is increased by matching the electroosmotic value to the mean particle mobility, Figure 10A.

The above considerations demonstrate that there is no simple method for choosing the optimum separation conditions. The resolution depends on many variables such as the mobility distributions of the species, the sample stream cross-section, the applied voltage, the value of electroosmosis in relation to the particle mobilities, and the channel dimensions. However, this theoretical analysis and prediction technique can be a valuable first step to obtaining the best separation conditions for a given set of restrictions. Even though we cannot define the absolute optimum conditions for every situation, these results do show the trends which should be considered in attempting to optimize the resolution.

COMPARISON OF EXPERIMENT WITH THEORY

The computer model allows a direct comparison between theoretical and experimental separations. Figure 11 shows the separation of the three polystyrene latexes in the Beckman CPE. Figure 11A presents the theoretical separations, assuming equal

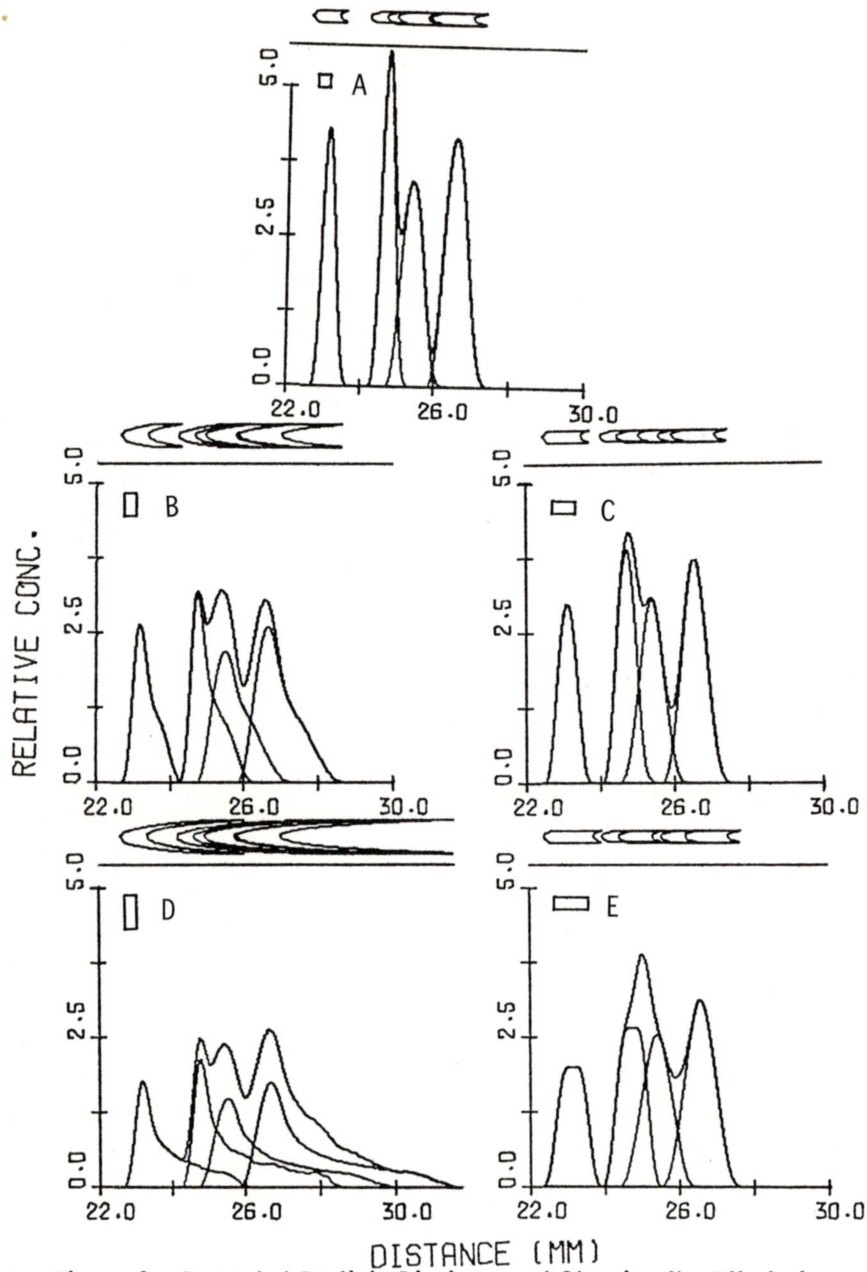

Figure 9. Computed Particle Displacement Showing the Effect of Increasing Band Volume in the R or θ Direction with U_{OS} = -1.00 μm cm/volt sec.; A. R = 0.2, θ = 0.2, B. R = 0.4, θ = 0.2, C. R = 0.2, θ = 0.4, D. R = 0.6, θ = 0.2, and E. R = 0.2, θ = 0.6.

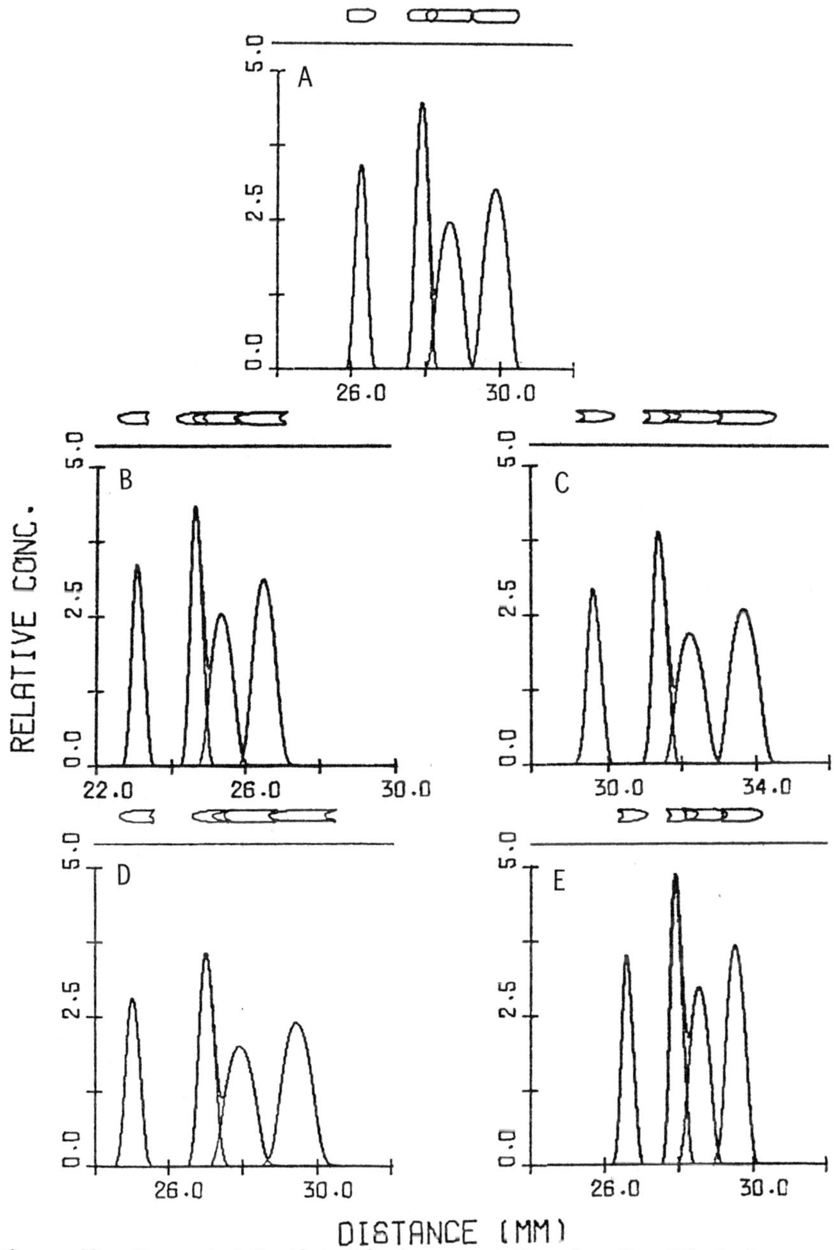

Figure 10. Computed Particle Displacement Showing the Effect of Electroosmosis and Adjusted Voltage Gradients with R = 0.2; A. U_{os} = -4.00 μm cm/volt sec, 100.0 volts/cm, B. U_{os} = -7.00 μm cm/volt sec, 100 volts/cm, C. U_{os} = -100 μm cm/volt sec, 100.0 volts/cm, D. U_{os} = -7.00 μm cm/volt sec, 80.0 volts/cm, and E. U_{os} = -1.00 μm cm/volt sec, 130.0 volts/cm.

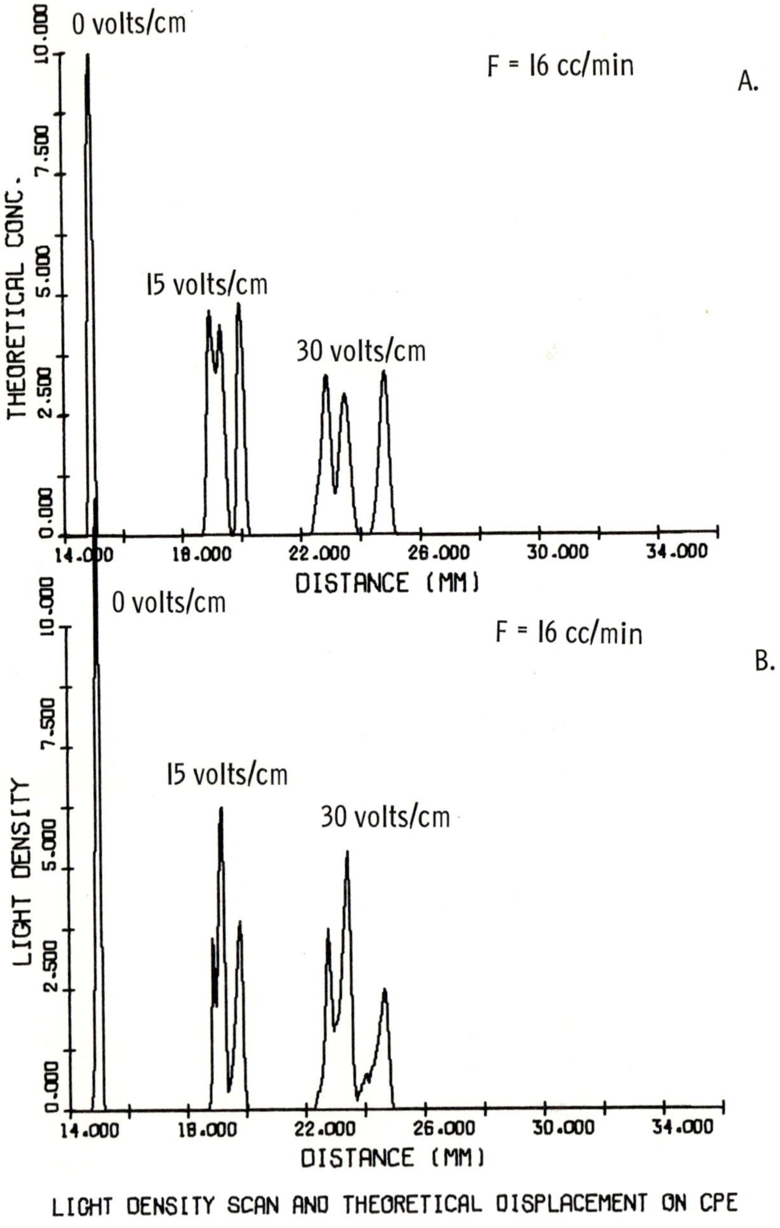

Figure 11. Comparison of the Theoretical and Experimental Separations of the Three Latexes Proposed for the SPAR CPE.

weight percents of each sample, and electrophoretic mobilities of 3.5 $\pm$ 0.1, 4.5 $\pm$ 0.3, and 6.7 $\pm$ 0.2μm cm/volt sec. The heights of the peaks are related to the spread of each particle distribution since the area under each peak is the same. Figure 11B shows the experimental results obtained on the Beckman CPE. In this case, the peak heights are related to the narrowness of the bands also, but more importantly to the wavelength of the UV light source and the degree of absorbance and reflection of each particle species. Although the peak heights do not match, the actual displacement distances and the spreads of the peaks show very good agreement.

Figure 12 shows a separation of the seven polystyrene latexes. These latexes ranged in size from 2.02μm to 0.088μm, and were run in the Beckman CPE in a barbital-sodium barbital buffer at 2 x 10^{-3} ionic strength. Again the concentration by weight of each species was equal in both the theoretical and experimental curves and the experimental peaks show a strong dependence on particle size. The experimental particle displacements are demonstrated to agree with theory.

Modeling of the CPE has allowed many possibilities, both in analyzing results and in predicting the theoretical optimum separations. By being very flexible and general, it allows many parameters to be varied to determine the specific effect of these parameters.

CONCLUSIONS

Resolution for particles of different mobilities in the Beckman CPE may be detected very well; collection of separated fractions, however, is not as well resolved due to the size limitations of the collection tubes. The use of monodisperse latexes of varying sizes, i.e., two latexes differing in mobility by as little as 0.15 mobility units, have been separated and detected but not collected. This has occurred with a very thin sample injection stream, under optimum electrolyte conditions, at voltage gradients from 30 to 60 volts/cm. This result is probably due to the coarseness of the collection system, the electroosmotic tailing of the bands, and some particle-particle interactions.

Modeling of CPE system on the computer has led to interesting revelations about the effect of the various parameters. Electroosmosis is the main consideration since it manifests itself in all the other parameters. Maximum resolution is obtained when the electrophoretic and electroosmotic mobilities match. Obviously, if two or more different particles are present in the CPE, only one can have a matching mobility. The larger the difference the more crescent shaped and elongated the bands become. The effect of electroosmosis may be minimized by keeping the sample injection stream very thin at the sacrifice of sample throughput rate, and by keeping the stream in the center of the column. If sample throughput rather than resolution is of prime importance, there may be an advantage to a rectangular injection port rather than a cylindrical port, since the rectangular port could be

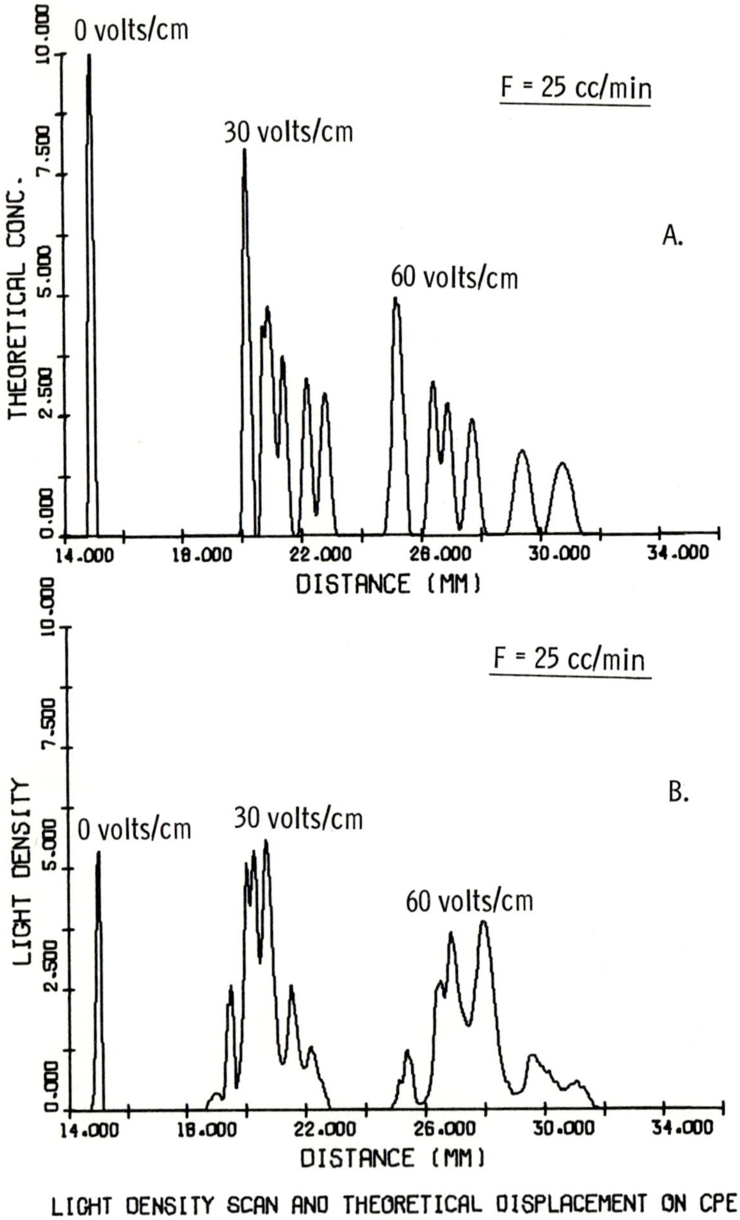

Figure 12. Comparison of the Theoretical and Experimental Separations of the Seven Latexes Ranging in Size From 2.02 μm to 0.088 μm.

increased in size in one direction only. The dimension that may be increased depends on the electroosmotic and electrophoretic mobilities. If the electrophoretic and electroosmotic mobilities match, then the width in the direction of the column wall could be increased without undue distortion. On the other hand, if the mobilities do not match, the thickness in the electrode direction would result in less distortion than increasing sample width. When resolution, rather than throughput is the primary consideration, there may be special conditions when a low value of electroosmosis will give better resolution than matching electroosmotic and electrophoretic mobilities. For instance, if the experimental conditions limit the migration distance, it is best to use that distance for electrophoretic migration rather than electroosmotic migration which does not contribute to the separation. However, in most cases where migration distance is not the limiting factor it is best to match mobilities. Therefore, the optimum conditions for any separation depend on a number of factors, the most important of which is electroosmosis.

REFERENCES

1. Hannig, K. (1961) Z. Anal. Chem. 181, 244-254.
2. Hannig, K. (1964) Hoppe-Seyler's Z. Physical Chem. 338, 211-227.
3. Hannig, K., Wirth, H., Meyer, B. H., and Zeiller, K. (1975) Ibid, 356, 1209-1223.
4. Strickler, A., Kaplan, A., and Vigh, E. (1966) Microchem J. 10, 529-544.
5. Strickler, A., and Sacks, T. (1973) Prep. Biochem. 3, 269-277.
6. Strickler, A., and Sacks, T. (1973) Ann. N. Y. Acad. Sci. 209, 497-514.
7. Hannig, K., and Heidrich, H. G. (1974) Methods Enzymol 21, 746-761.
8. Zeiller, K., Loser, R., Pascher, G., and Hannig, K. (1975) Hoppe-Seyler's Z. Physical Chem. 356, 1225-1244.
9. McCann, G. D., Vanderhoff, J. W., Strickler, A., and Sacks, T. (1973) Sep. & Purific. Methods 2, 153-198.
10. Micale, F. J., Krumrine, P. H., and Vanderhoff, J. W. (1976), J. Coll. & Interface Sci. Vol. IV, 247-258.
11. Krumrine, P. H. (1978), Ph.D. Thesis, Lehigh University, App. B., 143-155.

Electrokinetic Separation Methods,
P.G. Righetti, C.J. van Oss and J.W. Vanderhoff

CELL SEPARATION BY CONTINUOUS-FLOW ELECTROPHORESIS AND ISOELECTRIC FOCUSING

WILHELM W.JUST and GOTTFRIED WERNER
Max-Planck-Institut für Hirnforschung, Arbeitsgruppe Neurochemie, Deutschordenstraße 46, 6000 Frankfurt/Main, BRD.

INTRODUCTION

If one looks at the various organs of the mammalian organism and in particular at their manyfold functions, one may see that several individual, often highly specified, cell types cooperate for the maintenance of the overall activities of these organs. Again several subcellular organelles contribute to the unique functions of various cell types. To bring light into this sometimes mysterious interactions, methods for the separation of heterogeneous cell or cell organelle mixtures are necessary, which in turn enable studies on the particular properties of individual cell types.

Various highly elaborate separation techniques utilizing different underlying principles are readily available. They were described rather extensively in a recent compilation[1]. Electrophoresis, separating cells according to their surface charges, is one such method. The special techniques of continuous-flow electrophoresis (CFE) and of isoelectric focusing (IF) have been treated in this chapter. The main effort was to provide some information as to what extent CFE and IF are reliable tools for the biochemist interested in particle fractionation. Successful applications as well as the limitations of these techniques should be exhibited, placing practical aspects in the forefront.

CONTINUOUS-FLOW ELECTROPHORESIS

Different names such as vertical deflection electrophoresis, curtain electrophoresis, free-flow electrophoresis and free zone electrophoresis have been used to characterize the same underlying electrophoretic principle, the so called zone electrophoresis, in which anode, cathode and separation chamber compartments are filled with the same supporting electrolyte. In free solution separation will occur due to different electrophoretic mobilities of the sample ionic species introduced into the system.

Theoretical considerations

Since the content of this section was kept to that minimum neces-

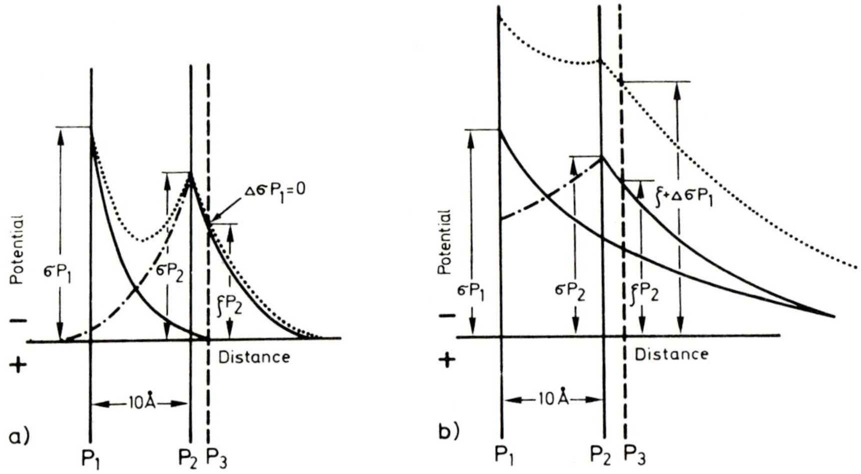

Fig.1. The resultant ζ-potential (.......) at the surface of shear for a multi double layer system, (a) for high ionic strength medium (>0.1) and (b) for low ionic strength medium (<0.01). From Heard and Seaman (ref.3), by the courtesy of Rockefeller University Press.

sary for the basic understanding of the practical experiments, the reader is referred to the theoretical chapters of this volume.

Considering a particle suspended in an electrolyte solution and bearing a net negative surface charge, positively charged ions will be attracted and will accumulate near the cell surface, thus forming an electrical double layer[2].

Quantitative considerations have shown that the electrophoretic mobility of the particle is governed by the potential at the surface of shear. This potential is called ζ-potential. With respect to the migrating particle the liquid within the surface of shear is stationary. In addition viscosity and dielectric constant are to be considered to be the same at the surface of shear and within the bulk liquid.

Heard and Seaman[3], taking into account the structure of the human erythrocyte membrane, suggested that the effective ζ-potential at the surface of shear derives from a multi double layer system. For that purpose (Fig.1) the authors hypothetically depicted, in addition to the surface of shear (P_3) and the phase boundary (P_2) which is given by the outermost ionogenic groups of the membrane, a third arbitrarily chosen plane (P_1) located deeper (10 Å) within the three dimensional membrane meshwork than P_2. On both planes P_1 and P_3 "diffuse double layers" may be generated, each with its own characteristic distribution of charge asymmetry. By superimposing both of them the resultant ζ-potential becomes evident. For the case of high ionic strength media (>0.1), shown in Fig.1a, the ζ-potential is not influenced by the course of the potential at P_1 whereas for the case of low ionic strength (<0.01)(Fig.1b) ionogenic groups situated at P_1 considerably contribute to the resultant ζ-potential. In addition the ionic strength

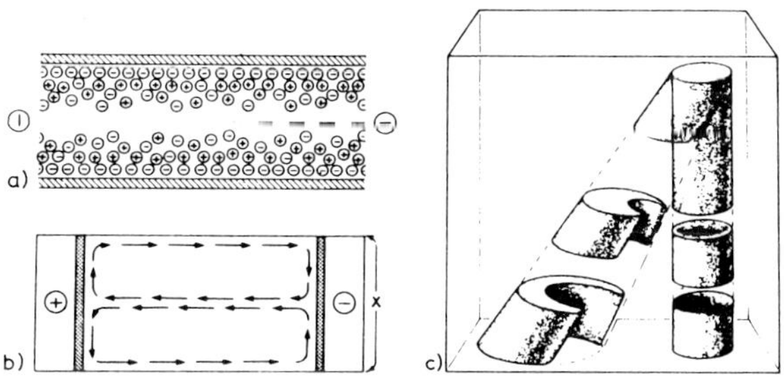

Fig.2. Electroosmotic flow in (a) an open and (b) a closed capillary system of diameter x. Influence of electroosmosis and of Poisseuille velocity flow (vertical zone) on the sample zone during CFE (cf.ref.6)(c).

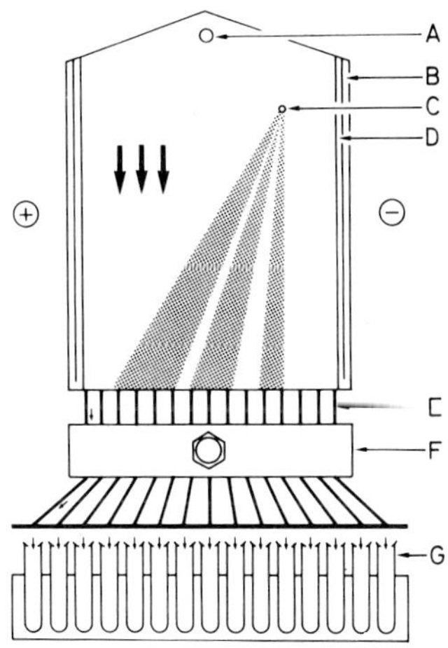

Fig.3. Principle of CFE. A detailed discription is given in the text.

of the medium was shown to be most important for the influence of size and shape of the particle on its electrophoretic mobility[4,5].

Although electroosmosis appears to be less pronounced in CFE and IF, under certain conditions it may lead to a considerable loss of band sharpness. In an open end glass capillary(Fig.2a) which is filled with an electrolyte solution, electroosmotic liquid transport toward the cathode takes place under the influence of an electric dc field due to the particular ion distribution on the capillary wall. The electroosmotic flow profile generated in a closed system such as the separation chamber of a CFE apparatus is shown in Fig.2b[6]. From this flow profile a deformation of the sample zones during the electrophoretic run is expected, as is shown in Fig.2c. However, the flow profile of the buffer curtain (Fig.2c) as well as the fact that biological membranes carry a net negative surface charge, both counteract the band broadening effect of electroosmosis[6-8]. During CFE electroosmotic flow might be visualized by the injection of either an uncharged dye or a solution of different refractive index than the chamber solution.

The current flow in a stationary electric field always will produce Joule-heat directly proportional to the electric field strength and the electrolyte concentration. For that reason CFE utilizes low conductivity buffer systems. Most physico-chemical parameters as viscosity, diffusion, electrical conductivity and electrolyte dissociation are temperature-dependent. Therefore, thermal inconsistencies, although they cannot completely suppressed, must be avoided. Several features which might considerably influence the CFE experiments such as the stability of pH and electrical conductivity within the separation buffer, the influence of substances used to balance the osmolarity of the buffer system, or the effect of buffer and sample flow rate on the band width, were recently discussed in a series of publications by Hannig et al.[7,9] and Zeiller et al[10].

Instrumentation

Since most of the experiments on cell fractionation have been performed using the commercially available instruments, the Hannig-type[11] FF-5 apparatus serves for the description of the basic operational principle (Fig.3).

Between two vertical parallel glass plates separated by a distance of 0.5-0.7mm, a vertical laminar buffer flow (A=buffer inlet) of constant but variable velocity is effected by the activity of a 90-fold pulse-free peristaltic pump (F) dividing the buffer curtain into 90 individual fractions (E,G). The electric field is applied at right angles to the buffer flow by means of platinum electrodes (B) which are contained in the electrode vessels separately rinsed by circulating electrode solutions. The electrode vessels are divided from the separation chamber by ion exchange membranes (D). A special device serves for the introduction of the sample through a front opening (C). The cooling system, one of the most important parts of the instrument, is connected to the rear glass plate and works according to the refrigerator principle by an electronically controlled Freon injection.

Optimal operational conditions may vary greatly from one experiment to another. However, for cell separation in general a field strength of approximately 100V/cm and a current of 100mA are recommended. A low buffer flow rate may be advantageous, but flow rates of less than 10cm/min may result in turbulances in the chamber.

A completely different, most interesting device for a CFE apparatus was introduced by Kolin[12], and is described in detail in chapter 10 of this volume. Utilizing also the principle of rotational stabilization against thermal convection, two other suitable devices were reported by Hjertén[13] and Aitchison et al.[14], the latter of which also allows continuous free-flow operation. This apparatus is displayed for production scale electrophoresis. It offers the possibility of collecting 30 fractions. Due to short exposure times of the sample material to the fractionation procedure, a cooling device was omitted.

Other more recent developments in the technology of CFE were reported by Mel[15], Strickler[16], Howery et al.[17], Pruzik[18] and Bier et al.[19] the last of which was especially designed for work at zero gravity. The apparatus developed by Seiler et al.[20] which was used for electrophoretic experiments in a continuously generated pH gradient is mentioned below in more detail.

Practical applications

Several aspects most important for the work on CFE or IF of biological particles could not be discussed as extensively as might be desirable. They are briefly mentioned in the following, and the interested reader is referred to the current literature.

A severe problem, which can largely be controlled, but which sometimes, however, may limit the application of the methods, is the tendency of biological particles to aggregate or clump. The underlying mechanisms are not completely understood yet. For practical work it seems necessary to solve the problem individually for each new type of separation experiment.

To determine the maintenance of cell viability throughout the separation experiment, dye exclusion tests are applied most commonly. However, the question is, to what extent are these tests relevant in determining cell viability. It seems also to be preferable to clarify cell viability individually, for example, by analysis of the uptake of radiolabelled tracers or by determining reproduction rates after cultering the cells.

When applying electrophoretic separation techniques following the preparation of cell suspensions by tissue dissociation, the common methods, such as use of hydrolytic enzymes or Ca^{2+}-chelating agents, although preserving cell morphology, may cause severe alterations in the cell surface charge.

Lymphocytes. Hannig and Krüsmann[21] reported for the first time the electrophoretic separation of a human leucocyte preparation. Both granulocytes and lymphocytes were found to reveal a biphasic distribution profile. In continuation Hannig and Zeiller[22], using a buffer system based on triethanolamine, potassium acetate and glycine of pH 7.4 reported the separation and characterization of immune competent cells from rat lymph nodes, which during CFE splitted into four major peaks (Fig.4). After immunization of the rats with sheep erythrocytes an increase in the total number of cells in peak II and peak III was observed, which in the case of peak III was due to an increase in the number of blast cells. Evidence of an organ specific distribution of rat lymphoid cells was provided by Zeiller and Dolan[23]. Cells belon-

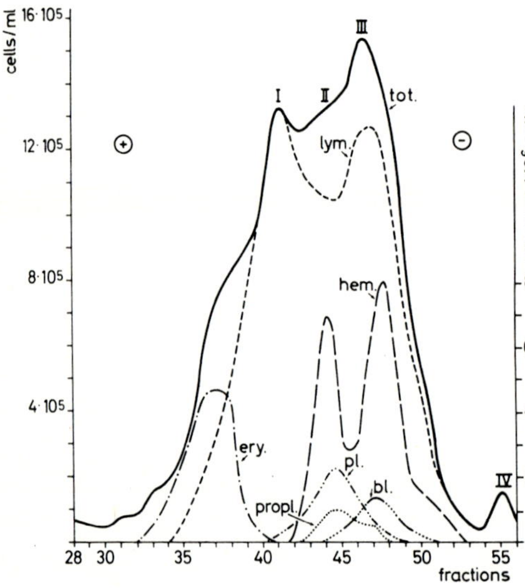

Fig.4. CFE of rat lymph node cells after immunization; tot= total cell distribution, hem= hemolysine producing cells (right ordinate), lym=lymphocytes, ery= erythrocytes, pl=plasma cellular elements, bl=blast cells, probl= proplasmocytes. From Hannig and Zeiller[22], by the courtesy of Walter de Gruyter.

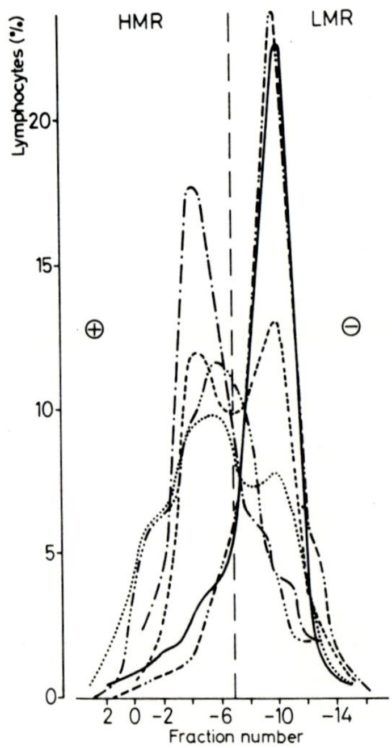

Fig.5. Distribution profiles of lymphocytes from various organs of the rat; thymus (——), spleen (—···—), lymph nodes (·······), thoracic duct lymph drainage 12h (—·—) and 48h (--- ---), bone marrow (--------); HMR, high mobility region; LMR, low mobility region. From Zeiller and Dolan[23], by the courtesy of Verlag Chemie.

ging to two electrophoretically defined regions of low and high mobility were detected (Fig.5). Following reinjection of previously separated lymphoid cells the analysis of the immune response strongly suggested that the cells of the high mobility region were T cells (thymus derived lymphocytes) whereas the cells of the low mobility region were B cells (thymus independent lymphocytes)[24]. This suggestion was confirmed by Nordling et al.[25], who characterized electrophoretically separated mouse lymph node cells by means of mouse specific antisera. The different electrophoretic mobilities of T and B cells were found to be consistent with striking differences in the membrane surface of these cells[25,26].

For the separation of different cell classes from lymphoid organs of the mouse von Böhmer et al.[27] developed a buffer system which minimized clumping of the cells. Most noticable in that respect is the observation of Zeiller and Pascher[28] that even totally clumped cell suspensions are easily deaggregated by short time incubation with deoxyribonuclease I. Therefore,the authors concluded that the factor causing clumping might be chromatin released by damaged cells.

The electrophoretic fractionation of lymphoid cells of different species mainly concerned with the evaluation of distinct maturation stages of these cells revealed T lymphocyte as well as B lymphocyte subpopulations[29-32]. These investigations may considerably contribute to the better understanding of immunological phenomena.

Bone marrow cells. Bone marrow cells from the mouse were electrophoretically separated using essentially the same buffer systems as for lymphocyte separations[33]. Separating bone marrow cells of patients with chronic myeloid leukemia, Schubert et al.[34] observed higher electrophoretic mobilities of the myeloma cells than of normal bone marrow cells, indicating a neoplastic change in the surface structure of the myeloma cells.

Subcellular organelles and membrane vesicles. When submitochondrial preparations were examined the distribution of membrane bound marker enzymes indicated that inner and outer mitochondrial membranes were separated from each other[35]. Following sonification of inner mitochondrial membranes, separation of outside-out and inside-out membrane fragments was achieved[36].

The isolation of a highly purified rat liver lysosomal preparation from a prefractionated rat liver homogenate was reported by Stahn et al[37]. The resulting fraction consisted of well preserved lysosomes 65-fold enriched over the total homogenate with respect to β-glucuronidase. In a similar manner Dew and Heidrich[38] recently purified renin containing granules from prefractionated rabbit kidney cortex. The authors achieved an 18-fold enrichment of renin activity compared to the kidney cortex homogenate. Using a Ca^{2+}-depleting perfusion medium for the isolation of rabbit kidney cortex cells, Heidrich and Dew[39] were able to separate renin containing cells from proximal tubule cells.

Hog gastric cell membranes which were prefractionated by isopyknic zonal centrifugation were further purified yielding a gastric mucosal membrane preparation containing high activities of K^+-ATPase as well as K^+-activated p-nitrophenyl phosphatase[40].

CFE was also applied to isolate subcellular particles from guinea pig brain cortex. Crude synaptosomal-mitochondrial preparations as well as synaptosomal vesicle fractions were electrophoresed in a Michaelis-Veronal buffer[41]. However, the degree of synaptosomal enrichment did not exceed that obtained by the use of common centrifugation procedures.

Other cell biological applications. To separate a mixture of two different types of bacteria, Strickler et al.[16] used a 3mM phosphate buffer of pH 6.1 and an electric field gradient of approximately 60 V/cm. The separability of various two-component mixtures out of five individual bacterial species were studied by Sinsabaugh and Normore[42]. Subfractions from the electrophoretic runs were found to be not composed of pure cultures. On the other hand, Ramsey[43] recently reported the successful separation of clinically significant microbes such as E.coli and Staphylococcus aureus and suggested the use of CFE as a method for species identification.

Studying the electrokinetic behavior and the separability of mixed phytoplankton suspensions Sheldon[44] fractionated mixtures of different algal genera.

The isolation of morphologically homogeneous spinach chloroplasts was reported by Klofat and Hannig[45]. Packer et al.[46] fractionated spinach chloroplasts into fractions of different size and different activity utilizing the stable flow free boundary sedimentation electrophoresis technique[15].

DENSITY GRADIENT ISOELECTRIC FOCUSING

In a remarkable treatise Kolin[47] recently demonstrated that IF may be considered as one special physico-chemical phenomenon out of a family of "isoperichoric focusing effects". Within that family IF is designated by the fact that an amphoteric substance introduced into a pH gradient increasing in the direction of the current migrates by the action of the electric field force towards its isoelectric position where it becomes isoelectrically focused (cf.chapter 21). It was the theoretical and practical work of Kolin[48], Svensson[49] and Vesterberg[50] which mainly contributed to the widespread use of IF as a powerful technique for the separation and characterization of amphoteric macromolecules. Therefore, as will be seen below, attempts were not lacking to use the technique for the fractionation of heterogeneous cell populations as well, although most authors entering that field did so with a good deal of scepticism.

Chemistry of carrier ampholytes.

Several approaches exist for the generation of a pH gradient for IF. Kolin[48] used for the functional demonstration a pH gradient which was engendered within an aqueous protein solution lying between an acidic and an alkaline buffer. The formation of pH gradients utilizing normal amphoteric as well as nonamphoteric buffer systems was recently investigated by Nguyen and Chrambach[51] and Chrambach and Nguyen[52] under the viewpoint of steady state stacking (isotachophoresis).

Troitsky et al.[53] reported on two other possibilities for the production of pH gradients. First, by the formation of a linear concentration gradient of glycerol, ethanol, etc. in an aqueous buffer solution, thus generating a pH gradient over an entire pH range of 1-1.5 pH units due to a gradual change of the dielectric constant of the solvent. The second method utilizes concentration gradients of polyols, e.g. mannitol, within borate buffers, resulting in the formation of pH gradients of an entire range of 2-3 pH units. Up to now this systems have not been used for the IF of cells or particles, although they might become important, because they allow ionic strength to be adjusted over a wide range.

Originally pH gradients were observed during stationary electrolysis of mixtures of amphoteres[54]. The complete separation of two amphoteres, however, may be achieved only if a third amphotere having an intermediate isoelectric point (pI) is present in the mixture[49]. Therefore, the separation of a mixture of several proteins requires a high number of amphoteres having distinct pIs. These amphoteres, which are called carrier ampholytes, determine the pH course of the gradient in the steady state. The carrier ampholytes should exhibit a good buffering capacity at their pI, and the pH gradient should be distinguished by having a uniform conductivity throughout. For cell separation in addition it is necessary that the carrier ampholytes do not exhibit cytotoxic side effects.

The first synthesis of a suitable carrier ampholyte mixture for IF was reported by Vesterberg[50]. By the copolymerization reaction of acrylic acid and an oligoethylene oligoamine Vesterberg was able to synthesize an undetermined but high number of amino carboxylic acid compounds ranging in their pIs between 3.5-10. Recently, several other attempts for the production of carrier ampholyte mixtures were reported[55,56] which all are based on Vesterberg's original idea.

The synthesis of amino-sulfonic acid carrier ampholytes were reported by Pogacar and Jarecki[57] and by Grubhofer[58]. The obtained pH gradients exhibited a gap between pH 3.5-5.8 which could be avoided only by the introduction of additional carboxylic acid groups.

Also in this laboratory attempts to synthesize suitable amino-carboxylic acid carrier ampholytes were initiated[59]. Fig.6 demonstrates some properties of the obtained carrier ampholyte mixtures. The pH gradient was separated into 48 fractions by means of CFE (see below) and the obtained fractions were analyzed for pH, buffering capacity, conductivity and UV/vis absorbancy. The synthesis was devised so as to allow also for the preparation of carrier ampholyte mixtures covering

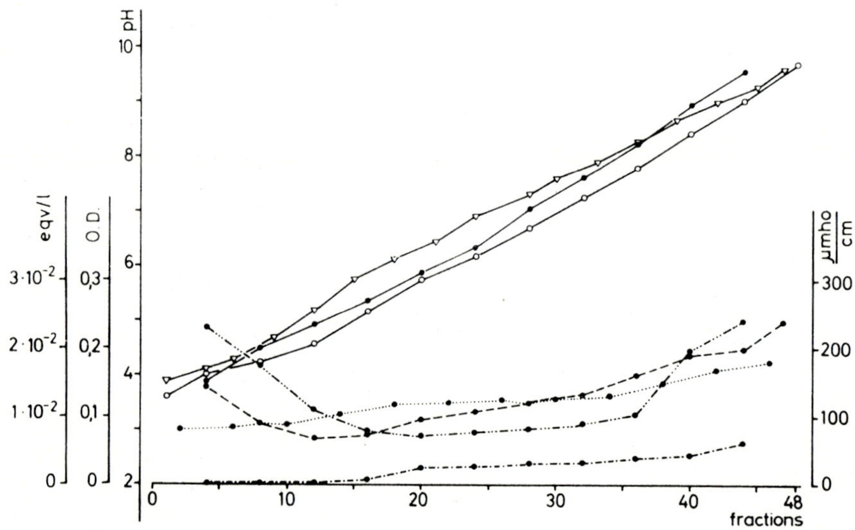

Fig.6. pH profile of three individually synthesized carrier ampholyte mixtures[59] established by CFE of a 1% solution. Only from the batch (—•—) buffering capacity (—...—), conductivity (-----) and absorbancy in the UV (......) and the visible region of the spectrum are depicted (-·-·).

restricted pH ranges. The mixtures thus obtained seem to be good media for microorganisms. Although this may sometimes be troublesome, it may suggest that this compounds lack strong cytotoxic effects.

Galante[60] indicated a general important property of carrier ampholytes: their ability to form chelates, especially with two-fold positively charged metal ions. The carrier ampholytes formed less strong complexes than EDTA with Mg^{2+}, Zn^{2+} and Fe^{3+}, but equally strong complexes with Cu^{2+}. Complex formation of carrier ampholytes with negatively charged macromolecules such as nucleic acids[61] and sulfated polysaccharides, e.g. heparin[62], was also discussed due to anomalous pI values observed following IF of these compounds. A completely satisfying explantion has not been given yet.

Cells and their isoelectric state

To obtain some information on the nature of ionizable groups situated in the outermost region of the cell surface it is a common procedure to titrate the cells using the electrophoretic mobility as an indicator[63]. Changing the ionic strength from 0.145 to 0.00218, Seaman[63] reported an increase of the pI of human red blood cells from 2.0 to 4.5. However, it was stressed by the author that below a certain pH and ionic strength the human erythrocyte membrane may suffer irreversible alterations, becoming no longer electrokinetically stable. Charged groups which are not "visible" at high ionic strength but exert their influence at low ionic strength as demonstrated in Fig.1, may offer an explanation for the pI shift unless other drastic changes in the membrane may be excluded. Electrokinetic instabilities of cells due to low pH media were also observed on other cells, e.g. human platelets[64] and human lymphocytes[65], whereas Greig et al.[66] reported a reversible electrokinetic behavior of various mammalian cells in that, after exposure of the cells to a medium of pH 2.8, the ζ-potential at pH 7 returned to the original value.

Nevertheless, the fact that a pH gradient consisting of carrier ampholytes is a medium of low "ionic strength"[67] (the term ionic strength strictly refers only to strong electrolytes) as well as the fact that the cells may exhibit low pIs, require careful attention to the maintenance of cell integrity applying IF for cell fractionation.

Instrumentation

As there is a current trend to take advantage of the high resolving power of IF on a preparative scale, several attempts for the design of suitable equipments were undertaken, which have been reviewed in a recent compilation by Righetti[68]. This section will only refer to those types of instruments actually used for IF of particles in free solution.

Most of the published work was performed using the commercially available LKB-electrofocusing column which originally was devised by Svensson[69]. Stabilization of the pH gradient against convection is performed by superimposing the pH gradient on a density gradient. It takes,however, about 24h to reach steady state IF. For cell fractionation it is therefore a real necessity to use a device which allows for the sample to be loaded into a preestablished pH gradient.

This necessity was taken into account in the modified electrofocusing column used by Boltz et al.[70] which allows the introduction of the sample at any desired pH after the pH gradient is formed.

Sherbet and Lakshmi[71], using the 110ml LKB-column, fused a side arm onto the middle of the column. The end of the side arm was sealed with a septum for sample injection.

Catsimpoolas and Griffith[72] recently demonstrated the versatility of the TRANS-analyzer (Bascom-Turner Instr., Newton, Mass., USA) for the analysis of heterogeneous cell populations. Six vertical glass columns are sequentially scanned by the vertical movement of a suitable light source, thus allowing the continuous recording of the time course of various cell separation experiments including IF.

Practical applications

Sherbet et al[73] used IF for the characterization of ionogenic groups on the cell surface of various mammalian cell types. The pH gradient, range 3-10 or 5-8, was supported either by a linear sucrose, glycerol, or Ficoll density gradient. A relatively sharply focused zone varying in pH between 5.6 and 6.85, dependent on the cell type analyzed, and a rather broad zone having a pI of about 5, irrespective of the cell type, was observed. Dye exclusion tests revealed that the broad peak mainly consisted of nonviable cells, whereas the cells found in the sharply focused zone were maily viable. HeLa cells loaded into a preestablished pH gradient were focused at pH 6.85 within 2-3h. The cells were cultured for 8 days with a plating efficiency of 55-75%.

It should be kept in mind that for the significance of the cell pI measured it is necessary to demonstrate steady state IF of the cells. Therefore, carrier ampholyte complexes with the cells as well as the influence of the supporting medium on the cell pI have to be excluded, preferably by using different media and carrier ampholytes of different pI ranges. Furthermore, the probability that the electrophoretic mobility of the cells in one direction is balanced by the sedimentation velocity in the opposite direction, or that the zero electrophoretic mobility of the cells is determined by their buoyant density rather than by their pIs should be considered as well. These effects may largely be excluded by conducting the experiments in duplicate under opposite electrode polarities.

The surface properties of brain tumor cell lines in relation to the degree of their malignancy was also recently investigated by Sherbet and Lakshmi[74]. Different human astrocytoma, meningioma and normal foetal brain cells were cultured in an appropriate medium following mechanical dissociation of the tissue. IF was performed in a 12ml capacity column and revealed a correlation between pI and the grade of the malignant state of the cells. No influence of the time of cell culturing on the cell pI was found.

Leise and LeSane[75] reported the application of IF to the separation of peripheral lymphocytes both of human and rabbit origin. The authors used 1% Ampholine, pI range 3-10 or 5-7 and Dextran 40 as the pH gradient supporting density gradient material, which was prefered over sucrose and Ficoll for maintenance of cell viability. Fig.7 demonstrates IF of human peripheral lymphocytes. Fractions focusing at pH 5.5, 5.4, 5.0 and 4.0 contained 10, 15, 9 and 5% of the total number of cells, respectively. Cells below pH 2 were found to be nonviable. Cells collected in other fractions exhibited high capability of survival, while only those cells focused at pH 4.18 showed significant replication during a 72h culturing period.

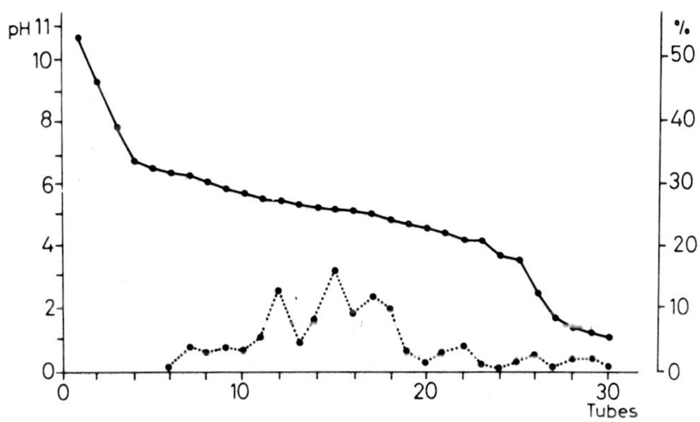

Fig.7. IF of human peripheral lymphocytes. From Leise and LeSane[75], by the courtesy of Marcel Dekker.

Hirsch and Gray[76] reported on the IF of peripheral lymphocytes of the rat in a dextran gradient in isotonic sucrose. Two major peaks of lymphocytes were obtained at pH 4.6 and pH 3.8. It was found that 75- to 85% of the cells in each fraction were viable and ^{3}H-1-alanine incorporation into cultured cells following IF was similar to that of fresh control cells.

The electrokinetic behavior of mouse spleen lymphocytes in a Ficoll sucrose gradient containing 2% ampholytes covering the pH range 2.5-8.0 was followed by Catsimpoolas and Griffith[72]. While the cells reached the focused state after approximately $3x10^{3}$sec they did not remain focused at that position, but instead became unfocused after $6x10^{3}$sec elapsed time of electrophoresis. The authors suggested that upon focusing the cells suffer membrane changes which finally lead to electrokinetic instabilities. No heterogeneity of the mouse spleen lymphocytes was detected during the experiment.

Another critical project to clarify the suitability of density gradient IF for cell fractionation was recently started by Boltz et al[70].

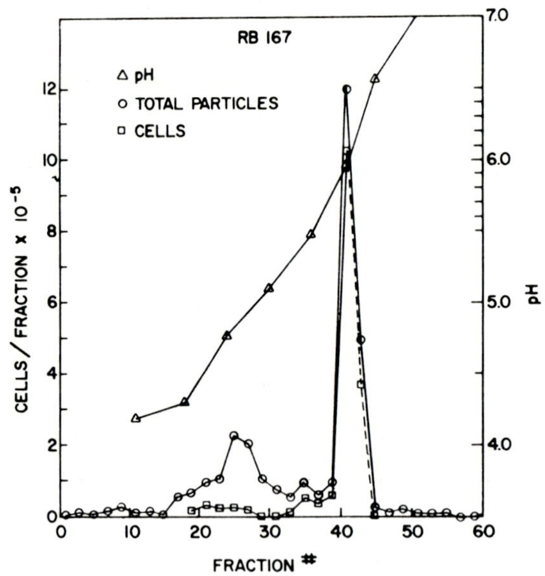

Fig.8. IF of cultured Chinese hamster fibroblasts (M3-1F3). From Boltz et al[70].

Using preformed pH gradients which were supported by a linear Ficoll density gradient made isotonic throughout by sucrose and glucose cell separation as well as changes in pI caused by various biological treatments were investigated. Fig.8 demonstrates IF of Chinese hamster fibroblasts (M3-1F3) which focused in a single sharp peak at pH 6.0, whereas after trypsinization the cells focused at pH 6.4 and 7.2. The authors stated the usefulness of IF and concluded that the technique provides information of physiological responses exhibited by the cell surface due to, e.g. hormonal changes, secretory activity, or virus infection. Due to the fact that cells subjected to IF frequently focused at higher pH than might be expected , the authors suggested the use of the term "isofocusing pH" for that pH at which the cells come to rest. The reasons for the unexpected high focusing pHs are still unclear.

Mammalian cells such as Ehrlich-Lettré mouse ascites tumor cells, as well as rat hepatocytes, were subjected to IF in a Ficoll-sucrose-Ampholine gradient by Manske et al[77]. Rat liver cells were collected at a focusing pH of 6.7 and Ehrlich ascites cells at a pH of 6.3-6.4. The short focusing times used resulted in cell viability of 92-95%.

Moore and Hibbitt[78] following IF in a 2-15% Ficoll gradient containing 1% Ampholine observed that the pI of spermatozoa from unoperated boars was 6.5 compared with 4.5 after the seminal vesicles were removed. It was stressed that the technique offers a means for studying spermatozoa during maturation and fertilization processes.

The pIs of rat liver mitochondria, of inner as well as outer mitochondrial membrane fractions were determined by IF in sucrose-Ampholine. The pIs obtained by IF correlated well with those obtained by cross partition within two aqueous polymer phases containing sodium salts[79]. Submitochondrial particles analyzed in the same way showed a pI and an isopartition point of approximately 6.6[80].

The technique of IF was also used successfully to determine the pI of membranes of human erythrocytes, platelets and lymphocytes[81]. The focusing pattern of lymphocyte membranes (pI=1.9), achieved in a pH gradient 1-3, was found to be changed after deoxyribonuclease treatment.

HeLa cell metaphase chromosomes were found to form a continuous distribution in the pH range 3.9-4.3[82]. By the application of velocity sedimentation prior to IF an appreciable enrichment of r RNA genes was observed in the chromosome fractions.

IF of bacterial cells was reported by Sherbet and Lakshmi[73] and by Langton et al[83]. In a Ficoll-Ampholine gradient Sherbet and Lakshmi found E.coli cells sharply focused in one single zone at pH 5.7. The authors pointed out the large discrepancy to the pI 2.2 obtained by cell electrophoresis. On the other hand, Langton et al., using sucrose as a density gradient material, obtained divergent results on the IF of E.coli cells, in that a biphasic distribution with maximal number of cells at either pH 4.2 or 6.4 was observed.

CONTINUOUS-FLOW pH GRADIENT ELECTROPHORESIS

The following paragraphs are intended to give some information about the experiences with IF applied in a CFE system. Since IF refers to a steady state process where the carrier ampholytes as well as the sample specimen reach a steady state focused position, one has to ascertain the attainment of that steady state. Otherwise, the term pH gradient electrophoresis may better describe an experiment cccuring in a transient state of the isoelectrofocusing process[84].

IF in a continuous-flow system may offer certain advantages over both the column and the normal buffer system. On the one hand, the pH gradient itself may influence the surface ionogenic groups of various particles differently. IF offers the use of narrow pH range carrier ampholytes to improve resolution. The use of carrier ampholytes results in a considerable decrease of current flow. Thus, applying the same field strengths used for CFE, less Joule-heat is produced and the thermal stability within the separation chamber is increased. In a representative experiment an electric field strength of 100V/cm will cause a current flow of only 10-20mA depending on the pH range used. On the other hand, the time the particle stay within the pH gradient is greatly reduced. There is no necessity to support the pH gradient by a density gradient assuring the constant osmolarity throughout the pH gradient. In the continuous-flow system a maximal number of $1x10^7$cells/ml may be separated within 1-1.5h. The cells thus stay in the separation chamber for approximately 10-20min.

Instrumentation

The apparatus used works in principle like a CFE-type (Fig3) instrument. The main feature of the apparatus is the U-shaped separation chamber which is built up by the elements shown in Fig.9a. The three main elements are fitted together by screw joints visible on the whole face picture of Fig.9b. At the outlet the chamber width is devided into 48 fractions (b). Accordingly, a 48-fold peristaltic pump governs the flow-through volume of the chamber solution, which enters the separation chamber via 11 inlets (a) on the top. At different heights between solution inlet and outlet there are possibilities to introduce the sample specimen. Constant temperature of the chamber solution is maintained by electronically thermostated liquid circulating upward through the cooling block (e).

The complete apparatus is contained in a thermostatically controlled cooling box (Fig.9b) adjustable to the desired temperature. A thermistor inserted at the upper part of the chamber allows temperature control of the separation solution. Two light sources on each side along the frontal separation field allow for the observation of the experiment by the Tyndall-effect.

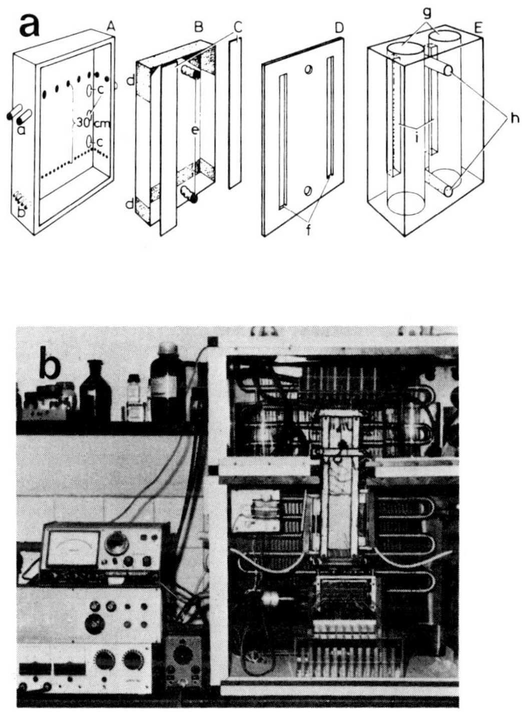

Fig.9. (a) Schematic drawing of the single elements forming the 30x12 cm separation chamber of the CFE apparatus developed by Seiler et al.[20] A, chamber block; B, cooling block made from aluminum and coated with an 0.25mm Teflon sheet; c,d, Teflon spacers and strips for maintenance of the chamber depth (usually 0.45mm); C, semipermeable membranes; D, rubber fitting; E, electrode block. A and E are made from plexiglass. (b) Whole face picture of the modified instrument used for pH gradient electrophoresis.

Practical applications

When the carrier ampholytes enter the separation chamber the molecules might approximate to their isoelectric position more and more during the downward flow. In this dynamic system several gadients exist, some of them correlating with the generation of the pH gradient. For the maintenance of the steady state they have to be kept constant. Horizontally, between the anode and the cathode there is the voltage gradient, the pH gradient and, as shown in Fig.6, a conductivity gradient as well. In the vertical direction, due to the progressive establishment of the pH gradient, there is a continuous decrease in the current density and the"ionic strength".

The variables which might be changed during the course of the experiment to obtain optimal separation conditions are flow-through time, electric field strength and the injection position at which the sample is introduced. In general, sample introduction into a somewhat preestablished pH gradient at about neutral pH was prefered.

Fig.10 shows the almost identical pH profiles obtained by CFE of a synthesized carrier ampholyte mixture (pH range 3.5-10) under a flow-through time of 30min at 110V/cm and that obtained in a polyacrylamide gel using markers to assure steady state IF. In the same figure the pH

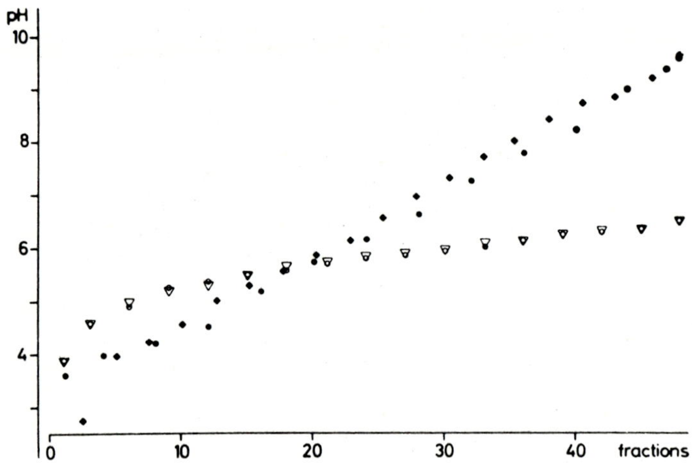

Fig.10. pH profile of carrier ampholytes developed by CFE (••••) and by polyacrylamide gel electrophoresis (▪▪▪); pH profiles analyzed after 1h (∘∘∘) and after 2h (▿▿▿) elapsed time of CFE.

profile of an Ampholine gradient, pI range 5-7, is shown after 1 and 2h of electrophoresis, demonstrating the time constant pH profile within the separation chamber. The electrode solutions used in most experiments were 0.1M H_3PO_4 and 0.1M NaOH for the anodic and the cathodic side, respectively.

Human red blood cells washed three times in 0.9% saline, 1mM EDTA, and resuspended in 0.25M sucrose, 1mM EDTA, were injected simultaneously at two different positions. Using an Ampholine gradient 3.5-10, an electric field strength of 90V/cm and a flow-through volume of 1.1-1.5ml/min the red blood cells combined into one single zone suggesting focusing of the cells[85]. Under similar conditions [86,87] a 1:1:1 mixture of mouse, human and rabbit red blood cells was separated into the individual components, both in a 3-10 and a 5-7 pH gradient (Fig.11). The use of the narrow range pH gradient resulted in a considerably improved resolution, as demonstrated by the absorbance of the hemolysate in the different fractions. Furthermore, the elution profiles of the cells resemble fairly well Gaussian distribution suggesting only a small influence of electroosmosis, although the separation chamber is composed of two different materials (plexiglass and Teflon).

On the other hand, the pH at which the cells are eluted in these experiments is inconsistent with their known pI and the composition of their membrane surface[63]. This might be due to insufficient time of electrophoresis and/or to electrokinetic instabilities of the cell membranes under the applied conditions. When human erythrocytes were

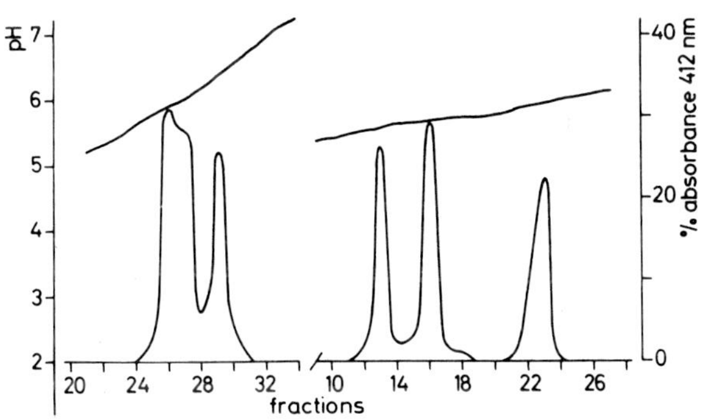

Fig.11. Separation of mouse, human and rabbit erythrocytes in a pH gradient 3.5-10 (left panel) and 5-7 (right panel).

washed in 0.9% saline, 0.3% glucose, without the presence of EDTA, there was no indication for an isofocusing pH of the cells down to pH 3.5[88]. Whether the above mentioned chelating properties of the carrier ampholytes itself may be a limiting factor in some applications of cell IF has to be further clarified.

Other experiments using continuous-flow pH gradient electrophoresis were concerned with the isolation of an extensively purified lysosomal fraction of rat liver. The light mitochondrial fraction (LMF), which is obtained by differential centrifugation of the rat liver homogenate and sediments between about 4000-12000 x g, approximately 10 times enriched in lysosomes, served as the starting material. The LMF in addition mainly consists of mitochondria, microsomes and peroxisomes.

Great difficulties deriving from aggregation and clumping of the LMF upon entering the separation chamber could be overcome by the addition of small amounts of heparin to the particle suspension[89]. The additional presence of small amounts of RNA aided the prevention of clumping[90]. Both substances exhibited a stabilization of the lysosomal membrane[85,91]. Washing the LMF with Tris-buffered sucrose largely prevented the tendency for clumping, suggesting that material adsorbed on the particle surface may at least partially be responsible for it.

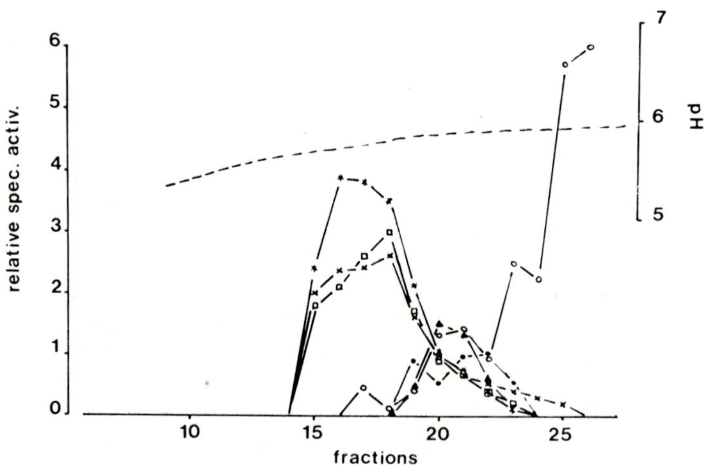

Fig.12. Separation of rat liver LMF in a pH gradient 5-7; N-acetyl-ß-glucosaminidase (—*—), acid phosphatase (—x—), aryl sulfatase (—□—), glucose-6-phosphatase (—•—), uricase (—o—) and cytochrom c oxidase (—▲—). The specific activities refer to the LMF.

A representative demonstration of the separation of the LMF is given in Fig.12. The distribution of several marker enzymes demonstrates the separation of mitochondria, microsomes and peroxisomes from lysosomes. With respect to acid phosphatase, a lysosomal marker enzyme, an approximately 50-fold enrichment of lysosomes over the total homogenate was achieved[85]. Electron micrographs of the LMF and of the lyso-

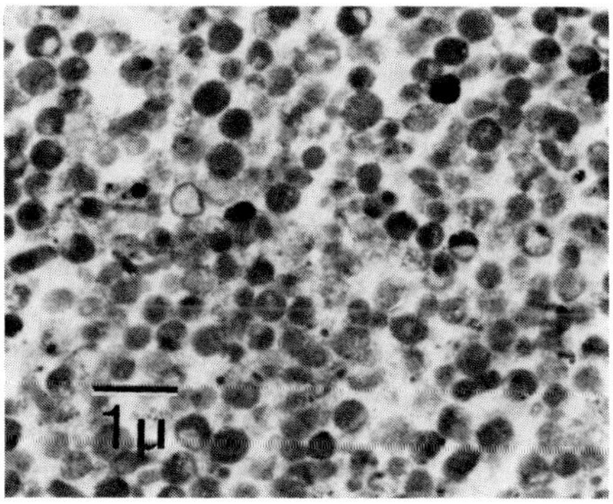

Fig.13. Electron micrograph of the lysosomal fraction obtained by pH gradient electrophoresis of rat liver LMF[85].

somal fraction, as shown in Fig.13, exhibit the high degree of purification and the attainment of fairly well preserved lysosomes.

CONCLUSIONS

The numerous experiments successfully utilizing the technique of CFE for particle fractionations clearly give evidence of the versatile applicability of this technique. The method works on both the analytical and the preparative scale with considerable resolving power. In addition reliable equipment is commercially available, the technology of which, although several-fold revised, is still being subjected to further improvements.

For preparative work prepurification of the sample mixture prior to CFE may be particularly recommended, regarding yield and final degree of purification of a certain fraction. For that purpose most commonly sedimentational techniques at unit or multiple gravity are used.

It was extensively shown that cell viability as examined by various techniques is largely preserved during CFE. However, recent observations on intranuclear electrophoresis occuring under similar conditions as applied for CFE[92] indicate that intracellular alterations may not be excluded. It might be stressed in that connection that the electrical double layer resembles a high resistance to the current, thus particles subjected to electrophoresis largely behave as if insulated[5].

On the other hand, it was indicated that several questions concerning the applicability of IF for cell or particle fractionations are still unanswered. Although it was shown repeatedly that certain cell types were successfully subjected to IF, the question whether membrane stability and hence cell viability can be maintained under isofocusing conditions should be further critically analyzed. The membrane stbility also may not be disregarded when discussing the method as a merely analytical one, since changes in the membrane structure, if they do occur, might lead to an artifactual heterogeneity of the separations.

Besides some other effects three properties of the carrier ampholytes seem to be important regarding cell viability and the observed differences between cell pI, as determined by cell electrophoresis, and isofocusing pHs. First, carrier ampholytes in the steady state focused position represent a medium of undetermined but very low "ionic strength". Second, the ampholytes are capable of chelating 2-fold positively charged metal ions. And third, they have been demonstrated to form complexes with polyanionic macromolecules. Although these properties are known, there is uncertainty as to what extent they are able to exert their influence on the cell membrane surface. Since some questions obviously derive from the particular character of the carrier ampholytes, the application of suitable buffer pH gradients should be considered. Furthermore, the recent development in synthesizing carrier ampholytes, allowing their radioisotopic labelling, also may be helpful in studying cell-carrier ampholyte interactions.

The results obtained by the application of IF to the fractionation of biological particles are by all means encouraging and certainly justify further extensive research.

ACKNOWLEDGEMENTS

The authors are indebted to Dr.John Sarvey for his valuable comments upon careful reading of the manuscript and to Miss Sigrid Thel for photographic and secretarial help.

REFERENCES

1. Bloemendal,H.(1977) Cell Separation Methods, North Holland, Amsterdam.
2. Overbeek,J.T.G.(1952) in Colloid Science, Vol.1, Kruyt,H.R.ed., Elsevier, Amsterdam, pp.115-193.

3. Heard,D.H. and Seaman,G.V.F.(1960) J.Gen.Physiol. 43,635-654.

4. Wiersema,H.P., Loeb,A.L.and Overbeek,J.T.G.(1966) J.Colloid Interf.Sci. 22,78.

5. van Oss,C.J.(1975) Sep.Purif.Methods 4,167-188.

6. Strickler,A. and Sacks,T.(1973) Ann.N.Y.Acad.Sci. 209,497-514.

7. Hannig,K., Wirth,H., Meyer,B.-H.and Zeiller,K.(1975) Z.Physiol.Chem. 356,1209-23.

8. Kolin,A.(1967) J.Chromatogr. 26,180-193.

9. Hannig,K., Wirth,H., Schindler,R.and Spiegel,K.(1977) Z.Physiol.Chem. 358,753-63.

10. Zeiller,K., Löser,R., Pascher,G.and Hannig,K.(1975) Z.Physiol.Chem. 356,1225-44.

11. Hannig,K.(1971) Methods Microbiol. 5B,513-548.
Hannig,K.and Heidrich,H.-G.(1974) Methods Enzymol. 21,746.

12. Kolin,A.(1967) J.Chromatogr. 26,164-179.

13. Hjertén,S.(1977) in Methods of Protein Separation, Catsimpoolas,N.ed., Plenum Publ.

14. Aitchison,G.F., Mattock,P., Steptoe,D., Thomson,A.R.and White,W.J.C.(1976) Tenth Intern.Congress Biochem., Hamburg.

15. Mel.H.(1964) J.Theor.Biol. 6,150.

16. Strickler,A., Kaplan,A.and Vigh,E.(1966) Microchem.J. 10,529-544.

17. Howery,D.G., Vermeulen,T.and Nady,L.(1972) J.Chromatogr.Sci. 10,557-559.

18. Prusik,Z.(1974) J.Chromatogr. 91,867-872.

19. Bier,M., Smolka,J.K., Kopwillem,A.and Ostrach,S.(1976) J.Colloid Interf.Sci. 55, 197-207.

20. Seiler,N., Thobe,J.and Werner,G.(1970) Z.Anal.Chem. 252,179-182.

21. Hannig,K.and Krüsmann,W.-F.(1968) Z.Physiol.Chem. 349,161-170.

22. Hannig,K.and Zeiller,K.(1969) Z.Physiol.Chem. 350,467-472.

23. Zeiller,K.and Dolan,L.(1972) Eur.J.Immunol. 2,439-444.

24. Zeiller,K., Holzberg,E., Pascher,G.and Hannig,K.(1972) Z.Physiol.Chem.353,101-110.

25. Nordling,S., Andersson,L.C.and Häyry,P.(1972) Sci. 178,1001-1002.

26. Mehrishi,J.N.and Zeiller,K.(1974) Brit.Med.J. 1,360-362.

27. von Böhmer,H., Shortman,K.and Nossal,G.J.V.(1974) J.Cell Physiol. 83,231-242.

28. Zeiller,K.and Pascher,G.(1973) Eur.J.Immunol. 3,614-618.

29. Zeiller,K., Pascher,G.and Hannig,K.(1976) Immunol. 31,863-880.

30. Shortman,K., von Böhmer,H., Lipp,J.and Hopper,K.(1975) Transpl.Rev. 25,163-210.

31. Zeiller,K., Pascher,G., Wagner,G., Liebich,H.-G., Holzberg,E.and Hannig,K.(1974) Immunol. 26,995-1012.

32. Shortman,K., Fidler,J.M., Schlegel,R.A., Nossal,G.J.V., Howard,M., Lipp,J.and von Böhmer,H.(1976) Contemporary Topics Immunobiol. 5,1-45.

33. Zeiller,K., Schubert,J.C.F., Walther,F.and Hannig,K.(1972) Z.Physiol.Chem.353,95.

34. Schubert,J.C.F.and Walther,F.(1973) Klin.Wschr. 51,327-332.

35. Heidrich,H.-G., Stahn,R.and Hannig,K.(1970) J.Cell Biol. 46,137-150.

36. Heidrich,H.-G.(1971) Febs Letters 17,253-256.

37. Stahn,R., Maier,K.-P.and Hannig,K.(1970) J.Cell Biol. 46,576-591.

38. Dew,M.E.and Heirdich,H.-G.(1975) Z.Physiol.Chem. 356,621-623.

39. Heidrich,H.-G.and Dew,M.E.(1977) J.Cell Biol. 74,780-788.

40. Saccomani,G., Steward,H.B., Shaw,D., Lewin,M.and Sachs,G.(1977) Biochim.Biophys. Acta 465,311-330.

41. Ryan,K.J., Kalant,H.and Llewellyn-Thomas,E.(1971) J.Cell Biol. 49,235-246.

42. Sinsabaugh,H.A.and Normore,W.(1971) Sep.Sci. 6,467-472.

43. Ramsey,W.S.(1976) 172.Meeting of the Amer.Chem.Soc., Abstr.No.59.

44. Sheldon,R.W.(1972) Limn.Ocean 17,481-498.

45. Klofat,W.and Hannig,K.(1967) Z.Physiol.Chem. 348,739-741.

46. Packer,L., Nobel,P.S., Gross,E.and Mel,H.(1966) J.Cell Biol. 28,443-448.

47. Kolin,A.(1977) in Electrofocusing and Isotachophoresis,Radola,B.J.and Graesslin,D. eds., Walter de Gruyter, Berlin, pp.3-33.

48. Kolin,A.(1954) J.Chem.Phys. 22,1628-1629.

49. Svensson,H.(1961) Acta Chem.Scand. 15,325-341.
Svensson,H.(1962) Acta Chem.Scand. 16,456-466.

50. Vesterberg,O.(1969) Acta Chem.Scand. 23,2653-2666.

51. Nguyen,N.Y.and Chrambach,A.(1976) Analyt.Biochem. 74,145-153.

52. Chrambach,A.and Nguyen,N.Y.(1977) in Electrofocusing and Isotachophoresis, Radola, B.J.and Graesslin,D. eds., Walter de Gruyter, Berlin, pp.51-58.

53. Troitsky,G.V., Zav'yalov,V.P., Kirjukhin,I.F., Abramov,V.M.and Agitsky,G.J.(1975) Biochim.Biophys.Acta 400,24-31.

54. Ikeda,K.and Suzuki,S.(1912) U.S.Pat., 1.015.891.

55. Vinogradov,S.N., Lowencron,S., Andonian,M.R., Bagshaw,J., Felgenhauer,K.and Pak, S.J.(1973) Biochem.Biophys.Res.Commun. 54,501-506.

56. Righetti,P.G., Pagani,M.and Gianazza,E.(1975) J.Chromatogr. 109,341-356.

57. Pogacar,P.and Jarecki,R.(1974) in Electrophoresis and Isoelectric Focusing in Polyacrylamide Gel, Allen,R.C.and Maurer,H.R.eds., Walter de Gruyter,Berlin,pp.153.

58. Grubhofer,N.(1977) in Electrofocusing and Isotachophoresis, Radola,B.J.and Graesslin,D. eds., Walter de Gruyter,Berlin, pp.111-120.

59. Just,W.W., in preparation.

60. Galante,E., Caravaggio,T.and Righetti,P.G.(1975) in Progress in Isoelectric Focusing and Isotachophoresis, Righetti,P.G. ed., North Holland, Amsterdam, pp.3-12.

61. Drysdale,J.W.and Shafritz,D.A.(1975) in Progress in Isoelectric Focusing and Isotachophoresis, Rhighetti,P.G. ed., North Holland, Amsterdam, pp.293-306.

62. McDuffie,N.M., Dietrich,C.P.and Nader,H.B.(1975) Biopolym. 14,1473-1486.

63. Seaman,G.V.F.(1975) in The Red Blood Cell, Vol.2, Surgenor,D.MacN. ed., Academic Press,New York, pp.1135-1229.

64. Stoltz,J.F., Stoltz,M.,Gineste,J., Streiff,F.and Larcan,A.(1970) Path.Biol.18,159.

65. Mehrishi,J.N.and Thomson,A.E.R.(1968) Nature 219,1080-1081.

66. Greig,R.G., Jones,M.N.and Ayad,S.R.(1976) Biochem.J. 160,325-334.

67. Vesterberg,O.(1968) Svensk.Kem.Tidskr. 80,213-225.

68. Righetti,P.G.(1975) Separation and Purification Methods 4,23-72.

69. Svensson,H.(1962) Arch.Biochem.Biophys.Suppl. 1,132.

70. Boltz jr.,R.C., Todd,P., Hammerstedt,R.H., Hymer,W.C., Thompson,S.J.and Docherty, J.(1977) in Cell Separation Methods, Bloemendal,H. ed., North Holland, Amsterdam, pp.145-155.

71. Sherbet,G.V.and Lakshmi,M.S.(1973) Biochim.Biophys.Acta 298,50-58.

72. Catsimpoolas,N.and Griffith,A.L.(1977) in Electrofocusing and Isotachophoresis, Radola,B.J.and Graesslin,D. eds., Walter de Gruyter,Berlin, pp.469-479.

73. Sherbet,G.V., Lakshmi,M.S.and Rao,K.V.(1972) Exp.Cell Res. 70,113-123.

74. Sherbet,G.V.and Lakshmi,M.S.(1974) Oncology 29,335-347.

75. Leise,E.M.and LeSane ,F.(1974) Prep.Biochem. 4,395-410.

76. Hirsch,R.L.and Gray,I.(1976) J.Cell Biol. 70,358.

77. Manske,W., Bohn,B.and Brossmer,R.(1977) in Electrofocusing and Isotachophoresis, Radola,B.J.and Graesslin,D. eds., Walter de Gruyter, Berlin, pp.497-502.

78. Moore,H.D.M.and Hibbitt,K.G.(1975) J.Reprod.Fert. 44,329-332.

79. Ericson,I.(1974) Biochim.Biophys.Acta 356,100-107.

80. Lundberg,P.and Ericson,I.(1975) Biochem.Biophys.Res.Commun. 65,530-536.

81. Jamieson,G.A.and Groh,N.(1971) Analyt.Biochem. 43,259-268.

82. Landel,A.M., Aloni,Y., Raftery,M.A.and Attardi,G.(1972) Biochem. 11,1654-1663.

83. Langton,R.W., Cole III,J.S.and Quinn,P.F.(1975) Arch.Oral Biol. 20,103-106.

84. Catsimpoolas,N.(1973) Analyt.Biochem. 54,66-78.

85. Just,W.W.and Werner,G.(1977) in Electrofocusing and Isotachophoresis, Radola,B.J. and Graesslin,D. eds., Walter de Gruyter, Berlin, pp.481-493.

86. Just,W.W., Leon-V.,J.O.and Werner,G.(1975) Analyt.Biochem. 67,590-601.

87. Just,W.W., Leon-V.,J.O.and Werner,G.(1975) in Progress in Isoelectric Focusing and Isotachophoresis, Righetti,P.G. ed., North Holland, Amsterdam, pp.265-280.

88. Just,W.W.and Werner,G.(1977) in Cell Separation Methods, Bloemendal,H. ed., North Holland, Amsterdam, pp.131-142.

89. Just,W.W., Leon-V,J.O.and Werner,G.(1974) Histochem. 41,161-166.

90. Just,W.W.and Werner,G.(1976) Z.Physiol.Chem. 357,266.

91. Just,W.W., Leon-V.,J.O.and Werner,G.(1974) Z.Physiol.Chem. 355,1565-1568.

92. Skaer,R.J., Whytock,S.and Emmines,J.P.(1976) J.Cell Sci. 21,479-496.

© 1979 Elsevier/North-Holland Biomedical Press
Electrokinetic Separation Methods,
P.G. Righetti, C.J. van Oss and J.W. Vanderhoff

ENDLESS BELT CONTINUOUS FLOW DEVIATION ELECTROPHORESIS

ALEXANDER KOLIN
Molecular Biology Institute, University of California, Los Angeles, Calif. 90024 (U.S.A.)

INTRODUCTION

The method described in this chapter deals with electrophoresis in a flowing liquid without stabilization against thermal convection by a porous medium or establishment of a density gradient in the fluid. Electrophoretic separations under such conditions cannot achieve the high resolution which can be attained in conjunction with chromatographic and filtration effects obtainable with gel electrophoresis. This drawback is, however, offset by the speed and ease of performance of the separations, the applicability to particulate components (such as cells and subcellular particles) for which gel electrophoresis is not suitable and the ease of collection of the separated components in continuous-flow preparative fractionations. Last but not least, the possibility of measuring electrophoretic mobilities in well defined homogeneous liquid media is an important advantage.

Endless belt electrophoresis is a method of deviation electrophoresis in which electrophoretic migration in the direction of the applied electric field is combined with entrainment of the electromigrating particles in a transverse laminar flow. Deviation of the particle path from the direction of the laminar flow is caused by the electromigration in the transverse electric field. Theoretical discussions which have been worked out in connection with endless belt electrophoresis have been subsequently applied to other (actually earlier) methods of deviation electrophoresis which have been reviewed elsewhere[1,2,3].

Among the unique features of endless belt electrophoresis are the following: 1) The use of a magnetic field along with the electric field, 2) pulsation-free magneto-hydrodynamic propulsion of the flowing buffer, 3) absence of membranes, 4) orbital motion of ions superimposed upon axial electromigration resulting in a helical path of particles undergoing electrophoresis, 5) periodically repeated separation steps in the established electric field, 6) superposition of a lateral flow in the electric field direction, 7) possibility of non-isoelectric accumulation of a fraction in a stationary orbit, 8) compact accommodation of a long migration path in a small space, 9) stabilization against thermal convection by fluid circulation about a horizontal axis, 10) rapid separations at field intensities in excess of 100 V/cm, 11) preparative and analytical capabilities at sample volumes of less than one tenth of a microliter, 12) cancellation of effects due to sedimentation of particles and of the streak itself along the direction of flow of the fluid buffer curtain.

THERMAL CONVECTION

Electrophoresis in an electrically conductive fluid requires the flow of an electric current in order to maintain an electric field. There will be thus a power dissipation (P) per unit volume of $P = J^2/\sigma$ where J is the current density and σ the electrical conductivity of the buffer.

In a confined fluid channel, the heat will leave the lumen by heat conduction through the confining walls which can be maintained at a constant temperature by a cooling arrangement. The temperature at the center of the confined fluid, which is heated by an electric current, will be highest. Due to establishment of a horizontal temperature gradient in such a configuration shown in Fig. 1, a fluid convection pattern will be established (above 4° C in water) in which the warmer central fluid of lesser density rises in the central region of the channel while the cooler and denser fluid near the walls descends. In a state of more severe thermal perturbation, a disorderly turbulent fluid motion may be engendered exhibiting smaller vortices. Fig. 2 shows an actual photograph of a transient convection in a thermally and gravitationally disequilibrated fluid column[4].

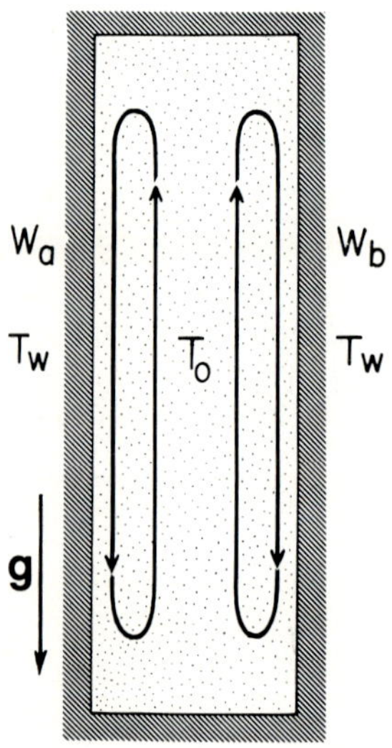

Fig. 1. Steady-state thermal convection caused by a horizontal temperature gradient between two closely spaced vertical walls W_a, W_b maintained at a low temperature T_w. The fluid between the plates is heated uniformly by an electric current which is perpendicular to the page. T_o: maximal central temperature; g: gravitational field vector.

It is obvious that turbulent thermal convection cannot be tolerated in a fluid in which an orderly sorting pattern of different species of particles is to be established for separate collection. Thermal convection even milder than that seen in Fig. 2 would thoroughly remix the separated fractions. It is therefore a prime objective of a method for electrophoretic separation in a homogeneous fluid to devise a method for suppression of small-scale thermal convection.

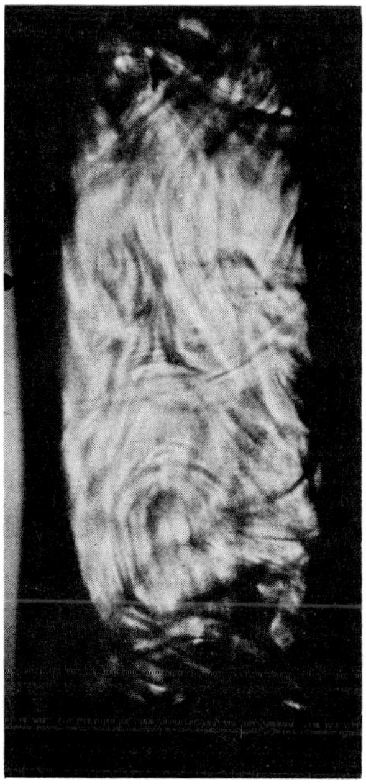

Fig. 2. Transient convection caused by instability in an inverted density gradient in the presence of a horizontal temperature gradient (from A. Kolin, Biochim. Biophys. Acta 32, 538, 1959).

In fluid curtain electrophoresis such suppression is achieved by using a very thin fluid layer between two parallel plates. The closely spaced plates promote suppression of thermal convection in the fluid trapped between them by taking advantage of fluid viscosity in a similar way as it serves to suppress thermal convection in beds of small closely packed glass beads between which the fluid is trapped.

The most perfect means of eliminating thermal convection would be by abolition of the field of gravity since thermal convection is a purely gravitational phenomenon[5,6]. It occurs only because of a difference in weight density between fluid elements differing in temperature. At $\vec{g} = 0$ all weight, and hence all weight density differences, would vanish[5]. Such conditions would prevail only in interstellar space far from celestial objects or in a freely falling system such as a space rocket or an orbiting satellite[5].

This, of course, is a very costly and normally not practical method of combatting thermal convection. A simple terrestrial approximation to zero-gravity conditions can be attained by a slow rotation of the separation cell[6,7]. A sedimenting particle will describe a small circle about an interior point fixed relative to the cell without sedimenting toward one of its walls. A thermally disequilibrated fluid volume can be compared to a sedimenting particle. It will "see" a rotating gravitational field vector in a rotating cell and will not move unidirectionally relative to its walls. The effectiveness of this method is such that large temperature gradients can exist in the separation cell without creating thermal convection[6]. Hjertén based a very effective method of electrophoretic separation in a rotating tube on this principle[7]. In both of the latter cases the apparatus had to rotate relative to the laboratory about a horizontal axis. The next section describes an approach in which rotation of the apparatus was obviated.

ELIMINATION OF GRAVI-THERMAL CONVECTION WITHOUT INSTRUMENT ROTATION

It is not as much macroconvection illustrated in Fig. 1 as disorderly thermal convection with small vortices such as seen in Fig. 2 that constitutes the main hazard in electrophoretic separations by a streak deflection apparatus. It is the horizontal component of fluid motion which can give rise to remixing of separated components. Large convection patterns, such as that shown in Fig. 1 can be considered as modifying the laminar flow velocity distribution by superposition of velocities.

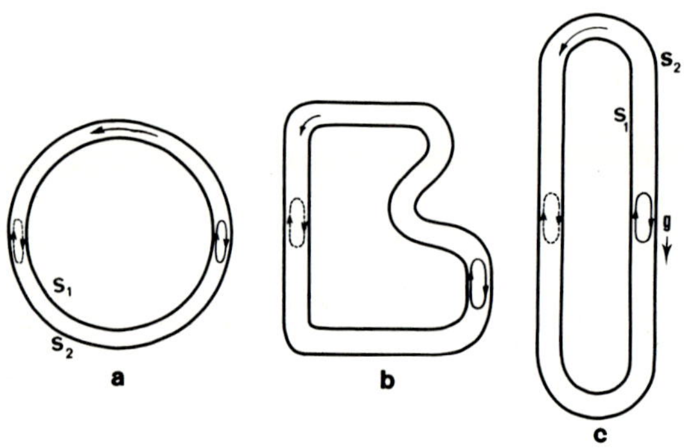

Fig. 3. Small thermal convection vortices in horizontal annular channels of different cross-section. The channel axis is perpendicular to the page and the inner wall S_1 is warmer than the outer wall S_2. The dotted vortex on the left of each channel represents the solid vortex shown in the right hand channel after its relocation by the counter-clockwise fluid circulation indicated by the curved arrows. (a): circular annular channel; (b): non-circular quasi-annular channel; (c): endless fluid belt annulus (vertical "race-track").

Fig. 3 illustrates how small thermally engendered vortices can be suppressed in a stationary apparatus by appropriate fluid circulation. S_1 and S_2 in Fig. 3 represent in cross-section the surfaces of two horizontal concentric cylinders whose axis is perpendicular to the page. The inner cylinder surface S_1 is maintained at a higher temperature than the outer surface S_2. If the fluid trapped in the annular space between the cylinders is water we assume both wall temperatures to be above 4° C. The horizontal temperature gradient thus produced in the annulus gives rise to a concomitant horizontal density gradient in the fluid. As a result, the warm (less dense) layers near cylinders S_1 will tend to rise and the cooler, denser layers adjacent surface S_2 will tend to descend. Thus, a thermally engendered vortical motion will be created which is illustrated by a small solid closed loop on the right hand side of the annulus. Arrows indicate the sense of fluid circulation.

Let us now perform an imaginary experiment. We assume the vortex considered above to be suddenly transferred to the diametrically opposed position illustrated by the dashed oval on the left hand side of the annulus. This will invert the directions of motion of the fluid in the vortex near the walls S_1 and S_2. The warm fluid near S_1 will move downward due to inertia, despite the net upward forces acting upon each of its volume elements due to predominance of buoyancy. The cool fluid near S_2, on the other hand, will (due to inertia) move upward despite its tendency to sink owing to its great density. Thus, the gravitational forces will act so as to oppose the inertial movement in the vortex and its rate of rotation will be retarded until it stops. After cessation of circulation, the fluid on the left hand side of the annulus will be set in motion again due to the horizontal density gradient. But now the sense of revolution in the left-hand vortex will be opposite to that indicated by the arrows, since the dense fluid near S_2 will descend and the less dense fluid near S_1 will rise.

It is easy to see now that a slow counterclockwise circulation in the annulus (as indicated by the top arrow) will inhibit the formation of small thermal convection vortices which require a certain amount of time to build up their rotational kinetic energy. They will be subject to an alternating retarding torque as they periodically oscillate between two horizontally diametrically opposite positions.

It is obvious that this mechanism of inhibition of thermal convection vortices is not limited to fluid circulation in a circular annular channel as illustrated in Fig. 3A. It will be equally effective in an arbitrarily shaped circulation path shown in Fig. 3B. A vertical annular "racetrack" shown in Fig. 3C is of particular interest to us in connection with subsequent considerations.

It is thus possible to achieve the same result as was pursued by the above mentioned rotating-cell electrophoresis apparatus in a stationary instrument by maintaining quasi-orbital motion of the fluid about a horizontal axis.

MAGNETO-HYDRODYNAMIC CIRCULATION OF THE ELECTROPHORETIC BUFFER COLUMN

Our next objective is to find a simple and stable way to maintain the fluid circulation considered in the preceding section. Fig. 4 illustrates a suitable mechanism.

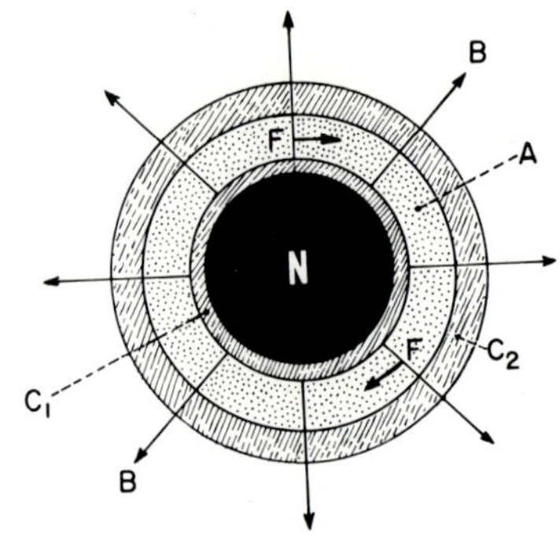

Fig. 4. Buffer-filled annulus between walls C_1, C_2 of two horizontal co-axial cylinders enclosing a cylindrical magnetic north pole N. B: magnetic field vectors. F: tangential electromagnetic forces generated by the interaction between the magnetic field B and the current which flows at right angles to the page between the cylinders C_1, C_2.

C_1 and C_2 correspond to the cylindrical surfaces S_1 and S_2 of Fig. 3. A is the buffer solution trapped in the annular space between the concentric horizontal cylinders C_1 and C_2. We imagine an electrical current to flow through the annulus A parallel to the cylinder axis (i.e. normal to the page). N is the idealization of a cylindrical isolated magnetic north pole whose radial magnetic field is depicted by the vectors B. The interaction of this magnetic field with the perpendicular electrophoretic current gives rise to tangential electromagnetic forces symbolized by the vectors F. This force-field, which is perpendicular to both the current and the magnetic field, sets up a revolution of the fluid about the cylinder axis which proceeds with the constancy of a clockwork at a constant current. The cross-section of the magnet N need not be circular. A normal magnetic field at its surface penetrating a narrow fluid channel closely conforming to the contour of the pole N will create fluid convection through such a closed channel filled with a conductive fluid traversed by an axial electric current.

While isolated north-magnetic field sources do not exist, a practically adequate approximation can be achieved by a configuration such as shown in Fig. 5. A soft-iron cylinder m is sandwiched between two bar magnets which face each other with like poles (N,N). As can be seen, the magnetic field is very close to being uniform and radial over most of the surface of the soft-iron cylinder m. Enclosing this magnetic sandwich into a plastic cylinder C_1 surrounded by a somewhat larger concentric plastic cylinder C_2 would thus provide us with the physical situation illustrated in Fig. 4.

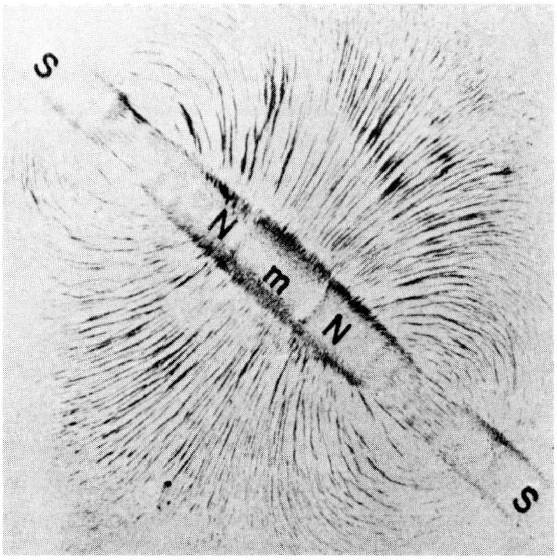

Fig. 5. Magnetic field distribution about a soft-iron cylinder m sandwiched between two cylindrical bar magnets facing each other with like poles. The field near the surface of m is substantially radial (from A. Kolin: J. Chromatogr. 26, 164, 1967).

THE CIRCULAR ENDLESS BELT APPARATUS

A simple demonstration instrument

Fig. 6 shows a simple instrumental implementation of the principles described above. The soft-iron cylinder M located between the two magnets N-S dwells in a snugly fitting plastic cylinder C_1. There is an annular space of about 1.5 mm between C_1 and the larger outer plastic cylinder C_2. The two electrode chambers EC_1 and EC_2 communicate with each other via the annulus between C_1 and C_2 which serves as a liquid bridge when the apparatus is filled with a buffer solution. When a voltage is applied to the electrodes E_1, E_2, an axial current flows through the annulus between the cylinders C_1, C_2. As a result of the interaction of this current with the radial magnetic field surrounding the iron cylinder M, the fluid in the annulus is set in rotation. The iron core M is milled out so as to permit transillumination through the window W thus created. This permits observation of electrophoretic components by light absorption or by dark-field illumination.

R is a reservoir containing the sample to be subjected to electrophoretic analysis. The material is introduced via a thin glass capillary injector IN into the center of the annulus in the direction of the buffer circulation. If the injected particles have no electric charge, they will accumulate in a circular orbit in the plane of injection. If they are negatively charged, they will electromigrate to the left toward the anode. The combination of this axial migration to the left with the orbital revolution results in a helical trajectory for the charged particles. While negative particles move to the left in a left-handed helix, the trajectory of positive particles moving to the right toward the cathode is a right-handed helix.

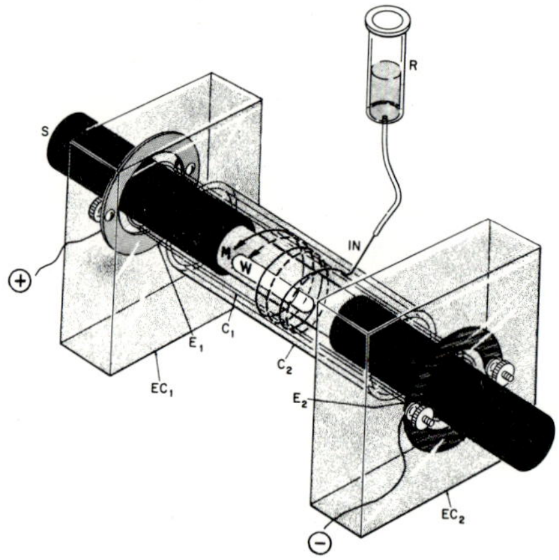

Fig. 6. Simple circular endless fluid belt demonstration separator. N-S: north and south poles of bar magnets. M: intermediate soft-iron cylinders with a glass-filled funnel W which serves as a window for trans-illumination. C_1, C_2: plastic cylinders which enclose the annular buffer column. EC_1, EC_2: electrode compartments. E_1, E_2: electrodes; N: sample injector; R: sample reservoir.

The pitch of the helix (i.e. the distance between adjacent helical turns) is the distance traversed by the particles due to electrophoresis during one revolution of the buffer column. The helical pitch is thus a measure of a particle's electrophoretic mobility. Fig. 6 shows two particle species indicated by a solid and a dashed helical path. Due to a difference in electrophoretic mobilities and, hence, of the helical pitch, the particle trajectories are separated widely enough at the end of the path of the particles for them to be collected in different compartments of a fraction collector by a suitable collector intercepting them after several helical turns.

Successive separation increments with each turn

A basic property of this separation principle is immediately apparent. Two electrophoretically distinct components gain equal increments in separation with every turn. Thus, if the minimum separation after one turn is k mm, the separation after n turns will be nk mm. The analogy with grating spectra of higher order suggests that it may not necessarily be advantageous to increase the number of turns indefinitely because there will be eventually an overlap between components of consecutive turns when the faster component of the n-th turn coincides with the slower component of the (n + 1)st turn. Such an overlap can, however, be easily removed by imposing lateral streaming in the direction of electrophoresis. For instance, if we add fluid to the right compartment EC_2 while removing it at the same rate from compartment EC_1, we superimpose an axial velocity upon all the particle species. This will increase the pitch of all helices and thus separate the slowest component of the (n + 1)st turn from the fastest component of the n-th turn by as large a distance as one pleases. While the helical pitch can thus be varied, the <u>distance</u> between two separated components remains unchanged by the lateral buffer flow.

Some initial simplifying assumptions and approximations

Our first simplified description will assume a uniform velocity distribution in the fluid flow in the annulus. We will further assume that the injected particle streak is of a negligible diameter and that it is placed at the center of the annulus. A further simplifying assumption will be constancy of the temperature throughout the annulus. While the temperature variation is only a few degrees C, the temperature gradient is great because of the thinness of the annulus. This temperature gradient is essential in order to remove the heat generated by the electrophoretic current. This is accomplished by placing the outer cylinder C_2 in a cooling bath and by circulation of cooling water through the iron core M. The instrumental implementation of cooling will be considered in more detail in connection with an improved apparatus described below. The simplifying assumptions regarding the velocity and temperature distributions will be dropped in a subsequent theoretical section. The initial description will also disregard electro-osmotic streaming which will be considered later.

Calculation of typical performance parameters

We assume in Fig. 4 the thickness of the fluid annulus surrounding the iron core M to be negligibly small as compared to the latter's radius and the magnetic field to be uniform and perpendicular to the annulus. For a current density $\vec{J}$ and magnetic field $\vec{B}$, the force density $\vec{f}$ (i.e. force per unit volume of the fluid in the annulus) will be

$$f = \frac{1}{10} |\vec{J} \times \vec{B}|, \qquad (1)$$

where $\vec{J}$ is in A/cm^2, $\vec{B}$ in gauss and $\vec{f}$ in $dynes/cm^3$ The fluid flow generated by this force-field in the annulus is governed by the differential equation

$$f = \eta \frac{\delta^2 u}{\delta z^2}, \qquad (2)$$

where η is the fluid viscosity in poise. Integration under appropriate boundary conditions leads to an equation for the velocity distribution[8,9]:

$$u = \frac{f}{2\eta} \left| z^2 - \left(\frac{h}{2}\right)^2 \right| . \tag{2a}$$

In terms of the maximum fluid velocity u_o at the center of the annulus ($z = 0$) we can write:

$$u_o = - \frac{f}{2\eta} \left(\frac{h}{2}\right)^2 \tag{3}$$

and

$$u = u_o \left| 1 - 4\left(\frac{z}{h}\right)^2 \right| . \tag{4}$$

z is the coordinate measuring the distance of a point from the mid-line of the annulus where $z = 0$. u is the tangential fluid velocity in the annulus corresponding to a given value of z. Equation (4) describes a parabolic velocity distribution for laminar flow between parallel plates separated by the distance h. This approximation is applicable to our case illustrated in Fig. 4 for $h \ll R$, where R is the mean radius of the cylinders C_1, C_2 and h the difference between their radii.

From equations (1) and (3) we get

$$|u_o| = \frac{h^2}{80\eta} |J \times B| . \tag{5}$$

For a typical circular endless belt cell we can use the following parameters: $h = 0.15$ cm; $\eta_{16^oC} = 1.11 \cdot 10^{-2}$ poise; $i = 200$ mA; $J = 0.061$ A/cm^2; $B = 340$ gauss.

These values yield from Equation (5) for the central velocity of fluid circulation in the annulus $u_o = 0.53$ cm/sec.

We can now compute the revolution time τ_o for an annulus circumference L of 20 cm

$$\tau_o = \frac{L}{u_o} \tag{6}$$

and we obtain in our example the revolution time $\tau_o = 37.8$ sec.

The helical pitch s_o for a centrally injected particle streak follows from

$$s_o = E \cdot \mu \cdot \tau_o , \tag{7}$$

where E is the electric field intensity and μ the electrophoretic mobility. For a particle of $\mu = 20$ TU electrophoretic mobility[10] (1 TU = Tiselius Unit = $10^{-5} \frac{cm}{sec} / \frac{V}{cm}$) we obtain at $E = 100$ V/cm the pitch $s_o = 7.6$ mm. Most of the circular endless belt apparatus had an annulus of about 10 cm circumference, whereas non-circular annuli were somewhat in excess of 20 cm in circumference.

Examples of performance of the circular endless belt electrophoretic separator

The following photographs will illustrate some of the characteristic features of the endless fluid belt apparatus, which in these illustrations comprised a horizontal buffer annulus 3 cm in diameter and 1.5 mm in annular thickness.

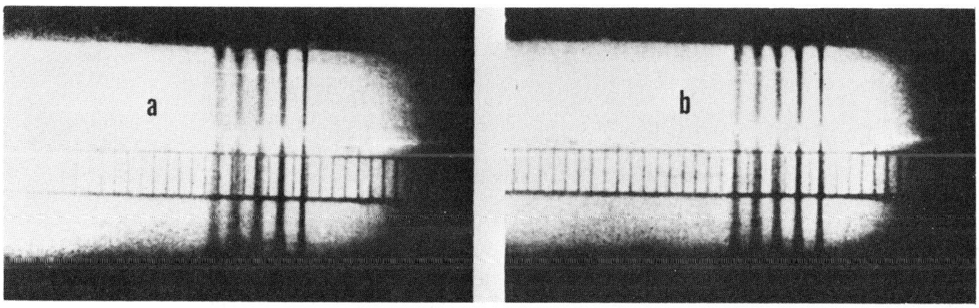

Fig. 7. Insensitivity of the helical pitch to variations in cell voltage. Four helical turns of the dye Evans blue are shown at a cell voltage of 75 V in (a) and of 150 V in (b).

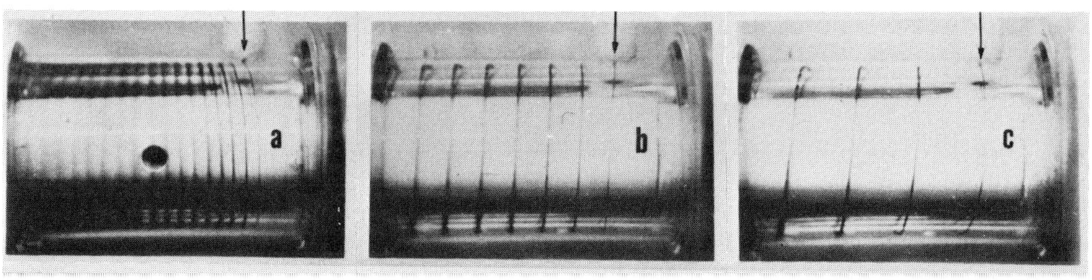

Fig. 8. Modification of the helical pitch by lateral flow from right to left. The point of sample entry is indicated by arrow. The rate of lateral flow is greatest in (c) and lowest (zero) in (a). (From A. Kolin: J. Chromatogr. 26, 164, 1967).

Fig. 7 illustrates the insensitivity of the helical pitch to the voltage applied across the separation cell. Fig. 7a shows a helix due to injection of the dye

Evans blue at a cell voltage of 75 volts. Fig. 7b shows the helical pitch after raising the cell voltage to 150 V. The pitch of the helical ion path has remained unchanged because the increase in voltage accelerated the ion electromigration by the same factor as the electromagnetic revolution of the buffer in the annulus.

In Fig. 8 we see how imposition of lateral streaming (from right to left) in the annulus can be used to vary the pitch of the helical path of the injected particles. In Fig. 8a electrophoresis proceeds without lateral streaming. In Fig. 8b the helical pitch is increased by a right-to-left flow and in Fig. 8c the distance between helical turns is increased further by an increase in the lateral flow velocity.

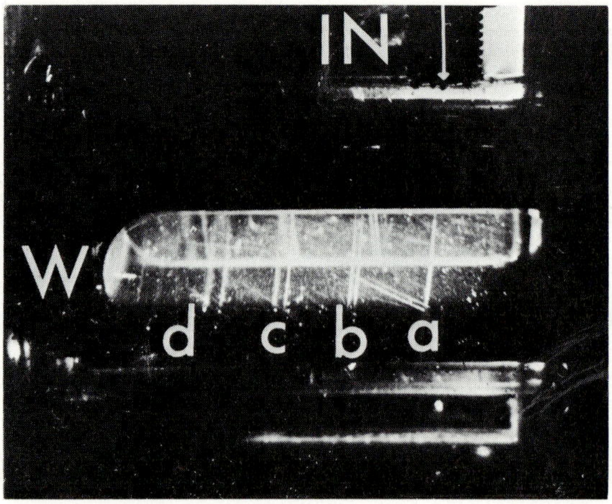

Fig. 9. Light scattering visualizes separation of two micro-organisms: Saccharomyces cerevisiae and rhodotorula in circular endless belt. a, b, c, d are consecutive helical turns. The arrow marks the point of sample injection. IN: injector; W: window. The light source is behind the window W and is directed toward the reader (the streak visualization is thus by forward scattering. From A. Kolin, J. Chromatogr. 26, 164, 1967).

A separation of two species of micro-organisms is shown in Fig. 9. A mixed suspension of two yeasts: Saccharomyces cerevisiae and rhodotorula is injected via injector IN at the arrow Light aiming toward the reader from a collimated light source behind the window W in the iron core visualizes the streaks of the light-scattering particles by dark-field illumination. We see four consecutive descending turns a, b, c, d. Electrophoresis is directed toward the left. A clear-cut separation is visible after a little more than one revolution (after about 20 seconds). Rhodotorula being the faster component, its streak lies to the left of the one corresponding to saccharomyces cerevisiae. We see how the separation grows with each consecutive turn by equal increments. This separation between the separated fractions is not affected by the magnitude of lateral streaming velocity.

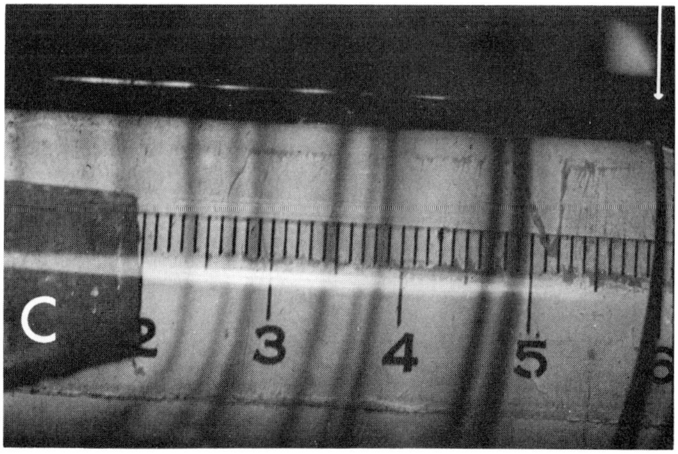

Fig. 10. Collection in the fifth helical turn in collector C of 3 dye components. From right to left: Evans blue, "Brush" recording ink, rose Bengal. The sample enters at arrow. Electrophoresis is from right to left. (From A. Kolin: J. Chromatogr. 26, 164, 1967).

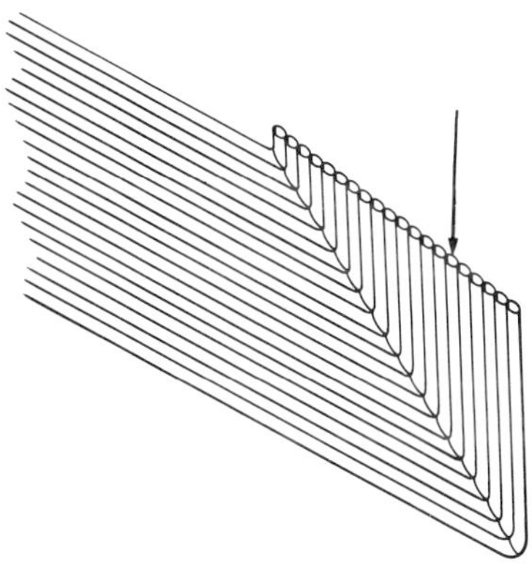

Fig. 11. A simple collector (as illustrated by C in Fig. 10) is made by cementing together fine parallel polyvinyl chloride tubes as illustrated. This flexible collector is introduced through the left electrode compartment into the annulus. The arrow indicates the direction of sample and buffer flow in the annulus.

Fig. 10 illustrates the separation and collection of three molecular components (the dyes are from right to left: Evans Blue, "Brush" recording ink, rose Bengal) in a circular endless belt after four helical turns. The dyes are injected into the annulus at the arrow. After four turns the separations between the streaks are considerably larger than the streak width so that collection without overlap becomes possible. The collector C is made of thin Polyvinyl chloride (PVC) tubing cemented with PVC cement into an L-shaped bundle shown in Fig. 11 . The arrow indicates the direction of the fluid motion in the endless belt pointing toward the tube openings. The streaks of the fifth turn are seen to enter the collector C. The rate of fluid withdrawal through the collector is adjusted to approximately equal the volume rate of fluid rotational flow in the annulus toward the collector over its entire width. In this case the streaks are nearly parallel as they enter C. If the rate of withdrawal is greater, they converge, and if it is smaller, they diverge. This buffer efflux through the collector generates an axial flow of suitable magnitude to separate the helical turns thus avoiding overlap between the different orders of separation.

THE NON-CIRCULAR ENDLESS BELT APPARATUS

Drawbacks of the circular endless belt which led to this modification

The circular endless belt scheme has the advantage of simplicity of construction. Its main drawback is the possibility of sedimentation of the injected streak upon the walls of the annulus. This is illustrated in Fig. 12. A streak of suspended particles is introduced from the injector IN. It is possible to adjust the density of the buffer by suitable additives so as to make it equal to the density of the injected particles to avoid sedimentation. This is, however, undesirable partly because the procedure is laborious and partly because such equilibration will as a rule succeed in equilibrating only one component of a particulate mixture while the other components will continue to sediment. But even in the case of a streak of molecular solutes whose individual particles do not sediment perceptibly, the density of the streak may differ from the density of the ambient buffer. In this case the streak as a whole will behave like a sedimenting object.

Fig. 12A illustrates a typical path of a particle whose density exceeds that of the buffer as it sediments while revolving about the iron core m in the circular endless belt. During the first quarter of the revolution between a and b the sedimentation brings it close to the central cylinder C_1 at b. Thereafter the particle path recedes from the inner wall on its way toward c reaching an intermediate position. As the particle journey continues toward d, further sedimentation brings it close to the outer wall C_2. On the way from d to a the particle returns to its level of origin within the annulus, due to sedimentation, when it reaches a. It is obvious that the excentric particle path may be such as to precipitate the particles upon the annulus walls in an extreme case. The same fate may affect a dense streak of a concentrated molecular solute.

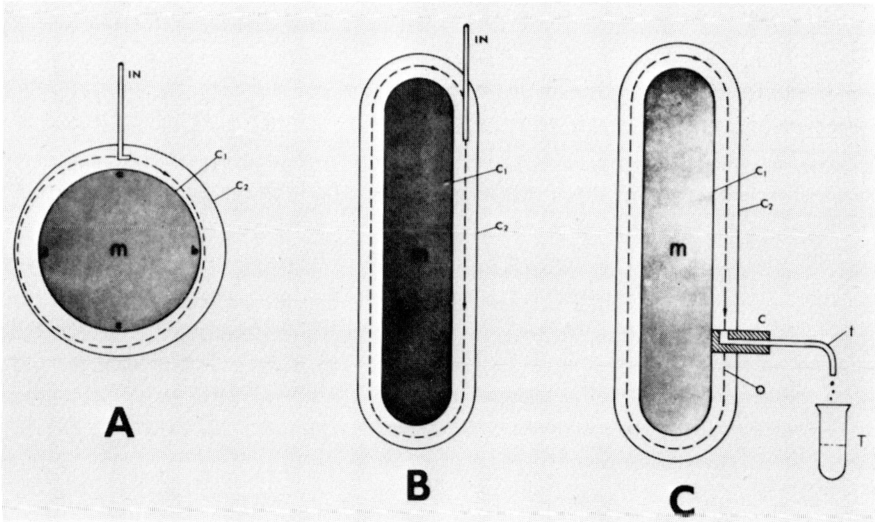

Fig. 12. Suppression of sedimentation by modification of the endless fluid belt contour. m: iron core inside annulus; IN: injector; C_1, C_2: annulus walls; C: collector; t: collector tubing; T: test tube. (A): a streak of sedimenting particles describes an excentric orbit which comes close to the walls C_1, C_2 of the annulus of the circular endless belt. (B): the sedimentation effect is minimized in the vertical "race track" which shows the injector IN. (C): placement of the opening of the collector C in the non-circular endless fluid belt. (Redrawn from A. Kolin, Proc. Nat. Acad. Sci. 56, 1051, 1966).

Remedy by the non-circular endless belt scheme

The undesirable sedimentation effects can be easily avoided by deformation of the inner iron core m into an elongated rectangle capped by semicircles as shown in Fig. 12B. Its surface C_1 will now be the inner boundary of the annulus. The outer annulus wall will be a surface C_2 surrounding the core m at a distance of 1.5 mm from it. The magnetic field will still be substantially normal to the surface C_1 and remain so throughout the annulus so that a current flowing at right angles to the page will engender tangential fluid motion indicated by the arrow.

A particle leaving the injector IN in the mid-plane of the annulus (actually preferably slightly to the left of it) will move without deviation with a vertical velocity affected only slightly by sedimentation. This velocity increase will be compensated by an equal decrease when the particle moves upward on the left side of the "race track" which has the shape of a belt sander. There is however a slight amount of downward sedimentation at the bottom of the annulus when the particle moves horizontally as it changes from a downward vertical path to an upward one on the opposite side of the fluid belt. This sedimentation deviates the particle slightly below the dashed midline so that it proceeds upward somewhat to the left of the midline belt. However, when the particle reaches the upward semicircular path, sedimentation brings it back toward and slightly below the midline so that its descent would be somewhat to the left of the midline on the injector side of the belt. Thus the dense particle (or streak) will oscillate about the midline of the fluid belt in a process akin to "gravitational focusing".

An additional advantage of this shape of the endless belt[12] is the possibility of significantly increasing its circumference without making the iron core m and the magnets excessively bulky and heavy. This leads to a wider separation of the electrophoretic components per helical turn. The particle path is now a non-circular helix.

Fig. 12C shows how a collector can be introduced to intercept the separated fractions. The rate of withdrawal of fluid can be adjusted to be equal to the volume rate of flow which would have existed in the absence of the collector. Buffer can be injected through a supply tube (CC of Fig. 20) under the collector at a rate equal to the collector outflow. Under these conditions the buffer circulates about the core m as it would have in the absence of the collector, i.e. as if the collector were hydraulically "transparent".

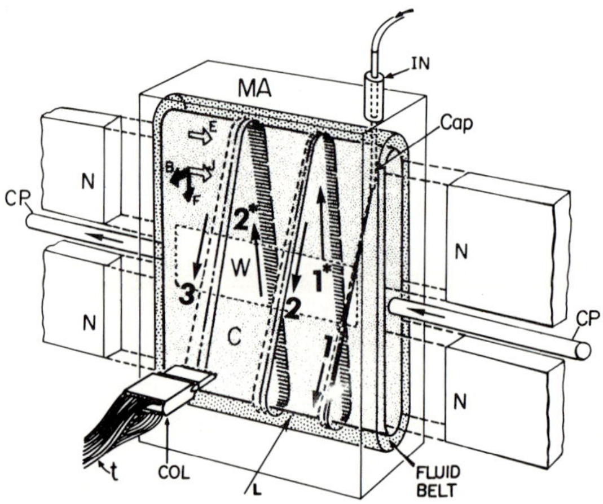

Fig. 13. Electrophoretic separation space of the vertical "race track" non-circular endless fluid belt apparatus. N: north poles of the (only partially visible) magnets (the magnets actually contact the core C). C: hollow iron core. CP: pipes conveying the coolant to and from the iron core. W: window in C. MA: water-cooled mantle surrounding the annulus. IN: injector. Cap: capillary for sample injection; COL: collector; t: collector tubing; E: electric field vector; B: magnetic field vector; J: current density vector; F: electromagnetic force vector; 1^*, 2^*: first and second ascending streaks (behind the core C); 1, 2, 3: first, second and third descending streaks (in front of the core C). L: light beam passing through the annulus below the core C.

The separation space of the endless belt apparatus

Fig. 13 shows a perspective view of the separation space of a non-circular endless fluid belt electrophoresis apparatus. C is the soft-iron core with a space W milled out to be filled with a quartz window. MA is a plastic mantle surrounding the core C so as to leave an annular space of 1.5 mm which is filled with a buffer solution which separates the core from the mantle. The latter contains two quartz windows, one in front and another one behind the core window W so that ultraviolet as well as visible light could be used for transillumination and visualization or photography of the separation pattern. The collector COL which could contain typically 20 openings over a distance of 2 cm feeding into 20 collection tubes t, can be plugged into the mantle through a gasketed aperture. The sample is injected into the central plane of the annulus through the injector IN (typically a glass capillary 0.2 mm i.d.).

The magnetic field can be provided by several bar magnets whose orientation and polarity are indicated in Fig. 13. The actual position of the magnets is such that they contact the core C (which they do not in the figure for clarity). The core is coated with a layer of about 0.15-0.2 mm thickness of epoxylite for electrical insulation. The magnets are separated by a thin rubber cushion from the core over the area of contact to avoid mechanical damage to the core insulation as a result of the forces of magnetic attraction. By interposing thicker padding between the magnets and the core, one can reduce the magnetic field in the annulus which can thus be regulated in simple fashion (one can also remove some of the magnets as well as replace them by a set of different intensity. A symmetrical arrangement of magnets is preferable to an asymmetrical one).

Proper cooling of the buffer in the annulus is of paramount importance to remove the heat which is generated in it at the rate of J^2/σ per unit volume (where J is the current density and σ the buffer conductivity). This is accomplished by maintaining the inner and outer walls of the annulus (C_1 and C_2 of Fig. 12) at a low temperature by circulating coolant fluid. The cooling of the inner surface is achieved by circulating the coolant via tubes CP through the core C of Fig. 13 which is hollow. The mantle MA is also hollow thus permitting circulation of cooling fluid which makes thermal contact with the outer surface of the buffer annulus through a thin (about 0.5 mm in thickness) inner wall of the mantle's front and back cooling chambers.

Fig. 13 illustrates the separation of three electrophoretically distinct components which electromigrate from right to left as they revolve around the core C as indicated by the arrows. We see the descending streaks 1, 2, 3 which pass in front of the core C as well as the ascending streaks 1^* and 2^* which move upward on the back side of the core. The particles under consideration have a negative charge since the electric field vector points to the right. The mutual orientation of the vectors of current density ($\vec{J}$), magnetic field intensity ($\vec{B}$) and electromagnetic force ($\vec{F}$) is shown in Fig. 13. Typical dimensions in cm are indicated in the diagram. L represents a light beam which permits visualization of the streaks in an end-on view. In this fashion the light traverses a much greater thickness of the particle cloud in the streak than is possible by tranverse illumination of the thin streaks through the window W. This increases the sensitivity of streak detection. In addition, this mode of observation reveals the cross-sectional shape of the streak seen in Fig. 14 the significance of which will be discussed below. (The view shown in Fig. 14 corresponds to looking over the top of the core rather than below it as suggested by the location of the light arrow L.)

Fig. 14. Distortion of the originally circular cross-section of a streak into a crescent. Electrophoresis direction is toward the left. The streak cross-section is viewed by looking horizontally through the annulus above the core C (see also Fig. 13). (From S.J. Luner: UCLA Thesis 1967.)

The flow scheme of the endless belt separator

Fig. 15 is a schematic representation of the arrangements for generation of flows other than the electromagnetic circulation in the Endless Fluid Belt apparatus. The process of collecting the separated fractions requires a certain amount of efflux from the collector COL which escapes via the collector tubing CT (0.6 mm i.d.) into the test tubes TT. The lost buffer must be replenished to maintain a steady state. The source for the replacement buffer is the Mariotte bottle MB. It maintains a constant pressure head while the fluid level in it is falling[11A]. The hydrostatic pressure at the bottom of the bottle is the same as it would have been in an open container filled with fluid up to the bottom level of the tube MT. The latter is an open tube establishing a communication between the buffer sealed in the bottle MB and atmospheric air. The pressure head can be changed coarsely by raising or lowering MB or fine-controlled by pulling the tube MB up or down.

The buffer is delivered to the electrophoresis cell via the distributor D from which fine PVC tubes (0.8 mm i.d.) lead to the buffer chambers BC. The syringe SC permits aspiration of the buffer from the chambers BC to remove air bubbles which may be trapped in the distributor D and the delivery tubes DT. The distributor D contains a coarse filter (a stainless steel perforated "Millipore" filter support) to prevent coarse suspended impurities from entering the apparatus.

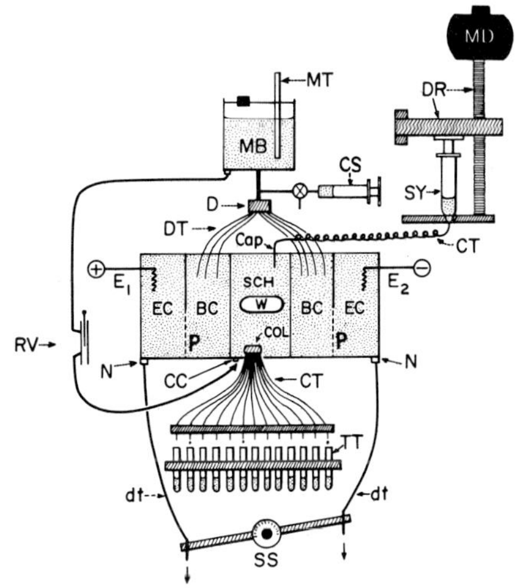

Fig. 15. Schematic diagram of the flow system of the endless fluid belt apparatus. SCH: separation chamber; W: window; Cap: capillary conveying the sample flow; COL: collector conveying the separated fraction to test tubes TT via the collector tubes CT. CC: collector compensator conveying buffer from Mariotte bottle MB to the cell via regulating valve RV. BC: buffer compartments. E_1, E_2: electrodes. N: nipples of rigid drainage tubes (D of Fig. 17, not shown here). dt: narrow plastic drainage tubes; SS: "see-saw" for fine-control of lateral flow. P: perforated plates across which centrifugal buffer flow is maintained; MT: tube of Mariotte bottle MB; CS: clean-out syringe; D: buffer flow distributor; DT: distributor tubes; CT: coiled tubing linking capillary to the sample syringe SY. MD: motor drive and DR screw drive for SY (from A. Kolin in Methods of Cell Separation, vol. 2, N. Catsimpoolas, ed., Plenum Publ. Co., New York, 1979).

When the collector is not activated (i.e. when the exit level of the tubes CT is raised to the fluid level in the compartments BC and EC so that fluid does not escape from the tubes) an even distribution of the tubes DT among the compartments BC on the right and left sides of the central section SCH will generate no lateral buffer flow in the annulus. Transferring a tube from the left BC chamber to the right one will generate a right-to-left buffer flow in the annulus. Similarly successive tube transfers will produce step-wise changes in lateral buffer flow in the annulus in either direction. A continuous fine-control of the lateral buffer flow is achieved by raising and lowering the level of the exit points of the drainage tubes dt (1.6 mm i.d.). The latter are mounted at the ends of a centrally pivoted "see-saw" SS. Changing the inclination of the "see-saw" raises the end points of one tube dt while simultaneously lowering the other one by the same amount. The rate of drainage through tubes dt can be altered by raising or lowering the pivot of the "see-saw" SS. The efflux from the collector COL contributes to the total drainage rate. In the steady state the rate of influx from the Mariotte bottle MB is equal to the total drainage rate.

As stated before, the buffer escaping from the collector tubes CT can be replenished at equal rate through collector compensator openings (not shown) below the collector COL in the annulus. RV is a regulator valve which permits the variation of this buffer flux delivered by the Mariotte bottle.

The electrode compartments EC which harbor the electrodes E_1, E_2 are separated from the buffer chambers BC by perforated plates P. The perforations play an important role as explained below[8,13]. The initial designs of the endless belt apparatus contained membranes separating the compartments BC and EC from each other hydraulically while providing electrically conductive continuity between them[11]. Like in the case of the "Free Flow" fluid curtain electrophoresis apparatus, such membranes cause spatial as well as temporal changes in the distributions of pH and electrical conductivity in the separation space[1,3,8]. Such drifts and gradients can be avoided by replacing the membranes by dielectric plates with 50 holes of about 1.6 mm i.d. through which a fluid flow of adequate magnitude is forced in the direction from the compartments BC toward the chambers EC. This flow must be rapid enough to prevent the electrolysis products from migrating from the electrode chambers toward the annulus. The idea of such a centrifugal buffer flow was first applied to an electrophoretic column[14] and was later adopted to the scheme of the fluid endless belt[13,15]. The maintenance of the centrifugal buffer flow accounts for the large rate of buffer consumption (6 liters/hour) of the apparatus. It is possible to collect and remix the buffer issuing from the two electrode chambers for re-use. It would be thus advantageous to reduce the buffer consumption by feeding the collected buffer back into the apparatus via a suitable filtering system. In this case, one would provide a constant pressure head by an overflow system rather than a Mariotte bottle.

The sample to be analyzed is injected by a motor-driven syringe Sy through a glass injection capillary Cap. MD is the motor drive in Fig. 15 and DR the screwdrive which pushes the syringe. CT is a fine teflon tubing (nr. 28 gauge) linking Sy with Cap. It is coiled as shown in order to suppress sedimentation of particles.

Increase in preparative resolution by fan collector

There is a practical limit to how closely the collector openings can be spaced. This sets a limit to preparative resolution in collecting the separated fractions. This limitation can be circumvented as follows by greatly increasing the number of collector openings to which the separation pattern can be guided from the electrophoresis cell[16]. Let us assume that it is difficult to space collector openings closer than 1 mm apart. A typical collector 2 cm in width would then feed its output into 20 test tubes. Fig. 16 illustrates how the fluid outflow from a collector can be spread in fan-like fashion to leave the apparatus through many more tubes than could be accommodated along the width of 2 cm available for the collector in the endless belt.

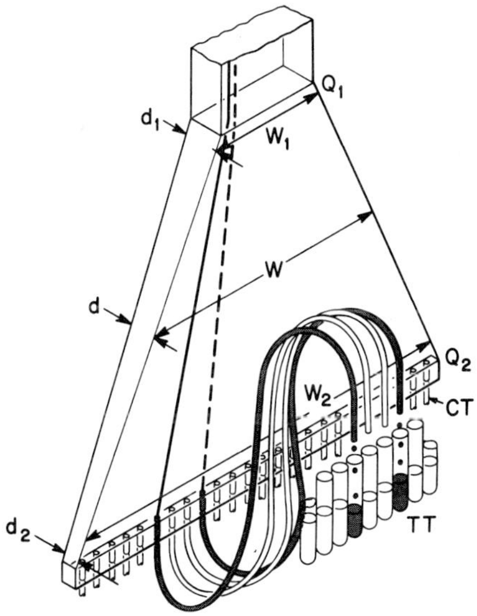

Fig. 16. Fan collector for increase in preparative resolution. The separated fractions (solid and dashed lines) leave the separator above level Q_1. The flow channel is deformed below in fan-like fashion. It expands laterally from width W_1 to width W_2 at the bottom. To maintain a constant linear rate of flow, the product of channel width (W) times channel thickness (d) is maintained constant. This flow pattern separates the streaks without further electrophoresis and reduces the hazard of fraction overlap in collector tubes CT which convey the fractions to test tubes TT (from A. Kolin, J. Chromatogr. 17, 532, 1975).

Let us assume that we replace the 2 cm wide collector COL in Fig. 13 by a rectangular channel FC of width W_1 = 2 cm as shown in Fig. 16. We can modify the cross-section of the channel as shown in Fig. 16. It grows wider toward the bottom (in the direction of fluid flow). The final width W_2 could, for instance, be five times as great as the initial width W_1. In this case, spacing the exit tubes as closely as possible (say, 1 mm apart) we would now discharge the collector output through 100 tubes, attached to nipples CT, instead of the 20 which can be accommodated along the width W_1. The solid and the dashed streaks illustrate the advantage of the fan collector. We see that the interspace between them increases as they migrate from the narrow section of width W_1 toward the wide terminal section of width W_2. The final separation between the streaks is enlarged by the factor W_2/W_1. (The width of each streak increases concurrently by the same factor.) To maintain constant flow velocity in the fan, the cross-sectional area (d • W) of the fan-like flow channel is kept constant as the distance d between the wings decreases while their width W expands in the direction of flow. The fan collector shown in the instrument of Fig. 17 is plugged into the mantle through an elongated O-ring gasket being thus removable for cleaning.

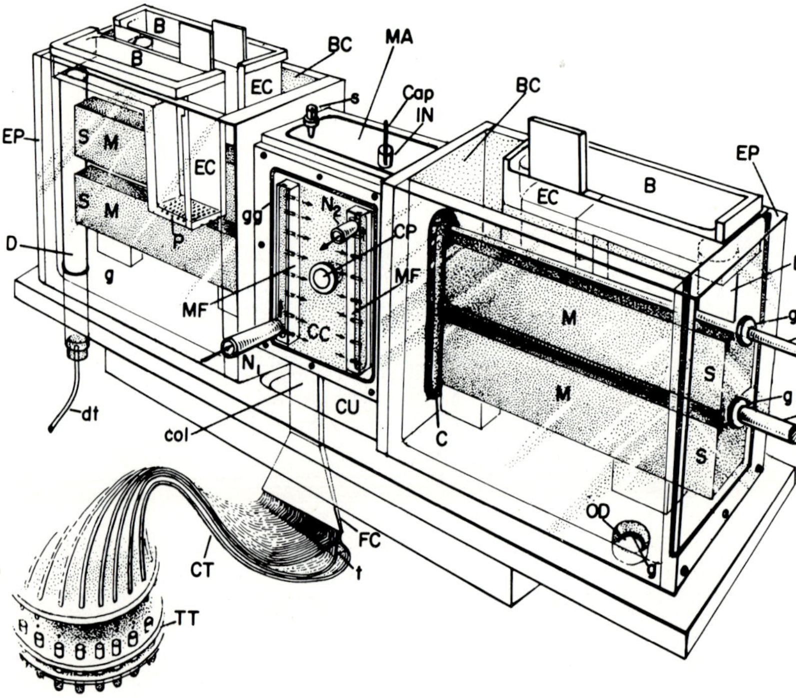

Fig. 17. Endless belt apparatus with submerged magnets. MA: mantle; S: tubing for removal of air bubbles from annulus. IN: injector holder; Cap: injector capillary; CC: cooling chamber in mantle; CP: center post in chamber CC. N_1, N_2: coolant supply nipples; gg: rubber gasket; MF: cooling manifolds. The flow in the cooling chambers is indicated by arrows. CU: cut-out in the base plate to allow the fan collector to pass. Col: plug-in collector; FC: "fan" of collector; t: terminal collector tubes; CT: tubing conveying collected fractions to test tubes TT. BC: buffer chambers; B: four "balconies" (the front right one has been removed in order not to obstruct the view of the structures behind it). EC: electrode compartments; P: perforated plates which separate the electrode compartments from the buffer chambers; D: drainage tubes; dt: terminal drainage tubes; g: rubber gasket; OD: gasketed openings for the drainage tubes D which act as balcony support columns; C: iron core; CP: coolant supply pipe of iron core; V: vent pipe for iron core (to permit escape of air); EP: removable gasketed end plates of buffer chambers; M: magnets; S: south poles of magnets whose north poles contact the iron core C. (From A. Kolin in Methods of Cell Separation, vol. 2, N. Catsimpoolas, ed., Plenum Publ. Co., New York, 1979).

APPARATUS WITH SUBMERGED MAGNETS

Construction of the apparatus

Difficulty in securing leak-proof seals may have been a primary reason which prevented wider use of endless belt electrophoresis. The apparatus described in this section eliminates the tricky gasketing system of the previous design by submerging the magnets in the buffer solution thus greatly simplifying the construction of the instrument[8]. This latest form of the Endless Belt Electrophoresis apparatus is shown in Fig. 17. It consists of three sections and exhibits almost perfect front-back and right-left symmetry. The central section is the separation chamber. The right and left adjacent chambers are plastic tubs which are initially empty as we begin the assembly of the instrument. The end plates EP are taken off by removing the screws which fasten them along the edges. A U-shaped gasket (visible at the end of the right buffer chamber BC) insures a leak-proof connection of the plates EP to the walls of each tub. The only difference between the right and left buffer chambers BC ("tubs") is that the plate in the left plate EP has only one opening for the cooling pipe CP (which conveys the coolant to the iron core in the annulus C), whereas the plate EP of the right BC chamber has two openings. The second opening passes the thin metal tube V which serves as a vent to permit air to escape from the core interior when it is filled with cooling fluid. All three openings for the brass tubes which are linked to the iron core are gasketed with O-rings g.

The iron core C is centered roughly by the location of the openings of the holes for the tubes CP in the plates EP. More reliable centering is achieved by means of teflon screws (not shown) which can be turned from the outside and which project into the annulus so as to clamp the core C between them.

While the plates EP are off, the core C is introduced through the annulus and is centered. Then the four magnets M are placed against the iron core C as shown and the end plates EP are screwed on at both ends of the buffer chambers BC after passing the tubes CP and V through the gasketed openings provided for them. There are niches milled into the end plates EP which support the outer ends (south poles S) of the magnets. A thin rubber sheet between the magnets' ends (N poles) and the epoxylite insulation of the core C protects the insulating layer against mechanical damage. The magnets are cast into Teets Denture Material for thorough electrical insulation (about 0.5 mm thick layer).

The only thing that remains to be done to complete the assembly of the cell is to insert the four "balconies" B. (The front balcony in the right BC chamber has been omitted in order not to obstruct the view of the magnets.) There are two balconies in each BC chamber, one on each side of the magnets. Each balcony consists of a horizontal flat tub section most clearly seen in the right compartment BC behind the magnets. There is an opening at the bottom on the right hand of the above tub to which a vertical drainage tube D is cemented. Each balcony has such a drainage tube which passes through a gasketed opening OD at the bottom of the buffer chamber

BC. At the opposite end of the balcony is cemented a vertical electrode chamber of rectangular cross section (EC) which is most clearly seen in the front balcony of the left BC chamber. The bottom of each EC chamber is perforated with about 50 holes 1.6 mm in diameter. Each BC chamber contains a Pt electrode (not shown in the figure).

The central iron core C is surrounded with a lucite mantle MA so that an annular space 1.5 mm thick remains between the core and the inner walls of the mantle. The top and bottom of the mantle are secured by nylon screws to the plastic frame of the mantle with suitable gasketing. The (identical) front and back panels of the mantle are also secured by nylon screws to the middle frame and can be easily removed. A gasket gg insures a leak-proof seal. The removal of these portions of the middle section is an essential convenience which greatly facilitates cleaning of the separation space of the instrument.

The removable front and back sections of the mantle harbor the cooling chambers (CC). The inner wall of such a plug-in chamber facing the annulus is made of glass 1.5 mm in thickness. This glass plate is cemented with RTV 112 silicone rubber cement to the lucite frame of the cooling chamber. The remainder of the cooling chamber is made of lucite. The cooling fluid enters through the lower nipple N_1 the left manifold MF from which it issues as a horizontal turbulent stream from the manifold openings. The coolant leaves the chamber CC through the right manifold MF via upper nipple N_2. The back cooling chamber is identical to the front one. To prevent membrane-like bulging of the inner glass plate with changes in pressure in the cooling chamber interior, which would alter the width of the annulus, a cylindrical aluminum post passes through an opening in the outer wall of the chamber CC. The inner flat surface is cemented to the glass plate with RTV silicone rubber cement which also seals and cements the protruding stump CP to the front surface of the CC chamber.

IN is a cylindrical injector stump with a central hole through which passes a snugly fitting glass capillary Cap so as to be precisely centered within the front section of the annular fluid "race track" (comp. Fig. 12C). S is a nr. 13 gauge syringe needle through which fluid can be aspirated from the top of the annulus to remove air bubbles which may get trapped during the process of filling the cell.

Col is a collector which is plugged into an O-ring-gasketed oval opening so as to intercept the fluid stream descending in the front section of the annulus over the final 2 cm width of the fluid belt. This plug-in arrangement has the convenience of making it possible to interchange different types of collectors and to replace a collector with a plug which leaves the end section of the fluid belt unobstructed. The collector section has a fan section FC with terminal hypodermic tubes t bent alternately forward and backward feeding the collector output to 100 teflon tubes CT 0.3 mm i.d. which guide the collected fractions to the test tubes TT in a circular rack. CU is a cut-out in the bottom plate for passage of the collector.

This particular Cell has no windows in the core and mantle. The separated streaks can be seen and photographed as they pass through the transparent collector above or inside the fan section FC. There is an advantage in shifting the streak observation to the collector section so as to avoid windows. This simplifies the construction of the core and the outer cooling chamber and makes the magnetic field as well as the cooling more intense and more uniform. The course and width of the streaks can still be observed within the annulus by providing horizontal white stripes on the core section against which colored streaks of injected tracer dyes can easily be seen.

Operation of the apparatus

In reading this description, one should refer concurrently to Fig. 15 and Fig. 17. The Endless Belt cell is filled by pouring buffer into the right BC compartment while the left compartment is raised. This inclination avoids entrapment of air on top of the annulus as buffer flows through it to fill the left BC chamber. The cell is filled adequately when the liquid level is about 1 cm above the floor of the balconies B.

When the buffer supply tubes DT are evenly distributed among the two BC chambers the centrifugal buffer flow toward the electrode chambers EC is activated by starting buffer delivery from the Mariotte bottle MB. The buffer enters the four electrode chambers EC at the bottom through the perforations P and flows upward until it reaches the balconies B. The upward buffer flow carries the small gas bubbles which are evolved by electrolysis at the electrodes, that dwell in the chambers EC, toward the balconies B where the slow horizontal buffer flow offers enough time for the bubbles to escape into the atmosphere before they reach the apertures at the ends of the balconies at which the drainage tubes D originate. These tubes carry the buffer toward small exit tubes dt. (Actually the two left as well as the two right drainage tubes D are joined to each other at the bottom by a horizontal tube (not shown) of somewhat larger diameter from which a single plastic tube dt leads to the sink.) The drainage tube D which fits into the gasketed opening OD in the front part of the right BC compartment and the attached front balcony are omitted to offer an unobstructed view of the magnets M.

The circulation of the cooling fluid is initiated and the current is turned on (ranging between 200 and 240 mA). The injector mechanism (comprising the components MD, DR, SY, CT, and Cap of Fig. 15) is activated with a tracer dye in the sample (usually brilliant blue) to trace the course of the helical streak. The collector tubing CT is lowered (with the test tube rack TT) to achieve the required rate of collector outflow and the desired lateral flow is established by appropriate transfer of tubes DT from one BC compartment to another and adjustment of the see-saw SS. This manipulation permits adjustment for the desired number of helical turns and placement of a given component of the separation pattern into a chosen test tube.

To stabilize the operation of the instrument, it should be turned on for about

15-20 minutes before preparative collection. For analytical observations such a waiting period is not necessary. The power supply used in our work is Sorensen Model DCR 3000-.5 and the cooling is accomplished by a Neslab cooler Model HX-50. The rate of separation and throughput is exemplified by a typical run processing 0.5 cc of a cell suspension of $5 \cdot 10^7$ to $5 \cdot 10^8$ cells/ml in 30 to 45 min. (these data are based on an instrument with a core with window as shown in Figs. 13, 19 and 20[13]).

A great variety of buffers can be used with the apparatus. The only requirement is a suitable value of electrical conductivity, typically in the order of 700 micromhos. The choice of the buffer depends on the requirements of the biological materials to be fractionated. The choice of the buffer is particularly critical in cell separations. An extensive discussion of suitable buffers is to be found in a paper by Zeiller et al.[17].

The following Table 1[8,15] illustrates typical operation parameters of an Endless Fluid Belt electrophoresis apparatus. It is based on measurements using an instrument of earlier design whose magnets are outside the buffer solution (Fig. 19)[13].

TABLE 1

EXAMPLES OF TYPICAL OPERATING PARAMETERS

i ma	E V/cm	σ μmho/cm	η cp	t °C	τ sec	μ_B -TU*
220	97.7	692	0.91	24	31.0	29
180	87.2	633	1.00	20	40.7	26
120	71.4	517	1.23	12	76.0	21

These operating conditions apply to a cell with a center section of the type shown in Fig. 13[13]. i: current in the annulus. E: electric field intensity in the annulus. σ: buffer conductivity (from current and voltage measurements). η: average viscosity in buffer belt. t: average temperature in buffer belt. (η and t calculated from average electrical conductivity.) τ: period of circulation of buffer belt around the core. μ_B: mobility of brilliant blue (by absolute measurement).

* $1\ \mathrm{TU} = 10^{-5}\ \frac{\mathrm{cm}}{\mathrm{sec}} / \frac{\mathrm{V}}{\mathrm{cm}}$.

Modes of observation and collection

The periodicity of the separation pattern in the Endless Belt system offers several different modes in which preparative and analytical separation can be conducted.

(A) Normal n-th order collection. Let us consider, for instance, three electrophoretic components A, B, C, spiraling around the iron core C of Fig. 13. If the distance between the collector openings is 1 mm, we would allow for enough helical turns before the collection to insure that the two closest components are at least 1 mm apart when they reach the collector to reduce the chances of collecting them in the same test-tube. This may require n helical turns prior to collection. In the absence of lateral streaming, it is neither possible nor advantageous to increase the number of turns indefinitely. Beyond a certain number of turns the fastest component of the n-th turn may come closer than 1 mm to the slowest component of the (n + 1)st turn and, in fact, these two turns may coincide. This coincidence can, however, be easily removed by imposing axial streaming of the buffer in the direction of electromigration. The different orders of separation can then be moved apart by distances determined by the rate of axial streaming as illustrated in Fig. 8 for a single streak pattern. Fig. 10 illustrates collection of the 5th turn for a three-component mixture with imposed axial streaming. While the axial streaming spreads the separated streak multipletts apart, it does not affect the distance between the component streaks within the individual orders of separation. This streak interval is determined only by the mobilities of the electrophoretic components, the electric field intensity within the annulus and the rate of buffer revolution (which is governed by the electrophoretic current and the magnetic field).

(D) Two-order collection. Let us assume that the slowest components B and C of the previous example are so close together after n turns that they enter three collector tubes and appear mixed in the middle test tube. The fastest component A may be quite well separated from the B and C doublet. It will be advantageous in this case to intercept A in the collector while permitting the B and C components to complete another turn before reaching the collector. This may increase their separation after (n + 1) turns to a point where they may no longer mix in an intermediate collector tube while collection of the A component after n turns remains adequate.

(C) Orbital accumulation. In principle, it should be possible to choose a buffer corresponding to the isoelectric pH of a protein component so that this component will go into a helical orbit of zero pitch, i.e. into a circular orbit corresponding to a component of zero mobility. This, however, presupposes absence of electro-osmotic streaming which would require a special effort to achieve.

It is much more practical to achieve the above condition without taking advantage of the isoelectric point under normally encountered circumstances. Fig. 18A shows the initial position of the first turn and four additional turns of india ink in a circular endless fluid belt annulus. The injection point is marked by an arrow. An axial streaming opposite to the direction of electromigration is then superimposed so as to place the injected particle species into a stationary orbit. The result is an accumulation of the particles in a circular orbit as shown in Fig. 18B under the arrow.

This mode of endless fluid belt electrophoresis permits detection of electrophoretic components present in small amounts by prolonged orbital accumulation. It also permits accumulation and purification of a selected component of a mixture while the undesired components are permitted to wander out of the annulus. Collection of the accumulated material can be initiated by shifting it by axial buffer displacement toward the collector.

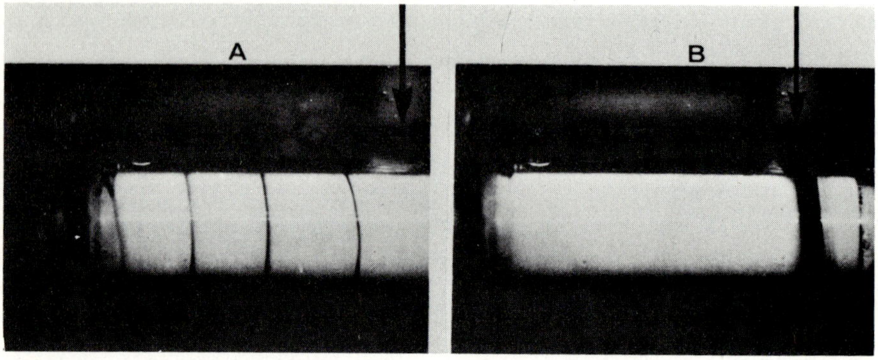

Fig. 18. Non-isoelectric accumulation in a stationary orbit. (a): helical path of india ink particles injected at arrow (part of the first run is not in the picture). (b): the lateral flow is adjusted so as to make the horizontal migration velocity component of the particles relative to the instrument equal to zero. The thick streak under the arrow shows the accumulation of india ink in a stationary circular orbit (from A. Kolin, J. Chromatogr. 26, 164, 1967).

(D) Orbiting zone electrophoresis. Microanalysis can be accomplished by a zone-injection approach. It is not necessary to supply the sample in a continuous stream as in the preceding descriptions. One can inject from a capillary (Cap of Fig. 13) of 0.1 mm i.d. a streak a few millimeters long into the center of the annulus. This streak will form an "orbiting zone" spiraling about the central core C. If it contains n electrophoretically distinct components, it will split into n streaks. By imposing an axial counter-flow so as to immobilize the slowest component in a stationary circular orbit, one may allow the separation pattern to spread over the entire width of the annulus. By viewing the separation pattern not transversely through the window W but end-on below or above the iron core C, as indicated by the line of sight L in Fig. 13, one can detect faintly visible zones with greater ease because of the greater absorption depth afforded by this mode of observation. Sample volumes in the order of 10^{-4} ml can be used in this analytical approach. Preparative collection is accomplished by activating the collector after a suitable zonal separation pattern has been achieved.

EXAMPLES OF ELECTROPHORETIC SEPARATIONS IN THE NON-CIRCULAR ENDLESS FLUID BELT

The instrument with central windows W in the core and mantle, as shown in Fig. 13, is more suitable for illustrations by permitting transverse streak photography of all helical turns in the annulus than the windowless apparatus of Fig. 17 in which the separation pattern can only be seen in the "fan" at the end of the particle path.

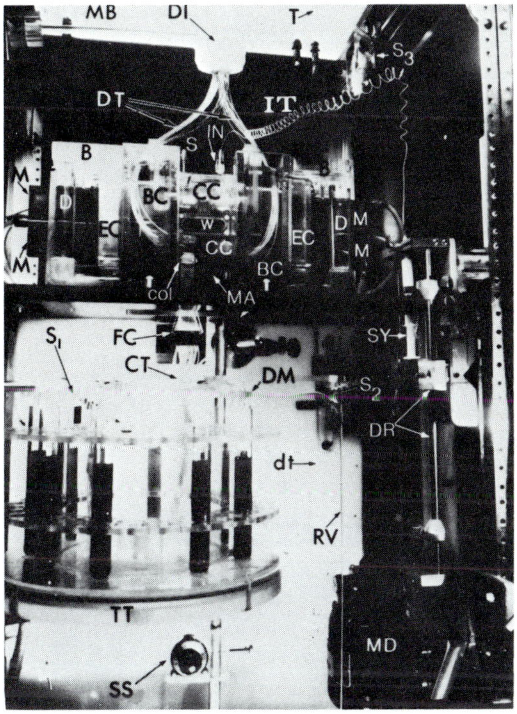

Fig. 19. Endless Belt apparatus with non-submerged magnets. MA: mantle; S: suction tube for air bubble removal; IN: injector; CC: cooling chambers; W: window; Col: collector; FC: fan collector; CT: collector tubes; DM: plate in which collector tubes CT terminate; dt: terminal draining tubes; S_1, S_2, S_3: stop cocks; RV: collector compensator regulating valve; BC: buffer chambers; EC: electrode chambers; B: "balconies"; D: large drainage tubes; M: magnets; DI: distributor; DT: distributor tubes; MB: Mariotte bottle; T: tubing linking Mariotte bottle to distributor DF. IT: coiled tubing for delivery of sample from syringe SY; MD: motor drive; DR: syringe drive; TT: test tube stand.

All of the below illustrations have been obtained with an instrument of the above design[13]. The photograph of Fig. 19 shows an overall view of a more recent form of an apparatus of this type. CC are the cooling chambers of the mantle MA in the central separator section. W is the window in the mantle behind which lie the windows in the core and in the rear section of the mantle. S is the air-removal syringe needle and IN the injector. DT are tubes delivering buffer to the buffer compartments BC from the Mariotte Bottle MB via the distributor DI. EC are the electrode compartments and B the balconies over which the buffer flows from the EC chambers toward the large vertical drainage tubes D from which it escapes via the thin flexible drainage tubes dt which terminate at the ends of the see-saw SS. Col is the point at which the collector plugs into the mantle MA. It terminates in the "fan" section FC from which tubes CT lead to the test tubes TT. SY is the syringe from which the sample is injected via the coiled teflon tubing IT into the glass capillary of the injector IN. DR is the screw drive propelling the syringe plunger and MD is the motor driving it. RV is the regulator valve of the collector compensator. S_1, S_2, S_3 are stopcocks (see legend). The magnets M are not immersed in the buffer solution in this design. They contact the ends of the iron core C which are also exposed to air. This requires a very precarious gasketing of the iron core which accounts for the main difficulty and weakness of this design which contributes to the intricacy of its construction. These drawbacks have been eliminated in the instrument with submerged magnets shown in Fig. 17.

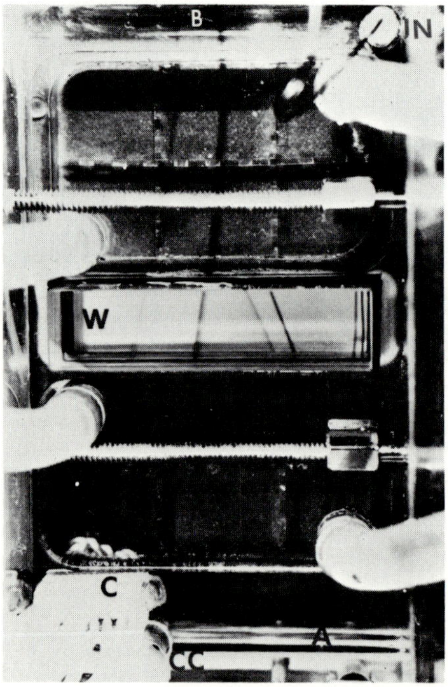

Fig. 20. Separation of small molecular weight components in the non-circular endless fluid belt. The slow component (on the right) is the dye brilliant blue and the faster one is the dye Evans blue as they appear in the window W. Electrophoresis is toward the left. The first descending streaks appear in the window about 5 sec. after injection and are clearly separated. The second descending streak slopes from the upper right to the lower left passing in part of the iron core, whereas the first ascending streak passing behind the core slopes from the lower right to the upper left. Thus, the ascending and descending streaks can be easily distinguished by their slopes. C: collector; IN: injector; CC: collector compensator (from A. Kolin and S.J. Luner, Anal. Biochem. 30, 111, 1969).

Fig. 20 shows a photograph of the front of the mantle. We see the separation of the dyes brilliant blue (slow component on the right) and Evans blue through the window W. The electromigration is from right to left. The sample is injected at IN and the two components are visibly separated at the right end of the window in the first descending streak about 5 seconds after injection. When the descending streaks reach A, they turn around behind the core C and rise in a path which is now seen rising from the lower right at A toward the upper left at B where the streaks pass over to the front side of the annulus and move toward the collector C in the second descending streak sloping from the upper right to the lower left. Thus, the ascending and descending streaks can be easily distinguished by their slopes. This photograph illustrates the narrowness and sharpness of the streaks as compared to their separation. The sharpest streaks are obtainable with microscopic particles which do not cause streak-broadening due to diffusion. As we shall see below, diffusion is not the only potential cause of streak widening which can occur even with particles of cellular dimensions[8,11]. The centering of the ascending streak between the first and second descending streaks is achieved by adjustment of the buffer delivery through the collector compensator CC whose output enters the annulus just below the collector C.

A selection of examples of separations of different types of biological materials is shown in Fig. 21[13]. We see in (a) a very rapid separation of two strains of a virus. The less abundant fast component is the U1 strain of tobacco mosaic virus (TMV). The slower component on the right is the U2 strain of TMV. Only the first descending and first ascending streaks are shown. The separation is clearly visible in the first descending streak about 4 cm below the injector capillary exit, which is about 7 seconds after injection. The separation is quite wide in the first ascending streak after less than one full helical turn. This picture has been obtained by ultraviolet photography.

Fig. (b) shows an ultraviolet photograph of the human serum proteins: albumin (left) and γ-globulin (right).

Visible light was used to photograph the streaks of the colored proteins: bovine hemoglobin (left) and horse heart cytochrome C (right) at pH 9.3 whose wide separation after about 1 1/4 turns is shown in Fig. (c).

Particles are most easily visualized and photographed by light scattering (darkfield illumination). Fig. (d) is a photograph of the streaks into which a mixture of the yeast rhodotorula and the bacterium Escherichia coli splits. E. coli, being the slower component, is in the streak on the right. Only the lower portion of the second descending streak of rhodotorula is clearly visible due to faulty illumination.

Fig. (e) shows a separation between human erythrocytes (left streak) and granulocytes (right streak) after 2 1/2 turns using the same type of illumination as in (d). Again the last descending granulocyte streak is barely visible due to poor lighting. The first descending streak is out of the view of the camera so that the first ascending streak appears on the right.

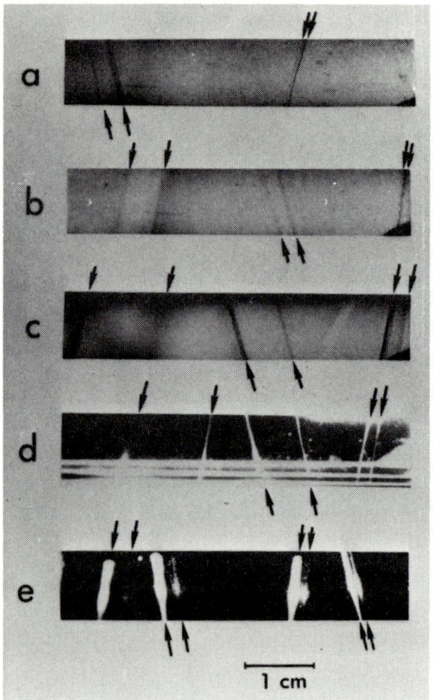

Fig. 21. Separation of different types of biological materials in the non-circular endless fluid belt apparatus. (a): rapid separation (within 7 sec.) of two strains (U1 and U2) of tobacco mosaic virus (only the first descending and ascending streaks are shown). (b): separation of human serum albumin (left streak) and gamma globulin (right streak). (c): separation of bovine hemoglobin (left) and horse heart cytochrome C (right) at pH 9.3. The separation is quite wide after about 1/4 turn. (d): separation of the micro-organisms rhodotorula and E. coli (slower component). (e): separation after 2 1/2 turns of human erythrocytes (left streak) and granulocytes (right streak) (from A. Kolin and S.J. Luner, Anal. Biochem. 30, 111, 1969).

Fig. 22 is a histogram for a typical collection pattern corresponding to the separation pattern of rhodotorula and E. coli shown in Fig. 21d[13]. The mixture injected into the annulus contained 1.7×10^{11} yeast cells and 1.2×10^{12} bacteria per milliliter. The injection rate was 0.13 ml/hr to obtain thin streaks for photography and mobility measurements. For preparative use the injection rate can be raised to 0.5 ml/hr.

Fig. 23 shows an interesting phenomenon which can be observed with some species of bacteria. A suspension of a pure strain of Escherichia freundii is shown to split into two widely separated streaks F and S[18]. This phenomenon has been further investigated[19] and it was found to occur with a number of other bacterial species. It was shown that one of the two streaks consists of piliated bacteria and the other streak of non-piliated ones. When the population of either streak is collected and cultured and subsequently re-injected into the Endless Belt apparatus, each of the cultures splits again into a piliated and non-piliated component.

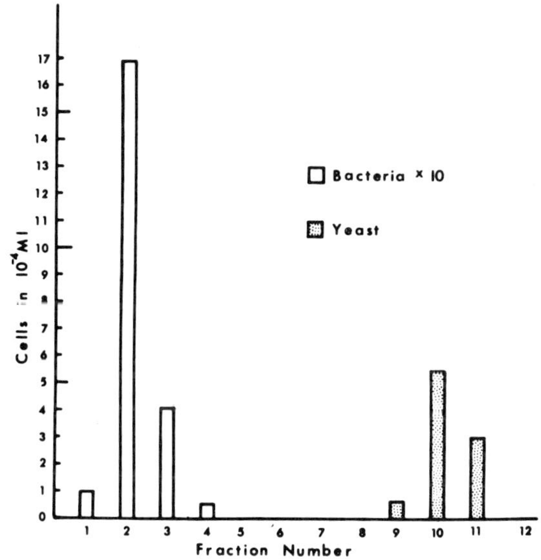

Fig. 22. Histogram for a collection pattern of rhodotorula and E. coli bacilli. The cell concentration in the sample was 1.7 x 10^{11} for rhodotorula cells and 1.2 x 10^{12} cells of E. coli per milliliter (from A. Kolin and S.J. Luner, Anal. Biochem. 30, 111, 1969).

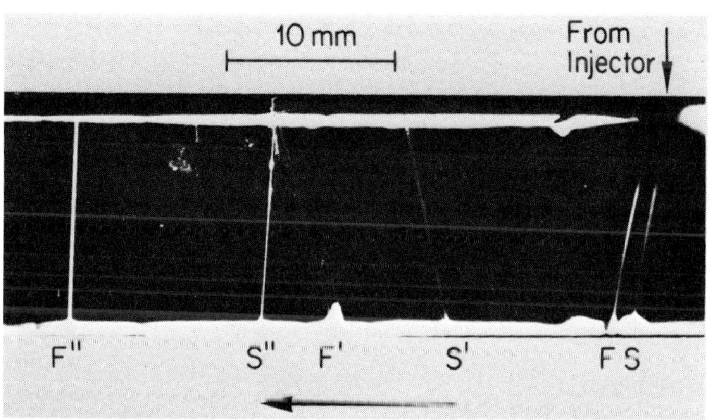

Fig. 23. Splitting of a pure bacterial culture (E. freundii) into two electrophoretically distinct cell populations (from A. Kolin, Proc. First Internat. Biophys. Congress, Baden, Austria, 481, 1971).

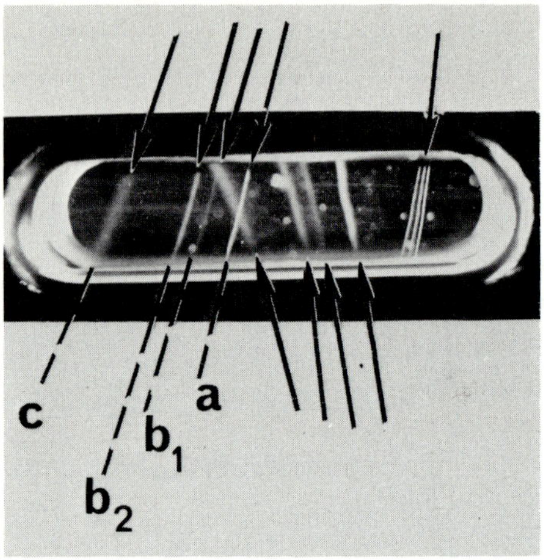

Fig. 24. Separation of a mixture of bacteria, labelled in the first descending streak: (a): E. coli, ML-35 strain; b_1, b_2: two electrophoretically distinct components of Proteus vulgaris; (c) E. coli BE-30 strain (from A. Kolin, in Methods of Cell Separation, vol. 2, N. Catsimpoolas, ed., Plenum Publ. Co., New York).

Fig. 24 illustrates the separation of a mixture of bacteria into four sharply delineated electrophoretic components[8]. It shows the separation pattern after adding to a suspension of E. coli-ML35 a suspension of the strain E. coli-BE30 and a suspension of Proteus vulgaris. The first downward arrow in Fig. B shows what appear to be three electrophoretic components in the first descending streak; but the four upward arrows marking the first ascending streaks show clearly the presence of four electrophoretic components which are labeled a_1, b_1, b_2 and c in the wider separation pattern of the first descending streaks. Like E. coli ML-35, the strain E. coli BE-30 gives rise to a single streak, but this streak (c) is clearly wider and more diffuse than the sharp streak (a) corresponding to E. coli ML-35. Proteus vulgaris, however, splits into two streaks b_1 and b_2.

Additional examples of cell separations in the Endless Belt apparatus can be found elsewhere[8].

STREAK DISTORTION DUE TO MISMATCHING OF VELOCITY PROFILES

The theoretical considerations of this and the following sections apply for the most part equally well to endless belt electrophoresis and to "free flow" fluid curtain electrophoresis. For simplicity we will consider flow between two vertical parallel closely spaced plates which could represent the fluid curtain or the vertical sections of the noncircular annulus of an endless fluid belt.

In the previous treatment we assumed the fluid velocity not to vary in the fluid ribbon. This is obviously incorrect since it is well known that under the conditions prevailing in the fluid curtain and the endless fluid belt the flow will be laminar with a parabolic velocity distribution. Nevertheless our approximation was justifiable because, at the center of the fluid curtain the velocity has a maximum and varies very little as we deviate slightly in either direction from the mid-plane of the fluid ribbon. Our approximation is thus valid for very thin streaks injected typically from capillaries in the range of 0.1-0.2 mm in internal diameter. This leads to sharp streaks which do not diverge very much in their course toward the collector as seen in the preceding illustrations. However, streaks emanating from a wider injector will diverge and grow noticeably wider as is pointed out below.

Let us consider a vertical fluid curtain descending with an average velocity $\bar{u}$ downward between the adjacent vertical plates P_1 and P_2 of Fig. 25. The fluid enters through inlet holes IH and exits through openings EH. M_1 and M_2 are electrically conductive membranes which confine the fluid ribbon. IN is an injector capillary located in the mid-plane between the plates P_1 and P_2. We shall ascribe to this plane the coordinate z = 0, where z measures the distance of a point in the fluid ribbon from the mid-plane.

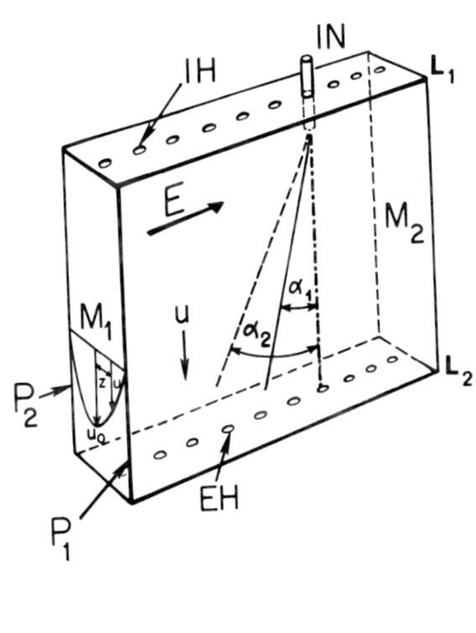

Fig. 25. Scheme of electrophoretic particle streak deviation in a flowing curtain. P_1, P_2: plates confining the fluid curtain. M_1, M_2: membranes; IN: inlet holes for vertical buffer flow; EH: efflux holes for vertical buffer flow; u: vertical buffer flow velocity; u_0: vertical fluid velocity at the curtain center; z: distance of a point from curtain mid-plane; α_1, α_2: angles of streak deviation from vertical path; L_1, L_2: two levels within the fluid curtain.

The vector $\vec{E}$ in Fig. 25 designates the direction of the electric field. Negative particles are carried by electrophoresis in the direction from right to left. As a result of the descent in the laminar flow and electromigration to the left, negatively charged particles in the central plane will be deflected from the vertical reference line of zero-deviation by the angle α_1. In a uniform velocity profile, all particles on the circumference of the injected streak will be deviated by the same angle so that the streak will remain well collimated when the electric field is applied. Actually however, the downward velocity profile is parabolic. A particle in the mid-plane is carried faster toward the exit holes EH than a particle near the wall. The latter will therefore dwell longer in the electric field before exiting and will be deviated by a larger angle α_2. Such a particle would be, for instance, the one located on the circumference of the streak nearest the reader (or its antipod). There will thus be an angle $(\alpha_2-\alpha_1)$ between the central minimum-deviation path and the path of the particles located farthest from the mid-plane. As a result the streak will appear to broaden in side view. We shall designate the divergence between the paths of the particles of fastest descent in the mid-plane and those of slowest rate of descent moving closer to the walls P_1, P_2 as "parabolic divergence" since it is due to the superposition of the uniform electrophoretic velocity with the parabolic profile of the downward laminar flow. We shall designate as "streak collimation" the reduction of the angle of divergence $(\alpha_2-\alpha_1)$ to zero. This would result in a streak in whose interior all particles would be deviated by the same angle.

Parabolic divergence is due to a mismatching between the velocity profiles of the vertical and horizontal motions of the particles in the streak. For a streak of initially circular cross-section, the latter will be distorted into a crescent in the course of electromigration as shown by the streak cross-sections C_1 and C_2 in Fig. 26. The levels L_1 and L_2 in the flow of Fig. 26 correspond to the highest and lowest positions of the particles emerging from the injector IN of Fig. 25. We have seen in Fig. 14 a photograph of such a crescent distortion of the streak cross-section.

The streak distortion is not harmful in measurements of the electrophoretic mobility. By measuring the displacement of the apex of the crescent one confines one's attention to particles in the central plane. This is equivalent to the use of an infinitely thin injector. The actual thickness of the streak can thus be disregarded in such measurements. In analytical observations, however, the apex of one crescent (C_3 in Fig. 26) can be hidden in the niche of an adjacent crescent (C_2 of Fig. 26) so that the side view may not reveal the presence of two separated components but, rather, create the impression of one wide diverging streak. In preparative collections the crescent distortion can be equally damaging. If we imagine a collector slot perpendicular to the plates P_1, P_2 of Fig. 25 to be located in the plane L_2 of Fig. 26 so as to intercept particles in the apex of crescent C_3, this exit slot will also collect particles located in the streak of the adjacent component by intercepting the contents of the "horns" of the crescent C_2. The removal of the streak distortion is thus a crucial objective of the quest for increased resolution in streak deflection electrophoresis. We shall see below how this can be achieved.

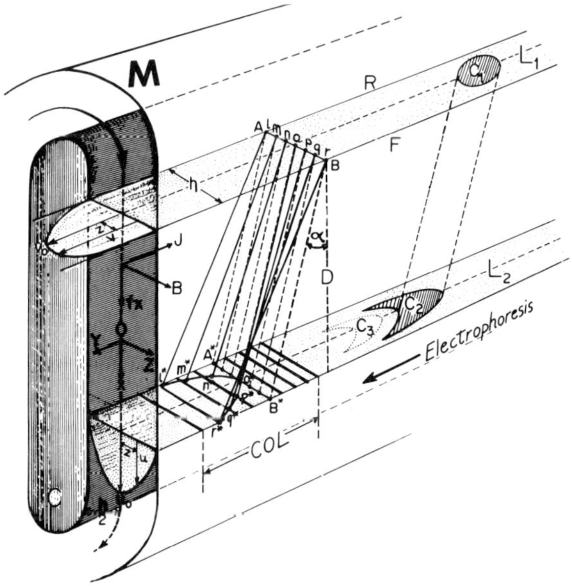

Fig. 26. Distortion of streak cross-section in fluid belt and curtain electrophoresis. C: iron core of fluid belt apparatus; z: distance of a point from center of annulus; v: horizontal flow velocity (v_o at the annulus center); u: vertical flow velocity (u_o at the annulus center); h: width of annulus; L_1, L_2: two levels within the vertical section of the endless fluid belt; X, Y, Z: coordinate axes; f_x: vertical force density; B: magnetic field vector; J: electric current density vector. COL: collector comprising slots parallel to A and B. AB: slot serving as an injection inlet for a ribbon-shaped sample streak. A^*B^*: the collector slot through which this injected sample would pass in the absence of streak distortion; l, m, n, o, p, q, r: particles along the line AB on level TL; l^*, m^*, n^*, o^*, p^*, q^*, r^*: location of these particles when they reach the level L_2 in the presence of streak distortion. D: path of undeviated particles; i: angle of deviation of particle path; C_1: cross-section of circular streak injected at level L_1. C_2: "crescent" into which the streak cross-section C_1 is distorted when particles originating in C_1 at level L_1 reach the level L_2. C_3: the "crescent" corresponding to a faster electrophoretic component of the level L_2. R, F: rear and front walls of the annulus confining the shown portion of the endless fluid belt (from A. Kolin, in Methods of Cell Separation, vol. 2, N. Catsimpoolas, ed., Plenum Publ. Co., New York, 1979).

AVOIDANCE OF STREAK DISTORTION BY MODIFICATION OF PARTICLE VELOCITY PROFILE BY A FORCE FIELD

We shall refer to Fig. 26 which illustrates horizontal and vertical velocity profiles in the fluid ribbon enclosed between the core C and the mantle M as well as between the horizontal levels L_1 and L_2. We can consider this fluid segment as being part of a vertical flowing curtain or the vertical front section of an endless fluid belt. The thickness of the fluid ribbon is h and the distance of the vertical walls from the mid-plane of the fluid belt is h/2. The magnetic field $\vec{B}$ is directed along the Z-axis, the curtain flow is parallel to the X-axis (proceeding with velocity u) and lateral (axial) streaming proceeds in the Y-direction with velocity v.

If the velocity of flow were uniform for the vertical flow, particles injected through a slot AB on level L_1 migrating electrophoretically in the Y-direction would be deviated by the same angle α to the left and would pass through a slot A^*B^* on the L_2 level. The same would happen, however, not only for a superposition of two mutually perpendicular uniform velocities. If, for example, the vertical and horizontal velocity distributions were are parabolic we would write for vertical velocity

$$u = u_o \left|1 - 4\left(\frac{z}{h}\right)^2\right| \tag{4}$$

and similarly for the horizontal velocity

$$v = v_o \left|1 - 4\left(\frac{z}{h}\right)^2\right| . \tag{7}$$

The superposition of these velocities would then yield for the angle of deviation α of a particle moving at a distance z from the mid-plane

$$\tan \alpha = \frac{v}{u} = \frac{v_o\left|1 - 4\left(\frac{z}{h}\right)^2\right|}{u_o\left|1 - 4\left(\frac{z}{h}\right)^2\right|} = \frac{v_o}{u_o} = \text{const.} \tag{8}$$

The terms dependent on z cancel out and all particles in the streak are deviated by the same angle. Thus, superposition of a lateral laminar flow in the endless belt apparatus upon the orbital endless belt velocity does not affect the sharpness of the streaks. In general, the particles in the streak remain sharply collimated as long as the two superimposed orthogonal velocity distributions (comp. Fig. 26) $u = f(z)$ and $v = F(z)$ are identical or have a constant ratio. Thus in order to avoid crescent distortion of the streak (i.e. parabolic divergence) we must find a way to modify one or both of the mutually perpendicular velocity profiles to fulfill the above condition for streak collimation.

Actually this condition does not prevail, as a rule, in endless belt electrophoresis, where the horizontal velocity of the particles is a result of superposition of the parabolic profile of the axial flow with the uniform electrophoretic velocity. Fig. 27 helps to visualize a few typical velocity profiles resulting from such superposition. F and R represent in top view the front and rear plates confining the fluid curtain which we imagine as flowing at right angles to the page away from the reader. The axial flow is in the Y-direction to the left. z measures the distance of a point in the fluid from the mid-plane. W represents a slit extending almost from wall to wall which acts as our injector through which a band of particles enters the curtain. We will temporarily disregard the particle motion along the X-axis (i.e. at right angles to the page).

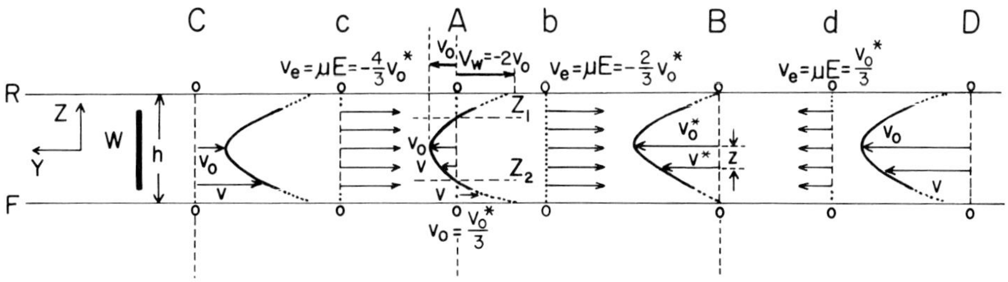

Fig. 27. Superposition of parabolic and uniform velocity distributions. A: electro-osmotic velocity profile; B: laminar flow velocity profile; C: velocity profile with central minimum velocity; D: velocity profile with central maximum velocity.

We begin with a row of particles along the line o-o shown under B. After one second they will be displaced to the left by the lateral laminar flow so that they will be located as shown under B along the parabola which thus depicts the particles' velocity distribution (v_o^* being the maximum velocity). We can now modify this velocity profile in various ways by imposing a uniform velocity v_e upon the particles by electrophoresis. The uniform velocity shown under d is $v_e = \frac{v_o^*}{3}$. The original velocity distribution is then altered to the velocity distribution seen under D by the superposition of this velocity v_e upon the velocity profile seen under B. If we impose electromigration with velocity - 4/3 v_o^* (shown under c) opposite to the direction of the laminar flow depicted under B, we obtain a velocity profile shown under C with a minimum velocity v_o at the center. The modification of greatest interest to us is the one shown under A. It is obtained by superimposing the uniform electrophoretic velocity $v_e = -2/3\ v_o^*$ opposite to the flow direction upon the parabolic laminar flow profile seen under B. The result is a velocity distribution in which the flow direction at the center is opposite to that near the walls R and F. There are two planes z_1 and z_2 parallel to the walls in which the fluid velocity is zero. The velocity distribution is parabolic as the original laminar profile under B but now the velocity does not vanish at the wall as shown under A.

We see thus that it is possible to modify the velocity distribution of particles introduced in a streak into a flowing curtain by imposing a force field, such as an electric field, which causes electrophoretic migration. We shall see below how such modification of the velocity profile of the horizontal motion of the particles can be used to match this velocity distribution to that of the transverse vertical curtain flow so as to satisfy the criterion for streak collimation (tan α = const. in equ. (8) for all values of z).

Restricted streak collimation by electro-osmotic convection

When we establish a horizontal electric field in a fluid curtain (Fig. 25) or endless fluid belt (Fig. 26), a horizontal fluid convection is generated by interaction of this electric field with the ionic double layer adjacent to the walls which confine the fluid belt. The fluid is set in motion parallel to the electric field near the walls and an oppositely directed fluid motion in the central region of the fluid belt is engendered so that the rate of horizontal net fluid transfer is zero.

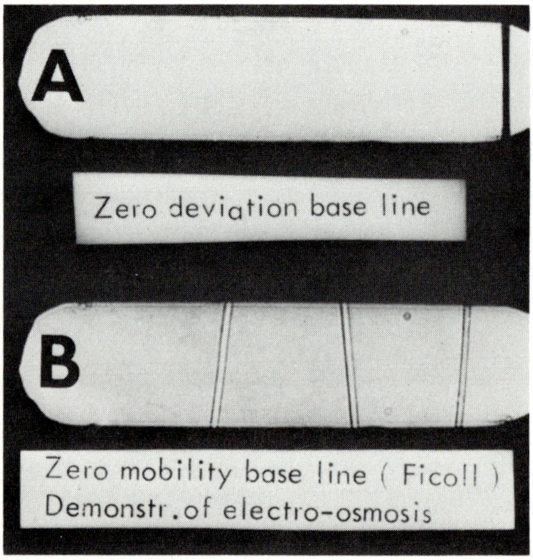

Fig. 28. Demonstration of electro-osmotic streaming and establishment of the zero-mobility baseline. (A): zero deviation baseline (wire passed through injector); (B): the first descending, first ascending and second descending streaks of a zero-mobility component deviated by electro-osmotic streaming (from A. Kolin, in Methods of Cell Separation, vol. 2, N. Catsimpoolas, ed., Plenum Publ. Co., New York, 1979).

The electro-osmotic convection can be demonstrated by injecting into the central plane of the fluid curtain a streak of zero electrophoretic mobility. The neutral yellow dye Apollon is suitable for this purpose. We used, however, Ficoll for better photographic results in Fig. 28. Fig. A shows the path of zero deviation obtained by pushing a straight wire through the injector capillary. Then, the electric current is turned on and a streak of a material of zero or negligible electrophoretic mobility is injected into the non-circular endless fluid belt. It is deviated from the path seen in Fig. 28A and moves in a helical path giving rise (from right to left) to the first descending, first ascending and second descending streaks seen in Fig. 28B. The pitch of this helix is determined by the ratio of the central velocity of the electro-osmotic flow to the maximum velocity of the laminar fluid curtain flow. This helical path can serve as the zero-mobility base-line in determinations of electrophoretic mobilities as discussed below.

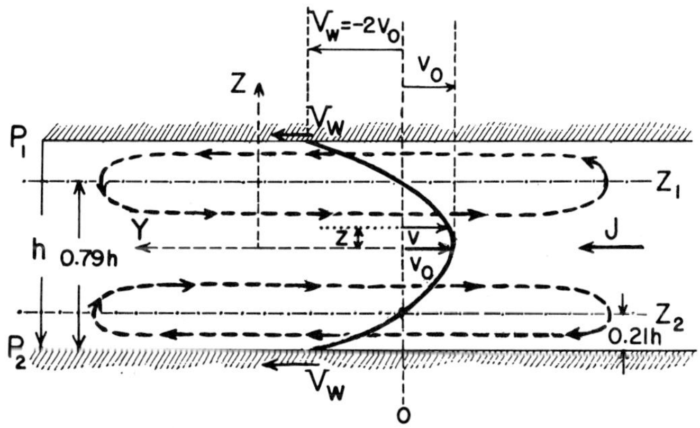

Fig. 29. Pattern of electro-osmotic convection. Parabolic distribution with non-zero velocity at the wall $v_w = -2v_o$ and central velocity v_o opposite to the wall velocity v_w. Z_1, Z_2: Smoluchowski zones of zero velocity. The convection pattern is indicated by arrowed dashed lines (from Kolin, in Methods of Cell Separation, vol. 2, N. Catsimpoolas, ed., Plenum Publ. Co., New York, 197).[9]

Fig. 29 illustrates the electro-osmotic convection pattern[8]. The velocity distribution is obtained by integrating twice the differential equation

$$f_y = \eta \frac{\partial^2 v}{\partial z^2} \qquad (9)$$

making use of the condition that the velocity assumes the value of v_w at the curtain walls for evaluation of the constants of integration. (f_y is the horizontal force density and η the fluid viscosity.) The resulting electro-osmotic velocity distribution is[3,20,21,22]:

$$v = \frac{f_y}{2\eta} \left| z^2 - \left(\frac{h}{2}\right)^2 \right| + v_w . \qquad (10)$$

It is a parabolic velocity profile in which, unlike the laminar flow profile, the velocity does not vanish at the walls. The wall velocity is twice the central velocity and points in opposite direction. Smoluchowski[22] has shown that there are two planes Z_1 and Z_2 at a distance of 0.21 h between the walls in which the velocity is zero. The resulting streaming pattern is a double circulation illustrated by the dashed lines in Fig. 29.

It is interesting to note that the velocity distribution of Fig. 29 is identical with the one shown in Fig. 27 under A, except that the flows are in opposite directions. We could now perform the same modification of the velocity profile in Fig. 29. Let us assume that we start with a row of ions deposited from an injector slit along the dashed line passing through 0. After a second, the electro-osmotic flow will have distorted this row into the parabola seen in Fig. 29. If we superimpose upon the electro-osmotic motion an electrophoretic migration of these particles to the right with the velocity v_W, the particles' velocity at the walls relative the plates P_1, P_2 will become zero and the particles in the wide streak filling the entire width h of the curtain will move with a velocity profile identical with that of the vertical laminar curtain flow. Since the two orthogonal velocity distributions are described by the same parabolic function, the condition for particle collimation is fulfilled according to equ. (8), the angle α of streak deflection being independent of z.

Since only one of the particle species in an electrophoretic mixture may have a suitable mobility to make its electrophoretic velocity equal to the electro-osmotic wall velocity v_W, it will be the only particle whose streak would be collimated. A clever approach has been made to modify the effective value of the wall zeta potential to match the mobility of each individual component of a mixture (one at a time) in order to collect in succession collimated components of a mixed sample[23]. But this would not permit to collect simultaneously all of the components of a mixture under optimal conditions.

The condition for streak collimation by means of electro-osmotic streaming follows from equ. (2) and the equation which describes the horizontal particle velocity v^* resulting from electrophoretic and electro-osmotic motion[3,20]:

$$v^* = \frac{f_y}{2\eta} \left| z^2 - \left(\frac{h}{2}\right)^2 \right| + E(W - \mu) \, . \tag{11}$$

(E is the electric field intensity; μ, the electrophoretic mobility and W the electro-osmotic mobility, i.e. the streaming velocity v_W at the wall at unit electric field strength.)

The angle of streak deviation is given by[3,20]:

$$\tan \alpha = \frac{v^*}{u} = \frac{f_y \left| z^2 - \left(\frac{h}{2}\right)^2 \right|}{f_x \left| z^2 - \left(\frac{h}{2}\right)^2 \right|} + \frac{E(W - \mu)}{f_x \left| z^2 - \left(\frac{h}{2}\right)^2 \right|} \tag{12}$$

(f of equ. (2) was replaced by f_x).

The functions of z cancel out in the left hand term. The right hand term vanishes for $W = \mu$, i.e. for particles whose electrophoretic velocity is equal to the electro-osmotic wall velocity, and the angular streak deviation becomes independent of z[3,20]:

$$\tan \alpha^* = \frac{f_y}{f_x} . \tag{13}$$

For this particular particle species the injector can be made as wide as the fluid curtain[20].

Possibility of simultaneous collimation of all streaks by thermal convection:

Qualitative description. Even if it were possible to collimate with ease one preselected streak for optimal collection of only one electrophoretic component at a time (wasting the rest of the ingredients), simultaneous collimation of all components of an electrophoretic spectrum would be much more desirable even if it were possible only over one half or one third of the width of the fluid curtain. The physical principle of such a possibility is outlined below.

Electro-osmotic convection is normally considered as a major disturbance interfering with electrophoretic analysis. We have seen, however, in the preceding section how it can be used to suppress streak deterioration by parabolic divergeance. The success was, however, limited to one component in a mixed sample and required a clever cell design to vary the cell wall zeta potential for one streak at a time. We will now show how thermal convection, an even more serious potential disturbance, can be harnessed to permit simultaneous collimation of all electrophoretic components in a mixed sample without alteration of cell design[8,24,25].

We started our review of streak deviation electrophoresis with a greatly simplified assumption that flow in a fluid curtain proceeds with a uniform velocity throughout its width. In such a case the injector diameter may be as large as the thickness of the curtain since electrophoretic deviation of ions will be independent of the z coordinate. We now ask ourselves the question if there is a physical possibility to achieve in a viscous fluid a uniform velocity distribution which we would normally expect to find only in a superfluid of zero viscosity. The answer is yes with respect to the horizontal flow in our case and a partial affirmation with respect to the vertical flow.

Fig. 29 shows the parabolic velocity profile for electro-osmotic convection. The maximum central velocity is v_o to the right and the maximum velocity to the left $v_w = -2v_o$ occurs at the walls. The velocity is zero only in the planes Z_1 and Z_2 at a distance ± 0.21 h from the walls. Let us now superimpose a laminar flow to the left with the central velocity $v_o = -3v_o$ upon this flow pattern. The superposition of these velocity distributions will result in a uniform horizontal flow to the left with the velocity $-2v_o$ at all values of the z coordinate. If we could obtain a similar rectification of the vertical velocity profile of the laminar curtain flow, we would have the ideal flow conditions we set out to create. This can be achieved with a useful degree of approximation over approximately half the width of the fluid curtain.

Let us imagine the fluid sheet trapped between the plates P_1 and P_2 in Fig. 29 rotated 90° so that the maximum of the velocity profile points upward. We replace the vector $\vec{J}$ by the gravitational field vector $\vec{g}$. We assume the fluid to be heated uniformly at a constant rate. The heat flows horizontally toward the plates P_1, P_2 which are maintained at a constant temperature and act as heat sinks. As a result of the horizontal temperature gradient thus engendered and the concomitant density gradient, thermal convection is generated which is qualitatively represented by the dashed lines. This streaming pattern resembles the one of Fig. 1 illustrating thermal convection. It would have been ideal if the thermal convection pattern were identical with that of electro-osmotic convection seen in Fig. 29 turned 90° because then it could be rectified by a downward laminar flow. Unfortunately, this is not the case because the fluid velocity vanishes at the walls in thermal convection. It remains to be seen if mathematical analysis will result in a velocity distribution for thermal convection whose superposition upon a laminar flow would yield a reasonably flat velocity profile over an appreciable portion of the curtain width h.

Mathematical analysis. The analysis of thermal convection in a uniformly heated fluid between two parallel vertical plates has been carried out under the simplifying assumptions that the distance between the plates is negligible as compared to their horizontal and vertical dimensions and that convection near the ends of the plates can be ignored when considering the flow near the central region[24,25]. Due to absence of horizontal flow in this region, there is no horizontal convective heat transfer and the temperature distribution follows from the differential heat equation:

$$\nabla^2 T = -\frac{q}{k} \,. \tag{14}$$

By integration and evaluation of the constants of integration[24] we obtain a parabolic temperature distribution:

$$T = T_w + \frac{q}{8k}(h^2 - 4z^2) \,. \tag{15}$$

The density as a function of temperature is

$$\rho = \rho_o[1 - \alpha\,(T - T_w)] \,. \tag{16}$$

The equation of motion is

$$\left(\frac{\partial}{\partial t} + \vec{u}\nabla\right)\vec{u} = -\frac{\nabla p}{\rho_o} - \vec{g}\alpha\,(T - T_w) + \eta\nabla^2\vec{u} \,. \tag{17}$$

With the assumption of steady-state flow, $\frac{dp}{dx}$ = const., $\vec{u}$ = u(z) and neglecting the temperature-dependence of viscosity η, the solution of equation (17) becomes*:

$$u(z) = \frac{g\alpha q}{80k\eta}\{h^2z^2 - \frac{10}{3}z^4 - \frac{h^4}{24}\} . \tag{18}$$

(The calculation is reproduced in greater detail elsewhere[24]).

Discussion of the solution

Like in the case of the electro-osmotic convection (Fig. 29) there are in the thermal convection pattern two planes of zero velocity (analogous to z_1 and z_2 in Fig. 29). While the coordinates of z_1 and z_2 are $h/(2\sqrt{3})$ and $-h/(2\sqrt{3})$ respectively, the corresponding coordinates for the stationary zones in thermal convection are $h/(2\sqrt{5})$ and $-h/(2\sqrt{5})$, i.e. the latter zones are closer to the central plane.

Unlike the velocity profile of electro-osmotic convection, the thermal convection velocity vanishes at the walls. The upward velocity maximum is at the center and the downward velocity maxima are located at $z = (h/2)\sqrt{\frac{3}{5}}$ and $-(h/2)\sqrt{\frac{3}{5}}$. The ratio between the maximum downward velocity u_d and maximum upward velocity u_o is $u_d/u_o = -5/6$. The magnitude of the maximum upward velocity at z = 0 is

$$|u_o| = \frac{g\alpha q}{80k\eta}\left(\frac{h^4}{24}\right) . \tag{19}$$

Rectification of the vertical velocity profile by superposition of thermal convection

We can obtain a velocity distribution of reduced curvature in the central region by superimposing a downward laminar flow given by equation (4) (curve w(z) in Fig. 30) upon the thermal convection described by equation (18) (curve u(z) of Fig. 30). The required maximum velocity for the laminar flow must be chosen so as to minimize the curvature-determining z^2-term of the equation R(z) = u(z) + w(z). Neglecting the z^4-term because of its smallness in the central region, one obtains (24) for the rectified velocity profile of the downward flow

$$R(z) = u(z) + w(z) = \frac{g\alpha q}{24\eta k}\left|\left(\frac{h}{2}\right)^4 - z^4\right| . \tag{20}$$

This velocity distribution is shown in Fig. 30 (solid curve). One sees that it is nearly uniform over a wide range of the central region. The value of the central velocity of the laminar downward flow which accomplishes this rectification of the velocity profile is

$$w_o^* = \frac{g\alpha q h^4}{320\eta k} . \tag{21}$$

* I am indebted for this derivation of the velocity distribution in thermal convection to Prof. E. Busse of the UCLA Institute of Geophysics and Planetary Physics.

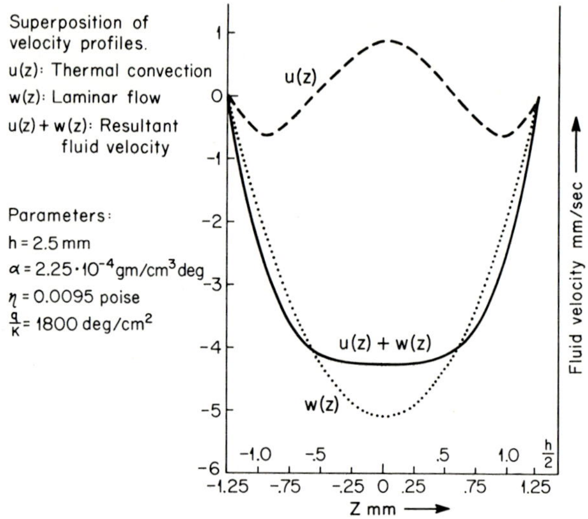

Fig. 30. Central rectification of the vertical velocity profile by superposition of a laminar flow (dotted curve w(z)) with a thermal convection velocity distribution (dashed curve u(z)). The resulting solid curve R(z) = u(z) + w(z) is flat over an appreciable fraction of the width of the annulus where the velocity profile can be assumed to be uniform with a good degree of approximation. The central temperature was assumed to be 22°C in this diagram (from A. Kolin and Ellerbroek, to be published in 1978).

The angular divergence of the streak in a rectified velocity profile shown in Fig. 30 (curve R(z) = w(z)) can be calculated from equation (30)[24]. It will depend, of course, on the width of the streak. The calculation for an h = 2.5 mm annulus width shows that the streak width may be as great as h/3 if we accept a streak divergence of about 1% (measured by the angle of divergence). At a streak width of 0.45 h the divergence becomes 5%[24]. Thus, increasing the injector width from d = 0.2 mm to about 1.1 mm in a 2.5 mm annulus, the throughput would increase by $(1.1/0.2)^2$ = 30 while the streak divergence would remain below 5% in a rectified flow. In a parabolic curtain flow the angular divergence for the same injector diameter (1.1 mm) would have been 25%.

In analytical work rectification of the velocity profile can be of equal importance by increasing resolution through diminution of streak divergence at a small injector width. Thus, for an electrophoretically homogeneous component of streak width h/10, the calculation yields a streak divergence of 1% in parabolic curtain flow and of 0.01% in rectified flow[24], which represents a hundredfold increase in resolution.

Applicability of the collimation principles to the fluid belt and flowing curtain

Streak collimation by matching one electrophoretic mobility to the cell wall zeta potential can be achieved with the "endless belt" system as well as the "free flow" curtain system. This collimation principle is based on making both the vertical and horizontal velocity profiles parabolic by superimposing electro-osmosis upon electrophoresis. The modification of the effective cell wall zeta potential[23] will be probably much easier to achieve in the fluid curtain system.

On the other hand, collimation by rectification of the vertical velocity profile by superposition of thermal convection and of the horizontal one by superposition of laminar flow and electro-osmosis can only be done in the "endless belt" system and not in the "free flow" curtain system because the flow channel is sealed off by membranes in the latter and does not permit maintenance of a horizontal laminar flow.

Collimation by thermal convection is, in principle, applicable to the normal mode of multi-turn operation of the "endless belt" system as long as an odd number of half-turns is used because the thermal convection collimation effects will cancel out in the ascending and descending paths of the endless fluid belt. However, this mode of operation is not favorable because the streak deterioration after several turns will have to be corrected in one single half-turn. It would be more favorable to use only the first descending half-turn and to make up for the loss in travel time and distance by making the height of the endless belt much greater. This mode of operation would only superficially resemble the "free-flow" curtain system. It would differ from it radically by relying on electromagnetic fluid propulsion rather than on a pressure gradient. There is thus no need for membranes and a horizontal laminar flow can be maintained.

Multi-turn operation with collimation is nevertheless possible by maintaining the fluid in the descending half of the endless belt above $4^{o}C$ and in the ascending half below the temperature of the density maximum ($4^{o}C$). In this case the central thermal convection velocity u_o will point <u>upward</u> in the descending half-turn (opposite to the direction of laminar flow) and <u>downward</u> in the ascending half-turn of the endless belt (again opposite to the direction of laminar flow). Collimation by thermal convection could thus occur throughout the full revolution of the endless belt.

RESOLVING POWER

The broadening of streaks in laminar curtain flow due to mismatching of velocity profiles sets a limit to the resolution of macro-molecular and particulate components where diffusion plays no role. This is illustrated in Fig. 26. We shall consider two adjacent components of mobilities μ_1 and μ_2 as resolved when the angular deviation of the trailing edge of the faster component streak surpasses the angular deviation of the leading edge of the slower component streak. We define as resolving power R the ratio of the mean mobility $\bar{\mu}$ of two adjacent components and their mobility difference $\Delta\mu$[11]:

$$R = \frac{\bar{\mu}}{\Delta\mu} \tag{22}$$

For an endless fluid belt or a flowing curtain apparatus, the resolving power can be expressed in terms of the instrumental parameters: curtain width h and internal injector diameter d (assuming the streak diameter to be equal to the latter[11]:

$$R = \frac{\bar{\mu}}{\Delta\mu} = \left(\frac{h}{d}\right)^2 . \tag{23}$$

We see that in the absence of velocity profile matching or rectification, the resolution deteriorates quadratically as the streak diameter d increases. On the other hand, one can obtain very sharp streaks and thus obtain high resolution by choosing a very small injector diameter. This is achieved, of course, at a sacrifice in throughput. Increase in the scale of separation can only be achieved at a sacrifice in resolution. The importance of the idea of rectification of the horizontal and vertical velocity profiles lies in that it permits a great increase in preparative throughput essentially without a sacrifice in resolving power.

MEASUREMENT OF ELECTROPHORETIC MOBILITIES

Streak broadening (crescent distortion) does not impair the measurement of electrophoretic mobilities in the endless belt apparatus. By measuring the deviation of the apex of the crescent, one measures in effect the electrophoretic displacement of a centrally located infinitely thin streak.

It is possible to make absolute as well as relative mobility measurements in the endless belt. In both cases, it is essential to avoid errors due to electro-osmosis. It has been shown in Fig. 28 how a zero-mobility base line can be established by observation of electro-osmotic displacement of a streak of neutral particles. The displacement of a streak of charged particles against this reference streak is a measure of electrophoretic migration.

In <u>absolute</u> mobility measurements we must know the electric field intensity in the annulus. We can determine it by measuring the voltage between two points, of known distance, such as the syringe needle S of Fig. 19 and a wire pushed into the annulus through the opening of the injector IN. The ratio of the voltage and distance gives us the electric field intensity dV/dy in V/cm.

The revolution time in the annulus can be measured by pulse-wise injection of dye. If we now inject the particle species whose mobility is to be measured together with a neutral component, such as the dye "Apollon" and measure the separation of the two streaks after one revolution, the ratio of this distance and the revolution time gives us the absolute velocity of the charged particle species. Dividing this velocity by the measured electric field intensity gives us the absolute mobility.

It is more convenient to make <u>relative</u> mobility measurements with reference to a chosen mobility standard. We have used mainly the diazo dye "brilliant blue" as a mobility reference standard. Its ionic charge remains constant in the range between pH 3.3 and pH 9.3. Its mobility has been established as a function of temperature[15] and it is thus possible to use this dye not only as a standard for mobility measurements but also as a "thermometer" to measure the temperature of the buffer in the annulus in the course of mobility measurements. The electrophoretic mobilities are thus determined at a known temperature[8,13,15].

The relative mobility measurement of an electrophoretic component simply involves simultaneous injection of this component in a streak with a neutral dye (e.g. "Apollon") and with a reference dye (e.g. brilliant blue). The separated streak triplet is photographed after n helical turns. The ratio between the distance of the streak of unknown mobility from the neutral streak and the distance of brilliant blue from the neutral reference streak gives us the relative mobility of the component of interest.

CONCLUSION

The orbital motion of particles about a horizontal axis in the endless fluid belt exposes them in effect to a rotating gravitational field in which unidirectional sedimentation is suppressed. For the same reason formation of small thermal convection cells is inhibited. Like "free flow" curtain electrophoresis (F.F.E.), "endless belt" electrophoresis (E.B.E.) is a method of streak deflection electrophoresis. The two methods differ in that fluid propulsion in E.B.E. is accomplished magneto-hydrodynamically without establishing a pressure gradient, whereas such a gradient is required for fluid propulsion in F.F.E. As a result of this difference, the separation space must be enclosed by membranes in F.F.E. but not in E.B.E. The omission of the membranes in E.B.E. avoids shifts in pH and conductivity which occur in the separation space of F.F.E. but not in E.B.E. Omission of the membranes also permits maintenance of lateral flow parallel to the current in E.B.E. but not in F.F.E. This capability makes the E.B.E. system suitable for velocity profile rectification by electro-osmosis and thermal convection with the consequent possibility of multiple streak collimation for high preparative output at high resolving power.

ACKNOWLEDGEMENTS

Endless Fluid Belt Electrophoresis was developed originally under support from the Office of Naval Research. Subsequent improvements were supported by Medical Testing Systems, Inc. of Fountain Valley, California, and experimentation with biological materials by the American Cancer Society.

I am indebted to Professor F. Busse of the UCLA Institute of Geophysics and Planetary Physics for the derivation of the velocity distribution in thermal convection.

Thanks are also due to Proceedings of the National Academy of Sciences, Analytical Biochemistry, Journal of Chromatography, Elsevier Scientific Publications and Plenum

Publishing Company for permission to reproduce illustrations credited to them in the legends.

LIST OF SYMBOLS

Symbol		Units
J	electrical current density	A/cm^2
f	force density (force per unit volume)	$dynes/cm^3$
B	magnetic field intensity	gauss
η	viscosity	poise
u	vertical fluid velocity	cm/sec.
u_o	vertical fluid velocity at the center	cm/sec.
v	horizontal fluid velocity	cm/sec.
v_o	horizontal fluid velocity at the center	cm/sec.
w	vertical velocity in laminar flow	cm/sec.
$R(z)$	resultant velocity due to laminar flow and thermal convection	cm/sec.
z	distance from central plane of fluid belt	cm
h	width of fluid belt	cm
i	current	A
τ	revolution time	sec.
μ	electrophoretic mobility	$\frac{cm}{sec}/\frac{V}{cm}$
V	voltage	Volt
TU	Tiselius unit of electrophoretic mobility	$= 10^{-5}\ \frac{cm}{sec}/\frac{V}{cm}$
T	temperature	^oC
T_o	temperature at center of fluid belt	^oC
p	pressure	$dynes/cm^2$
g	gravitational field intensity	cm/sec^2
q	power generation per unit volume	$cal/sec/cm^3$
k	thermal conductivity of fluid	$cal/sec/cm/^oC$
t	time	sec
s	helical pitch	cm
σ	electrical conductivity	$(ohm/cm)^{-1}$
ρ	density	gm/cm^3
T_w	temperature of the vertical walls confining the fluid sheet	^oC
α	temperature coefficient of density	$(^oC)^{-1}$

REFERENCES

1. Hannig, K. (1964) Eine Neuentwicklung der trägerfreien kontinuierlichen Elektrophorese, Hoppe Seyler Z. f. Physiol. Chem., 338, 211-227.

2. Hannig, K., Wirth, H., Meyer, B.H. and Zeiller, K. (1975) Free flow electrophoresis I, Hoppe Zeylers Z. f. Physiol. Chem., 356, 1209-1223.

3. Kolin, A. (1979) Continuous flow deviation electrophoresis: fluid curtain ("free flow") and endless fluid belt apparatus, in Electrophoresis: a survey of techniques and applications, Deyl, Z. ed., Elsevier, Sc. Publications, Amsterdam.

4. Kolin, A. (1959) Suppression of thermal convection and of similar types of fluid streaming by an electromagnetic force field, Biochim. et Biophys. Acta, 32, 538-543.

5. Kolin, A. (1964) Kinematic stabilization of continuous-flow electrophoresis against thermal convection, Proc. Nat. Ac. Sci., 51, 1110-1117.

6. Kolin, A. (1954) Method for elimination of thermal convection, J. Appl. Physics, 25, 1442.

7. Hjertén, S. (1958) Free zone electrophoresis, Arkiv Kemi, 13, 151.

8. Kolin, A. (1979) Cell separation by endless fluid belt electrophoresis, in: Methods of cell separation, vol. 2, Catsimpoolas, N. ed., Plenum Publ. Co., New York.

9. Dryden, H.L., Murnaghan, F.D. and Bateman, M. (1956) Hydrodynamics, Dover Publ. Co., New York, p. 184.

10. Catsimpoolas, N., Kolin, A., Hjertén, S. and Porath, J. (1976) Tiselius unit of electrophoretic mobility, Nature, 259, 264.

11. Kolin, A. (1967) Preparative electrophoresis in liquid columns stabilized by electromagnetic rotation. A: The apparatus; B: Artifacts, stability and resolution, J. Chromatogr., 26, 164-179 + 180-193.

12. Kolin, A. (1966) Helical path electrophoresis in vertical fluid sheaths, Proc. Nat. Ac. Sc., 56, 1051-1058.

13. Kolin, A. and Luner, S.J. (1969) Continuous electrophoresis in fluid endless belts, Analyt. Biochem., 30, 111-131.

14. Bergrahm, B. (1967) Apparatus for zone electrophoresis in vertical column, Science Tools, 14, 34.

15. Luner, S.J. (1969) Electrophoretic studies of isolated mammalian chromosomes, Thesis, University of California, Los Angeles.

16. Kolin, A. (1965) Magnification of resolving power of collectors in free flow electrophoresis, J. Chromatogr., 17, 532-537.

17. Zeiller, K., Löser, R., Pascher, G. and Hannig, K. (1975) Free flow electrophoresis. II. Analysis of the method with respect to preparative cell separation, Z. Physiol. Chem., 356, 1225-1244.

18. Kolin, A. (1971) Fractionation and characterization of cells and microbiological particles by means of endless fluid belt electrophoresis, Proc. First International Biophysics Congress, Baden, Austria, pp. 481-485.

19. Owen, R.M. (1972) Microbiological evaluation of endless belt electrophoresis, Calif. State College Thesis.

20. Kolin, A. (1960) Continuous electrophoretic fractionation stabilized by electromagnetic rotation, Proc. Nat. Ac. Sc., 46, 509-523.

21. Brinton, C.C. and Lauffer, M.A. (1959) Electrophoresis of viruses, bacteria and cells and the microscopic method of electrophoresis, in Electrophoresis, Vol. I, Bier, M. ed., Academic Press, New York, pp. 428-487.

22. Smoluchowski, M. (1921) in Handbuch der Elektrizität und Magnetismus, Vol. 2, Barth, Leipzig, p. 366.

23. Strickler, A. and Sacks, T. (1973) Focusing in continuous-flow electrophoresis systems by electrical control of effective cell wall zeta potentials, in Isoelectric focusing and isotachophoresis, Catsimpoolas, N. ed., New York Academy of Sciences, pp. 497-514.

24. Kolin, A. and Ellerbroek, B. (1978) Theory of Simultaneous Multiple Streak Collimation in Continuous-Flow Electrophoresis by Superposition of Electro-osmosis and Thermal Convection. (Submitted for publication.)

25. Ellerbroek, B. and Kolin, A. (1978) Thermal Convection in a Uniformly Heated Fluid Between Vertical Parallel Plates. (Submitted for publication.)

Electrokinetic Separation Methods,
P.G. Righetti, C.J. van Oss and J.W. Vanderhoff

ELECTROPHORESIS AND ISOELECTRIC FOCUSING IN FREE SOLUTION

RUDOLF QUAST
Herminenstr. 17 f, P.O. Box 13 14, D-3062 Bückeburg

In order to organize the large and heterogeneous amount of published data relating to electrophoresis and isoelectric focusing (IEF) in free solution, the different techniques will be arranged according to the technical principles involved and the type of apparatus used.

Related topics dealt with in other parts of this book, e. g. microelectrophoresis (chapter 6), electrophoresis in density gradient stabilized columns (chapter 12), or electrophoresis under zero gravity conditions (chapter 14), will not be treated here.

ELECTROPHORESIS IN FREE SOLUTION

Moving boundary electrophoresis in a U-tube

The first practical apparatus for free electrophoresis was realized by Tiselius[1] in his moving boundary electrophoresis which later became the reference technique for analytical electrophoresis. Here, the sample proteins created themselves the stabilizing density gradient in the U-tube during the electrophoretic run. Although other techniques for analytical electrophoresis, e. g. in gels, in membranes, in thin-layers of porous granules, and in density-gradients, have replaced the moving boundary in most analytical applications, the principle of the self-stabilizing density gradient of sample molecules gained more and more interest during the last two decades.

Electrophoresis in open troughs

In 1959, Kalous and Vacik[2] published details of an apparatus for preparative electrophoresis in an open, horizontal chamber, which may be regarded as the prototype of the horizontal, open-trough models later employed for isoelectric focusing (IEF). In this apparatus, a series of parallel equidistant ridges separated a horizontal trough into parallel trenches. The lid of the trough carried also a set of ridges dipping into the buffer solution in the trenches of the trough. In effect, the longitudinal section of the assembled chamber resembled in the direction of the electric field a meandering tube with horizontal axis and vertical bends. Recovering of separated sample molecules was facilitated by the fact that raising the lid lowered the level of buffer solution in the trenches below the ridges thus separating automatically fractions of solutes of different mobility. Resolution was limited by the number of trenches, however.

Electrophoresis in a rotating tube

A very different approach for the solution of the problem of stabilizing zones of sample molecules against gravitational mixing was chosen by Hjertén[3]. He designed an apparatus in which analytical electrophoresis took place in a horizontal rotating tube. Excellent resolution of various samples was obtained. Sedimentation of condensed sample zones is not prevented in this apparatus, but the effects of remixing and blurring of resolved bands were successfully circumvented by the rotation of the separation-chamber-tube around its length-axis. Sample molecules, cell particles or small volumes of concentrated solution move in small circles during the revolution of the tube, the circle being shifted continually in the direction of the electrophoretic migration (see Fig. 1). The resulting path of migration thus resembles a small spiral.

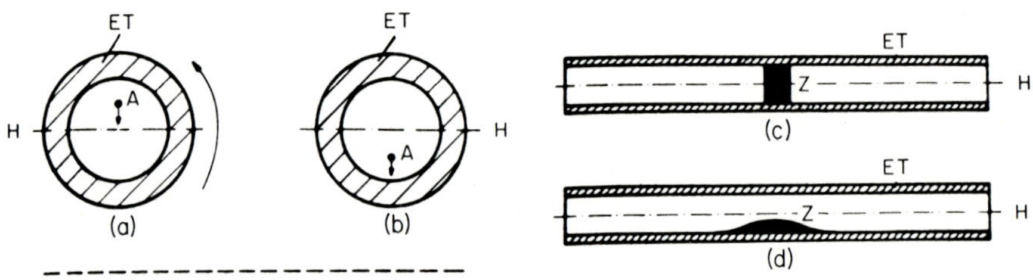

Fig. 1. Principle of zone stabilization in free zone electrophoresis. The horizontal electrophoresis tube (ET) is rotated around its long axis. H is the horizontal plane and A is a liquid element having a higher density than the surrounding buffer solution. If we assume that gravity and friction are the only forces acting on the liquid element, the direction of its movement must always be downward. Therefore, the liquid element moves away from the tube wall when it is in the upper half of the cross section (a) and toward the wall when in the lower half (b). According to this scheme, the liquid element is forced by rotation to move to-and-fro relative to the tube wall, which results in zone stabilization. If rotation is stopped, the sharp zone Z in (c) will spread rapidly, as shown in (d) (from Hjertén[3]; source: Elsevier Sci. Publ. Co.).

ISOELECTRIC FOCUSING IN FREE SOLUTION

IEF in multimembrane apparatus

Isoelectric focusing first was performed in instruments where several compartments were separated by semipermeable membranes (cf. Williams and Waterman[4]). Still, multi--compartment apparatus with dialysis-membranes draw considerable attention[5,6], although techniques employing density-gradients or gels for stabilization purpose are much more in use at present.

Soon it was realized, that sample molecules concentrated against a dialysis membrane sediment to the bottom of the chamber, if the content of the chamber is not sufficiently remixed by stirring. This effect was called electrodecantation.

Preparative Electrodecantation

An interesting application of electrodecantation of non-isoelectric contaminant proteins during purification of enzymes was described by Winchester *et al.*[7]. In short, a horizontal battery of dialysis cells was perfused from bottom to top by the sample solution. The pH of the solution was adjusted so that the enzymes of interest were at their respective isoelectric points. An electric field passing through the battery of cells was applied horizontally at a right angle to the stream of sample solution. All non-isoelectric molecules migrating in the electric field were pulled against one or the other dialysis membrane of a cell, eventually being concentrated there and decanted. Only the isoelectric enzyme molecules of interest could pass through the cell and were collected from the top.

IEF in open troughs

Valmet introduced his principle of zone convection in 1969[8,9]. Technically, the apparatus resembled that used by Kalous and Vacik[2] for electrophoresis in free solution. A series of parallel trenches of triangular cross-section constitutes the horizontal trough. Parallel ridges of the lid interlock with the trenches when the apparatus is assembled (Fig. 2). The volume of the solution is adjusted so, that the content of the trench is separated from that of neighbouring trenches as soon as the lid is raised, thus facilitating the recovery of separated components. The resolving capacity of the instrument was demonstrated by Valmet in experiments with albumin and gamma-globulin.

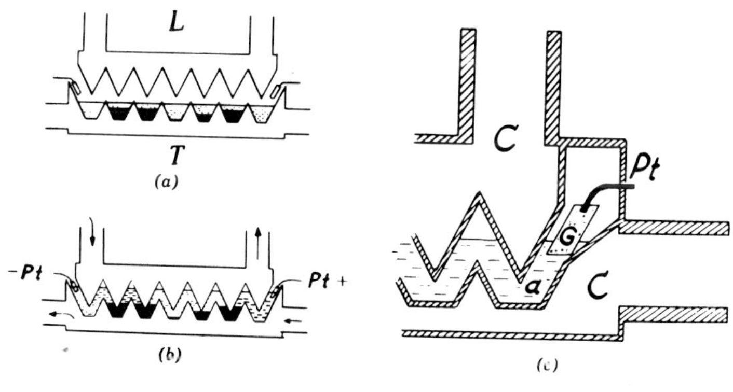

Fig. 2. Schematic drawing of the zone convection electrofocusing apparatus. (a): the lid (L) and the trough (T) before or after the experiment; (b): lid and trough put together, as during the experiment; (c): detail of one of the electrode compartments. The arrows indicate circulation of cooling water, Pt are the platinum electrodes, G is a sintered glass filter, a the ampholyte solution and C the cooling water (from Valmet[8]; source: LKB Produkter AB).

Later, Bodwell and Creed[10] reported on experiments with this technique in which a number of different proteins were used as samples in order to explore the specific difficulties of the zone convection technique.

A very similar design of instrument was used by Troizkii *et al.*[11] for the comparison of the microheterogeneity of human serum albumin found by IEF in a pH-gradient generated from Ampholine and by IEF in a pH-gradient generated from a 0,01 M boric--acid - borate-buffer pH 8,5 in 0,5 to 10 % solution of mannitol. The range of the pH-gradient from boric acid-borate in mannitol was from pH 6,0 (0,5 % mannitol) to pH 4,0 (10 % mannitol). The results of the experiments demonstrated nearly identical focusing patterns of human serum albumin in both types of pH-gradients used.

Other horizontal chambers used for IEF-experiments were described by Fawcett[12] and by Talbot and Caie[13]. The former instrument resembled the design of Kalous and Vacik[2], while the latter chamber was machined from a block of perspex and employed a different mechanism for separating the solution in the various trenches after the run (Fig. 3). Good results were obtained with that instrument in analyzing protein components of foot-and-mouth-disease virus[14].

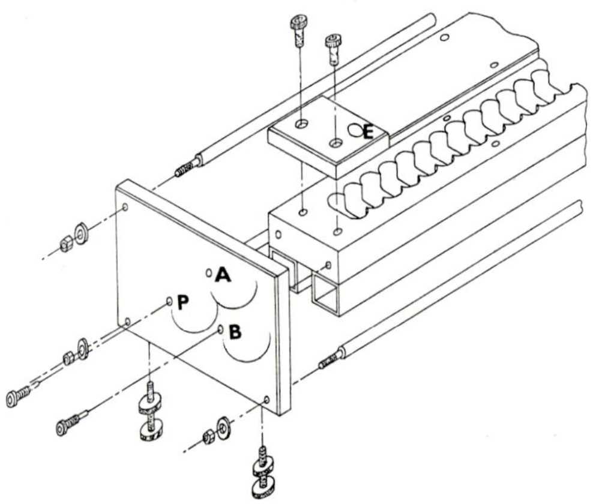

Fig. 3. Exploded diagram of part of the horizontal trough apparatus for isoelectric focusing. T: separation trough; L: lid; E: electrode locating plate; P: end plate; R: stainless steel tie rods; S: levelling screws; C: pivot pin; A and B: locking screws. During operation, the filled trough is titled 45°, so that the solution bridges the corrugations in the bottom. For autofraction, the trough is returned to horizontal position (from Talbot and Caie[13]; source: Butterworths).

IEF in closed U-tubes

U-tubes connected in series represented the principal feature of a closed apparatus for preparative IEF in free solution described by Rose and Harboe[15]. The authors called it 'sea-serpent shaped apparatus'. The separation chamber consisted of a glass tube bent in one plane in a meandering fashion. The glass tube was immersed in a cooling bath and on both ends connected to separate electrode compartments. Using human serum dialyzed against a diluted acetate buffer, separation was achieved in a run of 8 days duration.

Since the number of U-turns in such a design determines the resolving capacity of the apparatus, further improvement was achieved by Macko and Stegemann[16] in using a coil of polyethylene-tubing wound around a core of copper tubing for analytical IEF of potato tuber proteins. After the run they quickly immersed the polyethylene-coil in liquid nitrogen, cut the tubing into short pieces and thawed the fractions. The volume of solution to be separated was limited by the internal diameter of the polyethylene tubing of 4 mm. A similar design was used by Bours[17] for analytical IEF of lens-crystallins (Fig. 4).

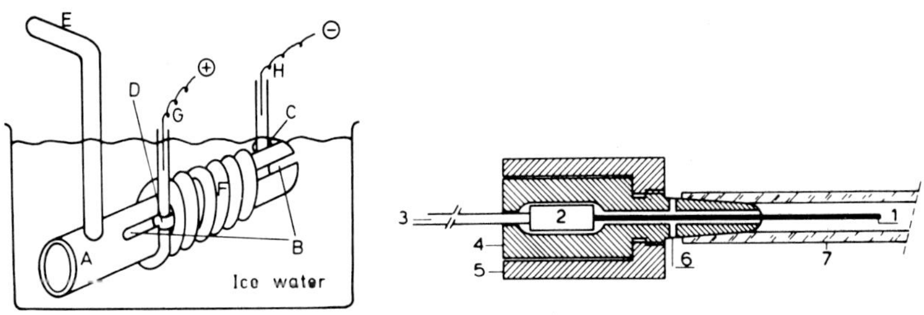

Fig. 4. Left: apparatus for free isoelectric focusing of proteins. A: brass tube (25 x 170 mm); B: slit (4 x 120 mm); C: fixed slot for tube F; D: movable slot for tube F; E: supporting bar; F: polyethylene tube (4 mm i.d., 6 mm o.d., length 1 m); G and H, platinum electrodes (from Chilla *et al.*[18]; source: Academic Press). Right: schematic drawing of platinum electrode used in free electrofocusing. 1: platinum wire; 2: soldered connection; 3: cable to power supply; 4: Perspex holder; 5: Perspex protective collar (18 x 25 mm); 6: gas outlet and 7: polyethylene tubing (from Bours;[17]; source: Academic Press).

Potentials of a horizontal coil of glass tube for free preparative IEF were explored by me when working in the department of hygiene at the university of Marburg[19-21]. The main difficulties with such instruments arise from:

a) the low field strength caused by the great length of the tube, e. g. about 250 cm for a 40 turn coil, resulting in a very long time of run, and

b) the formation of precipitates when using high concentrations of sample proteins.

While an increase in overall voltage applied to the coil might successfully reduce the time needed for establishing a pH-gradient and for subsequent separation of the sample molecules, at the same time the danger of local overheating increases.

IEF in open U-tubes

In contrast to the closed tube designs[15-17,19], an open coil design allows sampling of fractions after the run without destroying the tubing[16,17] and without excessive remixing of separated components[19] (Fig. 5).

Apparatus of the open tube design[20-22] tend to be sensitive to electroosmotic flow. The force of the electroosmotic flow depends, besides on fixed charges lining the interior of the glass tube and on the ionic strength of the solution, on the

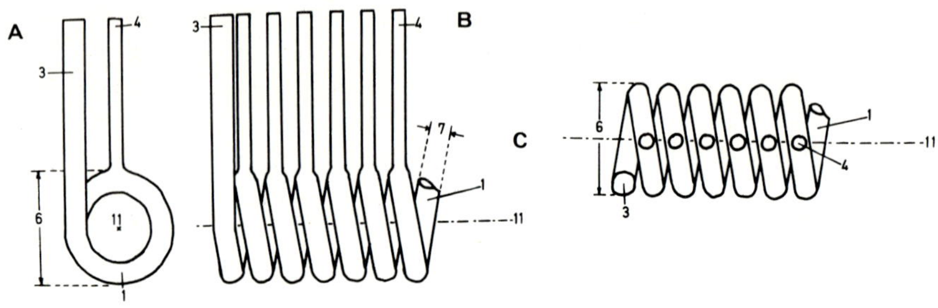

Fig. 5. Helical glass tube for preparative zone convection electrofocusing. A: end-on view of the glass helix; B: side-on view of the glass helix; C: top view of the glass helix (from Quast[20]; source: W. de Gruyter).

field strength. Hence with high field strengths a difference in electroosmotic force might occur in different turns caused by differences of conductivity in the various ranges of the pH-gradient forming - carrier ampholytes[25]. The differences do not cancel out as in columns[6] or in closed coil systems, but lead eventually to small differences in fluid level in the uppermost part of the turns. Hence the resistance increases in such parts and consequently local overheating may result.

Denckla[22] pointed out the importance of preventing coacervate formation in experiments with free preparative IEF. He used series of large volume U-tubes and high-voltage power supply in order to purify pituitary hormone preparations. He also discussed important practical aspects for the application of systems for IEF in free solution like influence of the Helmholtz-layer in U-tube-instruments of various dimensions, drift of pH-gradient, and various methods for the preparation of samples for IEF in order to prevent coacervate formation.

Conclusions

When considering advantages and draw-backs of IEF in free solution it becomes clear that, although various approaches have tested the feasibility of the experimental design, the optimum conditions have not yet been found. Further experiments using relatively high ionic strength buffer systems[24], may be of the boric-acid borate mannite type[11], and taking care of coacervate forming conditions, will eventually lead to an increasingly wider acceptance of preparative IEF in free solution.

REFERENCES

1. Tiselius, A. (1930), Nova Acta Regiae Soc. Sci. Upsal., 7, 4-14.
2. Kalous, V. and Vacik, J. (1959) Chem. listy, 53, 35-37.
3. Hjertén, S. (1967) Chromatogr. Rev., 9, 122-219.
4. Williams, R.R. and Waterman, R.E. (1929) Proc. Soc. Exptl. Biol. Med., 27, 56-59.
5. Rilbe, H. (1970) in Protides of the Biological Fluids (Peeters, H., ed.) Vol. 17, pp. 369-382, Pergamon Press, Oxford.

6. Rilbe, H. (1977) in Electrofocusing and Isotachophoresis (Radola, B.J. and Graesslin, D. eds.) pp. 35-50, de Gruyter, Berlin.

7. Winchester, B.G., Caffrey, M. and Robinson, D. (1971) Biochem. J., 121, 161-168.

8. Valmet, E. (1969) Sci. Tools, 16, 8-13.

9. Valmet, E. (1970) in Protides of the Biological Fluids (Peeters, H. ed.) Vol. 17, pp. 401-407, Pergamon Press, Oxford.

10. Bodwell, C.E. and Creed, G.J. (1973) Abstr. 9th Internatl. Congr. Biochem., Stockholm, p. 25.

11. Troizkii, G.W., Aschizkii, G.J. and Malji, K.D. (1975) Lab. delo, 12, 708-712.

12. Fawcett, J.S. (1975) in Isoelectric Focusing (Arbuthnott, J.P. and Beeley, J.A. eds.) pp. 23-43, Butterworths, London.

13. Talbot, P. and Caie, I.S. (1975) in Isoelectric Focusing (Arbuthnott, J.P. and Beeley, J.A., eds.) pp. 74-77, Butterworths, London.

14. Talbot, P. (1975) in Isoelectric Focusing (Arbuthnott, J.P. and Beeley, J.A., eds.) pp. 270-274, Butterworths, London.

15. Rose, C. and Harboe, N.M.G. (1970) in Protides of the Biological Fluids (Peeters, He., ed.) Vol. 17, pp. 397-400, Pergamon Press, Oxford.

16. Macko, V. and Stegemann, H. (1970) Anal. Biochem., 37, 186-190.

17. Bours, J. (1973) Exp. Eye. Res. 16, 501-515.

18. Chilla, R., Doering, K.M., Domagk, G.F. and Rippa, M. (1973) Arch. Biochem. Biophys., 152, 235-239.

19. Quast, R. (1976) in Plasma Protein Turnover (Bianchi, R., Mariani, G. and McFarlane, A.S., eds.) pp.3-15, MacMillan Press Ltd., London.

20. Quast, R. (1977) in Electrofocusing and Isotachophoresis (Radola, B.J. and Graesslin, D., eds.) pp. 455-462, de Gruyter, Berlin.

21. Quast, T. (1978) Offenlegungsschrift 26 29 339, Deutsches Patentamt

22. Dencka, W.D. (1977) in Electrofocusing and Isotachophoresis (Radola, B.J. and Graesslin, D., eds.) pp. 423-432, de Gruyter, Berlin.

23. Davies, H. (1970) in Protides of the Biological Fluids (Peeters, H., ed.) Vol. 17, pp. 389-396, Pergamon Press, Oxford.

24. Chrambach, A. and Nguyen, N.Y. (1977) in Electrofocusing and Isotachophoresis (Radola, B.J. and Graesslin, D., eds.) pp. 51-58, de Gruyter, Berlin.

Electrokinetic Separation Methods,
P.G. Righetti, C.J. van Oss and J.W. Vanderhoff

DENSITY GRADIENT ELECTROPHORESIS OF CELLS IN A VERTICAL COLUMN

ROBERT C. BOLTZ, JR.[*] and PAUL TODD[**]

[*]Separations Branch, George C. Marshall Space Flight Center, Marshall Space Flight Center, Alabama 35812, and [**]Department of Biochemistry and Biophysics, The Pennsylvania State University, University Park, Pennsylvania 16802, U.S.A.

Detailed methods for the density gradient electrophoresis of cells are presented, including glassware design, buffer compositions, cell preparation, and density gradient formation. Electroosmosis and convection are not serious problems in density gradient electrophoresis, but droplet sedimentation in a 2.5 cm diameter column restricts the method to approximately 10^7 cells per separation unless larger columns are used. The method has been applied to cell-surface and cell-separation problems in immunology, pathology, and virology.

INTRODUCTION

Several methods of preparative cell electrophoresis have been developed over the past two decades: flowing-curtain electrophoresis[1,2], endless-belt electrophoresis[3], rotating-tube electrophoresis[4], stable-flow free-boundary density gradient electrophoresis[5], static density gradient electrophoresis[6], and variations on these basic notions. Nevertheless, preparative cell electrophoresis is not in widespread use, because most methods function reliably only in the hands of specific investigators. Cell biologists cannot be expected to become electrochemists and engineers in order to achieve a cell separation that constitutes only a small part of an important scientific experiment. The objective of this chapter is to describe the construction and operation of a preparative cell electrophoresis system that is within the budget and experimental capability of most practicing cell biologists and to describe some of the test procedures that establish the physical limits of the functioning of that system.

PRINCIPLES OF DENSITY GRADIENT ELECTROPHORESIS OF SUSPENDED PARTICLES

Most large molecules, small molecules, and suspended particles have electrophoretic mobilities within approximately the same order of magnitude, namely 0.5-5.0 μm-cm/V-sec. The electrophoretic separation of molecules in free fluid is affected by *diffusion*, which occurs at velocities comparable to electrophoretic velocities, whereas the electrophoretic separation of suspended particles, such as biological cells, is affected by *sedimentation*, because the Stokes sedimentation velocity is comparable to the electrophoretic velocity. Diffusion is of no consequence in cell separation, since the diffusion coefficients of such enormous particles are so small.

The forces acting on a particle in density gradient electrophoresis are thus the electric field, the gravitational field, and the buoyant force. Balancing these forces leads to a constant particle velocity given by

$$v = \mu E + \frac{2}{9}\frac{a^2 g(\rho - \rho_0)}{\eta} \qquad (1)$$

where μ is the electrophoretic mobility in μm-cm/V-sec (derived from velocity per unit field, μm/sec per V/cm), E is the electric field strength in V/cm, usually determined from

$$E = \frac{I}{kA} \qquad (2)$$

in which I = current in amperes, k = conductivity of the density gradient in mho/cm, and A = cross-sectional area of the gradient column. The second term of equation (1) represents the Navier-Stokes sedimentation velocity for spheres of density, ρ g/cm^3, radius a cm in a fluid of density, ρ_0 g/cm^3, and viscosity η poise under the influence of the gravitational acceleration, g dynes/g.

A density gradient is a concentration gradient, so several factors in equation (1) are dependent upon column position x, including μ, which depends on Ficoll concentration.

It is sometimes standard to express electrophoretic mobilities, μ_0, at the viscosity of pure water, η_0. Thus the velocity of a particle undergoing vertical electrophoresis in a density gradient will not be constant. Equation (3) therefore expresses the x-dependence of particle velocity:

$$v(x) = \mu_0(x)\,\frac{\eta_0}{\eta(x)}\,\frac{I}{k(x)A} + \frac{2}{9}\frac{a^2 g\big(\rho - \rho_0(x)\big)}{\eta} \qquad (3)$$

This is the density gradient particle electrophoresis equation, and its components will be discussed individually in the later parts of this chapter.

GLASSWARE DESIGN FOR DENSITY GRADIENT ELECTROPHORESIS

Density gradient electrophoresis of particles was reported by van Regenmortel[7] who investigated viruses using a hydrostatically balanced U-tube of the type designed by Svensson[8]. Its basic design is shown in Figure 1, which is a drawing of the most primitive version of this concept as adapted for use with biological cells[6]. Subsequent versions eliminated hydrostatic balancing, improved on electrode and gradient configurations, provided inner and outer cooling, and added sample insertion features.

The fully jacketed U-tube design shown in Figure 2 provides continuous active cooling. It was shown[9] that this provision approximately doubles the current density (mA/cm^2) that can be applied to a column of fluid with conductivity in the vicinity of 10^{-3} mho/cm. Without adequate temperature control, convection disturbs the density gradient and destroys the band of cells.

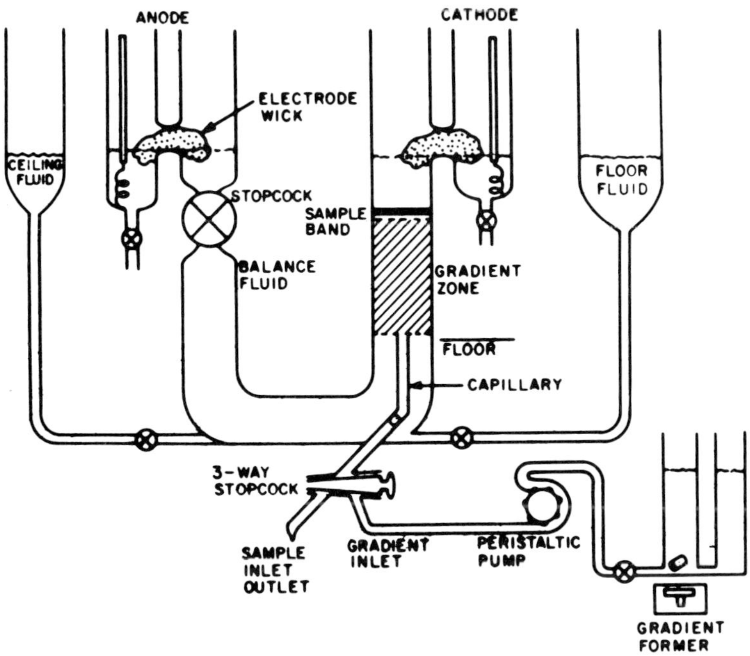

Fig. 1. Early version of density gradient electrophoresis apparatus for cells. Hydrostatic balancing was required, and electrodes were symmetrically placed on either side of the U-tube assembly, which could be disconnected at the bottom of the U. Revised diagram from Boltz et al.[6]

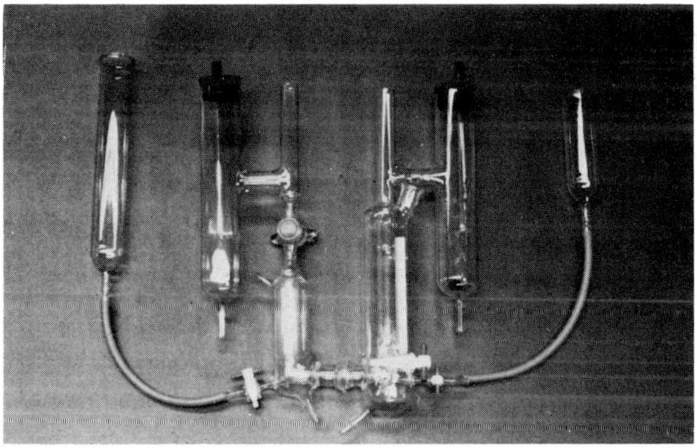

Fig. 2. Photograph of water-cooled U-tube density gradient electrophoresis apparatus for cells. The reservoir on the left is for "top solution" and the reservoir on the right is for "bottom solution" (see Table 1). All solutions enter and leave the U-tube at the bottom; cells and gradient enter through a capillary as in the apparatus described in Fig. 1.

It is awkward to build cooling jackets (and more so to build cooling fingers) for the U-tube configuration. This problem is managed in commercial electrophoresis and isoelectric focusing columns by eliminating electrode side-arms and hydrostatic balancing and constructing annular electrode compartments that support the liquid column at the bottom and contact it at the top[10,11]. However, the use of side-arm electrodes allows easy access to the center of the column at top and bottom. Figure 3 depicts a water-jacketed column with side-arm electrodes, which support and contact the gradient column through 15% polyacrylamide gel plugs which are made by the user and can be used for a number of separations and can be replaced[12]. This configuration is designed to allow gradient and sample insertion at the bottom and gradient harvesting at the top. The configurations of Figures 4 and 5 accommodate a central cooling finger, which may enter the column from the bottom (A) or top (B). The modular column designs shown in Figures 4 and 5 have several advantages: they can be modified into a wide variety of assemblies; the column length, diameter, and cooling provisions can be altered by interchanging columns; the side-arm gel plugs support the liquid column, thereby eliminating the need for hydrostatic balance; and complete access is available at both the top and bottom of the column. The procedure for assembling the column in preparation for cell electrophoresis is described in detail in a later section.

The T-joints that form the connections between the side-arm electrodes and the main column and its inlet and outlet components are machined blocks of plastic. Teflon blocks may be machined with precision so that "Precision Bore" pyrex tubing, from which the side-arms are blown and which is available in the U.S. from Corning Glass Co., fits snugly at room temperature. The Teflon-glass contact area is sealed with household silicone caulking compound and the entire apparatus is placed at 4°C to allow differential contraction to seal the joints tightly. An alternative and preferred method of constructing T-joints is to use either Teflon or Lexan polycarbonate with two "O"-ring channels milled in each of the receiving holes. This construction reduces the precision required of the glass tubing inserts by effecting a glass-to-rubber seal against the "O" rings captured in the channels. The use of "O" rings for this purpose is shown more clearly in Figure 6. Teflon and polycarbonate are the plastics of choice for this purpose because they do not melt or distort seriously upon autoclaving.

An especially versatile version of the device is illustrated in Figure 6. It provides inner and outer temperature control, inlet and outlet options, and a sample insertion system that permits loading of cells anywhere in the gradient. The central tube, through which cooling water passes, penetrates the upper and lower plastic blocks through sliding "O"-ring seals; it can thus be positioned vertically by the operator. The sample inlet tube descends through the center of this tube and terminates midway down in four outlets into the gradient. The sample suspension is injected into the gradient at the desired vertical position through these four outlets, as depicted in Figure 7. This version of the apparatus is especially useful for the isoelectric focusing of cells or molecules in a pre-formed pH gradient[13].

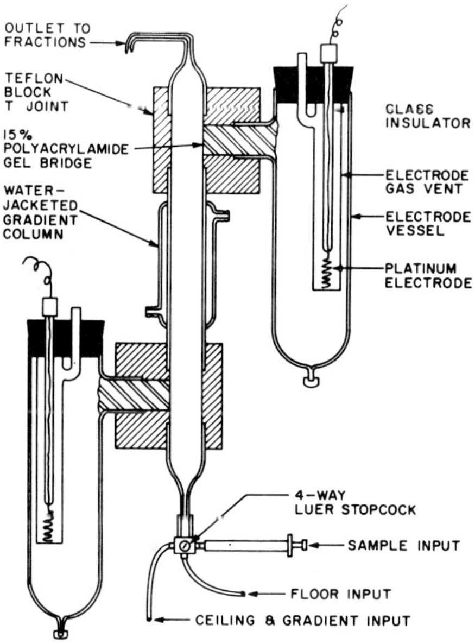

Fig. 3. Modular glass-and-Teflon density gradient electrophoresis apparatus. The polyacrylamide gel bridges support the mass of the fluid column and provide conductive contact to the electrodes. Any glass modules (column, top, bottom, side-arm) with the correct diameter may be inserted into the Teflon block to modify the geometry of the apparatus. Details of the stopcock system, gradient maker, and electrode assemblies are given in the text. Revised diagram from Boltz et al.[12]

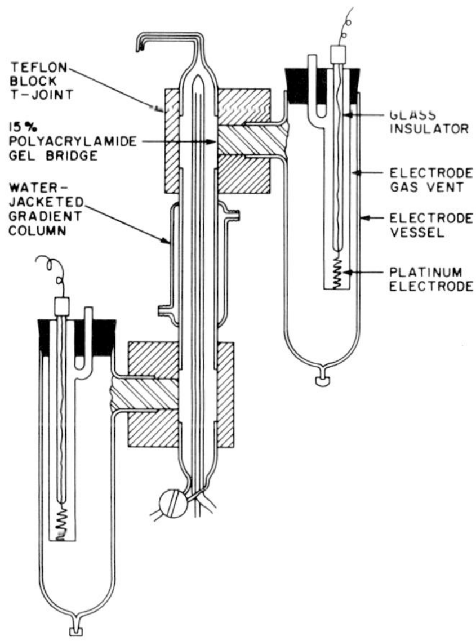

Fig. 4. Modular glass-and-Teflon density gradient electrophoresis apparatus with two-way cooling finger inserted at the bottom. Sample, gradient, and other solutions enter through the stopcock at the bottom. The upper tip of the cooling finger tends to distort cell bands during fraction collection at the top of the column.

It is also possible to construct an all-glass system which is nearly as versatile. Figure 8 shows a drawing of electrophoresis glassware in which the plastic joints have been replaced by blown glass joints. This arrangement completely eliminates any exposure of solutions to plastic surfaces, but the formation of side-arm gel plugs is slightly more difficult; it requires the inflation of balloons within the electrode vessels for the creation of a surface against which to pour the unpolymerized acrylamide solution, whereas this is accomplished by the simple insertion of a Teflon rod into the plastic connectors. The top and bottom of the column are completely open; it is thus possible to

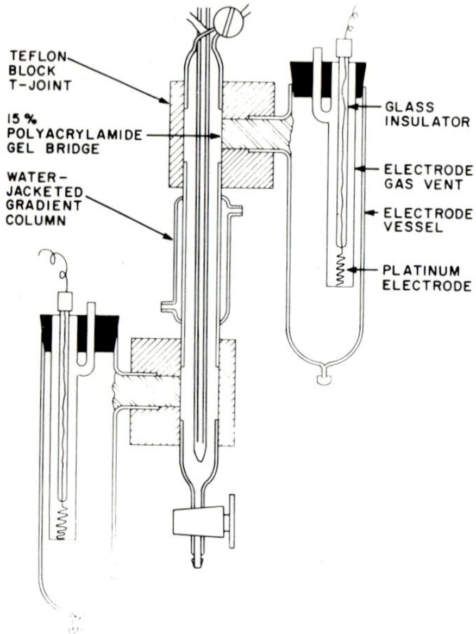

Fig. 5. Modular glass-and-Teflon density gradient electrophoresis apparatus with two-way cooling finger inserted at the top. All solutions enter through the bottom inlet. The conventional stopcock shown in this diagram is inferior to the simple taper shown in Fig. 3. In either case, a 4-way metal stopcock is attached by a short length of tubing. The combined presence of a stopcock and cooling finger at the top outlet tends to distort cell bands during fraction collection at the top of the column.

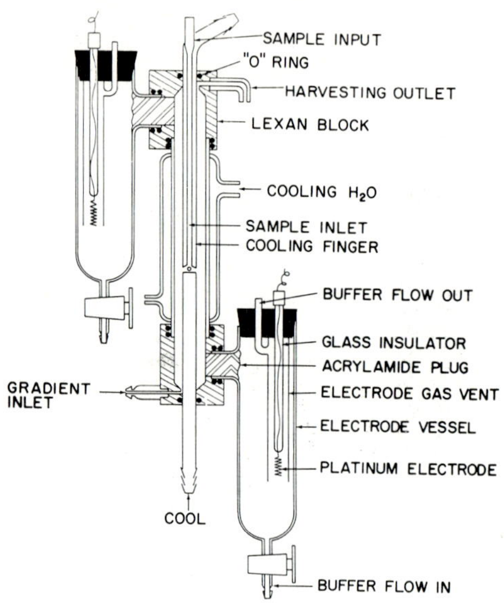

Fig. 6. Modular glass-and-Lexan density gradient electrophoresis apparatus with movable one-way cooling tube inserted straight through the column. The sample input capillary is inside the cooling tube and admits sample to the gradient column through four openings half-way down the cooling tube, which can be positioned vertically by sliding through the top and bottom "O" rings. This arrangement is ideal for isoelectric focusing of cells in pre-formed pH gradients. Two "O" rings are used to seal each glass-to-plastic connection involving the gradient column and side arms. The plastic blocks differ from those in Figs. 3-5 by having internal tapers ending in "O" ring seals at top and bottom. Gradient inlet and harvesting outlet channels are drilled in plastic and machined to accept glass capillaries by press fit. When recirculating of electrode buffer solutions is desired (as in certain types of isoelectric focusing), electrode buffer enters through the bottom electrode-vessel stopcock ("buffer flow in") and leaves through the electrode gas vent chimney through the electrode-vessel stopper ("buffer flow out"). Revised diagram from Boltz et al.[13]

explore a wide variety of internal configurations (sample insertion and cooling devices) by using conventional rubber stoppers or polyacrylamide gel plugs as closures at top and bottom. The open top and bottom are also useful for performing electrophoretic separations in a conventional (for molecules) or reversible (for cells) gel as the zone material. A plunger can be inserted into one of the open ends for extrusion of the gel out the other end.

The construction of glassware for the density gradient electrophoresis of cells can be accomplished in almost any laboratory. Of the versions listed here, those described in Figures 3 and 6 have been found most useful for routine use.

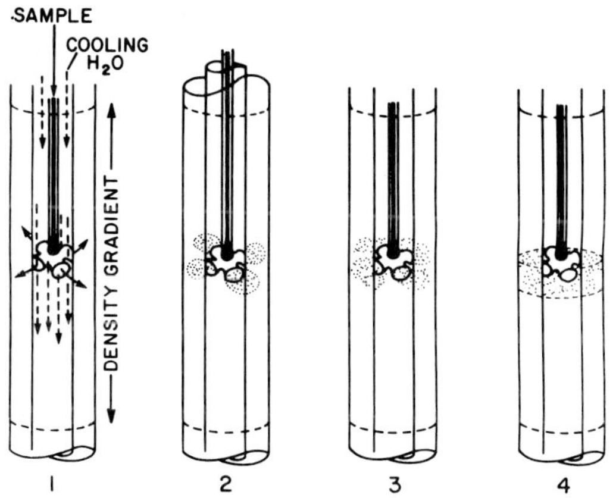

Fig. 7. Four stages of sample insertion using the apparatus diagrammed in Fig. 6. The outer cylinder is the density gradient column (jacket not shown). The middle cylinder is the cooling tube; dashed arrows indicate cooling water flow. The innermost cylinder is the sample inlet capillary; its interior is darkened to indicate the presence of sample (cell suspension). The solid arrows indicate four ports through which sample enters the gradient column. Diagrams 2, 3 and 4 indicate early, late and final stages of sample insertion by this method.

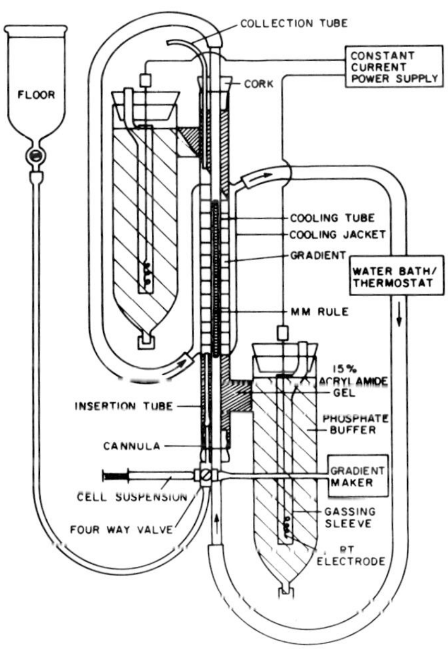

Fig. 8. All-glass density gradient electrophoresis apparatus. In this configuration it is similar, in principle, to the glass-and-Lexan apparatus of Fig. 6, except that cells are inserted beside the inner cooling tube at the bottom of the gradient. Silicone rubber stoppers are used at both ends, and, in this example, portions of the "ceiling" and "floor" regions are occupied by polyacrylamide gel, which is molded around the insertion and collection tubes at the top and bottom, respectively. Work of R. A. Gaines, The Pennsylvania State University.

ELECTRODE DESIGN FOR DENSITY GRADIENT ELECTROPHORESIS

The use of a polyacrylamide gel bridge permits a wide variety of electrode choices, since the invasion of the column by electrolysis products is minimized. It is still desirable, however, to keep electrolysis products away from the gel plug to avoid their possible ionophoresis into the column. The electrode vessel shown in Figure 3 provides various ways of accomplishing this.

The unplatinized platinum wire electrode is housed within an "electrode gas vent". This vent allows gaseous electrolysis products to escape without disturbing the liquid interface between the saturated NaCl in the bottom of the electrode vessel and the phosphate buffer in the top half of the electrode vessel. The lower half of the electrode vessel, which is 5.0 cm in diameter, is filled with 175 ml of saturated aqueous NaCl; the electrode reaction is thus zero order (less than 0.2% of the ions are converted to H_2 or Cl_2), and this segment of the circuit is highly conductive. The upper half of the electrode vessel contains 125 ml of 0.01 *M* phosphate buffer, pH 7.2 ("top solution", Table 1). This volume is adequate to prevent ionophoresis of Na^+ and Cl^- into the polyacrylamide gel plugs, thereby permitting up to several weeks of use of the same column without forming new gel plugs and without introducing extra ions into the central column.

The electrode gas vent may also serve another purpose. In cases where flowing or recycling electrode fluids are desired, the electrode fluid may be pumped through the opening at the bottom of the electrode vessel (with its stopper sealed in place) and out through the top of the electrode gas vent, as indicated in Figure 6.

TABLE 1

COMPOSITION OF SOLUTIONS USED IN DENSITY GRADIENT CELL ELECTROPHORESIS

The final concentration of each component is given in g/l

Component	Phosphate buffered glucose	Top solution	Bottom solution	Ficoll floor
KCl	0.20	0.20	0.20	0.20
$MgCl_2 \cdot 6H_2O$	0.10	0.10	0.10	0.10
Na_2HPO_4	1.15	1.15	1.15	1.15
KH_2PO_4	0.20	0.20	0.20	0.20
Glucose	10.0	10.0	10.0	10.0
Sucrose		68.0	51.0	42.5
Ficoll			100.0	150.0

SELECTION OF DENSITY GRADIENT SOLUTIONS

The application of density gradient techniques in cell biology requires the use of non-toxic gradient material. In electrophoresis applications the density solute must be as electrokinetically inert as possible and must have sufficient density at low osmolarity to buoy cells against sedimentation during migration in the electric field. The desiderata of low conductivity, low viscosity, low osmolarity per unit mass, and low cytotoxicity are difficult to fulfill in a single substance. It was considered[6] that the following substances were worthy of testing: polyvinylpyrrolidone (MW = 400,000), Ficoll (MW = 400,000), sodium 3,5-diacetamido-2,4,6-tri-iodobenzoate ("Hypaque", MW = 636), dextran (MW = 10,000), colloidal silica ("Ludox") 70 Å diameter particles, and sucrose. Osmolarity, density, viscosity, conductivity, and cytotoxicity were measured as a function of concentration in 0.01 *M* phosphate buffer, pH 7.4, and 1% glucose (second column, Table 1). An example of a data set is shown in Figure 9, in which density of the solution is used as independent variable, and osmolarity (in this case) is evaluated in the region around density = 1.05 g/cm^3. Sucrose does not qualify because the osmolarity exceeds 300 mOs*M*, which is considered suitable for living cells. Other compounds were disqualified on various grounds: Hypaque is imbibed by cells thereby increasing their density; polyvinylpyrrolidone is too viscous; colloidal silica is toxic. Dextran and Ficoll, polymers of saccharide derivatives, have similar properties, and Ficoll allows very high viability of cultured cells[6].

Dextran had been shown[14] to increase the zeta potential of suspended cells, presumably due to its high concentration near the cell surface leading to a local change in the dielectric constant. Ficoll has a similar effect, and R. A. Gaines, in an unpublished study, found that 0, 5, and 10% Ficoll solutions in low-conductivity 0.01 *M* phosphate buffer, pH 7.4, resulted in mobilities of 0.98, 1.77, and 2.60 µm-cm/V-sec, respectively, for fixed chicken erythrocytes.

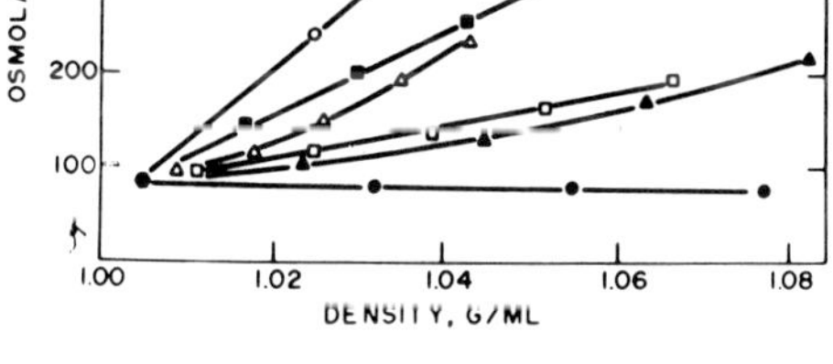

Fig. 9. Osmolarity as a function of density of various density gradient solutions. Similar data sets giving viscosity, conductivity, and cell toxicity were used as selection criteria for density gradient solutions[6]. Reproduced by permission of Marcel Dekker, Inc.

When cells travel upward through a Ficoll gradient, they enter zones of reduced viscosity and speed up [see equation (3)], but the reduced Ficoll concentration reduces their electrokinetic potential, and they slow down. The electrokinetic effect of Ficoll thus appears to counteract its viscosity to some extent, and this phenomenon may partly explain the constant migration velocities reported in test separations of cells in a Ficoll gradient[12].

The gradient in most common use[12,13,15] is an isoosmotic mixture of Ficoll and sucrose ranging in density from 1.035 to 1.055 g/cm^3. Its physical properties are summarized in Figure 10, which shows that the Ficoll gradient is also a reciprocal sucrose gradient.

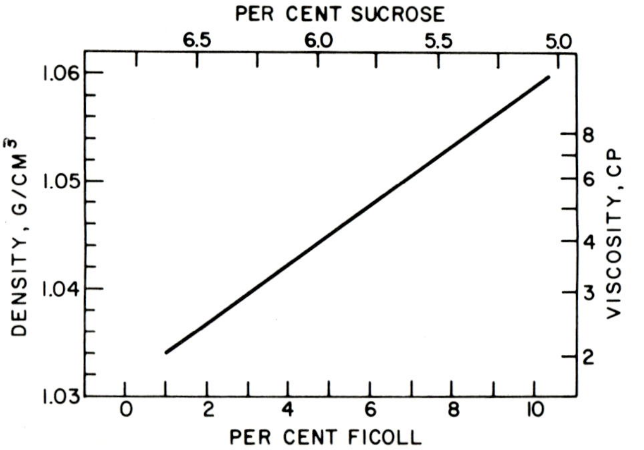

Fig. 10. Characteristics of Ficoll-sucrose density gradient used in cell electrophoresis. The osmolarity is constant at about 303 mOs*M*. The density increases linearly with Ficoll concentration, and sucrose concentration is reduced correspondingly to maintain isoosmolarity. Viscosity increases approximately exponentially with Ficoll concentration[6]. Reproduced by permission of Marcel Dekker, Inc.

ROUTINE PROCEDURE FOR CELL ELECTROPHORESIS

Apparatus assembly. Phosphate buffered 15% acrylamide is made up by adding 146 g acrylamide and 4 g N,N'-methylene bisacrylamide (Eastman Kodak chemicals) to the "top solution" (Table 1). The solution is heated to 37°C in a water bath. To 60 ml of this solution 1.0 ml of freshly prepared 10% ammonium persulphate and 0.05 ml of N,N,N',N'-tetramethylethylenediamide (temed) are added. The gel plugs are formed with the teflon blocks connected to the electrode vessel side arms. Tightly fitting teflon rods are substituted for the column and serve as a surface against which to pour 30 ml of the gel solution. The rods are removed after the gel has solidified, and the column is assembled as in Figure 3. A small amount of silicone sealer is applied to the glass-teflon interfaces, and the assembly is left overnight at 4°C. The cooling jacket is connected to a pump that circulates water from a thermostated bath. Saturated NaCl (about 175 ml) is layered gently beneath 125 ml of "top solution" in each electrode vessel until the level reaches the top of the gel plug. Cells are washed twice by centrifugation through phosphate buffered isotonic sucrose and resuspended at a concentration of 3 X 10^7/ml in 10% Ficoll, 5.1% sucrose phosphate buffer ("bottom solution", Table 1) in a disposable tuberculin (1 ml) syringe, which is connected to the B-D MS-15, 4-way metal stopcock, which is not yet attached to the bottom inlet of the column. One 50-ml disposable syringe is filled with "bottom solution" and another with "top solution" (Table 1). Some heavy fluid is flushed through the mixing chamber of a Beckman (catalog #35005Z) gradient former followed by a flush of light fluid. The 50-ml plastic syringes are inserted in the gradient former so that light fluid flow is controlled by the front drive, and heavy fluid by the rear. The mixing chamber is connected to the 4-way stopcock with a length of 1.1 mm diameter LKB silicone tubing and a 3083 BD male luer adapter. The reservoir containing floor solution (15% Ficoll, 4.25% sucrose, Table 1) is connected to the 4-way stopcock by a length of 3-mm diameter LKB silicone tubing and a 3096 male luer adapter. Small quantities of bottom solution, cell suspension, and top solution are flushed through the 4-way stopcock in that order to remove bubbles from the tubing and from the inlets of the stopcock. The 4-way stopcock is connected to the column at the ground glass luer fitting with a length of 5-mm diameter LKB silicone tubing. 30 ml of ceiling is inserted into the column using the top solution syringe. With the stopcock closed, the syringe is again filled with top solution and inserted into the gradient former. The stopcock is opened to the gradient input position, and a 25-ml gradient (range 2% Ficoll, 6.5% sucrose-6.4% Ficoll, 5.7% sucrose) is generated (settings 0-90 on the gradient former).

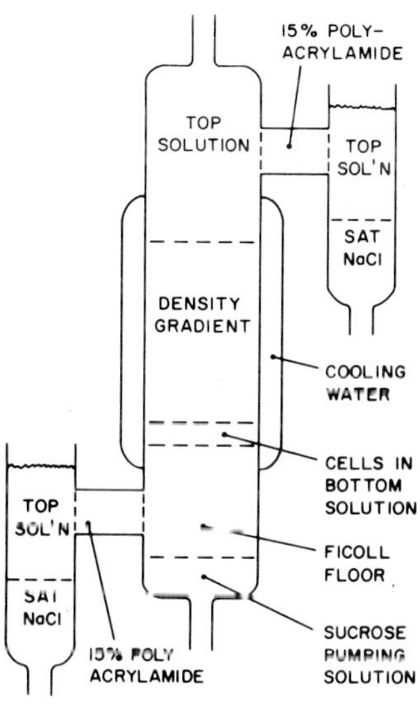

Fig. 11. Distribution of components at the start of a cell electrophoresis procedure using the apparatus shown in Fig. 3.

Next 0.3 ml of cell suspension is inserted followed by 40 ml of Ficoll floor solution (Table 1). The arrangement of solutions in the apparatus at the beginning of a separation is shown in Figure 11. A constant field strength of 4.0-5.4 V/cm at 15-20 mA is applied to the column for 2.5 hr. Close-up photographs of the migrating bands can be taken at suitable intervals.

By pumping 15% Ficoll or 30% sucrose ("pumping solution") beneath the floor, fractions are collected from the top of the column, checked for index of refraction and differential cells by staining and/or electronically counted with a Coulter counter.

TEST PARTICLES FOR DENSITY GRADIENT ELECTROPHORESIS

Test particles that are chemically stable and whose charged surface groups resemble those of living cells can be obtained by fixing erythrocytes of various animal species in dilute aldehyde solutions[16-18]. Fixed erythrocytes can be prepared, for example, from whole blood of Sprague-Dawley rats ($\mu = 1.10 \pm 0.06$), white leghorn hens ($\mu = 0.98 \pm 0.06$), or albino rabbits ($\mu = 0.62 \pm 0.05$ μm-cm/V-sec) following the procedures described by Tenforde[18]. Approximately 2.5 ml of blood are drawn and diluted immediately with 25 ml of 0.067 *M* Sørenson's buffer. They are then washed three times in 40 ml of this buffer by repeated centrifugation at 1700 *g* for 5 min at room temperature. Fixation is accomplished in 40 ml of this buffer with 2.5% electron microscope grade glutaraldehyde for 10 min at room temperature or 2% acetaldehyde for 20 days or 1.5% formaldehyde for 20 days or 1% osmium tetroxide for 1 min. The cells are rinsed three times in fresh 0.145 *M* NaCl and stored at 4°C in 0.145 *M* NaCl. It is worth noting that the isoelectric pH of erythrocytes can depend upon the fixative with which they are prepared[16,18,19]. Glutaraldehyde-fixed erythrocytes have been found to be electrokinetically stable over several months of storage in 0.145 *M* NaCl at 4°C.

ELECTROOSMOSIS IN DENSITY GRADIENT ELECTROPHORESIS

Most glass or plastic surfaces bear a net negative surface charge which gives rise to an electrokinetic (zeta) potential similar to that found on the surface of a colloidal particle[20]. This potential has the effect of polarizing the solvent molecules that lie within a few Å of the surface. The force exerted by this polarizing field depends on the dielectric constant, ε, of the solution. If an electric field parallel to the vessel walls is applied through the solution, the polarized solvent molecules will move in response to the applied field, E, with velocity, V_{os}, given by

$$V_{os} = \frac{\zeta E \varepsilon}{4\pi\eta} \quad (4)$$

as derived from hydrodynamic and electrostatic theory for a cylindrical tube under no pressure.

This electroosmotic velocity is comparable to the electrophoretic velocity in cylindrical[23,24] and rectangular[25,26] microscopic electrophoresis chambers and results in a parabolic distribution of fluid velocity across such chambers, since the force is exerted horizontally at the chamber walls. Very thin chambers, 0.1 cm, are used in microscopic electrophoresis, so the velocity parabolas are steep. Flatter velocity parabolas occur in wider chambers, such as those used in density gradient electrophoresis, 1.0-2.5 cm. It is thus important mainly to know the electroosmotic velocity at the chamber wall, as given in equation (4).

Under typical conditions of density gradient electrophoresis

$$\varepsilon \approx 5 \times 10^{-11} \text{ coul/V-cm}$$
$$\eta \approx 0.1 \text{ poise}$$
$$E \approx 5 \text{ V/cm}$$

and the ζ potential of pyrex glass, as determined in electroosmotic experiments by Ghosh and Ghosh[27], is about 0.1 V. These values, when applied to equation (4) give $V_{os} \approx 10^{-6}$ cm/sec, so that the fluid at the vessel wall flows about 0.1 cm toward the cathode in the course of a 3-hr separation.

Neutral blue dextran dye, when inserted at the top of a density gradient and subjected to fields up to 6 V/cm for up to 5 hr, were not distorted by electroosmotic flow.

It is concluded that electroosmosis does not distort electrophoretic separations in a density gradient under standard conditions of separation. It should be noted, however, that equation (4) does not characterize the fluid velocity parabola, which is also affected by hydrodynamic variables, including the gravity head. In the absence of gravity, serious electroosmostic distortion of particle migration due to parabolic fluid flow can occur[28,29].

DROPLET SEDIMENTATION IN DENSITY GRADIENT ELECTROPHORESIS

One limiting factor in resolution, common to all density gradient systems, is spreading of the cell bandwidth during the course of an experiment. Commonly termed "overloading the gradient" in gravitational and centrifugal velocity separations, droplet sedimentation can greatly broaden the cell bandwidth over the course of an experiment. Droplet sedimentation occurs when cells no longer behave as individual particles suspended in a density gradient, but rather as contributors to the density of the volume element they occupy. A density instability arises. A tiny droplet of cells and fluid is formed, and while it has a density ρ_{drop} only slightly higher that that of the surrounding fluid, the droplet has a sufficient radius, R, to result in detectable sedimentation. The velocity of sedimentation of such a droplet, once formed, is described by the equation

$$v(x) = \frac{2}{9}\frac{R^2 g}{\eta(x)}\left[\rho_{drop} - \rho_0(x)\right] \qquad (5)$$

The equation gives no information concerning the formation of these sedimenting droplets, which depends on the particle, its size, shape, and density and the physical properties of the density gradient fluid. To optimize resolution it is necessary to define the concentration at which the cells no longer behave as individual particles and behave as density contributors.

According to the theoretical analysis of Mason[30], the sedimentation of particles by the streaming or "droplet" mechanism, rather than sedimentation as individual particles, occurs when the number of suspended particles per unit volume exceeds a certain limit, which is proportional to

$$\frac{1}{a}\sqrt{\frac{g}{D\eta(x)}\frac{d\rho_0(x)}{dx}} \qquad (6)$$

where D = diffusion coefficient of the suspended particle. Mason has presented this relationship in the form of plots of critical particle concentration as a function of particle radius and solute density gradient. Agreement of the theory within a factor of 3 with observations on cells by Miller and Phillips[31] and agreement within 25% with observations on hemoglobin in sucrose density gradients used for zone electrophoresis by Brakke[42] were reported.

A series of experiments was undertaken to define droplet sedimentation limits in Ficoll using fixed chicken erythrocytes. Chick red blood cells (RBC) were suspended in the Ficoll density gradient of Figure 10, and the sedimenting acceleration was held constant at 1 X *g* during upward electrophoresis at 4 V/cm. Figure 12 shows a series of column fractionation profiles obtained following the electrophoresis of varying chicken erythrocyte concentrations. It demonstrates the effect of concentration-dependent droplet sedimentation on final harvested cell bandwidth. Droplet sedimentation evidently continued until each volume element contained the cell concentration at which the cells acted as individual particles. An extrapolation of the plot of the peak width *vs*. cell titre inserted (Fig. 13A) reveals that a titre of 10^7 cells/ml does not increase the initial bandwidth of 1 mm. A similar analysis of the number of cells contained in the peak fraction (Fig. 13B) again yields $\sim 10^7$ cells/ml as the critical concentration. Under the usual conditions it appears unwise to insert cell suspensions at concentrations higher than 10^7/ml. The droplet sedimentation problem could be partially overcome by the use of a gradient tailored to the cell type to be separated, similar to the isokinetic gradient developed by Pretlow[32] for gravitational velocity separations. The limiting load values determined for fixed chicken erythrocytes ($a = 4.8$ µm, $\rho = 1.19$ g/cm^3) can be used as guidelines for separations involving other cell types.

ACHIEVABLE RESOLUTION IN DENSITY GRADIENT ELECTROPHORESIS

Fixed rat, rabbit, and chicken erythrocytes have mobilities of 1.10 ± 0.06, 0.98 ± 0.06, and 0.62 ± 0.05 µm-cm/V-sec, respectively, in phosphate buffered sucrose[12]. Thus, rat and chicken erythrocytes have a 10% difference in average

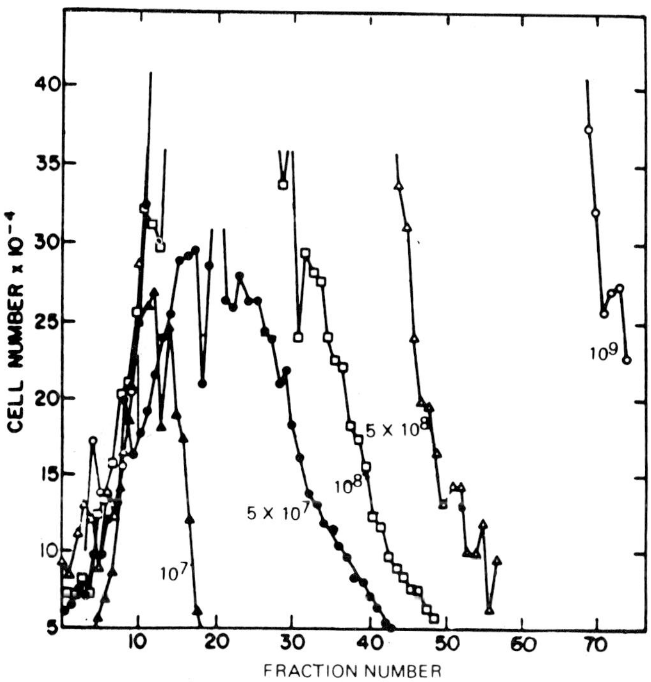

Fig. 12. The effect of droplet sedimentation on harvested bandwidth when the gradient is overloaded with cells (fixed chicken erythrocytes). The numbers associated with each curve are the numbers of cells inserted in a 1-mm band at the beginning of the electrophoresis procedure. Cell counts (per fraction) that were too high to allow presentation of these five fractionation profiles on a comparable scale have been omitted. Data from these profiles were used in plots of Fig. 13.

mobility. In addition, the rat erythrocyte appears as a biconcave disk under phase contrast microscopy. The chicken erythrocyte is ellipsoid with a dark nucleus; therefore, these cell types can be distinguished microscopically in fractions collected after density gradient electrophoresis. To quantify the resolution attainable with the modular apparatus of Fig. 3, rat, rabbit, and chicken erythrocytes, 1×10^7 total cells/ml, were electrophoresed at a field strength of 4 V/cm for 3 hr. The mixture was resolved into three visually distinct bands in the column (Fig. 14) when allowed to migrate for 3 hr at a field strength of 4 V/cm. The profile of cells per collected fraction exhibited three distinct peaks. The mid-peak contained 84% chicken erythrocytes upon microscopic examination, whereas the other two peaks were 94% free of chicken erythrocytes. These results indicate that subpopulations of cells differing in mobility by 0.12

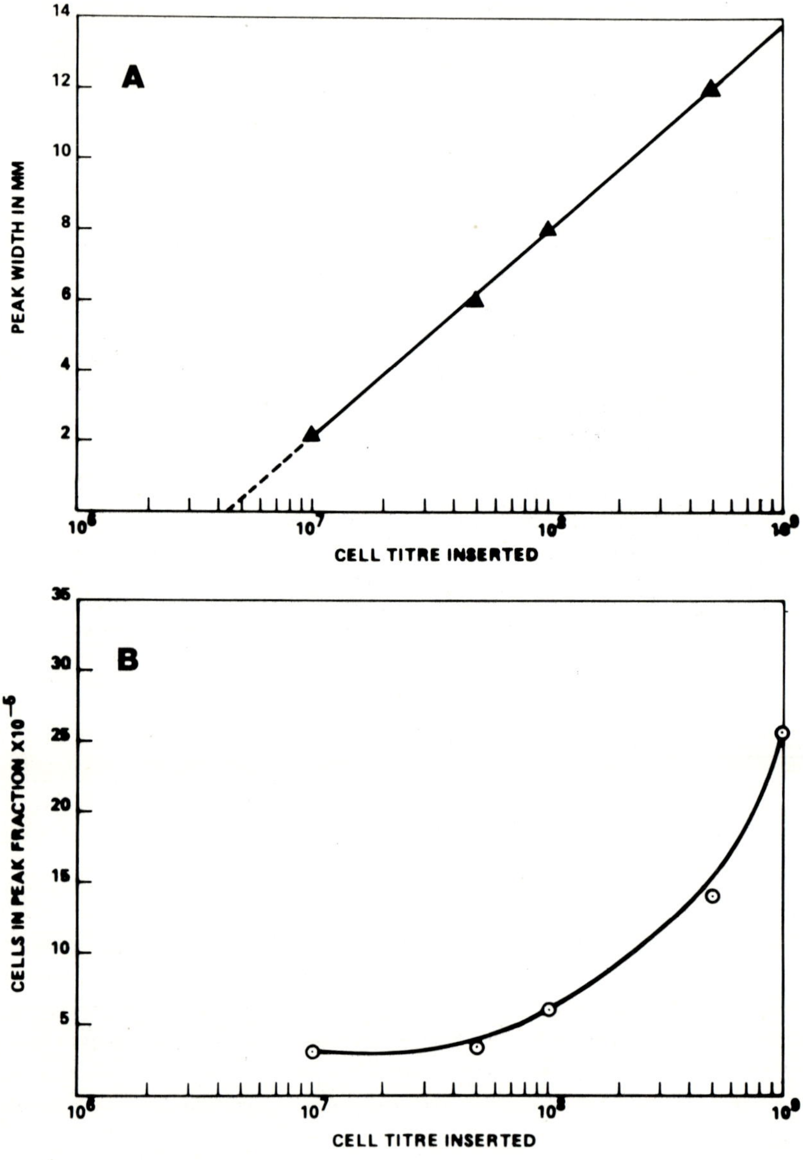

Fig. 13. Plot of the peak width (A) vs. number of cells inserted in a 1-mm band into the gradient. Peak width is defined as the number of fractions between those two fractions which contain half as many cells as the peak fraction. Plot of the number of cells in the peak fraction (B) vs. number of cells inserted in a 1-mm bank into the gradient. Band spreading due to droplet sedimentation prevents linear increase of peak height with number of cells inserted.

μm-sec/V-cm can be resolved under these conditions. Populations with smaller absolute differences might be resolved when allowed to migrate for longer periods of time.

Quantitative assessment of Figure 15 indicates that the two erythrocyte peaks were separated by twelve 0.25-ml fractions, giving an interfraction geometrical resolution of 0.01 mobility unit/fraction. It is possible to collect 0.05 ml fractions, giving 0.002 mobility unit/fraction, so that density gradient electrophoresis has geometrical resolution that exceeds biological precision by more than an order of magnitude.

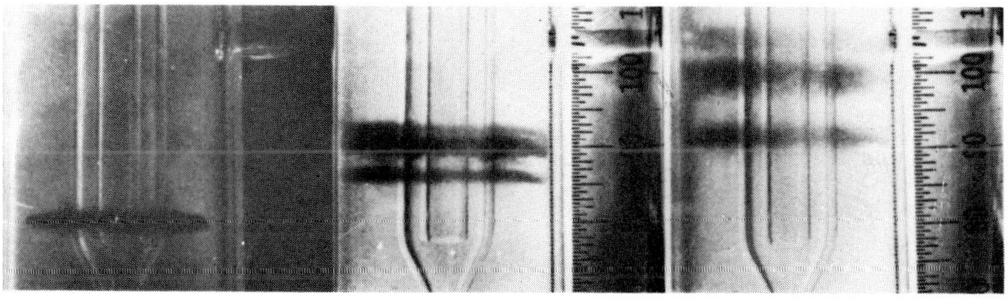

Fig. 14. Bands of mixed rat (top), chicken (middle), and rabbit (bottom) erythrocytes photographed after 0, 2 and 4 hr of electrophoresis in the apparatus shown in Fig. 5.

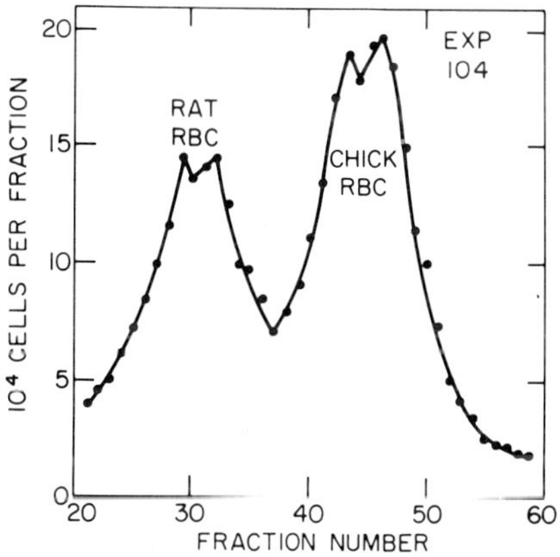

Fig. 15. Fractionation profiles of mixed fixed rat and chicken erythrocytes after electrophoresis at low cell density. These cell types differ in mobility by 0.12 μm-cm/V-sec, and the two peaks shown are separated by 10 fractions of 0.25 ml volume each. The flattening and apparent splitting of the peaks is considered to be caused by droplet sedimentation (see Fig. 12).

THE PREPARATION OF CELLS FOR DENSITY GRADIENT ELECTROPHORESIS

Several options are available when cells are prepared in suspension for density gradient electrophoresis. Cells dispersed from solid tissues or monolayer cultures may need to be treated with proteolytic enzymes and/or chelating agents, which may affect the mobility of cells. In some cases it may be desirable to allow cells to "regenerate" molecules lost from their surfaces by maintaining them for a few hours in suspension culture with complete nutrient medium. Some examples of methods that have been applied to particular cell types follow. The preparation of fixed erythrocytes as test particles was described earlier in this chapter.

Cells cultured in monolayer. Cells may be removed from their substratum by routine procedures such as trypsinization, EDTA treatment, etc. They are then suspended in complete nutrient medium (including serum at its usual concentration) with calcium omitted and maintained in suspension culture for 6-8 hr in a traditional tissue culture "spinner" flask. About 4 X 10^7 cells in 40 ml of medium is suitable. At the completion of incubation these cells are centrifuged 10-15 min at 300-500 X *g*, washed once or twice in "bottom solution" (Table 1) and resuspended in "bottom solution" for the application of about 5 X 10^6 cells to the density gradient electrophoresis column[6]. The electrophoresis of cells immediately after proteolytic resuspension has indicated a clear effect of trypsin and other enzymes on the surfaces of some cells but not of others[33].

Mammalian spermatozoa. Spermatozoa may be obtained by the collection of semen in an artificial vagina or by retrograde flushing of excised cauda epididymis with buffer[34]. The suspension in seminal fluid or buffer is cooled to electrophoresis temperature (4°C or 15°C) at a rate of 2°C/min. Cells are removed from the fluid by centrifugation at 300 X *g* for 10 min. The centrifuge tube is drained of all fluid by inversion on a paper towel and wiping of its rim with tissue paper. Cells are then resuspended in "bottom solution" for the application of about 10^7 cells to the density gradient electrophoresis column. After thè cells are harvested they may be returned to seminal fluid for motility testing.

Solid tissue cells. Cells from solid tissues can be dispersed with high viability by following a gentle procedure in which 0.1% crude trypsin (Difco 1:250) in Eagle's Minimum Essential Medium for Suspension culture containing 0.1% bovine serum albumin is used as incubation medium for the release of single cells[35]. This procedure has been used successfully for pituitary tissue[35] and modified for suspending bone medullary cavity cells[36]. It should be cautioned that when high concentrations of proteolytic enzymes are combined with mechanical forces to disperse cells, DNA is released to the medium, whereupon it promotes artifactious aggregation and changes in cell-surface properties. Adding 0.001% DNAase I to the proteolytic mixture eliminates this DNA. After satisfactory monocellular suspensions are obtained, cells are collected by centrifugation and resuspended in "bottom solution" for the application of 2-7 X 10^6 cells to column.

Exfoliated specimen cells. Proteolytic enzymes are highly destructive to exfoliated specimen cells. These may be collected in cold Hanks' Balanced Salt Solution, which is later brought to 5.0 mM in dithiothreitol to free cell junctions[37]. After 30 min at room temperature the cells may be dispersed by 20 passages through an 18-gauge needle at 5 lb/in^2 pressure[38]. The resulting single cells may be collected by centrifugation for 10 min at 300-500 X *g* and resuspended in "bottom solution" for the application of about 10^6 cells to the column.

Hematologic and ascites tumor cells. These cell types are already suspended in body fluids, and simple collection of cells by centrifugation is adequate unless a pre-purification step[13,35] is desired.

SOME ACCOMPLISHMENTS OF DENSITY GRADIENT ELECTROPHORESIS OF CELLS

The abundance of biological problems that has so far been attacked by the method of continuous flow electrophoresis[39] indicates that a similar variety of problems is available for investigation by density gradient electrophoresis. Although applications in immunology, hematology, virology, embryology, parasitology, endocrinology, pathology, and genetics are possible, applications to date, as indicated in Table 2, have been limited to immunology, virology, and pathology. Table 2 lists the chronology of these accomplishments using static density gradient electrophoresis.

TABLE 2

SOME ACCOMPLISHMENTS IN THE DEVELOPMENT AND USE OF STATIONARY FREE FLUID COLUMN ELECTROPHORESIS IN A DENSITY GRADIENT

Achievement	Year	Ref.
Application of moving boundary electrophoresis to plant virus purification	1947	40
Application of moving boundary electrophoresis to biological particle analysis	1953	41
Introduction of density gradient method as applied to molecules and viruses	1955	42
Separation of type I poliomyelitis virus	1958	43
Establishment of sucrose density gradient electrophoresis as a virus preparation routine	1961	44
Purification of several plant viruses free of host plant antigens	1964	7
Use of photometric analysis for analytic density gradient zone electrophoresis of viruses	1964	45
Preparative electrophoresis of viable cultured mammalian cells in a Ficoll gradient	1973	6
Test separation of human and rabbit erythrocytes	1975	15
Test isolation of viable mouse spleen and thymus lymphocytes	1975	15
Test separation of rat and chicken erythrocytes	1976	12
Analysis of herpes virus infected cultured cells	1976	12,26
Separation of various cell types from human biopsies and gynecologic specimens	1976	12
Separation of radioisotopically labeled mouse lymphocytes	1976	46
Separation of B and T mouse spleen lymphocytes	1976	47
Separation of human peripheral lymphocytes with differential ^{51}Cr uptake	1976	48
Separation of human blood leukocytes into monocytes, T lymphocytes, and different immunoglobulin-producing lymphocytes	1976	49
Use of photometric analysis for analytic density gradient electrophoresis of lymphocytes	1977	50
Separation of estrogen-responding avian femoral medullary cavity cells	1978	36

ACKNOWLEDGMENTS

Research at The Pennsylvania State University reported in this chapter was supported by Contract N01-CB-43984 and grants R01-CA-12589 and HD-05859 from the U.S. Public Health Service. We especially acknowledge the experimental contributions of Messrs. R. A. Gaines, M. K. Louie, and J. Kiefer, whose studies on column technology, electroosmosis, and droplet sedimentation, respectively, are presented.

REFERENCES

1. Zeiller, K., Loser, R., Pascher, G. and Hannig, K. (1975) Hoppe-Seyler's Zeit. Physiol. Chem., 356, 1225.
2. Strickler, A. and Sacks, T. (1973) Prep. Biochem., 3, 269.
3. Sturgeon, P., Kolin, A., Kwak, K. S. and Luner, S. J. (1972) Haematologia, 6, 93.
4. Hjertén, S. (1967) Chromatog. Rev., 9, 122.
5. Mel, H. C. (1964) J. Theoret. Biol., 6, 159.
6. Boltz, R. C., Jr., Todd, P., Streibel, M. J. and Louie, M. K. (1973) Prep. Biochem., 3, 383.
7. Van Regenmortel, M.H.V. (1964) Virology, 23, 496.
8. Svensson, H. and Valmet, E. (1955) Science Tools, 2, 11.
9. Gaines, R. A., Boltz, R. C., Jr. and Todd, P. (1974) Biophys. J., 15, 245a.
10. LKB Corp. (1974) Publication No. 8100(A)-E3. Ampholine Electrofocusing Columns.
11. Nelson, J. W., Allington, W. B. and Aron, C. B. (1974) ISCO Applications Res. Bull. No. 15. Isoelectric Focusing with ISCO Density Gradient Electrophoresis Equipment.
12. Boltz, R. C., Jr., Todd, P., Gaines, R. A., Milito, R. P., Docherty, J. J., Thompson, C. J., Notter, M.F.D., Richardson, L. S. and Mortel, R. (1976) J. Histochem. Cytochem., 24, 16.
13. Boltz, R. C., Jr., Todd, P., Hammerstedt, R. H., Hymer, W. C., Thompson, C. J. and Docherty, J. J. (1977) in Cell Separation Methods, Bloemendahl, H. ed. Elsevier/North-Holland, Amsterdam, pp. 143-155.
14. Brooks, D. E. and Seaman, G.V.F. (1973) J. Colloid Interface Sci., 43, 670.
15. Griffith, A. L., Catsimpoolas, N. and Wortis, H. H. (1975) Life Sci., 16, 1693.
16. Vassar, P. S., Hards, J. M., Brooks, D. E., Hagenberger, B. and Seaman, G.V.F. (1972) J. Cell Biol., 53, 809.
17. Walter, H., Krob, E. J., Brooks, D. E. and Seaman, G.V.F. (1973) Exp. Cell Res., 80, 415.
18. Tenforde, T. S. (1970) Adv. Biol. Med. Phys., 13, 43.
19. Haydon, D. A. and Seaman, G.V.F. (1967) Arch. Biochem. Biophys., 122, 126.
20. Bikerman, J. J. (1958) Surface Chemistry, 2nd ed., Academic Press, New York.
21. Bier, M., ed. (1959) Electrophoresis, Academic Press, New York.
22. von Helmholtz, H. (1879) Ann. Physik Wiedemann, 7, 337.
23. Seaman, G.V.F. (1965) in Cell Electrophoresis, Ambrose, E. J. ed., Little, Brown, Boston, pp. 4-21.
24. van Oss, C. J., Fike, R. M., Good, R. J. and Reinig, J. M. (1974) Anal. Biochem., 60, 242.
25. Fuhrmann, G. F. and Ruhenstroth-Bauer, G. (1965) in Cell Electrophoresis, Ambrose, E. J. ed., Little, Brown, Boston, pp. 22-25.
26. Thompson, C. J., Docherty, J. J., Boltz, R. C., Jr., Gaines, R. A. and Todd, P. (1978) J. Gen. Virol., 39, 449.

27. Ghosh, B. N. and Ghosh, A. K. (1958) Kolloid Z., 161, 110.

28. Snyder, R. S., Bier, M., Griffin, R. N., Johnson, A. J., Leidheiser, H., Jr., Micale, F. J., Vanderhoff, J. W., Ross, S. and van Oss, C. J. (1973) Sep. Purif. Meth., 2, 259.

29. Allen, R. E., Rhodes, P. H., Snyder, R. S., Barlow, G. H., Bier, M., Bigazzi, P. E., van Oss, C. J., Knox, R. J., Seaman, G.V.F., Micale, F. J. and Vanderhoff, J. W. (1977) Sep. Purif. Meth., 6, 1.

30. Mason, D. W. (1976) Biophys. J., 16, 407.

31. Miller, R. G. and Phillips, R. A. (1969) J. Cell. Physiol., 73, 191.

32. Pretlow, T. G., Weir, E. E. and Zettergren, J. G. (1975) Int. Rev. Exp. Pathol., 14, 91.

33. Milito, R. P. and Todd, P. (1976) Biophys. J., 16, 218a.

34. Hammerstedt, R. H. (1975) Biol. Reprod., 13, 389.

35. Hymer, W. C., Evans, W. H., Kraicer, J., Mastro, A., David, J. and Griswold, E. (1973) Endocrinology, 92, 275.

36. Furcinitti, P. S. and Hunter, S. J. (1978) in Electrophoresis 78, Catsimpoolas, N., ed., Elsevier/North-Holland, in press.

37. Escobar, G. J., Todd, P. and Sattilaro, R. F. (1977) J. Histochem. Cytochem., 25, 513.

38. Wheeless, L. L., Jr. and Onderdonk, M. A. (1974) J. Histochem. Cytochem., 22, 522.

39. Just, W. and Werner, G. (1978) in Electrokinetic Separation Methods, Righetti, P. G., van Oss, C. J. and Vanderhoff, J. L., eds., Elsevier, Amsterdam.

40. Lauffer, M. A. and Price, W. C. (1947) Arch. Biochem., 15, 115.

41. Hartman, R. E. and Lauffer, M. A. (1953) J. Am. Chem. Soc., 75, 6205.

42. Brakke, M. K. (1955) Arch. Biochem. Biophys., 55, 175.

43. Polson, A. and Kramer, R. (1958) Biochim. Biophys. Acta, 29, 187.

44. Cramer, P. and Svensson, H. (1961) Experientia, 17, 49.

45. Brakke, M. K., Allington, R. W. and Langille, F. A. (1964) Anal. Biochem., 25, 30.

46. Griffith, A. L., Catsimpoolas, N., Dewanjee, M. K. and Wortis, H. H. (1976) J. Immunol., 117, 1949.

47. Platsoucas, C. D., Griffith, A. L. and Catsimpoolas, N. (1976) J. Immunol. Meth., 13, 145.

48. Catsimpoolas, N., Griffith, A. L., Skrabut, E. M., Platsoucas, C. D. and Valeri, C. R. (1976) Cell Immunol., 25, 317.

49. Ault, K. A., Griffith, A. L., Platsoucas, C. D. and Catsimpoolas, N. (1976) J. Immunol., 117, 1406.

50. Catsimpoolas, N. and Griffith, A. L. (1977) Anal. Biochem., 80, 555.

Electrokinetic Separation Methods,
P.G. Righetti, C.J. van Oss and J.W. Vanderhoff

VERTICAL ASCENDING CELL ELECTROPHORESIS

CAREL J. van OSS and PAUL M. BRONSON

Immunochemistry Laboratory, Department of Microbiology, State University of New York at Buffalo, Buffalo, New York 14214 USA

INTRODUCTION

There is definite scope for a simple methodology for stationary (i.e. non-continuous flow) preparative electrophoresis of living cells at ambient gravity. In the preceding chapter such a methodology is described, utilizing density gradients obtained with polymer solutions. However, high concentrations of dissolved polymers tend to have variegated and often undesirable effects on cells,[1,2] and may even cause changes in their ζ-potentials.[1,3,4] We earlier described a method for preparative cell electrophoresis, utilizing a packed column of solid glass beads (of 0.1 mm diameter). The interstices between the beads were just small enough to prohibit sedimentation of erythrocytes due to ambient gravity as well as spreading through Brownian movement, but large enough to permit the migration of red cells under the influence of an electric field.[5] Complete separation was achieved between human and chicken erythrocytes, and between papain-treated and untreated human erythrocytes.[5] Upon further experimentation, however, it became clear that, while this method is eminently suitable for erythrocytes, it could not be used for the electrophoretic separation of other cell types; e.g., lymphocytes could at best be separated into very small, medium-sized and large cells, without much (if any) discrimination between electrophoretic mobility classes (van Oss *et al.*, unpublished results). Descending cell electrophoresis into gradients has been described by Boltz *et al.*[6] and by Griffith *et al.*,[7] and ascending cell electrophoresis has been reported on more recently by Boltz and his colleagues.[8] However, all these methods utilize high molecular weight polymers.

Preparation for 0 gravity stationary tube electrophoresis of living cells, to be done during the Apollo-Soyuz Flight Project, made it necessary to develop a method for ambient gravity cell electrophoresis by which a variety of non-polymer containing buffers could be tested prior to the actual flight experiments.[9] With the apparatus developed for that purpose (a cooled, vertical tube), cultured human lymphocytes were electrophoresed downward into a shallow, low molecular weight (Mw = 10,000) dextran gradient, as well as upward, straight into a non-gradient buffer. The descending method yielded hardly any enrichment of the (slower) B-lymphocytes at the cathodal end, and a significant enrichment but not an important purification of the (more negatively charged T-lymphocytes on the anodal side. The ascending method was more promising: although still no significant enrichment or purification of the slower B-lymphocytes appeared obtainable, the fastest fractions contained almost pure T-lymphocytes.[9] This method was further developed as it promised at least to yield the most highly charged cell fractions in a virtually pure state, on a preparative scale, by a rather simple procedure.[10]

However, the very high glucose or sucrose concentrations used in the cushion upon which the cells are layered[10] can be detrimental to the cells,[11,12] and even though low ionic strength buffers used in these methods also contain sizeable glucose concentrations, it is thought to be a distinct advantage to be able to eliminate at least the very high (>30%) sugar levels, by replacing them with D_2O. Also, if field strengths of >5 V/cm are to be sustained for any length of time, thermal convection is difficult to avoid in wholly unstabilized free liquid columns. However, fairly shallow density gradients made with D_2O/H_2O mixtures allow undisturbed ascending free liquid electrophoresis at up to 12 V/cm.[13]

METHODOLOGY

For the preparative ascending electrophoresis of $\approx 3 \times 10^8$ erythrocytes (or ≈ 1-2×10^7 lymphocytes) or other leukocytes or dispersed tissue cells a cylindrical tube (0.75 cm diameter), or a rectangular column (2.5 x 0.3 cm) of about 15 cm height is used. The bottom 4-6 cm of such a column is filled with a 2% agarose gel (set in the electrophoresis buffer). The columns are water-jacketed for adequate cooling (at 4°C jacket water temperature). A refrigerated water bath (LoTemptrol 154, Precision Scientific, Chicago, IL) is used for circulating cooled water through the water jacket. Wicks for the electrophoresis tubes consist of plugs of 2% agarose (Fisher Scientific, Rochester, NY) (dissolved in the same buffer as used for the electrophoresis) at the bottom and a filter paper bridge to the anode chamber at the top. About 0.5 ml of buffer (ρ = 1.124) prepared in D_2O is pipetted above the agarose plug. The cells suspended in 0.2 - 0.5 ml of a mixture of equal parts D_2O/H_2O are then layered on the D_2O buffer cushion. The rest of the tube is filled using 1 ml layers containing decreasing amounts of D_2O. This gives a density gradient of $\rho = 1.072$ at the point of sample insertion to $\rho = 1.020$ at the top (anode) of the electrophoresis tube. The wicks are connected immediately and an electric field of 9-12 V/cm is applied vertically across the column with the anode on top for up to 1 hour; the amperage should not exceed 7 mA. At the end of the electrophoresis separated zones of cells are visible. Samples of 0.5 to 1.0 ml each are taken from the top down by means of a 9 inch siliconized Pasteur pipet. Up to twelve fractions can generally be recovered. Phosphate buffers of low ionic strength (μ = 0.013, pH 7.3) are used, prepared in distilled water or heavy water (D_2O, Aldrich Chemical Co., Milwaukee, WI), consisting of 0.75 g anhydrous Na_2HPO_4, 0.10 g KH_2PO_4, 0.75 g NaCl, 1.0 g Na_2 EDTA and 80.0 g glucose for a total volume of 2 liters. The D_2O based buffer was usually prepared in 20 ml portions.

Erythrocytes of different species can be completely separated. The separation generally is visible within 10 minutes, and separated fractions can be obtained after runs of not more than 20 minutes;[13] see Figure 1.

Fig. 1. Electrophoretic separation of glutaraldehyde-fixed human and horse erythrocytes in a rectangular column (2.5 x 0.3 cm), ascending into a shallow D_2O gradient (from 80% to 0% D_2O in 4 cm). At left the mixture of 1.20×10^8 human and 1.46×10^8 horse cells (in 0.2 ml, 80% D_2O) had just been deposited on a pure D_2O cushion. Twenty minutes later (at 12 V/cm, 7 mA) total separation had occurred (right).[13] By microelectrophoresis[15] (in H_2O buffer) the mobilities of these cells were: 2.55 ($\pm$ 0.20 S.D.) μ/sec/V/cm for the human and 3.54 ($\pm$ 0.24 S.D.) μ/sec/V/cm for the horse erythrocytes. The buffer was a low ionic strength buffer (see test), pH 7.1, μ = 0.013.

Human peripheral lymphocytes can be fractionated in 55 minutes. However, even the fastest fraction rarely consists of more than 91-95% non-immunoglobulin-bearing cells; in contrast to the non-gradient levitation method.[9, 10]

Guinea pig pancreas cells (7×10^6) can be fractionated into 12 different fractions in 1 hour at 5 V/cm (at μ = 0.04); (van Oss *et al.*, in preparation).

DISCUSSION

Although there is much similarity between the older "levitation" ascending electrophoresis method[9, 10] (in which, apart from the cushion on which the cells are deposited, no gradient is used) and the D_2O density gradient ascending electrophoresis approach described here, there are important differences:

The levitation method is by far the simplest, and can be used in the cooled tubes described above at field strengths < 5 V/cm, at < 5 mA. Above these values intractable thermal convection occurs. However, the levitation method has consistently yielded much purer T cells in the fastest electrophoretic fractions of human peripheral lymphocytes. This is no doubt due to the fact that superimposed on the electrokinetic fractionation is the added effect of absence of the larger cells (generally B-cells) from the top fractions, due to sedimentation.

For pure electrophoretic fractionation, the D_2O density gradient method clearly is preferable. The separations possible in D_2O density gradient tubes suffer much less from convective disturbances, and thus permit better yields and allow the retrieval of <u>all</u> electrophoretic mobility fractions without contamination by sedimenting large cells or cell clumps. Also, D_2O by itself does not cause osmotic disturbances or engage in biochemical interactions, nor does it affect cell aggregation through polymer-binding or other mechanisms.[1,2] D_2O is not very expensive: the D_2O needed for one preparative electrophoretic cell separation as described above costs significantly less than $1.00. For electrophoretic mobility estimations it is necessary to realize that D_2O differs from ordinary water in its viscosity and dielectric constant, and thus in its conductivity. Also, its temperature of maximum density is higher than that of water (11°C instead of 4°C). The temperature inside the electrophoresis tube (cooled outside to 4°C) is 5-7°C, which is close to the temperature of maximum density of the average D_2O/H_2O mixture. The D_2O gradient is extremely simple, and will allow the preparative fractionation of up to several times 10^7 cells into 12 or more consecutive electrophoretic mobility fractions in a short time. The erythrocyte fractionation (see Figure 1) illustrates that, at the present state of the art and with simple means, a better resolution can now be achieved than was obtainable with the Apollo-Soyuz electrophoresis experiment in 1975.[14]

ACKNOWLEDGEMENT

Supported in part by NASA Contract NAS 8-32947 of the Marshall Space Flight Center.

REFERENCES

1. Brooks, D.E. (1973) J. Colloid Interf. Sci., 43, 714-726.
2. van Oss, C.J., Mohn, J.F. and Cunningham, R.K. (1978) Vox Sang., 34, 351-361.
3. Brooks, D.E. and Seaman, G.V.F. (1973) J. Colloid Interf. Sci., 43, 670-686.
4. Brooks, D.E. (1973) J. Colloid Interf. Sci., 43, 700-713.
5. Fike, R.M. and van Oss, C.J. (1973) Prep. Biochem., 3, 183-193.

6. Boltz, R.C., Todd, P., Streibel, M.T. and Louie, M.K. (1973) Prep. Biochem., 3, 383-401.

7. Griffith, A.L., Catsimpoolas, N., and Wortis, H.H. (1975) Life Sci., 16, 1693-1702.

8. Boltz, R.C., Todd, P., Gaines, R.A., Milito, R.P. Docherty, J.J., Thompson, C.J., Notter, M.F.D., Richardson, L.S. and Mortel, R. (1976) Histochem. Cytochem., 24, 16-23.

9. van Oss, C.J., Bigazzi, P.E., Gillman, C.F. and Allen, R.E. (1974) Proc. AIAA 12th Aerospace Sci. Mtg., 74, 211-213.

10. Gillman, C.F., Bigazzi, P.E., Bronson, P.M. and van Oss, C.J. (1974) Prep. Biochem., 4, 457-472.

11. van Oss, C.J. (1971) Infect. Immun., 4, 51-59.

12. van Oss, C.J., Gillman, C.F. and Good, R.J. (1972) Immunol. Commun., 1, 627-636.

13. Bronson, P.M. and van Oss, C.J. (1978) Prep. Biochem., in press.

14. Allen, R.E., Rhodes, P.M., Snyder, R.S., Barlow, G.H., Bier, M., Bigazzi, P.E., van Oss, C.J., Knox, R.J., Seaman, G.V.F., Micale, F.J. and Vanderhoff, J.W. (1977) Sep. Purif. Meth., 6, 1-59.

15. van Oss, C.J., Fike, R.M., Good, R.J. and Reinig, J.M. (1974) Anal. Biochem., 60, 242-251.

© 1979 Elsevier/North-Holland Biomedical Press
Electrokinetic Separation Methods,
P.G. Righetti, C.J. van Oss and J.W. Vanderhoff

ELECTROPHORETIC SEPARATION OF BIOLOGICAL CELLS IN MICROGRAVITY

J. W. VANDERHOFF AND C. J. van OSS

Center for Surface and Coatings Research, Lehigh University, Bethlehem, Pennsylvania 18015 (USA) and Department of Microbiology, State University of New York at Buffalo, Buffalo, New York 14214 (USA)

INTRODUCTION

The electrophoretic separation of biological cells has long been of interest in the biological and medical fields. The different characteristics of individual members of a given cell population often are manifested by differences in surface characteristics and, hence, electrophoretic mobilities, so that, in principle, these individual members may be separable by electrophoresis. The differences in electrophoretic mobility can be detected by analytical electrophoresis (e.g., microcapillary electrophoresis), but in only a few cases have significant proportions of a pure cell fraction been separated by preparative electrophoresis (e.g., continuous electrophoresis using the Beckman or Hannig instruments). Even in these cases, the resolution of separation has been poorer than desired and the amounts separated smaller than desired. Although the interest in electrophoretic separation has stimulated the development of several instruments, the goal of a routine method combining good resolution and adequate throughput has not yet been met.

Thus cell electrophoresis is presently at the same stage of development as was the electrophoresis of serum proteins in 1953, before the development of immunoelectrophoresis. At that time, only about a half-dozen different major electrophoretic mobilities were recognized, but few, if any, techniques were available to analyze the composition of each of these mobility fractions. With the development of immunoelectrophoresis[1,2] which allows the individual characterization of overlapping mobility fractions, the half-dozen known serum proteins grew to more than 30 in 10 years[3] and to more than 40 today[4].

At present, microelectrophoresis of human leukocytes detects no more than two or three mobility groups, but provides no capability for analyzing which cells belong in these groups, because of the small number of cells (10^3-10^5) examined. In comparison, more than 10^6 cells per fraction are needed to characterize the cells that each fraction comprises. Moreover, the electrophoretic separation of 10^7 or more cells is restricted by space and time requirements so that sedimentation of the cells is a limiting factor in addition to the usual thermal convection.

The NASA program to carry out scientific experiments in space provides an exciting prospect for electrophoretic separation of biological cells. The problems that beset ground-based separations would be moderated or eliminated by carrying out the separations in microgravity. The tendency of the cells to sediment and the thermal convection arising from Joule heating would be eliminated, leaving only such second-order effects as surface tension-driven convection to interfere with the separation. These advantages promise significant improvements in the resolution of separation and thus justify the difficult and expensive development of the flight hardware and ground-based supporting facilities to carry out electrophoretic separations in space. Therefore, NASA has initiated a program for electrophoretic separation of biological cells in microgravity with a resolution better than can be attained on earth and in quantities adequate for the intended scientific application.

RATIONALE FOR ELECTROPHORETIC SEPARATION OF CELLS IN MICROGRAVITY

The principal advantage of electrophoretic separation in microgravity is weightlessness, which obviates two problems that arise during electrophoretic separations of cells: sedimentation and fluid disturbances caused by local differences in temperature (and hence density). These fluid disturbances which arise from Joule heating by passage of the electrical current will be referred to as 'thermal convection', although the temperature variation of other parameters not affected by weightlessness, e.g., surface tension and viscosity, may also give rise to thermal convection.

The elimination of sedimentation in microgravity is important in the electrophoretic separation of cells that would sediment significantly on earth during the time required for their separation.

The elimination of thermal convection in microgravity is also important because it would allow the use of higher voltage gradients than would be permissible on earth. Of course, the use of higher voltage gradients (resulting in higher temperatures) is limited by the possible thermal denaturation of the biological cells and such second-order effects as surface tension-driven convection, electrical and magnetic effects, which are not influenced by microgravity.

The elimination of these two problems of electrophoretic separation would justify the expense and difficulty of carrying out the electrophoretic separations in microgravity.

Since sedimentation is a major problem of electrophoretic separation on earth than can be obviated in microgravity, the biological cells selected for separation in microgravity should: (i) sediment on earth during the time required for separation; (ii) merit electrophoretic separation. All particles or biological cells with particle diameters greater than 1μm and with densities of 1.05-1.10 gm/cc tend to sediment a significant distance (more than 1cm) in aqueous media during the time required to separate them by preparative electrophoresis. This category includes virtually all living cells. Several types of living cells are known --- and many more can be reasonably expected --- to have different electrophoretic mobilities linked to different biological properties[6-9]. A significant number of these merit preparative electrophoretic separation for research or production purposes, e.g., lymphocytes, stem cells (bone marrow), sperm cells, β cells of the islets of Langerhans (insulin-producing pancreas cells), kidney cells (producing urokinase and erythropoietin), and pituitary cells (producing various hormones such as somatotropic hormone)[9].

Particles smaller than 1μm in size do not sediment significantly during the time required for separation, and polymer molecules smaller than 0.1μm can be electrophoresed in gel phase without fluid disturbances arising from thermal convection. Thus there is no compelling reason to consider proteins or other soluble biopolymers as candidates for electrophoretic separation in microgravity because they do not sediment and can be protected from convection disturbances by gels or other porous media.

Three methods of preparative electrophoresis have been used for the separation of cells in ground-based experiments: (i) continuous flow (steady state) electrophoresis[10-18] (see Chapter 8 for a more complete description of this method), in which gravitational and thermal effects are obviated by the constant-rate flow of aqueous medium between the electrodes perpendicular to the electric field; this method is technically feasible, but difficult to carry out; moreover, the resolution of separation at a reasonable throughput rate is not as good as desired; (ii) vertical density-gradient electrophoresis, which is hampered by the fact that the polymers used to make the density gradient also tend to clump the cells[19,20]; (iii) vertical levitation electrophoresis, which permits the separation of only the fastest-moving cells[21]. None of these ground-based methods, at least in their present form, can offer the combination of excellent resolution of separation and adequate throughput rate than can be achieved with electrophoretic separation in microgravity.

Two methods of preparative electrophoretic separation can be used in microgravity: (i) free-fluid (static or stationary) electrophoresis in a closed cylindrical channel; (ii) continuous electrophoresis with the electric field perpendicular to a continuous flow of medium between two parallel plates.

Free-fluid electrophoresis was selected for the Apollo 14[22], Apollo 16[23,24], and Apollo-Soyuz[5] experiments because of: (i) its simplicity; (ii) its capability of allowing consecutive or simultaneous experiments during one flight; (iii) the advantage of having standby channels ready in case of malfunction; (iv) the absence of hydrodynamic problems; (v) its insensitivity to voltage gradient fluctuations. The disadvantages of this method are that the sample size is limited and the resolution can be improved only by increasing the length of the channel and the voltage gradient.

Continuous electrophoresis was selected for an experiment carried out on the Apollo-Soyuz mission[25]. The advantages of this method are: (i) relatively large sample size; (ii) relatively short time of electrophoresis; (iii) good resolution of separation combined with adequate throughput. The disadvantages are: (i) complexity of design and operation; (ii) sensitivity to variations in the hydrodynamic flow pattern; (iii) sensitivity to fluctuations in voltage gradient.

ELECTROPHORETIC SEPARATIONS DURING THE APOLLO 14 AND APOLLO 16 MISSIONS

The first electrophoretic separations in microgravity were carried out during the Apollo 14 mission[22]. The separation of mixtures of red and blue dyes, and of hemoglobin and DNA, was attempted using free-fluid (static) electrophoresis. The apparatus comprised three parallel channels 0.65cm in inside diameter and 10cm in length, with the electrodes separated from the chamber by semi-permeable membranes and rinsed continually to remove any decomposition products. The samples to be separated were confined in sample chambers with impermeable membranes. To start the experiment, the impermeable membranes were removed to release each sample to its channel, and the voltage gradient was applied. The dyes separated as expected, but the protein solution was altered by the long time between loading the apparatus and starting the experiment.

The Apollo 16 experiment[23] used a similar apparatus to separate 0.23μm and 0.80μm-diameter monodisperse polystyrene latex particles; one channel contained a mixture of the two different-size latexes and the other two channels, each latex alone. Upon electrophoresis, all three samples formed elongated parabolic bullet-shaped profiles (see Figure 5 of Chapter 5); for the faster-moving 0.80μm-size particles, the tip of this bullet-shaped profile approached the opposite end of the 10cm-long channel by the time all of the sample cleared the sample chamber. The tips of the profiles for the 0.80μm-size particles and the particle mixture moved at about the same rate, and that for the 0.23μm-size particles moved more slowly, as expected. The elongated bullet-shaped profiles were the result of a strong backward electroosmotic flow of the medium along the wall of the channel. For the channel containing the particle mixture, the separation was obscured, but presumably comprised two nested parabolic cones, with that of the slower-moving particles nested inside that of the faster-moving particles.

In comparison, ground-based experiments carried out under analogous conditions produced particle profiles that persisted for only a few seconds before the samples became completely roiled. Thus electrophoretic separation in space produced well-defined particle profiles (albeit elongated and bullet-shaped) whereas ground-based separation produced complete mixing by thermal convection.

These results were analyzed using the flight photographs and ground-based measurements[24]. This analysis was complicated by the appearance of bubbles in the electrophoresis channels, which obviously reduced the nominal voltage gradient of 26 volts/cm. The electrophoretic mobilities of the 0.23μm and 0.80μm-diameter particles measured in the same medium used for the experiment were 5.8 and 9.2μm cm/volt sec, respectively. From these and other ground-based measurements, it was deduced that the electroosmotic mobility of the glass channel wall in the medium used was 10μm cm/volt sec. Using these values, and the actual values of the voltage gradients (12-15 volts/cm for the three channels according to the individual bubble diameters), the two nested parabolic particle profiles were calculated as a function of separation time. To check these calculated profiles, a color densitometric scan was made of the black-and-white flight photographs, which showed clearly the parabolic cone of 0.23μm-size particles nested inside that of the 0.80μm-size particles, in good agreement with the calculated particle profiles.

This Apollo 16 experiment demonstrated the feasibility of electrophoretic separations in space and the correlation of the experimental results with theoretical predictions. Unfortunately, the separation of the different-size particles was not in a practical form because of the strong electroosmotic flow, and it was obvious that this electroosmotic flow would have to be moderated or eliminated for such a free-fluid electrophoretic separation to yield recoverable separated fractions.

DEVELOPMENT OF LOW-ELECTROOSMOTIC-MOBILITY COATINGS (see Chapter 5 for a more complete description of this subject)

The foregoing experiments demonstrated the need for elimination of electroosmosis if free-fluid electrophoretic separations in microgravity were to succeed. Electroosmosis is analogous to electrophoresis: it arises from the charge on the channel walls as compared with the charge on the particles themselves; the difference lies in the fact that all particles are free to move in an electric field according to their charges, but the channel walls cannot move, so the liquid medium moves along the channel wall. This movement is in the opposite direction to the electrophoretic migration and, in a closed system, reverses itself at the end of the channel, to move in the center of the channel in the same direction as the particles.

To obviate this electroosmotic flow requires that the charge on the channel wall be eliminated or covered up. Three possible approaches to solve this problem are: (i) prepare the electrophoresis channels from some material which bears no charge; (ii) cover the channel wall with an uncharged material which will render the wall hydrophobic and therefore not wettable by the aqueous medium; (iii) cover the channel wall with an uncharged material that will render the wall hydrophilic and uncharged. The first approach is undesirable because it limits the choice of materials from which the electrophoresis channels may be constructed. In the event, three different materials---Pyrex glass, Lexan polycarbonate, Plexiglas polymethyl methacrylate---were considered for the electrophoresis channels for the Apollo-Soyuz experiments, and it was necessary to develop a low-electroosmotic-mobility coating for each. Both of the other approaches were tried and found to be successful in eliminating electroosmosis[26].

For the first of the two approaches, γ-aminopropyltriethoxysilane (A-1100,

Union Carbide Corp.) applied on a Pyrex glass substrate gave an electroosmotic mobility close to zero. However, this approach was not developed further, as the hydrophobic substrate it presented might serve as a substrate for adsorption of ionic surface-active materials that may be contained in the sample to be separated.

The second approach was developed and eventually used for the Apollo-Soyuz Test Program (ASTP) MA-011 experiments[5]. The coating used comprised coating the glass electrophoresis channel with a solvent solution of γ-glycidoxypropyltrimethoxysilane (Z-6040; Dow Corning Corp.), followed by an aqueous solution of methylcellulose (Methocel; The Dow Chemical Co.) and baking to cause reaction of the glycidoxy and hydroxyl groups. The channels were then rinsed thoroughly with water to remove unbound methylcellulose. The proposed mechanism of binding comprises the chemisorption of the γ-glycidoxypropyltrimethoxysilane on the glass surface, condensation of adjacent silanol groups to form a two-dimensional "skin" of chemisorbed silane containing glycidoxy groups, reaction of the glycidoxy groups of the chemisorbed silane with hydroxyl groups of the methylcellulose, and removal of the unbound methylcellulose. The removal of unbound methylcellulose is an important part of this process; its presence was detected by allowing monodisperse polystyrene latex particles to stand in the coated channels for various times; the unbound methylcellulose desorbs from the channel wall and adsorbs on the latex particles causing their electrophoretic mobility to decrease, eventually to zero. Earlier experiments using the physical adsorption of methylcellulose crosslinked with aldehydes provided a less-permanent coating[27].

Other experiments[26] demonstrated that low-electroosmotic-mobility coatings could be applied to Lexan polycarbonate electrophoresis channels using the same process described above, and to Plexiglas polymethyl methacrylate channels by washing with aqueous sodium hydroxide to hydrolyze surface ester groups to the corresponding carboxyl groups, followed by reaction of the carboxyl groups of the Plexiglas surface with the hydroxyl groups of the methylcellulose.

The ground-based experiments were carried out using microelectrophoresis cells with replaceable center sections[26]. For the experiments using Pyrex glass substrates, the replaceable center sections were glass capillary melting point tubes which were coated in various ways and inserted in the cell for measurement. The electroosmotic mobility was determined by measuring the electrophoretic velocity of monodisperse polystyrene latex particles of known mobility as a function of depth in the tube and plotting this velocity as a function of the square of the distance, to form two intersecting lines that extrapolate to the electroosmotic mobility at the cell wall and interpolate to the electrophoretic mobility at the stationary level. For experiments using Lexan polycarbonate and Plexiglas polymethyl methacrylate substrates, the replaceable center sections were rectangular channels fabricated from sheet plastic, which were coated in various ways and inserted in the cell for measurement, and the electroosmotic mobility was measured as described above.

Using these methods, all three materials of construction can be coated to provide an effective low-electroosmotic-mobility coating.

ELECTROPHORETIC SEPARATIONS DURING THE APOLLO-SOYUZ MISSION

A series of electrophoretic separation experiments were carried out on the Apollo-Soyuz mission. One series (MA-011) comprised the free-fluid electrophoretic separation of human B and T lymphocytes, human kidney cells, and mixed human, rabbit, and horse fixed red blood cells, and the free-fluid isotachophoretic separation of human and rabbit red cells[5]. Another series (MA-014) comprised the continuous electrophoretic of a mixture of human and rabbit erythrocytes, rat bone marrow cells, spleen cells and lymphnode cells[25].

For the free-fluid electrophoretic separation of human B and T lymphocytes and human kidney cells, the apparatus design used for the Apollo 16 experiment was modified so that the electrophoretic separation chamber was replaceable; duplicate chambers containing the frozen cell sample were inserted, thawed, and subjected to electrophoresis; after separation, the samples were refrozen for return to earth, where the frozen sample was sectioned to give the separated cell fractions. To facilitate the freezing, a cooling chamber was placed over the electrophoresis chamber, which prevented the photographing of the separation.

The two samples of human B and T lymphocytes failed to give a satisfactory separation, in both cases, because of instrument malfunction. Of the two samples of human kidney cells, one failed to separate because of instrument malfunction, but the other separated satisfactorily to give fractions that were successfully cultured and grown after return to earth. One of these fractions is producing urokinase at a rate greater than has been observed before. This experiment must be repeated to confirm this exciting result, but it gives rise to the expectation that the electrophoretic separation of biological cells in microgravity can lead to ground-based production processes.

The electrophoretic separation of mixed human-rabbit-horse fixed red blood cells was intended as a demonstration experiment; therefore, the experiment was carried out without the cooling mantle, and the separation was photographed. This experiment comprises a rigorous test of the electrophoretic separation apparatus because the electrophoretic mobility distributions of these three samples overlapped, even though the average values (1.6, 1.9, and 2.4μm cm/volt sec for rabbit, human, and horse fixed red blood cells, respectively) were different. Obviously, the overlapping regions could not be separated, but it was hoped that the separation of reasonably pure fractions could be demonstrated. In the event, of the two experiments tried, one failed because of instrument malfunction, and the other worked satisfactorily.

Analysis of the flight photographs by microdensitometry and comparison with theoretical predictions based on the initial distribution of electrophoretic mobilities and the electroosmotic mobility of the electrophoresis channel showed that the general movement of the cells along the electrophoresis channel was in excellent agreement with theory, but that the separation of the individual peaks was not as well-defined as predicted[28]. One possible explanation advanced for the poorer-than-expected separation was too-high a concentration of the red blood cells, which gave rise to particle-particle interactions that interfered with the separation[28]. Nevertheless, the close agreement between the observed movement of the particle mixture and the predictions of electrophoresis theory shows that it is possible to predict the electrophoretic separation of a cell sample from the measured distribution of electrophoretic mobilities and the conditions of separation.

The two free-fluid isotachophoretic separations of fixed and live human-rabbit mixed red blood cells failed to give significant separation, presumably because the electrolyte intended to be interposed between the two fractions failed to function in this manner[5].

The continuous electrophoretic separation of human and rabbit cells failed to give a record of the separation. It is believed that the separation was satisfactory, but the ultraviolet lamp that was to have provided detection of the separated particle streams illuminated the sample with such intensity (such lamps burn more brightly in microgravity than on earth) that it overpowered the detector, giving only very faint traces of the particle streams[25].

It should be emphasized in evaluating the results of these electrophoretic separation experiments that these experiments can be performed only once in microgravity and that the experiment must work perfectly on that first attempt, a fortunate occurrence which is not often observed with ground-based experiments. This situation allows little margin for error, and it is not unexpected that some experiments fail because of unforeseen circumstances, as shown above. Nevertheless, the experiments carried out thus far demonstrate the potential of these electrophoretic separations in microgravity. Moreover, the excellent correlation between the experimental results and predictions based on the theory of electrophoresis demonstrate that the separations observed are in accord with theory and, therefore, that future separations in microgravity can be predicted with considerable accuracy.

RESOLUTION OF ELECTROPHORETIC SEPARATION IN MICROGRAVITY

The free-fluid electrophoretic separations carried out on the Apollo 16 and Apollo-Soyuz missions could not have been carried out on earth; ground-based experiments using apparatus of the same dimensions and the same conditions produced complete roiling of the sample within a few seconds. Ground-based experiments could be carried out in this apparatus using a very small voltage gradient and a very long separation time; however, the resolution of such an experiment would suffer because of Brownian diffusion of the particles, which becomes significant at very long separation times. Therefore, since separations in this apparatus can be carried out practically only in microgravity, it is useful to determine the resolution of separation that can be attained.

There are many ways to define the resolution of separation in electrophoresis; however, the most useful and pragmatic approach, which allowed comparison of one method of electrophoretic separation with another, is to assume that the sample comprises particles of two different electrophoretic mobilities and that the resolution of separation is equal to the minimum difference in electrophoretic mobility $\Delta U_{e_{min}}$ that can be separated without overlap[29].

For an ideal free-fluid electrophoretic separation, the value of $\Delta U_{e_{min}}$ is defined as[29]:

$$\Delta U_{e_{min}} = (U_e + U_{os})(\theta/L) + 2R^2 U_{os} \tag{1}$$

where $\underline{U}_e$ is the electrophoretic mobility, $\underline{U}_{os}$ the electroosmotic mobility of the channel wall, θ the thickness of the sample plug in the direction of electrophoresis, $\underline{L}$ the length of the electrophoresis channel, and $\underline{R}$ the ratio of sample plug diameter to channel diameter. The term "ideal" means the best resolution that can be attained under these conditions, i.e., all aberrations will decrease this calculated resolution of separation.

As can be seen, the value of $\Delta U_{e_{min}}$ depends upon the ratio of sample plug diameter to the electrophoresis channel diameter as well as the length of the electrophoresis channel (or time of separation). For biological cells of electrophoretic mobility 2.0μm cm/volt cm, the resolution of separation in this apparatus is 0.04μm cm/volt sec. The resolution of separation could be improved by decreasing the sample plug diameter/channel diameter ratio (at the expense of sample throughput) or by increasing the length of the channel or time of separation (the more likely approach).

The sample size separated in this apparatus is sufficient to recover fractions large enough to be cultured and grown on earth, i.e., sufficient for practical industrial applications.

Continuous electrophoresis was developed to obviate the effects of gravity. An electrolyte curtain is pumped downward through a thin flat cell (e.g., 0.15 cm depth, 5 cm width, 50 cm length for the Beckman CPE) between the electrodes (e.g., 30 cm length for the Beckman CPE) and out the bottom into the collecting tubes. Thermal convection in such an electrophoresis channel would tend to make the electrolyte rise, i.e., flow upwards against the electrolyte curtain flow. Therefore, the electrolyte curtain flow rate is increased to the point where the upward flow due to thermal convection is obviated. Only under these conditions of stable laminar flow can an electrophoretic separation be carried out successfully. (It is of interest that it has been proposed to invert the continuous electrophoresis channel and allow the thermal convection to determine the particle flow profile[30], an interesting but impractical possibility).

For an ideal continuous electrophoretic separation, the value of $\Delta U_{e_{min}}$ is defined as[29]:

$$\Delta U_{e_{min}} = \{U_e + (U_{os}/2)\}(\theta/L) \qquad (2)$$

where θ is the diameter of the sample stream, $\underline{L}$ the separation distance of the electrophoresis channel, and $\underline{R}$ the ratio of sample stream diameter to channel thickness.

As can be seen, in this method, too, the value of $\Delta U_{e_{min}}$ depends upon the ratio of the sample stream diameter to channel depth. For biological cells of electrophoretic mobility 2.0μm cm/volt cm and an electroosmotic mobility of the same value (the most favorable case), the resolution of separation in the Beckman CPE at a sample flow rate of less than 1 cc/hour is 0.04μm cm/volt cm. The resolution of separation can be improved by decreasing the sample stream diameter/channel depth ratio (again at the expense of sample throughput) or by decreasing the electrolyte curtain flow rate or increasing the voltage gradient.

This last approach demonstrates a way in which the resolution of separation of continuous electrophoresis can be improved substantially in microgravity. For example, with typical electrolyte media in the Beckman CPE, an electrolyte curtain flow rate of 10 cc/min combined with a voltage gradient of 100 volts/cm represents a limit of operation; if the electrolyte curtain flow rate is decreased further or if the voltage gradient is increased further, thermal convection interferes with the stable laminar flow. If the separation were carried out in microgravity so that thermal convection would be absent, the resolution of separation would be improved substantially by a decrease in electrolyte curtain flow rate (which would allow the particles to remain between the electrodes for a longer time) or an increase in voltage gradient (which would cause the particles to migrate more rapidly). This approach to improve the resolution would be limited only by the second-order effects not obviated by microgravity (e.g., surface tension-driven convection, electrical effects, magnetic effects).

Of course, as mentioned above, the resolution of separation of continuous electrophoresis can be improved by decreasing the sample stream diameter/channel depth ratio. When this ratio is decreased keeping the channel depth constant, the sample throughput would be decreased substantially; when this ratio is decreased keeping the sample stream diameter constant, the stable laminar flow cannot be maintained, at least on the ground. Such an approach could be used for a continuous electrophoretic separation instrument designed to operate in microgravity. This is the approach taken in the development of the General Electric continuous electrophoretic separator designed to operate in microgravity[31]. This instrument uses a channel 0.5 cm in depth, 5 cm in width and 10 cm in length. In principle, it can give the same throughput as the Beckman CPE at a substantially improved resolution of 0.01μm cm/volt sec or a 10-fold greater throughput at the same resolution of 0.04μm cm/volt sec. This instrument is tentatively scheduled to be tried in a sounding rocket flight during 1980.

In summary, the microgravity environment is a promising one to improve the resolution of separation, sufficiently so that it justifies the difficulty and expense of the development of the flight hardware.

CANDIDATES FOR SEPARATION IN MICROGRAVITY DURING SPACELAB/SPACE SHUTTLE MISSIONS

NASA has now initiated the Spacelab/Space Shuttle program, which will put a scientific laboratory in orbit routinely for periods of about one week. These missions will offer an incomparable opportunity for electrophoretic separation of biological cells. Both free-fluid (e.g., adaptations of the Apollo-Soyuz apparatus) and continuous (e.g., further developments of the General Electric instrument) electrophoretic separation instruments could be used for these separations.

The number of biological samples that can be subjected to electrophoretic separation in microgravity is limited, at least for the near future. Therefore, critera must be developed for the selection of candidate samples. The criteria proposed for this selection are: (i) the importance of the separation to the biological and medical fields; (ii) extensive ground-based separations supporting the importance of the separation, yet demonstrating that ground-based separations do not give the desired resolution; (iii) the expectation that an improved resolution of separation would yield separated fractions of scientific importance; (iv) the availability of ground-based testing methods to determine the efficacy of separation; (v) the expectation that the separated fraction would be used by the scientific community.

The importance of a given cell separation can be determined from the interest given it by the scientific community. Examples of worthwhile separations were mentioned earlier and are described in detail in Tables I and II. Other examples would certainly be proposed by the scientific community once the possibility of such separations in microgravity becomes known and the first successful electrophoretic separations have been achieved.

The second criterion could be satisfied by the recommendations of a representation or committee drawn from the scientific community, that would examine the scientific literature for worthwhile candidates and select the most promising candidates for trial in space. Such a program was initiated in 1976 by the Biological Separations Committee of the Universities Space Research Association (USRA)[9], which was established to advise NASA on the technical aspects of this program. The expectation that improved resolution would prove to be of importance for a given separation is, of course, a matter of scientific judgement. Nevertheless, the informed judgement of a committee drawn from the scientific community would give reasonable assurance that a given ground-based separation could benefit from transfer to microgravity. Such judgements must be made using the available ground-based data, but it is likely that the most promising candidates for separation in microgravity could be winnowed out without difficulty.

The ground-based tests and facilities to determine the efficacy of separation are also needed to determine experimentally if the separation in microgravity is demonstrably superior to the analogous separation on earth.

Similarily, the expectation that the separated fractions would prove of value to the scientific community is also a matter of judgement. There is little doubt, however, that the separation of a fraction rich in the antibody-producing human lymphocytes, urokinase-producing kidney cells, or insulin-producing β cells of the islets of Langerhans would be of importance to the scientific community and mankind in general. Other examples are certain to be developed as the program proceeds.

The foregoing criteria emphasize the need for organization of the search to elicit promising candidates for separation in microgravity. A committee should be established, or the present Subcommittee of the USRA Biological Separation Committee[9] should be empowered, to search out the most promising candidates for separation in microgravity, make sure that the requisite ground-based experiments have been performed, prepare the most promising candidates for separation in microgravity, and ensure that the post-flight testing and application protocols are established and carried out. Such an organized program would ensure that no promising candidate would be overlooked and that those candidates selected would be separated under optimum conditions, and also that the evaluation and commercial applications of the separated fractions would proceed as planned.

Promising candidates for electrophoretic separation during the Spacelab/Space Shuttle missions include, not only human lymphocytes and kidney cells, which show a distribution of electrophoretic mobilities that can be related to differences in surface characteristics, but also tissue cells with different functions, which, too, may have different electrophoretic mobilities. Thus many types of endocrine cells as well as endocrine tissue cells would merit consideration as candidates for electrophoretic separation in microgravity. Table I lists various cell candidates and Table II, tissue cell candidates[9]. Among these, the most promising candidates for separation on early Spacelab/Space Shuttle flights are β cells on the islets of Langerhans, human kidney cells, human B and T lymphocytes, and sperm cells.

TABLE I

Candidate Monodisperse Cells for Electrophoretic Separation in Microgravity

CELL CLASS	CELLS	SEPARATION	SCALE	APPLICATION	
				Clinical Research	Clinical Application
Blood Cells	Lymphocytes	Separation into B & T Cells and into subclasses of these	Preparative	Study of Biological Properties of Narrow Electrophoretic Fractions	Grafting of Pure Fractions in Various Immune Disorders and Leukopenias
	Monocytes	Separation into Electrophoretic Subclasses	Preparative & Analytical	Study of Biological Properties of Electrophoretic Subclasses	
	Granulocytes	Separation into Electrophoretic Subclasses	Preparative & Analytical	Study of Biological Properties of Electrophoretic Subclasses	Possible Transfusion of Pure Fractions in Patients with Leukopenias
Reticuloendothelial System	Macrophages	Separation into Electrophoretic Subclasses	Preparative & Analytical	Study of Biological Properties of Electrophoretic Subclasses	
Sperm Cells	Spermatozoids	Separation into Male & Female Generating Cells	Preparative		Selection of Sex of Offspring, through Artificial Insemination
Bone Marrow	Stem Cells	Separation from Other Bone Marrow Cells; Separation into Electrophoretic Subclasses	Preparative & Analytical	Study of Biological Properties of Electrophoretic Subclasses	Grafting of Pure Fractions in Various Immune Disorders and Leukopenias
	Megakaryocytes	Separation into Electrophoretic Subclasses	Preparative & Analytical	Study of Biological Properties of Electrophoretic Subclasses	
	Plasmacells	Separation into Electrophoretic Subclasses	Preparative & Analytical	Study of Biological Properties of Electrophoretic Subclasses	Grafting of Pure Fractions in Humoral Immune Deficiencies

TABLE II

Candidate Tissue Cells for Electrophoretic Separation in Microgravity

TISSUE	CELLS	SEPARATION	SCALE	APPLICATIONS	
				Clinical Research	Clinical Application
Pancreas	β Cells of Islets of Langerhans	Separation of Insulin-producing Cells	Preparative		Insulin Production by Tissue Culture Cells; Transplantation of Pure β Cells
	Exocrine Cells	Separation into Different Fractions	Preparative	Isolation and Characterization of Different Cells Producing Essential Digestive Enzymes	
Kidney	Kidney Cells	Separation of Urokinase-producing Cells	Preparative		Urokinase Production by Tissue Culture Cells
		Separation of Erythropoietin-producing Cells	Preparative		Erythropoietin Production by Tissue Culture Cells; Transplantation of Erythropoietin-producing Cells
Anterior Pituitary Gland	Pituitary Cells	Separation of, e.g., Somatotropic Hormone (STH)-producing Cells (e.g., from Acromegalics)	Preparative		STH Production by Tissue Culture Cells; Implantation of Pure Cell Fractions into 3rd Ventricle of the Brain of STH-deficient Patients
(MANY OTHER CELL-TYPES PRODUCING VARIOUS OTHER FACTORS NOT MENTIONED HERE ALSO BELONG IN THIS TABLE)					
Placenta	Syntrophoblasts	Separation from Other Cells; May on Occasion also Occur in Low Numbers in Circulating Blood of Pregnant Women	Preparative and Analytical	Study of Properties of Pure Syntrophoblasts	Diagnosis of Choriocarcinoma

SUMMARY

Electrophoretic separation of biological cells in microgravity offers an exciting prospect for the separation of fractions with specific biological functions. The weightlessness of microgravity would eliminate the sedimentation of the cells and the thermal convection of the electrophoretic separation process. The potential benefits are a resolution of separation far better than can be attained on earth combined with a throughput rate adequate to support the culturing and growth of the separated cell fractions. Preliminary experiments carried out during the Apollo 14, Apollo 16, and Apollo-Soyuz missions demonstrate the feasibility of such separations using both free-fluid and continuous flow electrophoresis. Criteria for the selection of candidates for separation are reviewed, and suggestions are made for future separations during the Spacelab/Space Shuttle missions.

REFERENCES

1. Grabar, P. and Williams, C. A. (1953) Biochim. Biophys. Acta, 10, 193-194.
2. Grabar, P. and Burtin, P. (1964) Immuno-Electrophoretic Analysis (Application to Human Biological Fluids), Elsevier, New York.
3. Schultze, H. E. and Heremans, J. F. (1966) Molecular Biology of Human Proteins (With Special Reference to Plasma Proteins), Vol. 1, Elsevier, New York, p. 173 ff.
4. Putnam, F. W. (1975) in The Plasma Proteins, F. W. Putnam, ed., Vol. 1, Academic Press, New York, p. 57 ff.
5. Allen, R. E., Rhodes, P. H., Snyder, R. S., Barlow, G. H., Bier, M., Bignazzi, P. E., Van Oss, C. J., Knox, R. J., Seaman, G. V. F., Micale, F. J., and Vanderhoff, J. W. (1977) Sep. Purif. Methods, 6, 1-59.
6. Ruhenstroth-Bauer, G. (1965) in Cell Electrophoresis, E. J. Ambrose, ed., Little, Brown & Co., Boston, 66-75.
7. Wioland, M., Sabolovic, D., and Burg, C. (1972) Nature New Biol., 237, 274-275.
8. Seaman, G. V. F. (1975) in The Red Blood Cell, D. M. Surgenor, ed., Academic Press, New York, Vol. 2, 1135-1229.
9. Van Oss, C. J. et al. (1976) Report of Subcommittee on Candidates for Separation at Zero Gravity, USRA Committee on Biological Separations.
10. Hannig, K. (1961) Z. Anal. Chem., 181, 244-254.
11. Hannig, K. (1964) Hoppe-Seyler's Z. Phys. Chem., 338, 211-227.

12. Strickler, A., Kaplan, A., and Vigh, E. (1966) Microchem. J., 10, 529-544.

13. Strickler, A., and Sachs, T. (1973) Prep. Biochem., 3, 269-277.

14. Strickler, A., and Sachs, T. (1973) Ann. N. Y. Acad. Sci., 209, 497-514.

15. McCann, G. D., Vanderhoff, J. W., Strickler, A., and Sachs, T. (1973) Sep. Purif. Methods, 2, 153-198.

16. Hannig, K., and Heidrich, H. G. (1974) Methods Enzymol., 21, 746-761.

17. Hannig, K., Wirth, H., Meyer, B. H., and Zeiller, K. (1975) Hoppe-Seyler's Z. Phys. Chem., 356, 1209-1223.

18. Micale, F. J., Krumrine, P. H., and Vanderhoff, J. W. (1976) Colloid Interf. Sci. 4, 247-258.

19. Van Oss, C. J., Mohn, J. F., and Cunningham, R. K. (1977) Proc. 5th Int. Convoc. Immunol., Karger, Basel, 56-64; ibid. (1978) Vox Sang. 34, 351-361.

20. Brooks, D. E. (1973) J. Colloid Interf. Sci., 43, 714-726.

21. Gillman, C. F., Bigazzi, P. E., Bronson, P. M., and Van Oss, C. J. (1974) Prep. Biochem., 4, 457-472.

22. McKannan, E. C., Krupnick, A. C., Griffin, R. N., and McCreight, L. R. (1971) NASA TMX-64611, Electrophoresis Separation in Space - Apollo 14.

23. Snyder, R. S., Bier, M., Griffin, R. N., Johnson, A. J., Leidheiser, H., Micale, F. J., Vanderhoff, J. W., Ross, S., and Van Oss, C. J. (1973) Sep. Purif. Methods, 2, 259-282.

24. Micale, F. J., Vanderhoff, J. W., and Snyder, R. S. (1976) Sep. Purif. Methods, 5, 361-383.

25. Hannig, K., Wirth, H., and Schoen, E. (1977) Apollo-Soyuz Test Project, Vol. 1, NASA SP-412, Washington, D. C., 335-352.

26. Vanderhoff, J. W., Micale, F. J., and Krumrine, P. H. (1977) Sep. Purif. Methods, 6, 61-87.

27. Hjertén, S. (1967) Chromatog. Rev., 9, 122-219.

28. Vanderhoff, J. W., Micale, F. J., and Krumrine, P. H. (1978) Sep. Purif. Methods, 8, (2).

29. Micale, F. J., and Vanderhoff, J. W. (1976) letter to the USRA Committee on Biological Separations, March.

30. Ostrach, S. (1976) ESA Spec. Pub. No. 114, 141-147.

31. McCreight, L. R. (1977) NASA TMX-58191 JSC-11582 Bioprocessing in Space-Electrophoresis for Biological Production, 143-158; McCreight, L. R., Griffin, R. N., and Locker, R. J. (1976) ESA Spec. Pub. No. 114, 175-179.

ACKNOWLEDGEMENT

The authors gratefully acknowledge the support of the George C. Marshall Space Flight Center by NASA Contract NAS8-32947.

© 1979 Elsevier/North-Holland Biomedical Press
Electrokinetic Separation Methods,
P.G. Righetti, C.J. van Oss and J.W. Vanderhoff

ANALYTICAL POLYACRYLAMIDE GEL ELECTROPHORESIS IN MULTIPHASIC BUFFER SYSTEMS ("DISC ELECTROPHORESIS").

ANDREAS CHRAMBACH
Endocrinology and Reproduction Research Branch, National Institute of Child Health and Human Development, National Institutes of Health, Bethesda, MD 20014.

INTRODUCTION

Nearly two decades ago, polyacrylamide gel electrophoresis originated in two forms: one associated with continuous buffers[1], the other with discontinuous buffer phases[2]. Increasingly, it is the latter form which has gained in popularity for reasons of allegedly better resolution[3], of automatic sample concentration and of the availability of a sharply defined dye front. This review will therefore be mainly concerned with PAGE in discontinuous buffers ("disc electrophoresis") or, to use the more general and descriptive term, PAGE in multiphasic buffer systems ("multiphasic zone electrophoresis" (MZE)[4]).

PAGE started as a qualitative separation method, with subjective criteria of recognition for individual zones and pattern analysis by "band counting". Frequently, rapidly moving zones were erroneously associated with small, slowly moving zones with large molecular weights. Complex gel patterns were subjectively interpreted by pinpointing the presence or absence, increase or decrease of particular zones. Procedurally, these "dark ages of PAGE" were characterized by use of a single gel concentration (7.5%), one or two pHs (10 and 4), and lengthy staining-destaining. Much of the present-day practice of PAGE still follows these original lines, in particular the clinical analysis. The original, rudimentary PAGE technique has gained new importance and popularity in the application of the one-(pH), one-(gel concentration) technique to molecular species differing only in size, not in charge density. These are either the SDS-derivatives of proteins or nucleic acids. Since pH under those circumstances does not come into play, and since for all molecules having the same free charge density (y-intercept on the Ferguson plot) the ratio of migration distances between protein molecular weight standards and unkowns remains the same, the original PAGE technique, at a single pH (any pH above neutrality) and at a single gel concentration yields information regarding molecular weight. Hence, the fractionation of nucleic acids and SDS-proteins according to molecular weight by "original PAGE" has become widespread to the point of exclusion, in the minds of many biochemists, of any other form and applicability of PAGE.

Over the last 10 years, a new and vastly more informative PAGE has arisen. This new, "quantitative " PAGE[5] for the first time does exploit charge differences between macromolecules effectively by operating at pH values as close to their isoelectric points (where differences between species are maximized) as the requirement for sufficient mobility will tolerate. Furthermore, instead of employing a sole, arbitrary gel concentration, the "new PAGE" uses the Ferguson plot (mobilities at several gel concentrations which, when plotted semi-logarithmically, yield a straight line[6,7]). Thus, instead of yielding "a band", PAGE is now capable of yielding, on the basis of precisely measurable migration distances at the various gel concentrations, the molecular size and net charge of any macromolecule in its native state, without loss of activities, and without having to isolate it from the multicomponent mixture in which it may be contained[8]. More important yet, conceivably, is the possibility to calculate, in simple fashion, from the Ferguson plots of any set of two macromolecules the particular gel concentration at which the two separate most effectively[9]. As a corollary to this calculation, we also learn, on the basis of objective criteria, in which cases to resolve the two species by PAGE (*i.e.* by molecular sieving, exploiting size differences) and in which to resolve by electrofocusing (a pure "charge fractionation"). In summary, the "new PAGE", by providing objective criteria for optimizing resolution, defines objectively a particular fractionation strategy for each separation problem, and paves the way for the preparative application of PAGE under optimally resolving conditions.

Another departure as yet from the original technique, increasingly popular over the last few years, has been the two-dimensional approach to macromolecular mapping, shown to be startling in its resolving capacity by O'Farrell[10]. The rationale for combining a "charge fractionation", electrofocusing, with a "size fractionation", SDS-PAGE", in two dimensions is not new[11]. However, the success of the method depends on an even zone spreading of a multicomponent mixture of macromolecules along each of the two dimensions. Each of the two gels, in conventional dimensions, will only be able to contain 30-50 equally concentrated zones in regular intervals, due to purely geometric limitations constraining a system of overlapping band distributions[12]; necessarily, these must represent the 30-50 most concentrated components in any complex mixture of macromolecules. Separated 2-dimensionally, an even distribution of 30-50 zones along each of the two dimensions must result in 900-2,500 spots, and this is precisely what O'Farrell first demonstrated. Thus, the simple combination of one of the "charge fractionation" methods with PAGE promises to improve the resolving capacity of gel electrophoresis by a factor of 10 to 30. But this holds only under conditions of "even zone distribution in each dimension" which is not easily achieved, and at present achieved mainly by trial-and-error approaches which are laborious and applicable to a single fractionation problem only. Nonetheless, 2-D gel electrophoresis appears as the wave of the future in the resolution of complex mixtures.

Finally, the last decade has seen an instrumental development which in even longer perspective than the 2-D method promises to replace present routine procedures of gel electrophoresis. This is PAGE with continuous optical scanning during the course of electrophoresis, developed by Catsimpoolas[13], with continuous automated data collection and, at least, the capacity for continuous automated data processing.

Of course, within either "original PAGE", or "quantitative PAGE", or "SDS- and nucleic acid PAGE", or 2-D-PAGE", or "PAGE with continuous optical scanning", there have been many procedural and instrumental variations and developments over the last decade which it is not possible to review in detail for reasons that have both to do with the ignorance of the author and his inability to absorb the thousands of publications in those fields, and with the volume limitations of a broad review of gel electrophoretic techniques. For instance, there have appeared immunological and affinity detection techniques in any of the 5 classes of PAGE which are very promising, very practical but exceed the scope of this review devoted mainly to fractionation strategy, optimization of conditions, choice between methods and potential for resolution. Similarly, apparatus design problems have not been discussed, although hundreds, and possibly thousands of PAGE apparatus models have appeared both in the literature and on the market over the last decade.

We will discuss "quantitative PAGE", SDS- and nucleic acid PAGE, 2-D-PAGE and PAGE with continuous optical scanning first since these seem to be significant innovations in analytical PAGE over the last decade, and conclude by some technical improvements in "original PAGE".

QUANTITATIVE PAGE

Page becomes "quantitative" when it aims at the measurement of migration distances, and derives from these 2 types of information not available from PAGE in the original form: 1) information regarding size, shape and net charge (surface charge density) of the molecule; 2) information leading to an objective definition of optimal fractionation conditions. Furthermore, quantitative PAGE implies also that the pH of PAGE analysis be objectively optimized rather that selected arbitrarily, and that the milieu of PAGE be rendered maximally suitable for the maintenance of macromolecules in their native, active forms.

This concept was fully developed and reviewed in 1971[8] and has recently been restated in conjunction with the apparatus requirements of the method[5]. It is perfectly clear, however, that the mere enunciation of the availability of buffer systems capable of stacking macromolecules (*i.e.* multiphasic buffer systems), and thus allowing for pH optimization of PAGE in these systems, was not enough to allow anyone to use the method - the buffer systems generated by the theory and program of Jovin[4,14] were available in form of expensive magnetic tapes and required deciphering by computer (of the FORTRAN language). Similarly, nearly all the apparatus needed for making polyacrylamide gels of reproducible pore size, for electrophoresis at controlled temperature and for the measurement of migration distances were commercially unavailable. Computer programs which translated migration distances to the desired parameters, molecular size and charge, and optimal gel concentration were only available as listings or in form of punched paper tape, by personal request, and they again required computing equipment (of the BASIC language). Small wonder, then, that to the present day, practically no one has been able to benefit from the wealth of information, and the automated way in obtaining it, that is inherent in the use of quantitative PAGE.

Let us briefly review the rational fractionation strategy of quantitative PAGE (Fig. 9 of ref. 8 and Fig. 1 of ref. 5), and then discuss fractionation-step by-step the progress in terms of availability, ease of procedure and instrumentation which has been made over the last decade and which finally, at this time, is beginning to become practical and feasible for large numbers of biochemists. The strategy for physically characterizing and isolating any macromolecule starts, for proteins, with a consideration of pH. What is the optimally resolving pH for the problem at hand? In principle, it's a pH sufficiently close to the isoelectric points of the proteins to be separated to maximize the fractional charge difference between them (e.g. 5/8 instead of 49/52) and, at the same time, to preserve adequate mobilities (equal or less than that of the trailing constituent of a multiphasic buffer system). In practice, finding of such a pH requires an experimental test of a series of Stacking Gels operative at pH's within the desired range. This, again, requires knowledge of the make-up and operative properties of the multiphasic buffer systems at those pH's, and is not meaningfully helped by reference to the 5,000 buffer systems of the Jovin-output[14]. This situation has been remedied in 3 ways: a) The Jovin buffer systems output is now available in inexpensive, and easily readable *microfiche* form[14], allowing anyone the access to buffer recipes, and the operative parameters of multiphasic (discontinuous) buffer systems at any pH. The Jovin program, and a computer, however are still needed for an explicit recalculation of the properties of any of the "subsystems" (Fig. 10 of ref. 5) comprising buffer systems of identical constituent composition but different pH from the "main systems" listed in the printed MZE buffer systems catalog (PB number 196090 of ref. 14). b) *Sample* buffer systems across the pH range have been published explicitly, at least for buffer systems operative in the anionic direction of migration, 0°C[5,15]. c) A very simple *apparatus* for the simultaneous testing of stacking pH's has been designed[16]. Although it is not commercially available as yet, it can easily be constructed by concentric sealing of Plexiglas cylinders into a conventional plastic gel tube apparatus, and insertion of a single circumferential electrode into each reservoir[16].

The second step in the general strategy of quantitative PAGE is the construction of a Ferguson plot (log R_f *vs* %T) for the macromolecule of interest. Experimentally, this implies PAGE at several gel concentrations, followed by measurement of migration distances either based on activity (*i.e.* slicing and assay of slice diffusates) or on staining. Gel concentrations for that purpose need to be reproducible, but not necessarily accurate - although any inaccuracy as may arise from incomplete polymerization will also give rise to increased irreproducibility of R_f[17]. Reproducibility of pore size depends on the standardization and control of polymerization conditions and purity of reagents[5]. Commercial reagents and instrumentation to conduct polymerization at a defined level of oxygen inhibition[18] are available at this time, as is an electronic mode of measuring migration distances with relative ease and accuracy[19]. For slicing and activity or radioactivity assay some models of the necessary transverse[20] and longitudinal[21] sectioning devices exist commercially for gel tubes. (Since the construction of a Ferguson plot by definition implies the simultaneous electrophoresis at several gel concentrations, tube rather than slab apparatus is required for this purpose). Any gel slicer can be used which is capable of preserving the origin and the front (moving boundary) in a recognizable form and thus allows for calculation of R_f values. Autoradiographic equipment for analysis of longitudinal slices is also available. Thus, while the price tag on analytical PAGE has greatly multiplied since the days of the "original PAGE", at least the tools needed to construct reproducible Ferguson plots are being manufactured today, although they are not available as yet as a functional unit or "kit", and thus have to be assembled by the investigator. Also not available as yet is a way to monitor the accuracy of the polymer, *i.e.* the % polymerization efficiency; methods have been described but are either laborious[17] or costly[22].

For the automated, rapid and error-free conversion of measured R_f values at various gel concentrations to the parameters characteristic of molecular size and surface net charge in a particular buffer milieu a series of computer programs designated as the "PAGE-PACK" is available in FORTRAN language[5,23], but this "availability" is not fully practical as yet: It requires a personal request, shipment of voluminous punched cards contained in 2 boxes, or the risk of introducing error in transcribing of a FORTRAN listing, availability of compatible computing facilities, funds for computations and preferably a remote terminal in the laboratory. These are a rare combination of conditions at present, although they should be commonly met as time progresses. In the meanwhile, a manual and graphic analysis of Ferguson plots is possible, but their adequate statistical evaluation is near-impossible for the average biochemist. Exactly the same considerations hold for the application of Ferguson plots to the problem of optimizing resolution between any set of two components by calculation of an "optimized gel concentration" in a particular buffer.

The third step in the general strategy of quantitative PAGE is the analytical application of fractionation conditions optimized with regard to gel concentration, pH, solvent and additives conducive to the maximal "happiness" of the molecule under investigation. Evidently, this does not pose any further procedural or instrumental requirements in those cases where molecular size differences between components to be separated dictate a PAGE fractionation. However, if resolving power depends primarily on net charge differences of components, and thus optimization leads into electrofocusing or isotachophoresis, further new procedural and instrumental requirements arise which at present may have to be met by improvisation or use of non-optimal conditions. Some of these problems have been discussed elsewhere[24], but they exceed the thematic scope of this review.

For the development of biochemistry into the organic and physical chemistry of macromolecules and their structure-function relationships the aim of fractionation is preparative. A "quantitative" science of preparative PAGE[25], electrofocusing and isotachophoresis[26] is not developed as yet beyond the optimization of pore size, the measurement of the first apparent diffusion coefficient in a gel for a single, homogeneous protein[26] and theoretical work on the subject[9]. Thus, in the most important aspect, the preparative one, PAGE is barely beginning to be "quantitative". Therefore, to avoid the inefficiency of trial-and-error sciences, it is felt that analytical PAGE scaled up to a preparative level under objectively optimized conditions also presents at present the simplest, risk-free and relatively non-laborious preparative route, once it is linked with suitable collection and purification procedures capable of yielding a homogeneous protein product in high yield and purity[27]. Evidently, this preparative aspect of analytical PAGE technique does exceed, when viewed rigorously, the bounds of a review on analytical methods. It must be included briefly, however, to give meaning and concise limitations to the very germane, central proposition that the most important thing that has happened to analytical PAGE over the past decade is that it has gained the potential, and the tools, for "quantitative" application.

SDS- AND NUCLEIC ACID PAGE

"Original PAGE" has been applied widely throughout biochemistry in two completely unexpected ways: in application to PAGE in SDS-containing buffers, and in application to polynucleotides. These two applications can be lumped together since they concern fractionation problems characterized by pure "size isomerism", *i.e.* they concern mixtures of molecules differing in size, but not significantly in molecular charge density. Furthermore, they concern families of macromolecules which are among themselves conformationally uniform, *i.e.* "random-coiled". Under these two conditions, it is possible to extract physical, "quantitative" information from PAGE carried out at a single pH and gel concentration, *i.e.* by the "original PAGE" technique.

A) Molecular weight:

Although it remains true that no available method other than sedimentation equilibrium can provide mass (dalton) measurements, while all other methods are sensitive to molecular size and shape[28], a comparison between standard molecular weight proteins of known mass (daltons) and unknowns of identical gross conformation (random coil) provides ratios of molecular volume identical to ratios of mass, allowing one therefore to correlate migration distance in PAGE with molecular weight. Thus, it has proved possible, and popular, to use PAGE at an uncontrolled, single gel concentration, anywhere in the neutral or alkaline range, for separation on the basis of "molecular weight", with assignment of characteristic "daltons" to each zone. A separate chapter (see Ch. 17) deals in detail with these methods. Within the framework of analytical PAGE as a whole, it seems only required in this review to reiterate clearly the conditions under which "original PAGE" is capable to yield "daltons":

1. Uniform molecular charge density: With SDS-proteins, this is achieved only in the absence of non-protein moieties in the molecule, such as carbohydrate or lipid or nucleotide moieties. It also depends on effectively opening up of tightly folded structures through reduction, heating and disaggregation requiring variable and unpredictable periods of reaction time, rather than the standard "5 minutes, 100°C". With nucleic acids, it requires absence of molecular rigidity (double strandedness) for both standards and unknowns. In either case it requires experimental verification of a uniform charge density among the component unknowns and standards, *i.e.* demonstration of a common y-intercept on a Ferguson plot in "quantitative PAGE" (see above). This is not easily demonstrated, since in practice the common point of intersection occurs near, but not at 0% gel concentration, and since the irreproducibility of y-intercepts (Y_o) is very large compared to slopes, for geometric reasons[6]. Considering now, that practically no one among the many protagonists and applicators of SDS-PAGE or nucleic acid PAGE determine such Ferguson plots, calculate meaningful statistics for their "daltons" or are even aware of the limitation of the method to populations of equal charge density and random-coiled conformation, many of the present molecular weight estimates by PAGE must be in error.

2. Correct choice of Stacking Limits, and of boundary tracking dyes: SDS-proteins and polynucleotides are far more electrophoretically mobile than proteins in general. Thus it is possible to raise the mobility of the Trailing Constituent of the stack (the so-called Stacking Limit[5]) to a value of 0.2-0.3 RM (mobility relative to Na[29]), thus separating at the stage of stacking protein that has not fully reacted with SDS or that is contaminating the nucleic acid. More important is the correct choice of a tracking dye which marks the front moving boundary (e.g. chloride-glycinate in the popular Tris-system) in SDS-PAGE[29]. Here, the problem is that micellar SDS continuously enters the stack in the Resolving Gel during electrophoresis, thus setting up an "extended stack"; within this stack, depending on gel concentration, the mobility of the tracking dye may fall behind that of the micellar SDS, with a corresponding shift of its position to the rear of the stack, leading to wrong R_f values[29].

3. Limitation to the linear range of sigmoidal standard curves or non-linear curve fitting: When charge densities among unknown and standards are equal or equalized by sufficient "cooking" of the proteins with SDS, MW values can be legitimately derived from the necessarily sigmoidal plots of log MW *vs* R_f or log R_f *vs* MW[28,31]. However, serious error may arise from either assuming that unknowns and standards fall into the linear range of the curves or from not using non-linear curve fitting techniques.

4. Avoidance of protein solubilization during staining: It is widely believed that alcohol and acid containing solutions of protein dyes are adequate protein fixatives. However, this is not true to the degree to which 10% TCA is a general protein fixative. In particular, hydrophobic proteins, being soluble in alcoholic solutions, are

not efficiently fixed in the conventional staining media for SDS-PAGE. The problem of incomplete fixation is exacerbated by the prolongued reaction time which is due to interference of SDS with the staining reaction, presumably through electrostatic repulsion of and competition with the dye. Thus, SDS needs to be removed from the gel prior to staining - and under conditions of effective protein fixation. Since SDS is not soluble in TCA solutions, this has not been possible to date with sufficient efficiency. At present, long periods of SDS-removal by 5-10% TCA prior to the short "no-background" staining of protein zones[84] after removal of SDS are required but at least give some assurance that proteins are not lost from the gel by solubilization. Even this cannot be held with any certainty, since the solubilizing action of SDS and the fixation by TCA are antagonistic[32].

B) Multicomponent analysis:

Gel length and bandwidth: Time dependent diffusion spreading of protein zones usually annuls the advantage to resolution that otherwise would obtain when the gel is elongated and separation between the peaks of zone distributions is correspondingly increased. However, the empirical finding that zones in SDS-PAGE or nucleic acid PAGE do not prohibitively increase in width with even very long migration path has lead to the practice, both in SDS-PAGE and in nucleic acid PAGE, of ever increasing gel length, with effective resolution in a single dimension of up to hundreds of zones. Thus, nucleic acid PAGE and SDS-PAGE seem to have a second advantage over the "original PAGE" in addition to their capacity, under appropriate conditions, to measure MW: that of being able to resolve multicomponent mixtures.

C) Pore size considerations:

SDS-PAGE has been applied in a) gels of different % crosslinking with the objective of obtaining conveniently flat Ferguson plots in the molecular size range of interest; b) gels of arbitrary gel concentration to determine molecular weight; c) pore gradient gels. Evidently, the choice among these 3 options with regard to pore size varies with the problem under investigation. a) Ferguson plots are needed to determine whether, under the conditions of sample preparation and electrophoresis used, uniform charge density among components obtains. By decrease of % crosslinking from 5 to 2, these plots can be obtained with relatively divergent gel concentrations and therefore more conveniently and accurately[29]. b) Single gel concentrations are the most convenient and accurate to apply as a molecular weight method, once the conditions for this application (see above) are secured. However, it may be difficult in some cases to secure an approximately even zone distribution in gels of single gel concentration. c) In these cases, pore gradient engineering can be used to achieve such a distribution of zones. This is of particular importance for 2-D-PAGE (see below).

Much of the present nucleic acid work on PAGE is done on conventionally sized gels in application to the products of hydrolysis by restriction enzymes[33]. In addition, bot RNA[34] and DNA[35] in the molecular weight range of 0.1 to 30 x 10^6 have been fractionated by gel electrophoresis. Routinely, 0.5 to 1.5% agarose are being used for the larger MW species in this range, conventionally crosslinked polyacrylamide for the small ones, and polyacrylamide-agarose "copolymer" for the intermediate range. Due to the lack of wall adherence of agarose, most of this work has been restricted to horizontal gel slabs and not employed stacking (since it is difficult to set up parallel Stacking and Resolving gels in the horizontal dimension). No advantage has been taken in that field, to date, of the possibility to extend the pore size of polyacrylamide gels indefinitely by increases in the % crosslinking (with Bis) beyond 5% and up to 50%[17].

D) Additional considerations regarding nucleic acid PAGE:

Problems in aggregation and binding of nucleic acids to the matrix have been overcome by conducting gel electrophoresis in formamide[35], urea[36] or methyl mercury[37]. Among detection methods, ^{32}P-labeling with or without autoradiography and the ethidium bromide sandwich technique have been pre-eminent, since both methods allow for excision and isolation of nucleic acid zones from the gel. A method for the purification of nucleic acid from polyacrylamide-like impurities in the diffusate from gel slices has been described[38]. Conformational differences among single and double stranded DNA species have been determined by Ferguson plot analysis, using agarose gels[39].

2-D-PAGE

The 2-D approach to the resolution of multicomponent mixtures, based on a "charge fractionation" followed by a "size fractionation", was first demonstrated in its full effectiveness by O'Farrell[10]. The geometric conditions upon which the success of this method rests have been pointed out above. Thus, it is ineffective, as widely tried, to apply the conditions carefully elaborated on *E. coli* for an even distribution of zones in both dimensions to other systems. To resolve systems of one to several thousand components, pH gradient engineering[24,25] and pore gradient engineering (see Ch. 16) have to be applied to each particular system until the geometric conditions for maximal resolution are met. This also implies the selection of a detergent or detergent mixture capable of solubilizing the large majority of proteins under investigation, preferably in such a way as to maintain protein activities[32,40,41].

The precise recognition of each component in a macromolecular map of several thousands, and its quantitation, comprising automated characterization of each component in terms of an x-y coordinate pair, with appropriate confidence limits, and automated detection of significant intensity increases or decreases, or spot appearance and disappearance from the map, in ways analogous to those elaborated for astronomy, remain unsolved problems of 2-D macromolecular mapping. At present, the method is at the mercy of subjective spot characterization and estimate of relative spot intensities.

PAGE WITH CONTINUOUS OPTICAL SCANNING

Technologically, the introduction by Catsimpoolas of PAGE with continuous optical scanning[13] has undoubtedly laid the foundation stone to a PAGE technology as it will be practiced one or a few decades henceforth. This is due to two reasons: First, the continuous UV-scanning technique requires very short migration paths for exceedingly accurate mobility determinations, of the order of 1 cm migration distance per up to 100 scans (less by one order of magnitude if zone width is measured, see below).Thus, it allows for PAGE fractionation at a very rapid rate. It also allows, for the first time, for a meaningful measurement not only of band position, but of zone width which provides quantitative information with regard to resolution and microheterogeneity. Second, the optical scanning technique is compatible with automated data collection (with regard to migration distance and zone width) and instant data processing. Thus, it is possible, in a single run with a commercially available 6-gel cassette (as yet limited to absorption in the visible range) containing gels at six different gel concentrations, to determine Ferguson plots for each component upon termination of a mere 1 cm migration distance, and to immediately derive from these plots values of molecular size and net charge, as well as objective information with regard to the optimal gel concentration for the resolution of each pair of the protein components from one another. In addition, the apparent diffusion coefficient[26] of each component in the particular gel can be calculated and also be plotted as a function of gel concentration. Thus, the continuous UV-scanning apparatus is capable of very rapid quantitative PAGE, in the most extended sense of the word, and provides error-free physical information for each component in a fully automated fashion. Thus, continuous scanning would be ideal for large numbers of PAGE analyses as accrue in clinical practice. What still inhibits the spread of the method is only the relatively high cost of the operation and the requirement for protein loads approximately one order of magnitude higher that those used in conjunction with staining techniques. But this seems to be a rather minor technical problem which undoubtedly is going to be solved once sufficient demand for apparatus of this type arises.

"ORIGINAL PAGE"- SOME TECHNICAL IMPROVEMENTS.

Gel

Pore architecture of polyacrylamide gels has remained relatively inflexible. The three polymarization catalysts introduced by Ornstein and Davis[2], persulfate, riboflavin and tetramethylethylendiamine, have remained in universal use. A few new crosslinking agents have been designed to allow for cleavage of the crosslinks through periodate oxidation (DADT, N,N-diallyltartardiamide)[42], reduction of disulfide bonds[43], even heating[44].DADT-cross-linked gels have been useful at the high %C range to provide mechanically stable open-pore gels[45]. However, compared with Bis-crosslinked high %C gels (in particular the %C range of 10-50), DADT crosslinked high %C gels seem more restrictive. Other commercially available crosslinking agents such as the (2-12) C-Bis homologues[44] are unexplored to date.

When Bis-crosslinked high %C gels, *i.e.* gels crosslinked between 10 and 50%, were first introduced, they were expected to free PAGE of an upper molecular size limit[46]. This expectation was based on the fact that among a few proteins tested, all appeared to exhibit 0 retardation by the gel, as $\%C_{Bis}$ approached 50%. However, intervening experience has shown the real behavior of high $\%C_{Bis}$ gels to be far more complex and problematic. First, high $\%C_{Bis}$ gels exhibit increasing wall separation problems with increasing %T. These lead to current leakage along the gel periphery. Second, the variation of Y_o with $\%C_{Bis}$ (Fig. 2 of ref. 46) raises the problem of the physical meaning of Y_o as a measure of molecular net charge. It is a problem that has been disregarded to date in all reports of molecular net charge based on PAGE. Third, it appears that mechanically more stable gels than those formed with Bis, viz. DADT crosslinked gels, do not exhibit the same continuous opening of effective pore size with increasing %C that has been demonstrated for Bis-crosslinked gels. However, a thorough investigation of this question, and of the effective pore size of high %C gels formed with any of the new types of crosslinking agents referred to above is still outstanding.

For electrophoresis of very large molecular weight compounds (in the range from 2 to 50 x 10^6) the competition between polyacrylamide at high %C, the agarose-polyacrylamide copolymer and agarose by itself remains undecided.Polyacrylamide at 30-50%C is probably the most mechanically unstable matrix of the three. Agarose has no wall adherence and gives rise to electroendosmosis due to its sulfate charge. The mixed polymer is not easy to form uniformly. Pevikon may be an attractive alternative to agarose[47].

Very little physical evidence of pore structure exists. Two reports on the scanning electronmicroscopic (SEM) analysis of polyacrylamide gel have been published[48,49]. The SEM data reveal cell-like pores enclosed by fibers of less than 1 nm thickness which could be the structures responsible for molecular sieving. However, the inside diameter of the "pores" themselves is too large to possibly be able to affect the migration of macromolecules. Also, the "pore" size varies in as yet indeterminate fashion as a function of the freezing rate during sample preparation for SEM.

Solvents

The polymerization of crosslinked polyacrylamide is compatible with solvents such as formamide[35,46] or chloral hydrate[50] but these have had application in practical analytical PAGE only very rarely. In contrast, the solubilization of hydrophobic proteins by detergents with maintenance of activity represents a significant widening in the applicability of analytical PAGE[32,40,41,51]. It is based on the realization that detergent properties may be "titrated" by mixing ionic and non-ionic detergents to give optimal solubilization and disaggregation without total inactivation[41]. It is also increasingly based not on arbitrary and subjective choice of detergents and detergent concentrations but on knowledge of the physical properties of detergents and detergent micelles[32].

Apparatus

To use PAGE efficiently and flexibly, a great variety of apparatus is required which has recently been listed in its entirety[5]: gel tube apparatus for PAGE at diverse gel concentrations and variable duration of time, slab apparatus for electrophoresis at constant gel concentration and time. Multichamber apparatus is useful for the determination of optimal pH[16]. Longitudinal and transverse slicers, autoradiographic equipment, R_f measuring devices, photographic equipment, deaeration devices etc.[5] are needed. An attempt has been made at least to list the various "apparatus modules" for the functional PAGE laboratory, but consensus on optimal apparatus designs remains to be established[5]. To date, a functional, multi-modular assembly of all this equipment remains unavailable commercially. Expensive and tedious assembly and, to a minor degree, construction, of all of the required apparatus remains burdensome and inhibitory to the widespread use of PAGE in a flexible and efficient way.

ZONE DETECTION AND CHARACTERIZATION

1) Fixation and staining:

a) Protein fixation and staining: it has been reduced to ½ h operation[84] since destaining is no longer required in PAGE (this is not as yet true for SDS-PAGE).

b) Prestaining of proteins: this has been accomplished with Drimarene Brilliant Blue K-BL[52] and is potentially useful in application to protein molecular weight standards. The dye reacts with amino groups on the protein and must therefore give a false value of Y_o on Ferguson plot analysis; but the molecular size of the protein remains uneffected (within the detection limits of PAGE) by the addition of dye, making it possible to obtain accurate values of K_R (C. Bui and A. Chrambach, unpublished data). Obviously, the method is applicable to SDS-PAGE. Since it avoids "ring-staining", at least this element of inaccuracy in quantitation is eliminated. Remaining defects of this method are that it appears difficult to reproduce the quality of reported data[52], and the degree of reported sensitivity, indicating that some procedural element has not been adequately defined as yet, and that it relies on a chemically unidentified, commercially unavailable dye, obtainable only as a sample from the manufacturer.

c) Prestaining of proteins with MDPF[53]: MDPF is a fluorescent reagent which conveys a stable fluorophore to protein zones in PAGE (in contrast to the previously used fluorescamine). The reaction product is stable under conditions of sample preparation for SDS-PAGE. Like the Drimarene dye, it covers up charged functional groups (in this case, $-COO^-$), but could be expected to exhibit an unmodified K_R in Ferguson plot analysis. Just as the dye, it is also available only as a sample by request, but not commercially.

d) Protein staining with ANS[54]: ANS is a fluorogenic reagent which lends itself to reaction with ammonium sulfate precipitated protein zones on polyacrylamide gel.Thus, it does not interfere usually with protein nativeness. The binding to the protein amino groups is non-covalent, *i.e.* reversible. The fluorescent interaction product can be obtained prior to and during electrophoresis as in the 2 above-cited methods; alternatively, reaction may proceed after electrophoretic zone separation has occurred. The technique involved wrapping the gel into filter paper saturated with reagent.

e) Dansylation of protein zones on gels[55]: Reaction of amino groups with dansyl chloride is analogous to prestaining with MDPF, except that the blocked functional groups are positively charged. It has the obvious advantage that it allows for N-terminal analysis on the extracted protein zones.

f) Detection of proteins on gels after hydrolysis: Proteins after staining may be hydrolyzed to their constituent amino acids and identified by amino acid analysis[56]. Sensitivity is increased when the detection of amino acids follows derivatization with fluorescamine[57]. Alkaline hydrolysis avoids the problem of amino acid contamination with a large concentration of ammonia derived from polyacrylamide.

g) Detection of proteins on gels by their phosphorescence: One of the most promising new methods of non-destructive protein zone detection rests on the natural(tryptophanyl) phosphorescence of proteins at low temperatures[58]. Sensitivity appears equivalent to that of staining.

h) Detection of proteins on gels after isotopic labeling on the gel: Labeling of protein from zones on polyacrylamide gel after elution, followed by enzymatic hydrolysis and fingerprinting of the protein has been reported[59,60]. More elegantly, iodination has been carried out on the gel slices, containing the protein, themselves[61,62], prior to fingerprinting.

i) Phosphoproteins: Highly phosphorylated proteins can be detected on polyacrylami-gels to a limit of sensitivity of 0.13 nanomole of phosphorus[63], corresponding to 40 ng of phosvitin. The method rests on the specific binding of phosphatidyl serine to trivalent metals, and it is six times more sensitive than previously described methods.

k) Carbohydrate stains: Several new methods for glycoprotein detection on polyacrylamide gel have been developed. Sensitivity of the classical PAS procedure has been increased to 2-3 μg of bound carbohydrate[65]. This allows for the detection of transferrin with approximately 1/10 the sensitivity of present protein stains, while a protein with high carbohydrate content like fetuin can be detected with about the same sensitivity. In addition, conditions for the quantitative densitometry of PAS stained gel patterns have been described[66]. Glycoproteins can also be detected by fluorescent stains[66,67]. The application of "Stains-all" has been expanded to include sialic acid rich glycoproteins[68].

2) Slicing and isotope analysis:

a) H_2O_2 oxidation: The simplest approach to isotope analysis of labeled macromolecules from gel slices is to disrupt the gel oxidatively by alkaline H_2O_2 incubation at or slightly above room temperature[69]. This procedure avoids the labeled water or CO_2 loss otherwise associated with this procedure. However, neither this nor any other study to date reports the time required for slice solubilization as a function of gel concentration. Since the sample is aqueous, counting efficiencies are relatively low.

b) "Amine-solubilization": Another approach is to solubilize the macromolecule and allow it to diffuse out of the gel slice under conditions of gel swelling and possibly some hydrolysis by strongly basic amines[70,71]. Since water is not involved, counting efficiency is high. Efficiency strongly depends on the choice of scintillation cocktails[71]. To the degree that these "amine-solubilization" techniques require heating, they invite chemiluminescence (phosphorescence) problems.

c) "Instagel" cocktail: Another approach is to disperse the gel slice into "Instagel" scintillation cocktail. High efficiency is claimed for this method[72]. It appears certainly attractive for work at high gel concentrations.

d) Staining prior to isotope counting: Gels may be stained and subsequently be analyzed for isotope. This has the advantage that stained bands and isotope content can be unequivocally correlated, that loss of resolution due to blind slicing is avoided and that isotope analysis may be limited to those parts of the gel which contain stained zones[73].

e) Determination of specific activity: Protein may be extracted from zones of polyacrylamide gel by SDS-urea buffer, with subsequent analysis for protein by the Lowry reaction and TCA precipitation of the protein on a nitrocellulose funnel followed by isotope analysis[74]. This technique provides specific activity data, and obtains a higher counting efficiency than previous methods of gel solubilization by H_2O_2.

f) Automated radioactivity scanning: Longitudinal polyacrylamide gel slices or slabs of no more than 3 mm thickness can be scanned for radioactivity directly, and within a few minutes, by means of commercial gas flow strip counters. Relevant techniques have been described both for ^{32}P[75] and, with relatively little sensitivity, for ^{14}C labels[76], but not as yet for tritiated or iodinated proteins, although some radiochromatogram scanners (e.g. Packard 7220) claim to possess this capability.

3) Autoradiography:

a) ^{3}H-labeled macromolecules: A simple fluorographic method for autoradiographic analysis of ^{3}H- containing gels has been described[77]. Similar methods have also been used to enhance the sensitivity of autoradiographic detection of the higher energy labels, ^{32}P and ^{125}I[78].

b) Double label autoradiography: Differential analysis for ^{3}H and ^{14}C on the identical gel by autoradiography has been described[79].

c) Increase in detection sensitivity: An expensive but effective way to increase detection sensitivity of ^{14}C-containing zones on polyacrylamide gels is the use of an image intensifyer[80].

4) Densitometric zone quantitation:

Quantitation of stained bands[81], or of UV-absorbing macromolecules[82] in polyacrylamide gels is widespread but problematic: a) Chromophore concentration on the macromolecule varies between species and as a function of staining-destaining conditions. b) Due to limits of detection sensitivity, densitometric estimates of protein decrease with increasing migration distance[83].

Band width increase due to the diffusion of the macromolecule from the time of termination of electrophoresis to the end of fixation is necessarily being neglected; the resulting zone flattening must, just like increases in migration distance, contribute to a lowering of apparent zone area.

Some of these problems can be avoided, although at great expense, by use of electronic image recognition systems[81].

REFERENCES

1. Raymond, S. and Nakamichi, M. (1961) Anal. Biochem., 7, 225-232.
2. Ornstein, L.L. (1964) Ann.N.Y.Acad.Sci.,121, 321-349.
3. Chen, B., Griffith, A., Catsimpoolas, N., Chrambach, A. and Rodbard, D. (1978) Anal. Biochem.,89, 609-615.
4. Jovin, T.M. (1973) Biochemistry, 12, 871-898; Ann. N. Y. Acad. Sci.,209,477-488.
5. Chrambach, A., Jovin, T.M., Svendsen, P.J. and Rodbard, D. (1976) in"Methods of Protein Separation" Catsimpoolas, N. ed., vol. 2, Plenum Press,N.Y.,pp.27-144.
6. Rodbard, D. and Chrambach, A. (1974) in "Electrophoresis and Isoelectric Focusing in Polyacrylamide Gel", Allen ,R.C. and Maurer, H.R. eds., W.de Gruyter, Berlin, pp. 28-61.
7. Rodbard, D. and Chrambach, A. (1970) Proc. Natl. Acad. Sci. USA, 65, 970-977.
8. Chrambach, A. and Rodbard, D. (1971) Science, 172, 440-451.
9. Rodbard, D., Chrambach, A. and Weiss, G.H. (1974) in "Electrophoresis and Isoelectric Focusing in Polyacrylamide Gel" Allen, R.C.and Maurer, H.R. eds., W. de Gruyter, Berlin, pp. 62-104.
10. O'Farrell, P.H. (1975) J. Biol. Chem., 250, 4007-4021.
11. Macko, V. and Stegemann, H. (1969) Z. Physiol. Chem., 350, 917-919.
12. Kapadia, G., Chrambach, A. and Rodbard, D. (1974) in "Electrophoresis and Isoelectric Focusing in Polyacrylamide Gel" Allen, R.C. and Maurer, H.R. eds.,W. de Gruyter, Berlin, pp. 115-144.
13. Catsimpoolas, N. (1976) in "Methods of Protein Separation" Catsimpoolas, N. ed., vol. 1, Plenum Press, N.Y., pp. 27-68.

14. Jovin, T.M, Dante, M.L. and Chrambach, A. "Multiphasic Buffer Systems Output", National Technical Information Service, Springfield, Virginia 22151, PB Numbers 196085 to 196091, 259309 to 259312, 203016.

15. Bui, C., Galea, V. a,d Chrambach, A. (1977) Anal. Biochem., 81, 108-117.

16. Newby, A., Matthews, G. and Chrambach, A. (1978) Anal. Biochem., 91, 473-480.

17. Chrambach, A. and Rodbard, D. (1972) Separ. Sci., 7, 663-703.

18. Chidakel, B.E., Ellwein, L.B. and Chrambach, A. (1978) Anal. Biochem.,85,3i6-320.

19. Peterson, J.I., Tipton, H.W. and Chrambach, A. (1974) Anal. Biochem., 62, 274-280; Tipton, H.W., Rumen, N.M. and Chrambach, A. (1975) Anal. Biochem., 69, 323-326.

20. Kohler, P.O., Bridson, W.E. and Chrambach, A. (1971) J. Clin. Endocrinol., 32, 70-76.

21. Bloemendal,H. and Vennegoor, C. (1969) in "Techniques in Protein Biosynthesis" Campbell, P.N. and Sargent, J.R. eds.,vol. 2, Academic Press, N.Y.,pp.187-197.

22. Chen, B. and Chrambach, A. (1979) J. Biochem. Biophys. Methods, submitted.

23. Rodbard, D. and Chrambach, A. (1971) Anal. Biochem., 40, 95-134.

24. Chrambach, A. and Nguyen, N.Y. (1979) in "Electrokinetic Separation Methods" Righetti, P.G., van Oss, C.J. and Vanderhoff, J.W. eds., North Holland/American Elsevier, Amsterdam and New York, Ch. 18.

25. Chrambach, A. and Nguyen, N.Y. (1977) in "Proc. International Workshop on Technology for Protein Separation and Improvement of Blood Plasma Fractionation" Sandberg, H. ed., Reston, VA, pp. 484-507.

26. Chen, B., Chrambach, A. and Rodbard, D. (1979) Anal. Biochem, submitted.

27. Nguyen, N.Y. and Chrambach, A. (1979) J. Biophys. Biochem. Methods, submitted.

28. Rodbard, D. (1976) in "Methods of Protein Separation" Catsimpoolas, N. ed., vol. 2., Plenum Press, N.Y., pp. 145-180.

29. Wyckoff, M., Rodbard, D. and Chrambach, A. (1977) Anal. Biochem., 78, 459-482.

30. Baumann, G. and Chrambach, A. (1976) Proc. Natl. Acad. Sci. USA, 73, 732-736.

31. Rodbard, D. (1976) in "Methods of Protein Separation" Catsimpoolas, N. ed., vol. 2, Plenum Press, N.Y., pp. 181-218.

32. Hjelmeland, L. and Chrambach, A. (1979) in preparation.

33. Maxam, A.M. and Gilbert, W. (1977) Proc. Natl. Acad. Sci. USA, 74, 560-564.

34. Rosenvold, E.C. and Honigman, A. (1977) Gene, 2, 273-288.

35. Staynov, D.Z., Pinder, J.C. and Gratzer, W.B. (1974) Biochemistry,13,5373-5378.

36. Rosenberg, M., Court, D., Shimatake, H. and Brady, C. (1978) Nature, 272, 414-423.

37. Bailey, J.M. and Davidson, M. (1976) Anal. Biochem., 70, 75-85.

38. Diener, X.X. (1973) Anal. Biochem., 55, 317-320.

39. Johnson, P.H. and Grossman, L.I.(1977) Biochemistry, 6, 4217-4225.

40. Newby, A.C., Rodbell, M. and Chrambach, A. (1978) Arch. Biochem. Biophys., 190, 109-117.

41. Newby, A.C. and Chrambach, A. (1978) Biochem. J., in press.

42. Anker, H.S. (1970) FEBS Letters, 7., 293-294.

43. Hansen, J.N. (1976) Anal. Biochem., 76, 37-44.

44. O'Connel, P.B.H. and Brady, C.J. (1976) Anal. Biochem., 76, 63-73.

45. Baumann, G. and Chrambach, A. (1976) Anal. Biochem., 70, 32-38.

46. Rodbard, D., Levitov, C. and Chrambach, A. (1972) Separ. Sci.,7, 705-723.

47. Muller-Eberhard, H.J. (1960) Scand.J.Clin.Lab.Invest.,12, 33-43.

48. Ruechel, R. and Brager, M.D. (1975) Anal. Biochem., 68, 415-428.

49. Gressel, J. and Rodbards, A.W. (1975) J. Chromatogr., 114, 455-458.

50. Ballou, B. and Smithies, O. (1977) Anal. Biochem., 80, 616-623.

51. Hjelmeland, L.M., Nebert, D.W. and Chrambach, A. (1978) Anal. Biochem.. in press.

52. Bosshard, H. and Datyner, A. (1977) Anal. Biochem., 82, 327-333.

53. Barger, B.O., White, F.C., Pace, J.L., Kemper, D.L. and Ragland, W.L. (1976) Anal. Biochem. 70, 327-335.

54. Neremberg, S.T., Ganger, C. and DeMarco, L. (1971) Anal. Biochem.,43, 564-574.

55. Kato, T. Sasaki, M. and Kimura, S. (1975) Anal. Biochem., 66, 515-522.

56. Houston, L.L. (1971) Anal. Biochem., 41, 81-88.

57. Boehlen, P., Stein, S., Stone, J. and Udenfriend, S. (1975) Anal. Biochem., 67, 438-445.

58. Isenberg, I., Smerdon, M.J., Cardenas, J., Miller, J., Schaup, H.W. and Bruce, J. (1975) Anal. Biochem., 69, 531-535

59. Bray, D. and Brownlee, S.M. (1973) Anal. Biochem., 55, 213-221.

60. Bonner, W.M. and Pollard, H.B. (1975) Biochem. Biophys. Res. Commun., 64, 282-288.

61. Christopher, A.R., Nagpal, M.L., Carrol, A.R. and Brown, J.C. (1978) Anal. Biochem., 85, 404-412.

62. Patterson, B. (1977) Cell, 12, 751-758.

63. Hegenauer, J., Ripley, L. and Nace, G. (1978) Anal. Biochem., 78, 308-311.

64. Kapitany, R.A. and Zebrowski, E.J. (1973) Anal. Biochem., 56, 361-369.

65. Matthieu, J.M. and Quarles, R.H. (1973) Anal. Biochem., 55, 313-316.

66. Eckhardt, A.E., Hayes, C.E. and Goldstein, I.J. (1976) Anal. Biochem., 73, 192-197.

67. Carson, S.D. (1977) Anal. Biochem., 78, 428-435.

68. Green, M.R. and Pastewka, J.V. (1975) Anal. Biochem., 65, 66-72.

69. Goodman, D. and Matzura, H. (1971) Anal. Biochem., 42, 481-486.

70. Paus, P.N. (1971) Anal. Biochem., 42, 372-376.

71. Hellung-Larsen, P. (1971) Anal. Biochem., 39, 454-461.

72. Leon, S.A. and Bohrer, A.T. (1971) Anal. Biochem., 42, 54-58.

73. Albanese, E. and Goodman, D. (1977) Anal. Biochem., 80, 60-69.

74. Buisson,M.,Reboud.,A.M. and Reboud,J.P.(1976)Anal.Biochem.,75, 656-659.

75. Rushizky, G.W. and Mozcjko, J.H. (1975) Anal. Biochem., 65, 540-542.

76. Magnusson, R.P. and Jackiw,A. (1979) J.Biochem.Biophys.Methods,submitted.

77. Bonner, W.M. and Laskey, R.A. (1974) Eur. J. Biochem., 46, 83-88.

78. Laskey, R.A. and Mills, A.D. (1977) FEBS Letters, 82, 314-316.

79. Gruenstein, E.I. and Pollard, A.L. (1976) Anal. Biochem., 76, 452 457.

80. Brisgunov, V., Rebentish, B., Gordon, I. and Debabov, V. (1975) Anal. Biochem., 66, 100-103.

81. Weidekamm, E., Wallach, D.F.H., Neurath, P., Flueckiger, R. and Hendricks, J.

(1974) Anal. Biochem., 58, 217-224.

82. Wadkin, J.E. and Miller, R.A. (1970) Anal. Biochem., 34, 424-435.

83. Fishbein, W.N. (1972) Anal. Biochem., 46, 388-401.

84. Diezel, W., Kopperschlaeger, G. and Hofmann, E. (1972) Anal. Biochem., 48, 617-620.

Electrokinetic Separation Methods,
P.G. Righetti, C.J. van Oss and J.W. Vanderhoff

ELECTROPHORESIS IN GELS OF GRADED POROSITY AND TWO-DIMENSIONAL TECHNIQUES.

ELISABETTA GIANAZZA and PIER GIORGIO RIGHETTI
Department of Biochemistry, University of Milano, Via Celoria 2, Milano 20133 (Italy)

INTRODUCTION

In gel gradient electrophoresis (GGE) the macromolecules run through a polyacrylamide gel of increasing concentration: this corresponds to a decreasing pore size (pore gradient electrophoresis, PGE) in the gel network, and an increasing sieving effect according to macromolecular size. The use of gradient gels allows a high resolution of the bands, since the migration through a medium of increasing viscosity sharpens the zone by retarding more the front than the rear of the band, thus counteracting diffusion effects. Under prolonged electrophoretic runs, the migration depends only upon the size (and shape) of the macromolecules: therefore, this separation procedure is useful for an (at least approximate) evaluation of native molecular weights (MW) without subunit disaggregation or loss of enzymatic activity. The sieving effect of the polyacrylamide depends in turn upon the gel concentration: in order to obtain a suitable resolution of any protein mixture either the limits or the shape of the gradient (linear, concave, sigmoid) can be varied.

Grossbach and Weinstein[1] first used a multi-layer gel of increasing acrylamide concentration for the separation in a single step of whole tissue RNA's species: with this artifice the higher MW molecules could enter the gel network while the smaller were prevented from running with the solvent front. This method was further developed by Slater[2-4] and by Margolis and Kenrick[5-7], who proposed a more versatile and simpler technique by the use of linear concentration gradients. The exploitation of this approach for MW determinations was first proposed by Thorun and Mehl[8] and by Slater[9]; Rodbard *et al.*[10] and Felgenhauer[11] have discussed the theoretical bases and the limitations of the method.

THEORY

The mobility m of a macromolecule in a polyacrylamide gel of concentration T (T% = g of acrylamide + g of N,N'-methylene bisacrylamide (Bis) per 100 ml solution[12]) is linked to the free mobility m_o by the Ferguson relationship[13]:

$$\log m_T = -K_R T + \log m_o \qquad (1)$$

in which K_R is the retardation coefficient. Equation (1) it theoretically derived from the random fiber model of Ogston[14] for the gel structure, and has been experimentally validated under a wide range of conditions (up to 15% acrylamide). By combining this equation with the gel concentration versus gel length relationship,

Rodbard *et al.*[10] derived the equations describing the behaviour of a macromolecule migrating in a gel gradient.

For a linear gradient the migration distance x at time t is given by:

$$x = \frac{\log_e\{2.303\ K_R(T_b-T_t)m_o t + \exp(2.303\ K_R T_t)\} - 2.303\ K_R T_t}{2.303\ K_R(T_b-T_t)} \qquad (2)$$

and the mobility m by:

$$m = \frac{m_o}{2.303(T_t-T_b)m_o t + \exp(2.303\ K_R T_t)} \qquad (3)$$

(T_t and T_b = gel concentrations at the top and bottom of the gradient, respectively). When t is very large, equations (2) and (3) become, respectively,

$$x = \frac{\log\{2.303\ K_R(T_t-T_b)t\}}{2.303\ K_R(T_t-T_b)} \qquad (4)$$

$$m = \frac{1}{2.303\ K_R(T_t-T_b)t} \qquad (5)$$

and

$$\lim_{t\to\infty} x = \infty \qquad (6)$$

$$\lim_{t\to\infty} m = 0 \qquad (7)$$

So, as t increases the migration distance becomes asymptotically proportional to the log of time; moreover m decreases and becomes inversely proportional to the time, but is not equal to zero for any finite gel concentration (for which the Ferguson relationship is validated) or time. Also, the distance between two bands (1 and 2 subscripts), expressed as:

$$\lim_{t\to\infty} \Delta x = \frac{1}{T_t-T_b}\left(\frac{1}{2.303K_{R_1}} - \frac{1}{2.303K_{R_2}}\right)\left(\log \frac{K_{R_1}}{K_{R_2}}\right) + \frac{1}{T_t-T_b}\left(\frac{1}{2.303K_{R_1}} - \frac{1}{2.303K_{R_2}}\right) \log t \qquad (8)$$

increases as the log of time, and the relative mean velocity (R_{12}):

$$\lim_{t\to\infty} R_{12} = K_{R_2}/K_{R_1} \qquad (9)$$

asymptotically approaches a constant related only to retardation coefficients. These equations are verified by the findings of Slater[9] which show that, after prolonged electrophoresis, the plot of migration distance versus VH (volts x hours) gives straight lines, which, by backward projection beyond the origin, meet in a single point (see Fig. 1). Thus the ratio between the migration distance of a protein to that of any other becomes a constant after the proteins have all entered a gradient

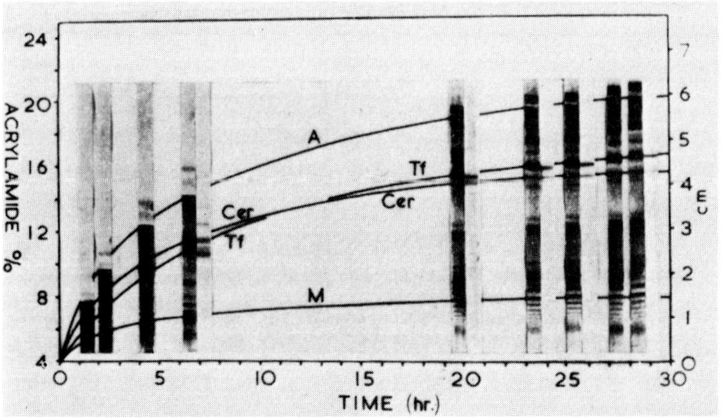

Fig. 1. 20 μl of serum mixed with 20 μl 20% sucrose solution used to produce a pattern at 26, 37, 42, 46, 50 and 61 hours. Electrophoresis in a cold room at 3±2°C with voltages ranging from 350 to 410. After 10×10^3 VH the proteins form stable patterns in which they continue to move slowly but their relative positions remain constant (from Slater[9]; source: American Chemical Society).

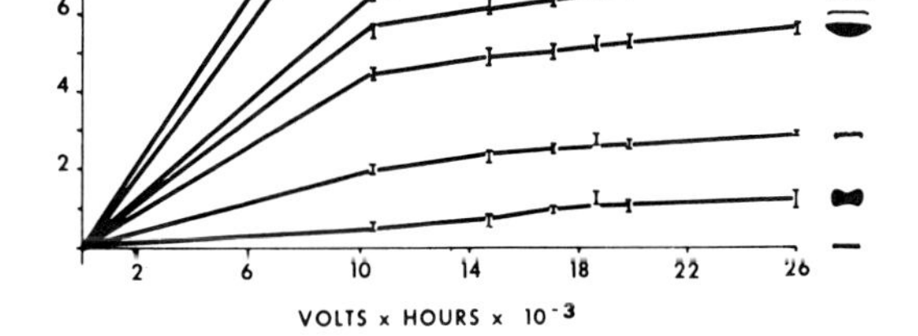

Fig. 2. Progress curves of plasma proteins electrophoreses at 280 V, 50 mA, 5°C in a 4-24% gradient slab. Successive 10 μl samples of (a) 50% plasma (haptoglobin type 2-2) with added hemoglobin and (b) mixture of 0.5 mg/ml each of transferrin and ceruloplasmin were applied at intervals to give running times indicated in the graph. The curves were drawn for albumin (A), transferrin (Tf), ceruloplasmin (Cer) and α_2-macroglobulin (M). Corresponding strips were cut from the photograph and superimposed on the curve. (from Margolis and Kenrick[7]; source: Academic Press).

region in which they are subjected to the sieving effect of the acrylamide matrix (Fig. 2). This causes the electrophoretic patterns to become constant after prolonged runs in a gel gradient. The gel concentration at which the migration rate of a protein becomes constant is called pore limit[9].

The retardation coefficient K_R depends upon the molecular radius R[15,12]:

$$\sqrt{K_R} = k(R+r) \qquad (10)$$

(where k=constant; r=radius of the polyacrylamide fiber; $R = (3MW\bar{\nu}/4\pi N)^{1/3}$, where $\bar{\nu}$=partial specific volume and N= Avogadro's number) and therefore upon the molecular weight. From equation (10), a simple relationship between log MW and log migration distance can be derived:

$$\log MW = -1.5(\log x - k) \text{ when } R \gg r \qquad (11)$$

$$\log MW = -2\ (\log x - k) \text{ when } R \cong r. \qquad (12)$$

The experimental data from Slater[9] and Kopperschläger *et al.*[16] validate these findings: the slope found by the former is, in fact, -1.75 and, by the latter, -2.

This technique, therefore, allows the evaluation of native molecular weights of proteins when run in parallel with a set of reference macromolecules. However, this experimental approach has some limitations. In aqueous buffers, only MW's above 50 000 daltons (d)[16] can be interpolated on a straight line, with a 3 - 20% gradient (with a 4 - 26% gradient, Lorentz[17] obtained a reliable MW evaluation for proteins of about 21 000 d). The use of 8M urea (which is known to reduce the effective gel pore size by about 50%) lowers the limit to about 20 000 d[18]. Moreover, it should be noticed that K_R depends directly upon R. For globular proteins, this means that the behaviour along the molecular sieve is determined by their hydrodynamic volumes: therefore, macromolecules with different volume to mass ratios (*i.e.* different frictional ratios) are scattered in a MW vs migration plot. For macromolecules with non-globular shape and large axial ratio, like fibrinogen, the penetration into the gel pores depends largely upon the minor diameter of the molecule. So, the reliability of MW determination by gradient gel electrophoresis depends on the homogeneity of shape and frictional ratio between the proteins used as standards and those under investigation[11]. Random coil formation in 8M urea and p-chloromercury benzoate (PCMB) causes proteins to exhibit larger MW's and to band less sharply. Fibrinogen still shows an anomalous behaviour. The use of sodium dodecyl sulphate (SDS) seems to be unnecessary, at least for globular proteins: in fact, the slopes of MW vs migration plots are similar for both, native and denaturated proteins[17], even though SDS obviously breaks down the oligomeric structure of quaternary proteins (see Ch. 17 by H. Stegemann).

In the first stages of electrophoresis along a gel gradient, the proteins move according to their charge to mass ratio, and only under prolonged runs (once the pore limit has been reached) the charge parameter becomes unimportant under the effect of molecular sieving. The relative importance of these two parameters on mobility is evident when the macromolecules are run perpendicularly (or transversely) to the gel gradient, instead of along it: the crossing of different protein traks indicates the gel concentration at which the electric charge is balanced by molecular size[7]. With a transverse run, for each protein several determinations of m_T can be obtained in a single experiment and, by plotting them according to the Ferguson relationship (1), m_o and K_R values can be extrapolated[5,10]. However, at high gel concentrations, the plot falls out of linearity: in these regions the behaviour of the macromolecules can be described by the Felgenhauer[11] relationship:

$$\log T = \log T_{EL} - K_R m_T \qquad (13)$$

which predicts a zero mobility above a limiting concentration, the exclusion limit ($=T_{EL}$). This parameter can be measured directly as the point of intercept between the track of a protein and the upper edge of the gel in transverse gradient electrophoresis. By comparing the data obtained in this way with the migration distances of the same proteins in a run along the gel gradient, it has been demonstrated that only the largest molecules approach their exclusion limit under the above conditions[7] (Fig. 3A and B).

Thorun and Mehl[8] used transverse electrophoresis data for MW evaluation: the ratio between migration distances, or mobilities, at two gel concentrations -which corresponds to the frictional ratio - is plotted against MW: in a medium favouring random coiling of proteins (such as acidic phenol mixtures), a reliable determination (±5%) in the range 10 000 - 100 000 is obtained.

EXPERIMENTAL

The gradients generally used are linear or concave: the latter improve the resolution of the lowest MW components: in fact, the sieving effect of the gel depends upon the pore size, and this, in turn, decreases as $T^{1/2}$, for low percentages of cross-link, and as $T^{1/3}$ for highly cross-linked gels[15]. Thus small proteins cannot in practice reach their pore size limit since they have to travel a long way through a medium of increasing viscosity[7].

Another problem is the highest T% which can be utilized: most workers use gradients ranging from 4 to 25% acrylamide concentration, but even 30%[9,19], 35%[20] and 40%[21] T_b are reported. However, such gels are brittle and difficult to stain and destain. In general, the cross link concentration is kept constant along the gel gradient, and often equal to 5%, as recommended by Rodbard *et al.*[10]. Margolis and Wrigley[22], however, use an increasing C% (from 3.8 to 6.5) at the highest gel concentrations, in order to improve the sieving effect in this region of the gradient, and to prevent the formation of gel turbidity around 10% concentration (T%). A linear cross-link gradient is used by Caton and Goldstein[23] and by Foissy[19]. When starting the experiment, the proteins are loaded in slots cast in a very soft gel of acrylamide (the use of agarose has never been proposed even though we feel it is a useful variant, since it prevents collapsing of the pocket walls); alternatively, a sample

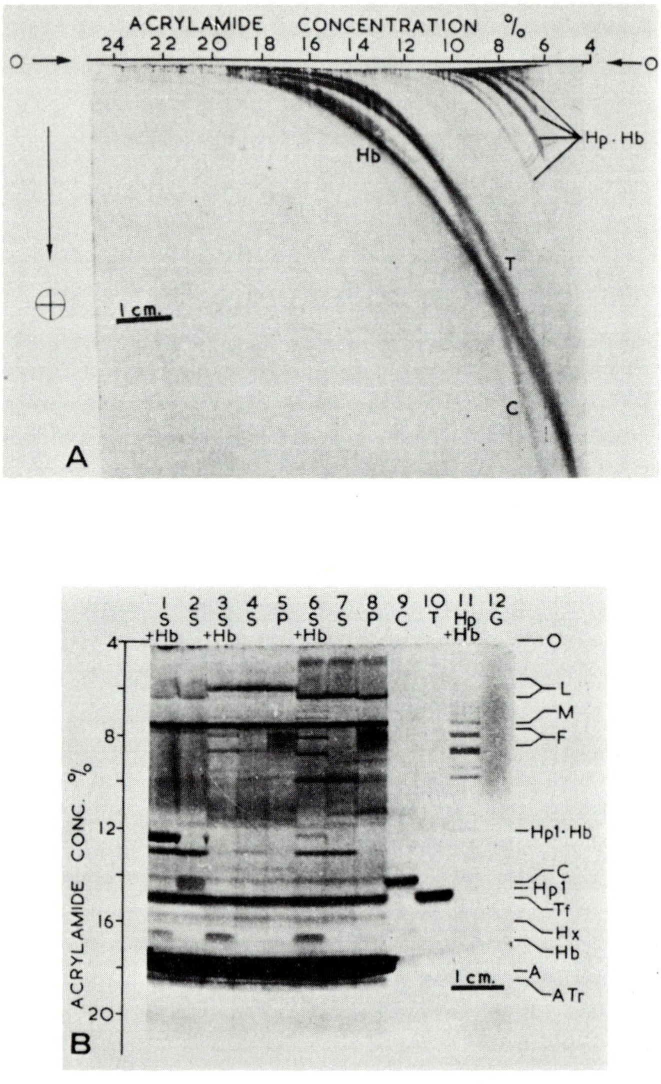

Fig. 3. (A) Electrophoresis across a 4-24% linear gradient gel. Mixture containing 0.5 mg/ml each of hemoglobin (Hb), ceruloplasmin (C), transferrin (T), haptoglobin polymers (Hp.Hb) and traces of other proteins were applied along the edge of the gel slab (0.2 ml/8cm). Electrophoresis: 4 hr at an average of 12°C, 220V, 90 mA. (B) Prolonged electrophoresis in a 4-24% linear gradient gel: 24 hr run at 6°C, 280V, 130-80 mA. Samples 1-8: 10 μl of 50% plasma (P) or serum (S). Tracks: (1-2): Hp 1-1; (3-5): Hp 2-2; (6-8): Hp 2-1; (9-11): 12 μg each of C, T and Hp-Hb, respectively; (12): 50 μg γ-globulins (G). Other proteins identified by letters: β-lipoprotein (L), α_2-macroglobulin (M), fibrinogen (F), hemopexin (Hx), hemoglobin (Hb), Albumin (A), and α_1-antitrypsin (ATr) (from Margolis and Kenrick[7]; source: Academic Press).

spacer in plastic material can be used[7]. During the gel preparation, to avoid a premature setting, the catalyst and promoter concentrations have to be carefully tested, the temperature controlled and the gradient delivered in a short time. In order to prevent layer mixing by thermal convection, the polymerization has to start at the top of the gel[5-7]. This happens spontaneously, so that earlier recommendations (formation of a gradient either of radical scavengers[5,16] -to prolong the induction time- or of catalysts[7,9], or of temperature[5]) can be disregarded. Moreover, the acrylamide gradient is often further stabilized by an additional density gradient of sucrose or glycerol[5,7,8,16,24]. In order to optimize the reproducibility from run to run, several gel plates or rods are prepared together[7,22,23]: in the early method of Margolis and Kenrick[5], a single block is polymerized, which is then sliced in slabs of appropriate thickness.

The actual slope of the gradient can be tested either from the weight (wet and dry) of gel segments punched at different heigths[7,22], or, in a simpler way, by a densitometric scan of the gel, prepared by adding a dye either to the mixing chamber or to the reservoir (bromophenol blue, which is also useful to stain some proteins, *i.e.* serum albumin[20], or 4-nitrophenol[17]). For gradient preparation, two types of devices are generally used: two communicating vessels (reservoir + mixing chamber) or a rectangular chamber with a sloping baffle (see, for instance, refs.[25-27]).

Linear gradients

These gradients can be obtained from two solutions corresponding to the desired top and bottom concentrations by mixing and delivering equal volumes from each of them. The simplest device for gradient formation is made of two cylindrical vessels, called reservoir (R) and mixer (M), connected at the bottom by a short tubing with a stopcock (Fig. 4). The outlet from the mixing chamber to the gel mold is closed

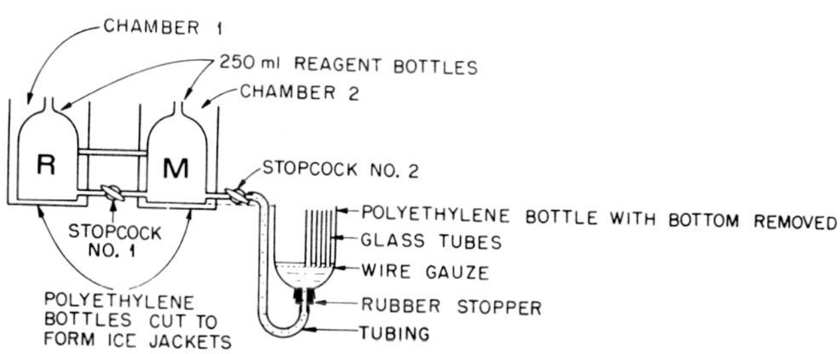

Fig. 4. Home-made apparatus for preparing linear gel gradients. R is the reservoir and M the mixing chamber (from Caton and Goldstein[23]; source: Academic Press).

by a second stopcock. The solution in M has to be continuously stirred; once the two locks have been opened, equal volumes flow from M and from R, since the two vessels are hydrostatically equilibrated (a small correction should be made to account for differences in density between the two solutions: $V_b = V_g d_t/(d_b + d_t)$; where the subscripts correspond to: b=bottom, t=top, g=gradient)[28]. A more refined device includes two syringes whose plungers, fixed to a support plate, are pushed down together (Fig. 5). This apparatus permits high rates of delivery (limited only by the turbulence in the gel chamber and by the rate of mixing in M). If the more concentrated solution is put in M, the receiver is filled from the bottom, (as in Fig. 4), otherwise the gradient is delivered from the mold top.

Another type of gradient-generating device consists of a rectangular cell divided into two compartments by a sloping partition (Fig. 6). The two outlets join in a small magnetic mixer or a narrow tube which forms several coils. The amounts of the two solutions flowing from the outlets are proportional to their surface areas at

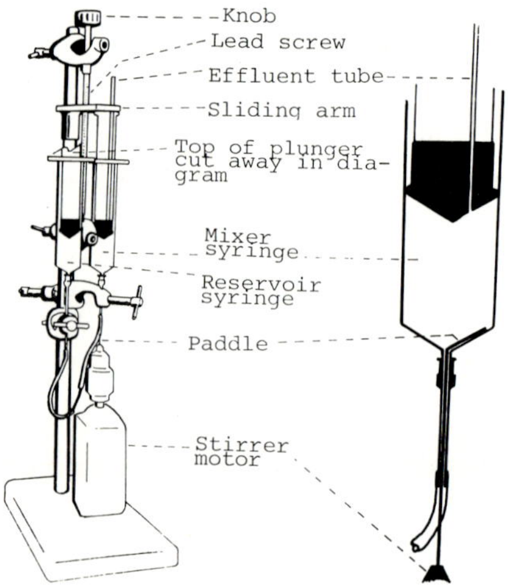

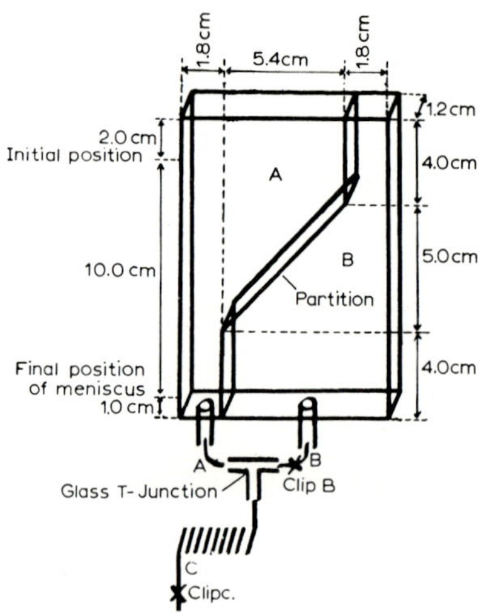

Fig. 5. (left). Gradient former made with two disposable plastic syringes and plungers. (A): general view, (B): enlarged diagram of mixer syringe. With this arrangement the two solutions do not have to be compensated for different hydrostatic pressure due to different densities. Notice that the gradient solution leaves the mixing chamber through an effluent tube embedded in the plunger (from Arcus[26]; source: Academic Press).

Fig. 6. (right). Gradient forming apparatus. The top of the vessel is open to the atmosphere. All measurements are internal dimensions neglecting the thickness of the perspex (3.2 mm). A and B are PVC tubes (internal diameter 4 mm), connection with the vessel being made via short glass tubes sealed in with rubber grommets. The lower part of the PVC tube C is omitted in the diagram (from Pratt and Dangerfield[24]; source: Elsevier Publ. Co.)

any moment, and the slope of the baffle within the vessel determines the shape of the gradient. For the preparation of large gel slabs, instead of using a single capillary outlet, the gradient solution is evenly distributed along the width of the mold from a perforated tubing (a spray nozzle) as large as the cell. This sprinkler is gently lifted from the bottom toward the top of the cell as the gradient solution is poured in[19]. Gel gradients in rods can be conveniently formed by layering two solutions of different acrylamide concentration in each tube, then by rotating the tubes clamped on a rack in a oblique position at a 5° angle with the level. After a 10-15 min rotation at 50 rpm, the gelation is allowed to proceed in a vertical position[17] (Fig. 7A and B).

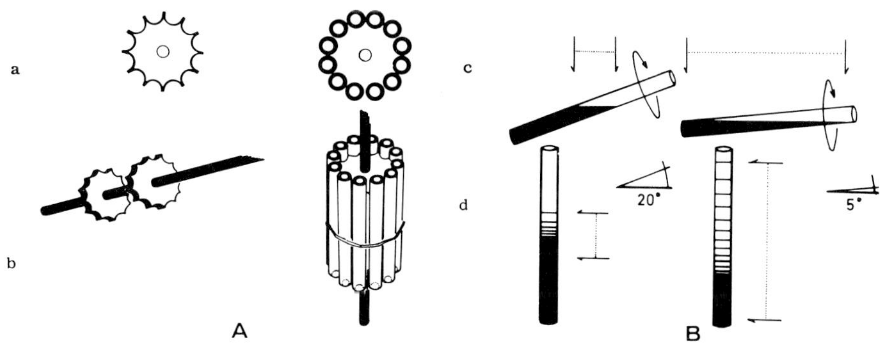

Fig. 7. Gradient formation on a rotating rack. (A) supporting rack. (a) 2 mm piece of Plexiglas, nearly 25 mm in diameter; (b) rotation rack without tubes, in oblique position; (c) cross-section and (d) view of the rack with tubes in vertical position. (B) gradient formation by rotation in oblique position and polymerization in vertical position. The arrows indicate the length of the gradient (from Lorentz[17]; source: Academic Press).

For capillaries of inner diameter smaller than 1 mm (1-10 µl capacity), the formation of the gradient can be achieved simply by allowing the reciprocal diffusion of two solutions, the first containing the catalyst, the second all the other gel constituents, which are drawn by capillarity in equal amounts into the tube[21].

Some typical separations

4-24% gradients for serum protein analysis. We describe here briefly one of the early systems reported by Margolis and Kenrick[7]. When preparing gel gradients, it is convenient to polymerize simultaneously several slabs or rods, so that all the gel matrices within a prepared stock will exhibit identical properties. To this purpose, these authors[7] have built a stack of 14 cells (4x100x100 mm inside) which rests on a strip of plastic net in a rectangular mold with a funnel-shaped bottom (see the lower drawing in Fig. 8). Since the gradient is fed from the bottom, the mixing chamber (M) contains the less dense solution. At first, a solution of 0.25% ammonium persulphate (AP) and 0.03% dimethyl amino propionitrile (DMAPN) is poured into the mold, in order to displace air bubbles from the cells and to insulate the gelling solution from atmospheric oxygen. The gradient components -302 ml each of 26% (in the reservoir) and 4% (in the mixer) acrylamide (C%= 5), containing 90 mM

Tris, 80mM borate and 2.5 mM EDTA buffer, pH 8.3, added with 7.75 ml of 10% AP and 30 μl (in R) and 90 μl (in M) DMAPN- are then delivered from two vessels fitted with linked piston floats. After delivering the entire gradient, the last portion left in the bottom funnel of the gel mold is displaced into the stack of 14 cells using a 30% sucrose solution. Once the gel has polymerized, one cell is transferred to an electrophoretic tank (see, for instance, Fig. 2 in ref. 7) and the sample applied with a spacer. The run usually lasts for 24 hours with a current density of 20 mA/cm^2. Examples of separations in this system are shown in Figs. 2 and 3A and B.

2.5-12% (T%) (C% = 5-2%) gradient for RNA separation[23]. For the preparation of 90 gel rods 5 mm ∅ x 90 mm height, 130 ml each of 15% and 2.5% acrylamide in 20 mM Tris, 10 mM Na acetate, 20 mM acetic acid and 1 mM EDTA buffer, pH 7.5, mixed with 100 μl of TEMED and 50 μg of AP/100 ml, are delivered with the gradient former of Fig. 4. The electrophoretic run is carried out in about one hour at a constant current of 4 mA/tube.

2-6.5% gradient for lipoprotein separation, in a pH gradient[24]. The concentrated gel is made of 6.5% T (C% = 5) in a 230 mM Tris, 55 mM HCl and 6 mM H_3PO_4 buffer, pH 8.5, containing 4% sucrose. The diluted gel is composed of 2% T (C% = 5) and 1% sucrose in 60 mM Tris, 15 mM HCl, 25 mM H_3PO_4 buffer, pH 7.23 (the gradient former is shown in Fig. 6). The cathodic buffer contains 150 mM glycine and 10 mM NaOH while the anodic buffer contains 150 mM Tris and 50 mM HCl. The run lasts for 6 hours at a total constant current of 50 mA.

Concave gradients

As Rodbard *et al.*[10] pointed out, MW determinations in PGE are satisfactory only in the "asymptotic case", where the zone velocity approaches zero. In practice, only for proteins with MW's well above 100 000 migration virtually ceases on prolonged electrophoresis, while migration of smaller proteins continues at a decreasing rate sufficient to interfere with size determination. To overcome that, concave gel gradients have been introduced. These types of gradients can be easily made by using any of the two-vessel apparatuses described above, provided the mixing chamber is sealed, so that its volume is maintained constant. In this case, the shape of the gradient is given by[28]:

$$C_V = C_R + (C_b - C_R)\ e^{-V/V_m} \qquad (14)$$

where C_V = gradient concentration when a volume V has been displaced. For any given set of V_g, C_b and C_t values, C_V (*i.e.* the shape of the gradient) depends upon V_m: by increasing the volume in the mixing chamber, the gradient becomes flatter.

Instead of the two-vessel equipment, a tank with a baffle of concave shape can be conveniently used[22], since the shape of the gradient can be varied to almost any gradient function simply by varying the slope of the baffle. The device shown in Fig. 8 is composed of an additional chamber, used to generate also a concave gradient

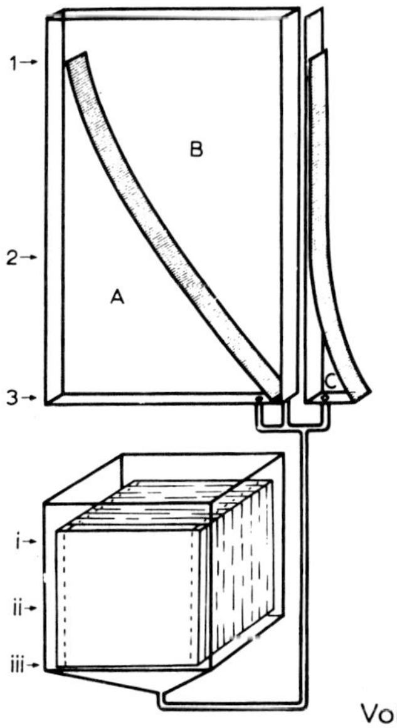

Fig. 8. Diagram illustrating the principle used to produce pore gradient gels with variable cross-linking. A contains a 30% acrylamide-Bis mixture (3.8% Bis and 96.2% acrylamide); B contains diluent buffer and C is filled with a 30% acrylamide-Bis solution (20% Bis, 80% acrylamide). Lower part: drawing of the moulding tower which contains a row of 14 empty glass cells. Each cell consists of two 82-mm squares of glass taped together and separated by two 3-mm-thick strips of glass. When the cells have been filled, the proportions of solutions A, B and C at levels 1, 2 and 3 of the gradient former correspond exactly to actual compositions at levels (i), (ii) and (iii) of the electrophoresis plates.

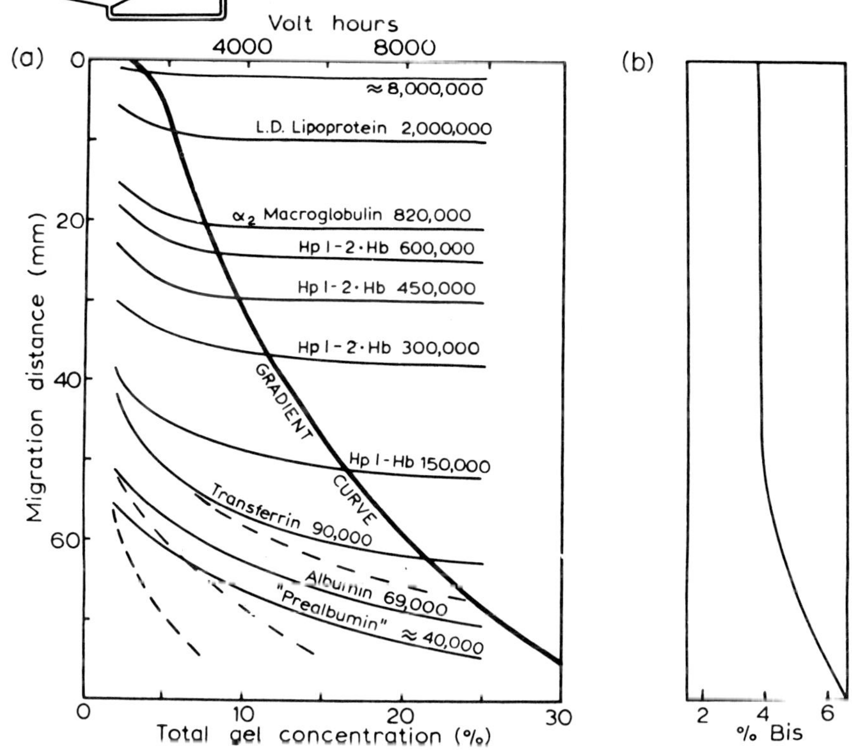

Fig. 9. Plots of the mobilities of serum proteins against time of electrophoresis using pore gradient gels with uniform 5% cross-linking (dashed lines) or variable cross-linking (solid lines), as indicated by side diagram (b). For both types of gel, the bold transverse plot indicates the variation in total gel concentration (acrylamide plus Bis). The curve indicating slowest mobility represents some very-low-density lipoproteins and γ M polymers with molecular weights in the region of 8 000 000 (from Margolis and Wrigley[22]; source: Elsevier Publ. Co.).

of cross-linkage (Bis) in the dense region of the gel. The mobility curves obtained in these gradients for serum proteins are shown in Fig. 9, in comparison with gels containing uniform cross-linking. While there are no differences between the two systems for proteins greater than 150 000 d, for smaller proteins the increased cross--link has effectively reduced the pore size in the denser region of the gel, since their plots (solid curves) come much closer to being asymptotic than equivalent plots (broken lines) in gels of uniform 5% cross-linkage. These concave (T% and C%) gels thus offer a much better basis for estimation of MW's by comparison with proteins of known sizes. An electrophoretic time equivalent to about 2000 VH is recommended, as a compromise between convenience in performing the experiment and approach to "pore limit" conditions. Examples on the use of these gradients can be found in Arosio *et al.*[29], for preparative ferritin purification, and in Margolis and Wrigley[22] for serum protein separation.

Precast gradient gels are commercially available from Gradient Pty, Oak Ridge, Tennessee, USA, or Lane Cove, Australia, and Pharmacia Fine Chemicals AB, Uppsala, Sweden. Universal Scientific Limited (Uniscil), London, UK, also produces equipment for concave, as well as linear, gradient gels. For a home-made apparatus, see Madjar *et al.*[30].

Conclusions

This brief survey illustrates the contrast between dynamic gel electrophoresis and pore limit electrophoresis. In the former case, the gel network offers frictional resistance for migration which is not, however, uniquely determined by protein size and shape, since protein charge also plays a significant role. On the other hand, pore limit electrophoresis represents the limiting case where macromolecules have penetrated a pore gradient so far that they become "jammed" in the gel random meshwork of fibers: migration distance is therefore a reflection solely of size and partly shape, but not charge. PGE is thus very useful in the resolution of complex mixtures of macromolecules exhibiting a wide range of MW's.

Gradient techniques have been highly successful in the field of nucleic acids, which exhibit only small differences in charge to mass ratio but high MW heterogeneities[23,31] (and which cannot be discriminated on the basis of their pI's with isoelectric focusing[32]). PGE has also been successfully applied to protein studies[33,34] and to the analysis of biological fluids[4,5,7,20,21,35,36] and tissue[8] or microbial cell extracts[19].

TWO-DIMENSIONAL TECHNIQUES

Notwithstanding the high resolving power of electrophoretic techniques (Giddings[37] has calculated that a number of theoretical plates of approximately 2000 can be achieved in electrophoretic fractionations using reasonable values of path length, voltage gradient and diffusion coefficient) no single electrophoretic method is able fully to resolve all components of a complex protein mixture. This is particularly critical in the case of body fluids, which contain molecules over a thousandfold range of MW's and relative concentrations, with a relatively small range of net charge or free mobility. Often, increasing sample load does not increase the ability to detect minor components, which remain as barely detectable shoulders or irregularities on the tail of the gaussian distributions of major components (overshadowing effect). The next logical step was thus the introduction of two-dimensional techniques (2-D) with the aim of spreading all the protein spots as far apart as possible and uniformly over the 2-D separation space. A "macromolecular map" (MM) would thus be generated, in which each spot will be uniquely defined and characterized by a set of coordinates, representing intrinsic macromolecular parameters of each protein species (e.g.: a set of pI-native MW or pI-subunit MW or native MW-subunit MW values, etc.). Ideally, if the various species of a multicomponent mixture have to be spread randomly over the 2-D plane, and not aligned along a diagonal in the second run, the fractionation method utilized for the second dimension of 2-D mapping should operate on a principle independent of that used for separation in the first dimension. This concept is very nearly achieved in paper electrophoresis/chromatography (finger-printing) of peptides, where molecules are fractionated on the basis of charge in one dimension and on the basis of their partition coefficients in hydrophobic solvents in the second dimension. The strategy, then, in a 2-D electrophoretic separation of proteins, would be to discriminate molecules on the basis of net-charge (or free mobility) in the first dimension, followed by fractionation according to size, in the second dimension, or *viceversa*. In this way, one might expect that every region of the 2-D gel slab would be covered with spots.

Smithies and Poulik[38] were among the first to apply 2-D techniques, by performing electrophoresis on filter paper followed by a run at right angles in a starch gel. The use of two acrylamide gels of different concentrations[39-42] and even of different pH was then proposed[42-46]. An improved resolution and a (rough) possibility of relating the position of a spot in a 2-D map to the macromolecular parameters of proteins was offered by gel electrophoresis followed by PGE as proposed by Margolis and Kenrick[5,7]. The true innovation, however, came when isoelectric focusing (IEF, which discriminates solely on the basis of charge), in one dimension, was combined with different electrophoretic techniques, discriminating either solely on size, or on size and charge, in the other dimension. We will therefore limit our discussion to the combined use of these techniques, as summarized in Fig. 10, since they meet the principle of spreading maximally the protein spots over the gel map. In Fig. 10, IEF has been depicted as the first dimension run, even though, in some cases, this combination can be reversed. However, it is customary to use IEF as the first step, also because it permits the use of dilute samples, not suitable for gel electrophoresis, and provides well-sharpened zones for the second stage of fractionation. Furthermore, use of electrophoresis as the second step offers the possibility of removing carrier ampholytes at that stage, thus simplifying the staining procedure. From a technical point of view it is often convenient to run the first separation in a gel rod, which

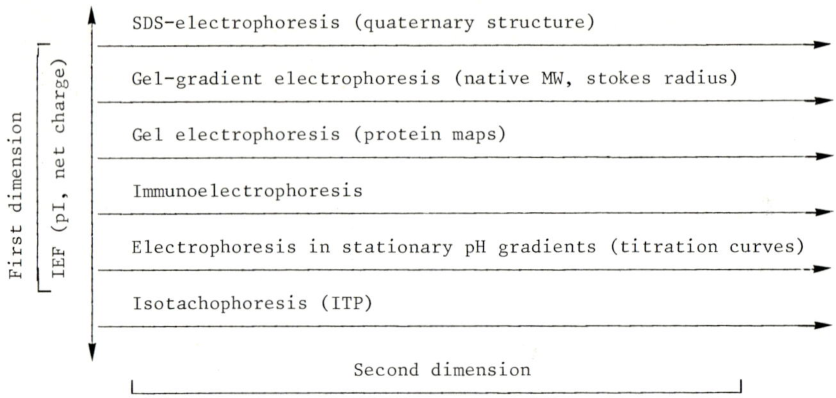

Fig. 10. Diagram illustrating various two-dimensional techniques in which one of the dimensions is isoelectric focusing (IEF).

is then embedded (for instance, with the aid of melted agar) on top of a gel slab in which the second stage separation will take place. (Fig. 11)[47]. If needed, the first gel can be equilibrated for a short time in an appropriate buffer, or incubated in SDS or urea, or detergents, prior to the second run. Alternatively, gel tubes can be used in both dimensions, in which case the initial gel rod is cut into sections each of which is applied to another gel tube for the second stage[48]. In another variant, very convenient when the second run is performed on the surface of a very thin, horizontal gel slab, the first run is performed in a gel slab too, which is subsequently cut into strips, which are then transferred to the second gel(laying-on technique[49]). Several of these 2-D techniques have been reported in the proceedings of recent meetings on IEF and ITP[50,51].

IEF-SDS gel electrophoresis. It allows the determination of pI in the first run and of subunit MW in the second step, and it is a true charge-size fractionation. The 2-D map shows spots spread all over the plane. A classical example is the 1100 spot map developed by O' Farrell[47] in the analysis of a *E. coli* lysate. With this method, detection and quantitation of protein species representing as little as 10^{-4} to 10^{-5} of the total cell protein was made possible. Barrett and Gould[52] and McGillivray and Rickwood[53] have applied this 2-D technique to the analysis of non-histone proteins in chromatin. The accuracy of the method was given as ± 3000 in MW and ± 0.2 pH units in pI determinations. The O' Farrell technique has also been adapted to monitor the synthesis of viral specific proteins in whole cell extracts of infected cells, to map lymphocyte surface antigens and to follow the rate of synthesis of steroid-inducible proteins in rat hepatoma cells[54]. This method seems to be ideally suited for the characterization of membrane proteins, such as those of the human erythrocyte ghost[55] or lymphocyte microsomal membrane proteins[56]. Applications to the mapping of plant proteins[57] and of pancreatic zymogen granules[58] have also been reported.

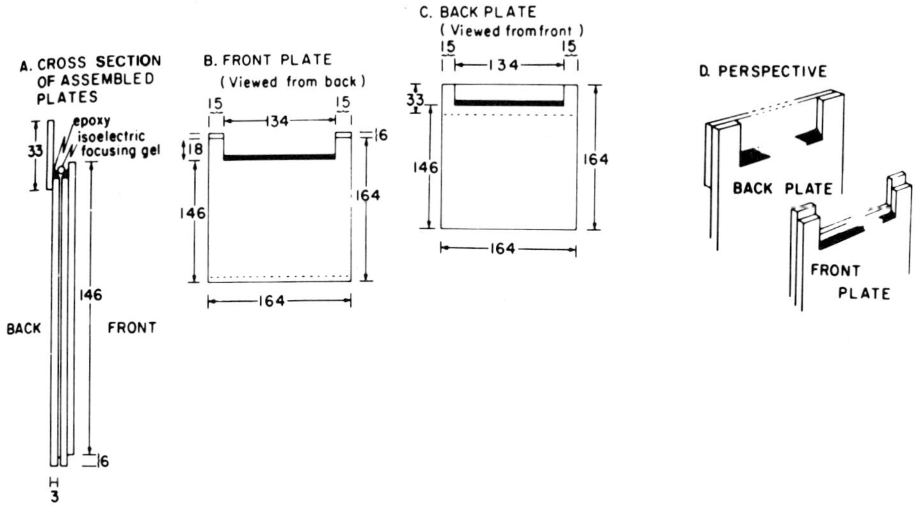

Fig. 11. Construction scheme of slab gel plate for use in 1- and 2-D techniques. After IEF in the first dimension, the gel rod is embedded on top of a polyacrylamide gel slab for SDS-electrophoresis in the second dimension. All measurements are given in mm (from O'Farrell[47]; source: Williams and Wilkins Co.)

Usually, the IEF run is performed prior to the SDS-electrophoresis. However, Miller and Elgin[59] have shown that proteins pretreated with SDS band at the same pI as the untreated proteins if focused in a 8 M urea gel. Apparently, the SDS-protein complex dissociates in urea in the electric field to give an SDS-free protein which behaves indistinguishably from the protein treated only with urea. Ames and Nikaido[60] obtain similar results by adding to the SDS-protein extract Nonidet P-40 and then by performing IEF in a gel containing 9.5 M urea and 5% detergent (NP-40). Danno[61] has run fluorescamine labelled proteins in SDS-electrophoresis and then performed the IEF run in 8 M urea gel tubes top loaded with 1-3 mm slices of the first run gel. A type of modified peptide mapping or fingerprinting of large peptides has been proposed by Epstein and Wolff[62]. The protein is first demolished in CNBr, the cleaved peptides subjected to performic acid oxydation and then analyzed first in IEF followed by SDS-urea gel electrophoresis.

IEF-Pore gradient electrophoresis. This also is a true charge-size fractionation which gives the pI in the first run and native MW in the second dimension. One of the first examples is the separation of serum proteins in a 3-10 pH gradient followed by electrophoresis in a concave 4.5-26% acrylamide gel slab, in Tris/EDTA/borate buffer, pH 9.3, for 20 hours[63]. Wright *et al.*[64] have applied the same technique to the analysis of serum, cerebrospinal fluid, urine and bacterial cell extracts. An interesting variation has been described by Felgenhauer and Pak[65]. They perform IEF in a granulated gel layer (Sephadex G-75) followed by electrophoresis in a linear gradient of 3-30% acrylamide. The use of granulated gels in the first dimension allows rapid equilibration of large proteins, such as ferritin, α_2-macroglobulin and β-lipoprotein, and quick detection of the focused bands by the paper print technique[66]. A micro-technique has also been described[67], by which IEF is performed in a 5 μl capillary followed by electrophoresis in a continuous slab gel gradient of 1-40% acrylamide of less than postage stamp size.

IEF-gel electrophoresis. This is indeed an hybrid method of charge versus charge-size fractionation in the second dimension. Dale and Latner[68] were the first to demonstrate the potential of this procedure in detecting qualitative and/or quantitative changes in protein maps from normal and pathological sera. This study was followed by the application of this technique to several pathological cases, including nephrotic syndromes, mono- and polyclonal gammopathies, multiple sclerosis, IgG myelomatosis, proteinuria[69], and typing of haptoglobin genotypes[70]. Wrigley[71,72], Wrigley and Shepherd[73],Macko and Stegemann[74] and Stegemann *et al.*[75] have used similar 2-D procedures to map wheat grain proteins and potato proteins, respectively. Generally speaking, the resolving power of the 2-D technique is greater than that of each single step, since nearly twice as many components are revealed in the 2-D map.

IEF-immunoelectrophoresis. Several variants of this technique have been reviewed by Catsimpoolas[76]. After IEF, the gel rod can be sectioned and the squashed slices used either for double diffusion, or for radial diffusion or for electro-immuno diffusion. Alternatively, the intact gel rod after IEF can be embedded in an antibody--free agar for diffusion or in an antibody-containing agar for crossed electrophoresis or incubated as such in an antibody solution for *in situ* immunoprecipitation. (see Fig. 1 in ref. 76). In the case of crossed immunoelectrofocusing, it is more advantageous to run IEF in a thin gel slab, cut a 5 mm broad polyacrylamide strip and transfer it to the surface of an antibody-containing agarose gel for electrophoresis (laying-on technique)[49,77]. This prevents gel swelling and water accumulation at the polyacrylamide-agarose interface, with blurring of protein patterns, typical of the moulding-in procedures.

IEF-electrophoresis in a stationary pH-gradient. This 2-D technique is carried out at right angles in the same gel matrix. We know of only one report so far[78], but the idea is fascinating and certainly deserves to be further exploited. At first, a gel slab is focused without applying the sample. Once the pH gradient is established, the electrodes are turned by 90°, the sample applied all along the pH gradient, in a pocket precast at the center of the gel slab, and electrophoresis continued at right angles to the direction of focusing for 1 to 3 hours. Since the carrier ampholytes are isoelectric, their migration is practically negligible during the second stage, which means that the sample applied migrates in a continuum of pH values (pH 3-10), which are kept stationary. This is the equivalent of performing a titration curve of a protein. This technique, appears to be very promising, especially in mapping protein polymorphisms(genetic mutants, post-synthetic modifications, etc.).

IEF-ITP. This too is a brand new technique described so far in the reverse combination ITP-IEF[79]. Crossed ITP-IEF should allow an investigation of the differences between the separation principles of these two methods. As presently described, ITP is first run in a thin layer polyacrylamide gel, then a strip is cut and transferred to the cathodic side of a thin layer (0.5 mm) IEF slab, to be used for IEF separation. Crossed ITP-IEF combines the high resolution of both techniques while overcoming their respective limitations. For instance, proteins with nearly the same isoelectric points can be separated by isotachophoresis at a pH at which their relative pH/mobility curves diverge. On the other hand, proteins with very similar pH/mobility curves at the prevailing pH in ITP fractionation, but with different pI's, will co-migrate in the ITP step, but be separated in the IEF step.

Concluding remarks. We hope this brief survey on 2-D techniques will give to the reader the feeling of the great potential inherent to these combined electrophoretic methods. Judicious use of different combinations of these 2-D techniques will often define several physico-chemical macromolecular parameters in a distinct manner for ready reference or comparison. Meanwhile, we are waiting for the three-dimensional techniques announced by Stegemann in Hamburg[57].

ACKNOWLEDGEMENTS

We are indebted to Dr. J. Margolis, Sidney, Australia, who, by declining at the last moment the invitation to write this chapter, made possible to us this wonderful journey through this fascinating branch of electrophoretic techniques. Supported in part by grants CT76.00311.06, 76.01488.04, CT77.01471.04 and CT77.00278.06 from Consiglio Nazionale delle Ricerche (CNR) Roma, Italy.

REFERENCES

1. Grossbach, U. and Weinstein, I.B. (1968) Anal. Biochem., 22, 311-320.
2. Slater, G.G. (1965) Fed. Proc., 24, 215-230
3. Slater, G.G. (1968) Fed. Proc., 27, 384-390
4. Slater, G.G. (1968) Anal. Biochem., 24, 215-217.
5. Margolis, J. and Kenrick, K.G. (1967) Nature, 214, 1334-1336.
6. Margolis, J. and Kenrick, K.G. (1967) Biochem. Biophys. Res. Commun., 27, 68-73
7. Margolis, J. and Kenrick, K.G. (1968) Anal. Biochem., 25, 347-362.
8. Thorun, W. and Mehl, E. (1968) Biochim. Biophys. Acta, 160, 132-134.
9. Slater, G.G. (1969) Anal. Chem., 41, 1039-1041.
10. Rodbard, D., Kapadia, G. and Chrambach, A. (1971) Anal. Biochem., 40, 135-157.
11. Felgenhauer, K. (1974) Hoppe-Seyler's, Z. Physiol. Chem., 355, 1281-1290.
12. Rodbard, D. and Chrambach, A. (1971) Anal. Biochem., 40, 95-134.
13. Ferguson, K.A. (1964) Metabolism, 13, 985-992
14. Ogston, A.G. (1958) Trans. Faraday Soc., 54, 1754-1759
15. Rodbard, D. and Chrambach, A. (1970) Proc. Natl. Acad. Sci. U.S.A., 85, 970-977.

16. Kopperschläger, G., Diezel, W., Bierwagen, B. and Hofmann, E. (1969) FEBS Letters, 5, 221-224.

17. Lorentz, K. (1976) Anal. Biochem., 76, 214-220.

18. Anderson, L.O., Borg, H. and Mikaelsson, (1972) FEBS Letters, 20, 199-202.

19. Foissy, H. (1975) J. Chromatogr., 106, 51-57.

20. Arcus, A.C. (1970) Anal. Biochem., 37, 53-63.

21. Rüchel, R., Mesecke, S., Wolfrum, D.I. and Neuhoff, V. (1973) Hoppe-Seyler's Z. Physiol. Chem., 354, 1351-1368.

22. Margolis, J. and Wrigley, C.W. (1975) J. Chromatogr., 106, 204-209.

23. Caton, J.E. and Goldstein, G. (1971) Anal. Biochem., 42, 14-20.

24. Pratt, J.J. and Dangerfield, W.G. (1969) Clin. Chim. Acta, 23, 189-201.

25. Bock, R. and Nan-Sin-Ling (1954) Anal. Chem., 26, 1543-1546.

26. Arcus, A.C. (1967) Anal. Biochem., 18, 381-384.

27. Margolis, J. (1969) Anal. Biochem., 27, 319-322.

28. Noll, H. (1969) in "Techniques in Protein Biosynthesis" Campbell, P.N. and Sargent, J.R. eds., Academic Press, London-N.Y., pp. 101-180.

29. Arosio, P., Adelman, T.G. and Drysdale, J.W. (1978) J. Biol. Chem., in press.

30. Madjar, J.J., Arpin, M. and Rebound, J.P. (1977) Anal. Biochem., 83, 304-310.

31. Schaup, H.W., Best, J.B. and Goodmann, A.B. (1969) Nature, 221, 864-865.

32. Galante, E., Caravaggio, T. and Righetti, P.G. (1976) Biophys. Acta, 442, 301-315.

33. Skakoun, A. and Daussant, J. (1971) C.R. Acad. Sc. Paris, 273, 733-736.

34. Niku-Paavala, M.L., Skakoun, A., Nummi, M. and Daussant, J. (1973) Biochim. Biophys. Acta, 322, 181-184.

35. Margolis, J. and Kenrick, K.G. (1969) Aust. J. Exp. Biol. Med. Sci., 47, 637-641.

36. Sheridan, J.W., Kenrick, K.G. and Margolis, J. (1969) Biochem. Biophys. Res. Commun., 35, 474-477.

37. Giddings, J.C. (1969) Separ. Sci., 4, 181-188.

38. Smithies, O. and Poulik, M.D. (1956) Nature, 117, 1033-1034.

39. Raymond, S. (1964) Ann. N.Y. Acad. Sci., 121, 350-365.

40. Raymond, S. and Nakamichi, M. (1964) Anal. Biochem., 7, 225-230.

41. Holmes, R. (1967) Biochim. Biophys. Acta, 133, 174-177.

42. Kaltschimdt, E. and Wittmann, H.G. (1970) Anal. Biochem. 36, 401-412.

43. Kaltschimdt, E. and Wittmann, H.G. (1970) Proc. Natl. Acad. Sci. U.S.A., 67, 1276-1282.

44. Wittmann, H.G., Stoffler, G., Hindennach, I., Kurland, C.G., Randal-Hazelbauer, L., Birge, E.A., Nomura, M., Kaltschimdt, E., Mizushima, S. Traut, R.R. and Bickel, T.A. (1971) Mol. Gen. Genet., 111, 327-333.

45. Fahnestock, S., Erdmann, V. and Nomura, M. (1973) Biochemistry, 12, 220-224.

46. Held, W.A. and Nomura, M. (1973) Biochemistry, 12, 3273-3281.

47. O' Farrell, P.H. (1975) J. Biol. Chem., 250, 4007-4021.

48. Wrigley, C.W. (1976) in "Isoelectric Focusing" Catsimpoolas, N. ed., Academic Press, New York, pp. 93-117.

49. Soderholm, J. and Smith, C.J. (1975) in ref. 50, pp. 99-114.

50. Righetti, P.G. (1975) Progress in Isoelectric Focusing and Isotachophoresis, North Holland/American Elsevier, Amsterdam and New York.

51. Radola, B.J. and Graesslin, D. (1977) Electrofocusing and Isotachophoresis, W. de Gruyter, Berlin.

52. Barrett, T. and Gould, H.J. (1973) Biochim. Biophys. Acta, 294, 165-170.

53. MacGillivray, A.J. and Rickwood, D. (1974) Eur. J. Biochem., 41, 181-190.

54. Ivarie, R.D., Gelfand, D.H., Jones, P.P., O' Farrell, P.Z., Polisky, B.A., Steinberg, R.A. and O' Farrell, P.H. (1977) in ref. 51, pp. 369-384.

55. Bhakdi, S., Knüfermann, H. and Wallach, D.F.H. (1975) in ref. 50, pp. 281-291.

56. Knüfermann, H. (1977) in ref. 51, pp. 395-404.

57. Stegemann, H. (1977) in ref. 51, pp. 385-394.

58. Scheele, G.A. (1975) J. Biol. Chem., 250, 5375-5385.

59. Miller, D.W. and Elgin, S.C.R. (1974) Anal. Biochem., 60, 142-148.

60. Ames , G.F.L. and Nikaido, K. (1976) Biochemistry, 15, 616-623.

61. Danno, G. (1977) Anal. Biochem., 83, 189-193.

62. Epstein, H.F. and Wolff, J.A. (1976) Anal. Biochem., 76, 157-169.

63. Kenrick, K.J. and Margolis, J. (1970) Anal. Biochem., 33, 204-207.

64. Wright Jr., G.L. Farrell, K.B. and Roberts, D.B. (1973) Biochim. Biophys. Acta, 295, 396-411.

65. Felgenhauer, K. and Pak, S.J. (1975) in ref. 50, pp. 115-120.

66. Radola, B.J. (1973) Ann. N.Y. Acad. Sci., 209, 127-143.

67. Rüchel, R. (1977) Anal. Biochem., 132, 451-468.

68. Dale, G. and Latner, A.L. (1969) Clin. Chim. Acta, 24, 61-68.

69. Latner, A.L. (1973) Ann. N.Y. Acad. Sci, 209, 281-298.

70. Latner, A.L. and Ames, A.V. (1975) in ref. 50, pp. 223-233.

71. Wrigley, C. (1970) Biochem. Genet., 4, 509-516.

72. Wrigley, C. (1968) J. Chromatogr., 36, 362-365.

73. Wrigley, C. and Shepherd, K.W. (1973) Ann. N.Y. Acad. Sci., 209, 154-162.

74. Macko, V. and Stegemann, H. (1969) Hoppe-Seyler's Z. Physiol. Chem. 350, 917-919.

75. Stegemann, H., Franksen, H. and Macko, V. (1973) Z. Naturforsch. 28, 722-732.

76. Catsimpoolas, N. (1973) Ann. N. NY. Acad. Sci., 209, 144-146.

77. Holmberg, K. and Wadström, T. (1977) in ref. 51, pp. 413-419.

78. Rosengren, A., Bjellqvist, B. and Gasparic, V. (1977) in ref. 51, pp. 165-171.

79. Brogren, C.H. (1977) in ref. 51, pp. 549-558.

© 1979 Elsevier/North-Holland Biomedical Press
Electrokinetic Separation Methods,
P.G. Righetti, C.J. van Oss and J.W. Vanderhoff

SDS-GEL-ELECTROPHORESIS IN POLYACRYLAMIDE, MERITS AND LIMITS

HERMANN STEGEMANN
Biochemie, Biologische Bundesanstalt, Messeweg 11, D-3300 Braunschweig

INTRODUCTION

Sodium dodecyl sulfate(SDS)* and other salts of this ester have become an indispensable tool in protein chemistry [1]. Four decades ago, its capability to strip off coat proteins from viral nucleic acids brought the first indications that the latter are responsible for genetic information. The pioneering work of Shapiro, Viñuela and Maizel [2,3] created a further burst of new applications. They observed that ME-reduced proteins, when heated with SDS, disaggregated into protomers and that their electrophoretic mobilities were inversely related to the logarithm of their MW if measured in crosslinked polyacrylamide, an inert carrier introduced by Raymond [4]. Since this estimation of MW is possible on a microgram-scale with almost any mixture, including proteins solubilized by SDS, without tedious prepurification and with simple equipment, the wide-spread use of this method is not surprising. The abundant literature has been partly reviewed in [1,5,6,7] with emphasis on animal and microbial proteins with less emphasis on plant proteins. Cathionic detergents as described by [8] and others are not considered.

Theoretical considerations are found e.g. in [9,10,11] and are not discussed here. It seems to me that not all the parameters of the acrylamide polymerization process are fully understood nor can the behavior of the mostly unknown individual proteins be put into a mathematical jacket. These calculations are even more difficult if small and hard-to-detect, nonpeptidic constituents are part of the molecule or if the amino acid composition, as well as their sequence, are unusual.

---

*ABBREVIATIONS AND DEFINITIONS

GE	= Gel-electrophoresis	(P)AA	= (Poly)Acrylamide
PAGE	= Polyacrylamide-GE	BIS	= N,N'-Methylen-bis-AA
PoroPAGE	= Porosity gradient-PAGE	ME	= Mercapto ethanol
PAGIF	= Focusing in PAA	MW	= Molecular weight(s) in sense of molecular size after SDS-treatment
AGGE	= Agarose-GE	2-D-	= Two-dimensional
PAGGE	= PAA/Agarose-GE	Mapping	= PAGIF and PAGE as 2-D-procedure
Protomer	= Protein monomer (subunit) Incorrect: polypeptide		
SDS-Protein	= Protomer with SDS		

Since all common proteins completely unfolded with SDS have a MW-dependent migration rate in SDS-PAGE, a common shape must have been produced. A rodlike[12] or an ellipsoidal[13] or even better, a random-coil[14] model fits the observations, either when 0.4 or 1.4 g of SDS is bound per gram protein. Therefore, one can follow certain rules and it is the intent of this review to compile as much valuable data as possible from recent literature as well as from our experiences with PAGE-procedures since 1960.

EQUIPMENT. TUBES OR SLABS ?

Either tubes or slabs are used for PAGE. Preparing gel rods in tubes involves very cheap, mostly home-made, equipment. Limited comparability of protein bands is due to a small porosity gradient on top of an otherwise homogeneous gel, formed when water has been layered successively on the monomer mixture. Layering with 2-methyl-propanol-1 has other drawbacks. Tubes are advantageous for large sample volumes especially with stacking gels[15] . The different salt concentrations of the samples will cause less band blurring and distortion in rods since the wall restricts unevenness in the horizontal electric field in contrast to the section below the slot in slabs. On the other hand, unknown ions in the sample may lead to faster or slower migration of bands in tubes, whereas in slabs, there is a warning distortion.

Equipment for casting slabs and running vertical PAGE at controlled temperatures are more expensive, usually in the order of $ 600 and not as easy to build in a meager workshop. However, precise comparison of many bands with MW-markers and particularly in genetic studies is possible only in slabs. Furthermore, the slot-forming process does not require any preventive measures for excluding oxygen from critical areas, it is faster, and all other variables during polymerization are more uniform. If not, the slab pattern will soon reveal these irregularities. Due to their overall advantages, we use slab gels predominantly. Finally, 2-D-methods are possible only with the slabs. Simple constructions like in[16] for horizontal slabs provide lower quality. While almost any apparatus can be used, there seems to be a new suggestion every month. Consult the buyers' guides for commercially available equipment with or without cooling. Good results are obtained with the PANTA-PHOR® (Labor-Mueller, D-3510 Hann.Münden, W.-Germany), a Plexiglas®-apparatus for vertical and horizontal PAGE under cooling, without or with an inlet of two glass plates when needed for better adherence of the gel. This version was developed in 1968 from the first vertical slab apparatus introduced to this field by Raymond[17] in 1960. Gel size is

13 x 18 cm by any thickness up to 2 cm. The PANTA-PHOR® is "student-proof" and allows for all operations (PAGE, including a preparative scale with stepwise elution of the separated proteins, PoroPAGE, Immuno-AGGE, PAGIF and 2-D-procedures) in one apparatus with simple auxiliary equipment. All figures shown here are results of separations done with this system. The rectangular attachment for preparative separations with a 20 cm^2 working surface is either made from glass for SDS-PAGE or from plastic for AGGE and PAGGE involving nucleic acid and virus separations. For larger slabs (18 x 22 cm) in 2-D-work, the MONO-PHOR® (Labor-Mueller) is used. To reduce diffusion, tap water or buffer circulated through ice may be used as coolant for the slab, or, better yet, a temperature controlled cryostat set at $2^{o}C$. Low temperatures allow for SDS-PAGE with lower concentrations of PAA and thus, shorter running time with equally sharp bands.

Power supplies for tube-PAGE should deliver a direct current of 200 V and 100 mA, for slab-PAGE and PAGIF, 1000 V and 300 mA. Constant voltage, current or wattage is fine but not essential.

CHEMICALS

For routine experiments, the purity of some chemicals is of less importance than generally considered. A good technical grade of AA will do the job if kept cool, dry, and in the dark. But even 5 year-old lots of 50 kg when stored dry at room temperature performed well. For gel blocks (12 x 8 x 1.6 cm) in preparative work, recrystallized AA is preferred since it swells less. AA always contains some PAA, acrylic acid, and ions which increase during storage and can be removed by mixed-bed ion exchange and recrystallization from hot acetone (1:4, W:V) or chloroform (1:5, W:V). Filter off insoluble (polymerized) material, then slowly chill to 0° and filter. A second filtration after storage at -20° is recommended. Avoid skin contact and use a hood. For further information see[10 6]. Since polymer formation and hydrolysis occur spontaneously even in the best preparation, buying costly AA does not necessarily point to an experienced worker. BIS can be recrystallized from acetone (1:60, W:V). Cyanogum 41®, a mixture of 95% AA and 5% BIS, cannot be recrystallized. It is often used by us for large-scale work (more than 80 samples per day per person). Activity of ammonium-persulfate is best checked before the first use by a test polymerization in a beaker since decay, composition of the buffer, concentration of AA etc. effect the gelling time. AA in buffers with urea needs only half the amount of persulfate. SDS occasionally contains considerable amounts of chains longer or shorter than C_{12} which strongly influence binding and staining of unfolded proteins[18] [19] [20], affect separations in - only multiphasic - buffer systems[21], and detection of nucleases[22]. Recrystallization from propanol-2 improves SDS-performance. Nevertheless, it is better to stick to reliable preparations. At the moment, Serva (D-6900 Heidelberg), Merck (D-6100 Darmstadt) and British Drug House (GB-BH12 4NN Poole) can be recommended. Buffer salts should be free of potassium ions to avoid precipitation of K-dodecylsulfate. The more soluble Li-salt can replace SDS. Urea is crystallized from propanol-2 (1:14, W:V). Supranolcyanin 6 B from Bayer (D-5090 Leverkusen) or one of the Coomassie blues is used for staining. All other chemicals are of standard purity. Lower quality TRIS (yellowish in concentrated solution) leads to a small increase in pH-values without harming the separation. MW-markers such as in TABLE I or from BDH (cross-linked with diethylpyrocarbonate) for the range from 14.3 to 71.5 and 53 to $265 \cdot 10^3$ daltons or from Boehringer (D-6800 Mannheim 31) are used in 5% PAA. Better fitting of the log-plot is achieved if 3.5% and 10% PAA are used for the lower and higher MW range respectively. 7.5% PAA is a compromise, recommended by BDH.

TABLE I

STANDARD SOLUTION FOR MW CALIBRATION

A. Markers heated with 1% SDS (100°/3 min) only (Gel load 15 µl)
B. Markers heated with SDS and ME (1%) as above (Gel load 8 µl)

MW dalton x 10^{-2}	Protein	Stock solution Amount mg	Buffer ml	µl from stock for final solution A (420 µl)	B (475 µl)
1500	Immunoglobulin (IGG)	1	1.0	160	-
975	Phosphorylase b (muscle)	8*	0.2	20	50
670	Albumin (bovine serum)	1	0.2	-	75
490	Fumarase	1	1.0	160	-
370	Alcoholdehydrogenase (yeast)	1.7*	0.2	-	150
257	Chymotrypsinogen A	1	0.2	40	100
143	Lysozyme	1	0.2	40	100

*corresponds to 1 mg pure enzyme

For details: see APPENDIX

BUFFER SYSTEMS

For continuous buffers, phosphate buffer pH 7.1 with 0.1% SDS in 5% PAA in the original or slightly modified versions is most frequently used[2]. Borate buffer, pH 8.9, is often used in the field of phytochemistry and virology[23]. A borate buffer, pH 7.9, is relatively insensitive to ionic contaminations of the sample[24]. For buffer compositions, see TABLE II. Continuous systems are more convenient to handle, are faster, and are less expensive: an important factor if numerous samples are screened. Patterns have sharp bands when sample volumes smaller than 30 µl are applied per 15 mm^2. Most of the figures in the text show separations done in this manner.

A successful discontinuous system was introduced[25 26 27] based on the Davis/Ornstein system. It has the advantage of giving very sharp bands, even with larger sample volumes[15] and works equally well with or without urea. About 95% of the separation problems can be solved using a few formulae. Other buffers may be tried[6]. Since all possible interactions (accumulation of hydro-phobic or -philic regions, borate complexes, voltage gradient etc.) cannot be predicted for an unknown protein, theoretical considerations serve only as guidelines. Discontinuous systems have to be applied cautiously due to the observation that SDS is stripped off from proteins. For detailed discussion, see section LIMITATIONS. If reoxidation of SDS-proteins during PAGE is suspected, which is one possibility for causing streaking, the buffer in the cathodic compartment should contain 0.01% thioglycolate to provide for a reducing environment.

SAMPLE PREPARATION

There is currently no method for unequivocal SDS-treatment of the crude sample. For complete disruption of the higher protein structures and binding of sufficient SDS, concomitant unfolding by SDS and reduction of S-S-bridges are needed. ME is mostly used[2], as well as dithiothreitol (1%) or thioglycolate (2%). The latter can be removed electrophoretically and the reduction checked for completeness[28]. The rate of unfolding varies from protein to protein. Since some proteases especially in plants, bacteria and fungi are very resistant to SDS and may digest other already denatured proteins[29], the boiling of small volumes (1 - 2 ml) with 1% SDS (at least 2.5 mg/mg protein) and 1% ME for 3 and 12 min respectively is recommended. SDS-PAGE patterns should be the same. Omitting SDS for solubilization in favor of non-ionic detergents with ampholytes and ME, in combination with a later SDS-charging, with or without boiling, seems to be a good alternative[30]. For structured material, see MICROMETHODS.

Boiling may cause a shift in the charge distribution as well as a loss or gain of bands or spots in the SDS-dimension unrelated to digestion. PAGIF-separated proteins incubated with SDS at room temperature (50 min) followed by SDS-PAGE give slightly different patterns when heated within the gel, and also when compared with the boiled so-

TABLE II

COMPOSITION OF BUFFERS FOR SDS-PAGE

Buffers for continuous (C) and discontinuous (D) systems.
Volumes given in D adjusted to gels, 3 mm thick, in the PANTA-PHOR® apparatus.

C Continuous Buffers, No. 1 to 6

	pH	Solutions	TRIS	Borate $Na_2B_4O_7 \cdot 10\ H_2O$	Boric acid	Phosphate $NaH_2PO_4 \cdot H_2O$	Phosphate $Na_2HPO_4 \cdot 2\ H_2O$	NaCl	EDTA .2 Na	SDS	from Stock ml	Water added to ml
C1	7.1	Stock	18.39 g		185.49 g					—	—	5000
		Final	10mM		200mM					3.0 g	1000	3000
C2	7.1	Stock				69.9 g (B)				—	—	1000
		Stock					89.0 g (A)			—	—	1000
		Final				25mM				1.0 g	(15.2 (B) / 34.8 (A))	1000
C3	7.9	Stock	151.25 g		183.00 g					—	—	5000
		Final	31mM		74mM					2.4 g	300	2400
C4	8.2	Final		6.60 g 3.5mM	11.78 g 38mM			2.78 g 10mM	—	—		5000
C5	8.2	Final	4.85 g 40mM		3.00 g 48mM				0.60 g	—		1000
C6	8.9	Stock	605.00 g		46.00 g					—	—	5000
		Final	125mM		19mM					2.4 g	300	2400

D Discontinuous Buffer System, Composites No. D_1 to D_4

	pH	Solutions	TRIS	Glycine	HCl 1N	SDS	Water to ml	Gel Concentr.	D_1 ml	D_2 ml	D_3 ml	D_4 ml	Water ml	10% AP µl	TEMED µl
D_1	Monomer: 300 g Acrylamide and 8 g Bisacrylamide with Water to 1000 ml							14.0% resolving	17.3	-	9.5	-	11.2	320	20
D_2	8.0	Stock / Final	60.6 g / 125mM	-	315 ml / 79mM	4.0 g	1000	6.3% spacer	2.0	-	2.4	-	5.2	50	5
D_3	8.8	Stock / Final	181.6 g / 375mM	-	315 ml / 79mM	4.0 g	1000	4.0% stacking	7.9	14.8	-	-	36.5	360	30
D_4	8.3	Final	6.0 g / 50mM	28.8 g / 384mM	-	1.0 g	1000	6.3% base	3.60 g AA and 0.18 g Bis			57.0	-	300	30

lution. Complete unfolding is retarded in gels and for rigid proteins is not achieved by short boiling nor by soaking with 8 M urea. Prolonged processing results in loss of proteins or formation of artificial spots. Similar observations were made when heating crude samples[31 32]. Unexplained patterns[33] and relationship to the temperature of the SDS-treatment occured[34]. It is noteworthy that SDS-incubation at a lower pH than usual (pH 5) results in complete unfolding without heating[35]. Presence of urea does not guarantee complete unfolding with SDS and ME[36] and if the sample is boiled carbamylation takes place. Inhibitor/enzyme complexes can resist dissociation[37]. Lipoproteins may become insoluble. Their irreversible precipitation is avoided by solubilization first with SDS alone in an ultrasonic bath allowing the temperature to rise from 20 to 60°C, and then adding ME[38]. Cystin oxidation by performic acid[39] at 0°C is an alternative route for complete elimination of disulfide groups even in insoluble proteins[40].

Samples with more than 0.02% potassium ions must be dialyzed to reduce the salt concentration. Total ionic strength of samples should be below that of the gel buffer. For SDS-PAGE in continuous buffers 1 µg of protein in a layer about 1 mm high when applied to a sample slot or tube will give sharp and easily visible bands. For dilute samples, it was found that of all procedures tested, concentration by collodium bags was the most reliable, convenient, and gentle method[41]. After treatment, sucrose or, if interference is expected or high density solutions have to be handled, iodinated sugar derivatives like Metrizamid® are added to create a 10% solution. Glycerol is not as good due to the higher osmotic force resulting when applied to the gel. Small samples when frozen quickly and stored at -20°C have yielded the same protomer distribution in SDS-PAGE 6 years later. Tracking dyes may be added. They are of limited value in monophasic systems since their migration rates relative to proteins depend upon a number of variables[42] including the voltage gradient. Bromophenol blue, pyronines and tropaeolin OO can be used. Pyronin B has been described as a misleading marker[43] which is not surprising.

PREPARATION OF THE GEL and EVALUATION

The polymerization of AA and BIS (usual ratio 20-40 : 1) or other crosslinking agents can be done by photopolymerization with riboflavin (or its faster soluble phosphate), which gives "softer" gels compared with the initiation by persulfate and tertiary amines. The two processes follow different routes, and the latter is strongly inhibited by oxygen. For better reproducibility, reduced lag time, and less unpolymerized material, we prefer the addition of dry Na-sulfite which is dissolved in the monomer mixture (trials done by Dr.V.Loeschcke) based on[44] and used with good results since then[45]. Gelling time is about four times longer without sulfite. If re-oxidation of reduced proteins is feared, the gel is washed with ME or thioglycolate or pre-electrophoresis is done with the latter. For mappings, the special slot former with slots on each side is inserted into the monomers for running, stacking etc. gels before polymerization (Fig.1 and 2). One slot is for markers and one for the sample not fractionated in the first dimension, leaving space for the gel with the proteins to be separated in the second dimension. Gels prepared with 6% propanol-1 keep their size particularly well and separate some SDS-proteins better than aqueous gels. Running time is shorter. Gels with borate buffer pH 7 to 8 can be stored at least 4 weeks before use. However, the migration rate is slightly diminished. Borate ions act as preservatives. After the run, visible bands are marked with a needle dipped into India ink.

At least three slabs with different PAA concentration (e.g. 5, 8, 12%) should be prepared to evaluate the samples in a log plot of the relative mobility of the markers and sample proteins (in mm) versus % PAA concentration (Ferguson plot). If the lines have no common intersection one should be cautious about the MW-determination. The MW is found by plotting relative mobility (mm migrated) vs. log of MW, using the MW given in TABLE I. See also[46] and the evaluation of proteins across a SDS-PoroPAGE[47] for MW up to 1 million.

DETECTION OF PROTEINS IN PAA

There are almost as many staining and destaining procedures as there are authors. We[48] use 0.025% Supranolcyanin Bayer (which is similar to the Coomassie blue family[49]) in methanol, acetic acid and TCA, with occasional trials of other published procedures. Supplementation with Supranol-echtrot Bayer stains some more acidic proteins. For details see the appendix. Simply chilling the gel after SDS-PAGE or incubation with alkyl(C_8-C_{18})benzyldimethylammoniumchloride[50] or acidic KCl-solution[51] and searching for turbid bands is fast but not very sensitive in our hands. Staining methods will detect about 0.1 µg of protein in a 5 x 0.5 x 1 mm cube after destaining by diffusion. Electrophoretic destaining may save some time but weak bands are often lost without them being noticed. Staining SDS-proteins needs more time (3 hours per mm slab thickness) in comparison to slabs not containing SDS. Staining intensities vary with the source of SDS, with Pierce-SDS giving low intensities (own observation and[21]). In gels with a PAA-content above 7% it may be advisable to first fix the proteins with our staining solution without the dye (to partially remove SDS) and then replace this solution with the complete mixture. If destained too long restaining is possible by addition of a drop of stain to the destaining mixture. It is most practical to do all procedures in stainless steel trays having rounded edges and a cover. They should be swayed gently.

Quantification of proteins with or without staining remains a problem in spite of many efforts, since the type and concentration of pro-

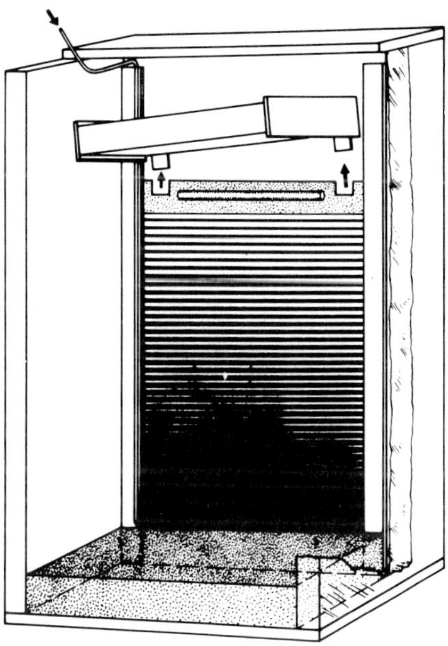

1

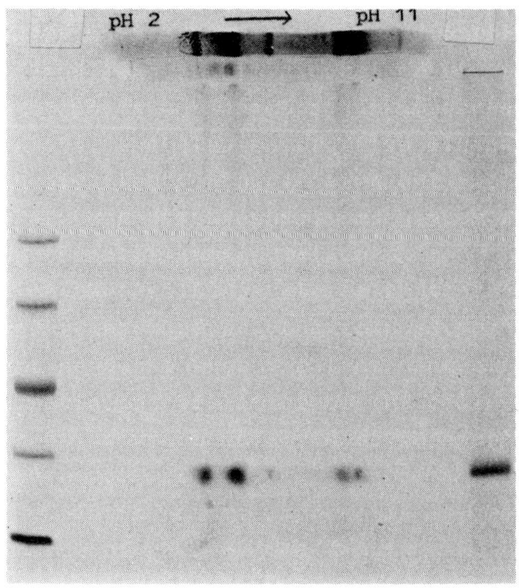

2

Fig. 1. Equipment for preparation of PoroPAGE slabs, inclusion of the gel rod (from e.g. PAGIF) and slot formation for reference samples in the second dimension. Cathode is on the top, as in all figures.

Fig. 2. in-situ-Incubation. Intact turnip yellow mosaic virus separated by PAGIF pH 2 - 11, incubated within the gel with SDS/ME for 5 min at 100°C and pH 7.1, followed by SDS alone (7 min/90°). SDS-PAGE in Tris/borate pH 7.1. Applying the virus on the cathodic site for PAGIF gave slightly different results when compared to anodic feeding. On top, gel rod after PAGIF not extracted into the second dimension. Left: MW-markers B as in TABLE I. Right: virus particles boiled for 3 min with SDS/ME, then applied to SDS-PAGE directly.

teins and of dyes, staining duration, and other parameters play a role. Coomassie blue is commonly used and for a late version see[52] and[53]. Complete elution from the gel has to be considered as an alternative and is described in PREPARATIVE SDS-PAGE or by[54] and[55]. Autoradiography can be done in slabs soaked first in the destaining solution containing 5% glycerol, dried on a clean glass plate and covered with Cellophan® or similar material[3] [56]. If dansylated proteins[57] were separated in highly purified PAA-gels UV-light is used for detection. SDS is determined via the chloroform-soluble basic fuchsin complex[58]; in gels it is detected by methylene- or toluidine-blue.

PREPARATIVE SDS-PAGE

Separation is done more efficiently in rectangular gel blocks than in roundshaped gels. The repeated comparisons of various methods, starting with our own cylindrical design (Stegemann and Rien (1963) Reports Med. Fschg. Anst. M.-Planck-Ges., Göttingen) strongly point to block gels for several reasons. The effluent proteins are flushed away during PAGE with increasing intervals between flushes[59]. Again, one can choose from several commercial instruments. The separation shown in Fig. 3 was done with the PANTA-PHOR® in monophasic borate buffer pH 7.1. In contrast to PAGE-blocks which take protein loads up to 3 g per 20 cm^2 (but not more than 400 mg per individual protein), the SDS-PAGE will take only about 10 mg per compound. 2 M urea in the gel and in the buffer for stepwise elution favors complete recovery. Asao[60] produced a good yield with comparable load per cm^2. The elution steps are regulated by a timer (Typ DCSW, Labor-Mueller, D-3510 Hann.Münden) allowing any time pattern above 3 minutes, periodic or not, for up to 5 days. This is achieved by punching holes into a black super-8-film. Usually a 5 or 10% increase of the intermission time in respect to the foregoing one is chosen. Thus, proteins are sufficiently concentrated in each fraction (6 ml) and can be spotted directly (50 µl) for re-electrophoresis or used for other purposes.

Cutting pieces from a gel block and squeezing them through a PVC-filter (or a piece of polyester-cloth) in a disposable syringe is suitable, if the extract is to be used for re-electrophoresis or, with some restrictions, for hydrolysis. It is not recommended for "fingerprinting" due to the considerable amount of oligomeric PAA, which would disturb paper- and TL-chromatography, and must be removed by gel filtration, with some loss of proteins.

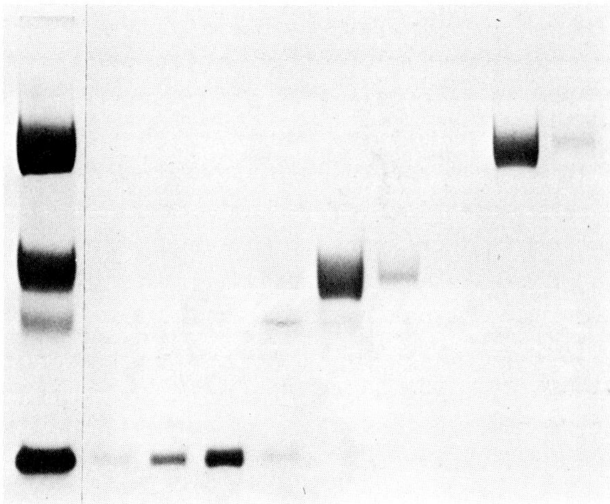

Fig. 3.

Fractions, checked analytically, from preparative SDS-PAGE in borate buffer pH 7.1 with 0.1% SDS and 2 M urea in the gel and in the elution buffer. Electrode buffers circulated. 10 mg each of crude myoglobin (whale), ovalbumin and bovine serum albumin, SDS loaded. From left: unfractionated mixture, fractions (6 ml) No. 9 to 15 and 17 and 19, 50 µl were applied directly to the analytical slab. 15 min intermission between the first and second fraction increased by 5% per step, being e.g. 33 min after the 17th step.

MICROMETHODS, in-situ-PROCEDURES

Neuhoff and his team [61] [62] [63] checked many PAGE-methods for their use in capillaries and reported for SDS-PAGE new insights and applications. Similar techniques have been published by others, e.g.[52] . Micromethods save chemicals and time, and they are unavoidable if there is a real need to deal with minute samples. However, 9 years ago, it was our experience with microslabs and capillaries that besides the inferior comparability of tubes, micromethods tended to be less reproducible and had limited separation ability in comparison to larger gels. This may no longer be valid although critical comments still persist [64] . PAA-pores in capillaries can be made particularly small since the heat of polymerization is dissipated quickly even with 50% PAA. Here it was first shown [62] that SDS is lost from the SDS-protein-complex in SDS-PAGE with discontinuous buffers and that "a great number of bands of only one protein may represent various SDS-charges". This was later confirmed by[54] Continuous buffer systems are less misleading in the interpretation of mappings . The stripping of SD^{35}S and the occurrence of multiple zones from albumin and its oligomers were demonstrated[62] . The use of lower percent PAA reduces the problem but leaves the chance of SDS-dissociation due to jumps in ionic strength.- If structured material is scarce and cannot be brought into solution without severe loss, or stems from a former separation in a gel, direct SDS-incubation on top of a capillary can be done as shown by[65] with a single insect gland, or with intact virus particles in a porous PAGIF-gel (Fig. 2).

IMMUNO-TECHNIQUES WITH SDS-PROTEINS

The gain from the solubilization of proteins obtained with SDS and after SDS-PAGE is counteracted by SDS interference in subsequent immunochemical analysis. Almost normal precipitation can be achieved after focusing[66] , by addition of non-ionic detergents[67] and potassium chloride, or by dilution through diffusion and SDS-stripping[68] by PAGE to reduce the SDS-concentration to less than 0.1%. Even dansylated and stained proteins will undergo precipitation, not, however, if stained with Coomassie blue. I assume that Supranolcyanin and Coomassie blue also stain by intermolecular bounds, unlike amidoblack, and hence render the protein inaccessible to the antibody. See also REMOVAL OF SDS.

REMOVAL OF SDS and PAGIF OF SDS-PROTEINS

Removal of SDS from proteins by anion exchangers or by electrodialysis (70 V/cm) with continuous removal of the dialysate[69] in the presence or absence of urea or nonionic detergents[70] was not very efficient in our experience due to too great a loss of protein. Electrophoretic discharge by PoroPAGE (up to 40% PAA) is possible[62] , if proteins are not too basic. A "two-stage" PAGE (first SDS-PAGE 10% PAA, then PAGE 15% PAA) is suitable for separation of neighbouring SDS-proteins as well as for freeing them from SDS[55] . We agree that the most promising and convenient way, with a good yield, is the removal by PAGIF (2% Ampholine® pH 3.5-10 and 8 M urea in 5% PAA) after washing SDS-gels with an urea/Ampholine® solution[66 71]. The total SDS-content should be below 0.03% before PAGIF is started. Some problems are encountered with proteins having isoelectric points around 3.5 but re-focusing in a narrower range, pH 2 - 4.5, will help to separate the SDS-band from proteins. No interference was seen in the PAGIF-pattern for viral proteins stripped off by SDS when compared with the same protein never in contact with SDS[72] . This implies that proteins loaded with SDS can be used for PAGIF if disruption of hydrophobic bonds and sufficient solubility is ensured by additives[73 33 74] .

The competition of nonionic and anionic detergents is well-known in lipid chemistry[75 76 77], has been used to pick up SDS by nonionic detergents competing with the protein[67 78 70 72], and provides another route for SDS-removal.

TWO-DIMENSIONAL PROCEDURES

Among the 2-D-separations most commonly used for proteins are the PAGE/PAGE-combinations at different pH-values and porosities[79] and the PAGIF/PAGE/PoroPAGE methods[80 81 82]. It is their unequivocal virtue to separate native proteins without possible artificial heterogeneity caused by an induced charge or incomplete unfolding created by SDS. This type of separation including a microversion[83], makes activity staining possible. We chose the term "mapping"[81] for these combinations of PAGIF with PAGE-techniques when applied to macromolecules in contrast to the Ingram-term "fingerprinting" for smaller molecules like oligo- and polypeptides separated by chromatography and high voltage electrophoresis.

In 1973 we combined the charge-dependent PAGIF with the size-dependent SDS-PAGE using a continuous buffer for investigations on reduced and unreduced fractions of potato[24] and other plant proteins[84 85]. More experimental details are available as a leaflet for a student course as mentioned in[24]. Those SDS-mappings using SDS- and SDS-PoroPAGE without urea proved to be an excellent tool for separation of SDS-loaded, sometimes not easily soluble, protomers in spite of the limitations to be discussed below. Klose used this approach with discontinuous buffers for fetal tissue obtaining nearly 500 spots[86 87] or almost 1300 in a later work[88]. A SDS-mapping including urea was independently published for chromatin non-histone proteins and produced some hundred spots[89]. Similar methods were published in 1974[90 91 92] or later[93]. O'Farrell introduced the same principle once again using nonionic detergent as well[33]. Danno[71] inversed the order, since in PAGIF (as the second step) bound SDS was completely removed from the proteins. This confirmed the finding that some SDS in extracts[72] does not interfere with PAGIF. See REMOVAL OF SDS. Protein elution rates from, and SDS-conversion in gels were found quite different when compared with proteins in free solution which were directly applied to the second gel. To do this on the same slab (6 mm thick) the slot formers are positioned on the same level as the gel rod to be tested. PAGIF, PAGE or SDS-PAGE is used in the first dimension, and in the second dimension, SDS-PAGE or -PoroPAGE is done in the PANTA-PHOR® apparatus. All proteins stay in their respective plane making it a 3-D-method (to be published). A strong solubilizing agent (phenol, aqueous urea, acetic acid) for the second dimension after SDS-PAGE[94] and organic-base dodecyl sulfate after chloral hydrate-PAGE[95] seems promising for membrane proteins.

See also: in-situ-PROCEDURES and LIMITATIONS and the APPENDIX.

ENZYMES IN SDS-PAGE

It is obvious that any enzyme will lose its specific activity if thoroughly unfolded by SDS. As a matter of fact measurement of residual activity is one method for checking for clusters of remaining native structure. In this case, two or more discrete bands (or smearing) will result from only one enzyme if a protein stain is used. With activity staining, only one band is seen but this will lead to different "MW" if compared with marker proteins at different PAA-concentrations. Peroxidases from potatoes are an example ([24] and unpublished results). Any remaining enzyme activity makes MW-determination doubtful. This problem has been further discussed[43]. Various enzymes can be regained (see REMOVAL OF SDS), preferably nucleases after SDS-PAGE in the presence of DNA or RNA if washed with SDS-free buffers. The presence of nucleic acid aids SDS-removal[22]. The detection limit for nucleases is of the order of 10^{-11} g. Nucleic acids incorporated during polymerization of PAA are tied within the matrix and do not migrate during PAGE. The same is true of nucleic acids if used as primer to detect polynucleotid phosphorylases[48] and polymerases[96], of pectic acids (pectinases)[45] or of the even more acidic chondroitin sulfates (hyaluronidases)[97]. The presence of acidic compounds should aid SDS-removal from these types of enzymes.

For fast protein identification, an additional tool to fingerprinting (after complete enzymolysis) is the gradual degradation by enzymes in the presence of SDS followed by SDS-PAGE in a discontinuous[98] or continuous buffer system, see Fig. 4 or by the 2-D-method PAGIF/SDS-PAGE (SDS-mapping, unpublished). Applied to potato proteins isolated by preparative PAGE it could quickly identify near-identical fractions and is less cumbersome though not yet as sensitive as fingerprints with dansylated peptides. In general, thermolysin and papain are more suitable for producing characteristic peptide patterns from the proteins tested compared with the more common serine-proteases.

LIMITATIONS

Only one quarter of the possible alterations of amino acids in the peptide chain can be detected by change in charge[99]. The chance of a MW-shift by mutation is very low. The merits of molecular differentiation associated with separations in an electrical field could be ex-

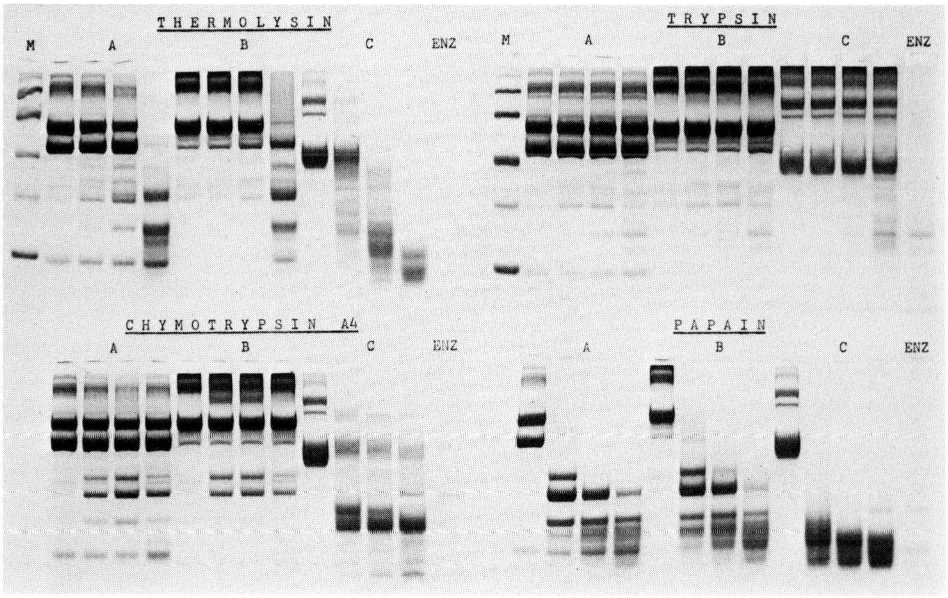

Fig. 4. SDS-polypeptides separated in phosphate buffer (pH 7.1) and 7.5% Cyanogum obtained from substrate A, B and C after limited proteolysis by enzymes as indicated. M = MW-markers, A and B = two different batches of serum albumin, C = ovalbumin, Enz = enzyme alone. 4 slots per substrate, starting from left to right with substrate alone followed by increasing amounts of added enzyme.

tended if a change of hydrophobicity was detectable as well. At the moment, it just confuses SDS-rules[100]. In 1974, we tried to benefit from the interaction of a dodecyl-chain with exposed hydrophobic regions of protein molecules by weaving in N-dodecylacrylamide into the PAA-network. Native proteins more hydrophobic on their surface should be retarded during PAGE but our experiments are not yet successful in showing this. The most important hydrophobic regions are buried in native proteins and are composed of different parts of the chain. Nevertheless, unfolded proteins should provide enough individual interaction of this sort to take advantage of a fixed hydrophobicity in the matrix. This field of inversed SD(S)-PAGE promises further developments.

Polypeptides and proteins of lower MW deviate in SDS-PAGE from the rules for medium MW proteins[101]. Fig. 5 illustrates not only the deviation but unpredictable scattering of "SDS-MW" for protomers with an

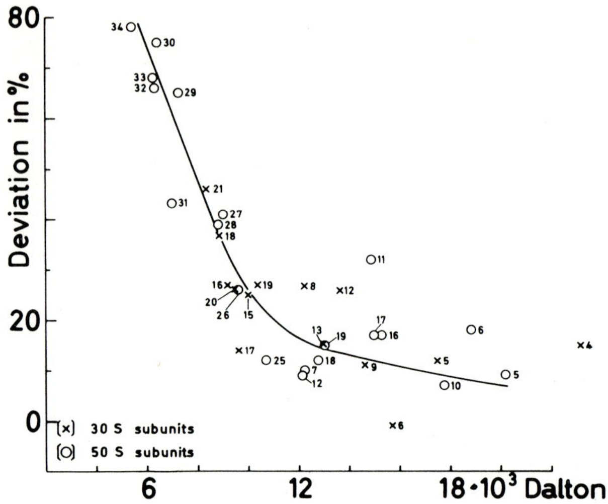

Fig. 5.
% Deviation of MW as found by SDS-PAGE from actual MW of ribosomal proteins (E. coli) as sequenced and numbered by Wittmann[102]. Abcissa: MW as calculated for the chemical composition.

exactly known amino acid sequence[102]. It may be worthwhile to correlate the type and accumulation of a given amino acid chain with the MW-aberration in SDS-PAGE. Other irregularities are observed in the common MW-range, some of them were mentioned in section BUFFER-SYSTEMS, SAMPLE PREPARATION and MICRO-METHODS. Whereas different PAGIF- and PAGE-procedures give readily reproducible results under a variety of conditions, the chosen SDS-method including the sample preparation must be standardized to provide this reproducibility. Fig. 6 illustrates the influence of complexing and non-complexing ions and EDTA on migration rates in SDS-PAGE with monophasic buffers in steady-state. We have observed this for a long time[103] and it has been discussed by [104]. For applications of SDS-PAGE in the field of genetics, it is important to realize that a single band or spot in one system will not necessarily appear under slightly different conditions. Therefore, pattern evaluation in relation to heredity should be done with great care.

Reproducible streaking of bands or spots in SDS-PAGE and SDS-mapping is frequent and signals either insufficient experimental conditions

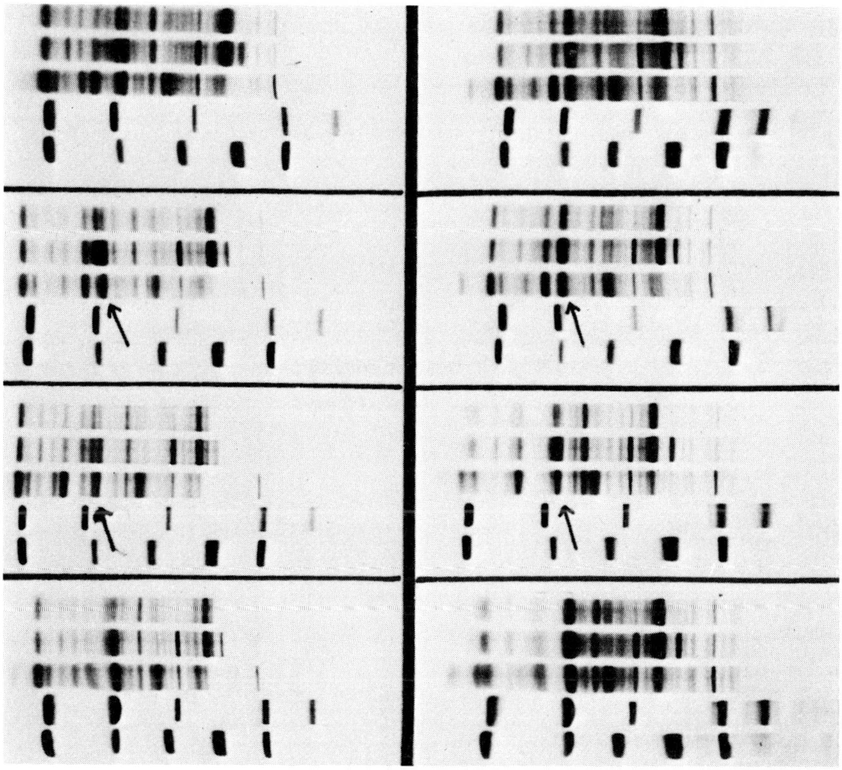

Fig. 6.

5 samples separated under 8 different conditions of SDS-PAGE (continuous buffer). Upper row pH 7.1 buffers: --- Phosphate without and with 0.2% EDTA; Tris/borate - and + EDTA. Lower row pH 8.9 buffers: --- Tris (0.125M)/HCl - and + EDTA; Tris/borate - and + EDTA. 0.1% SDS in all buffers. See TABLE II for composition. Samples: buffer extracts from milled kernels of 3 different maize cultivars, boiled with SDS/ME for 5 min (slot 1-3). MW-markers as in TABLE I, slot 4 solution A, slot 5 solution B.

(buffer system/partly aggregated, but not always precipitated, redissolving sample/slow release from one gel to the other in 2-D-methods etc.) or glycoproteins. Unlike proteins whose MW are exactly coded the carbohydrate moiety has to some extent a random chain length. A sharp band in PAGIF and a corresponding streak in SDS-PAGE may point to a glycoprotein if the sugar part carries no charge. Both methods should start with samples in free solution. If starting with one sample enclosed in a gel, as in a 2-D-method, one should run the reversed sequence as well, discriminating between possible slow release from the first gel or a range of MW. We favor doing PAGIF first, either followed by PAGE[81] or by SDS-PAGE[24] since the uncertainties of the SDS-step (unfolding, loading) are then not carried into the second dimension.

In SDS-PAGE one may use additives like urea, dimethylsulfoxid, formamide, the teratogenic dimethylformamide and non-ionic detergents, simultaneously or successively, mono- or multiphasic buffers, even or graded PAA-concentration. No system is "best". One has to watch for traps. Virus proteins boiled for 2 min in 2% SDS, 5% ME and 8 M urea (with surely partial carbamylation[41]) showed different patterns and MW depending on the content of urea during SDS-PAGE[105]. Transient discrete bands of SDS-complexes have been demonstrated for one pure, not completely unfolded protein[35]. Continuous buffers are more reliable than discontinuous ones especially if used in SDS-PoroPAGE, since the different ionic strenth can govern the binding level of SDS[106 103] and the pore size the partial removal of SDS[62]. Ampholytes in the sample solution, or worse, if unevenly distributed in gels used for SDS-mapping, capture SDS and lead to inadequate loadings. Multiple spots result if SDS is picked up later from the cathodic buffer reservoir, Fig. 7. Dif-

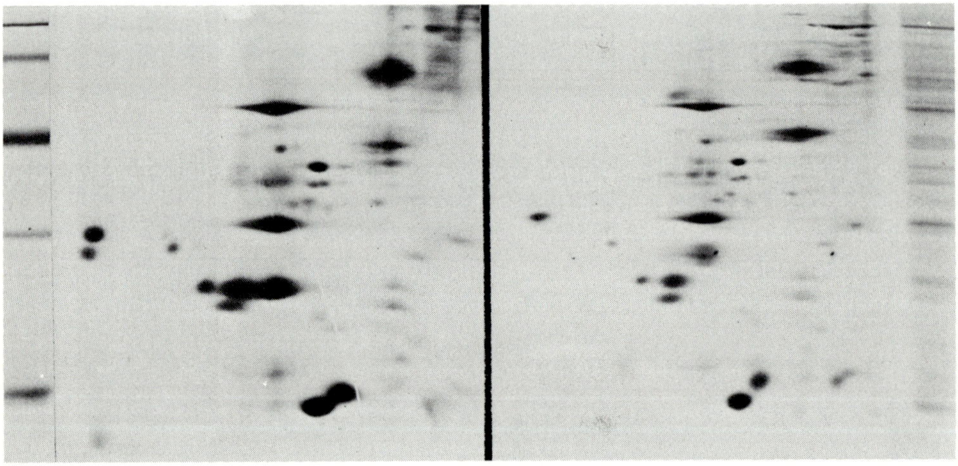

Fig. 7. 2-D-separation of a maize meal extract. Top: PAGIF pH 8 to 4 (left to right, not shown). The focused gel for the left slab (SDS-PAGE in system D, resolving gel, TABLE II) was insufficiently loaded with SDS. Right slab: the PAGIF-gel was not only incubated at 22°C, but also boiled for 10 min in 1% SDS/ME. Similar results, though not as pronounced, were obtained with urea containing slabs, where the focused gel was not heated. Left slot: markers as in TABLE I/B. Right slot: maize extract not focused.

ferent batches of SDS (with 0% and 25% carbon chains $>C_{12}$ respectively) did not alter protein patterns in SDS-PAGE with continuous buffers, while in discontinuous systems origin of SDS and type of protein affected the separation[21]. On the other hand, multiphasic buffers usually yield very sharp bands and therefore should be tried in addition, Fig. 8.

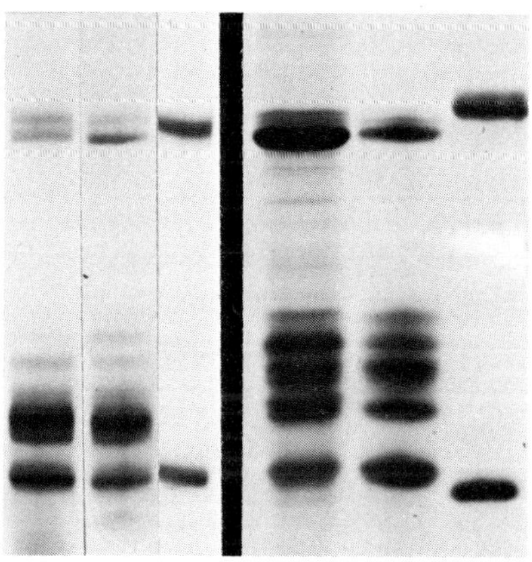

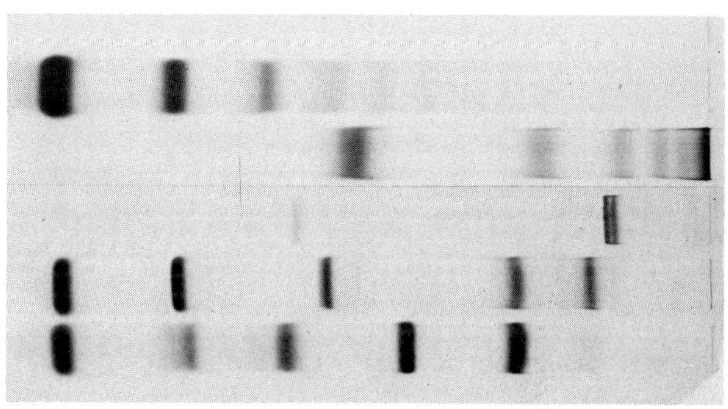

Fig. 8. Comparison of MW as found in monophasic SDS-PAGE with Tris/borate pH 8.9 (slot 1-3) and in a discontinuous system pH 8.8 (slot 4-6), see TABLE II. Samples are tuber proteins from potato cultivars 'Bintje' (slot 1 and 4) and 'Voran' (2 and 5), MW-markers (3 and 6): ADH from yeast, and lysozyme, see TABLE I.

Fig. 9. Markers for MW-determination in SDS-PAGE (see also CHEMICALS and TABLE I). From left: BDH-markers, MW range 14300 to 71500 (slot 1) and 53000 to 265000 (slot 2); Boehringer RNA-polymerase, MW 39000, 155000 and 165000 (slot 3); our set, solution A (slot 4) and B (slot 5). Conditions of SDS-PAGE: 5% Cyanogum, Na-phosphate buffer pH 7.1 (see TABLE II) with 1% SDS, running time about 2.5 hours, 200 V, in the PANTA-PHOR.

After reviewing the literature of the last years, I am not sure, however, how many extra bands and spots were "discovered" due to discharge and recharge of SDS and other influences related to the type of separation employed. Reproducibility of a pattern is not a criterion for the number of individual proteins originally in the sample before SDS-treatment. But this excellent reproducibility allows for a very good test to detect any compositional change of a sample compared with the reference, provided the experimental parameters are kept exactly constant. Labeling individual compounds in SDS-mappings, as has been started, combined with specific enzym staining in mappings and preparative PAGE for re-runs will lead to insights hitherto not possible with other methods.

The Ferguson plot of relative mobilities obtained at different gel concentrations (see EVALUATION) may reveal an anomalous behavior in SDS-PAGE, which makes MW-determination unreliable[107]. On the other hand, anomalous behavior was an elegant tool for aiding the identification of narcissus mosaic virus in the commercially grown ornamental plant 'Nerine'[108].

Despite certain aberrations, all kinds of SDS-PAGE have become an indispensible tool in protein chemistry. No other method (except PAGIF) can be combined as efficiently with other separation steps, nor work with such small amounts of proteins and still give a good estimate of MW, a characteristic molecular parameter. It is hoped that the limitations and aberrations discussed above are regarded as a sign of caution, reckoned as an additional characteristic and as encouragement to use the slab techniques more frequently since it is the only unequivocal comparison of protein patterns.

APPENDIX

Lack of space prohibited the inclusion of an appendix. A 25 page leaflet describing experimental details may be obtained from the author.

ACKNOWLEDGEMENTS

I would like to thank Dr. V. Loeschcke, H. Francksen, R. Knackstedt, E. Krögerrecklenfort and Ch. Legaye for valuable discussions and experimental help, K. Wieczorek for construction of several apparatus, R. Buhr for the typescript and layout, H. Schlobach for the drawings, BASF (D-6700 Ludwigshafen) and Bayer (D-5090 Leverkusen) for supplying chemicals free of charge. We are indebted to BASF for synthetizing N-dodecyl-acrylamide.

REFERENCES

1. Waehneldt, T.V. (1975) BioSystems 6, 176-187.
2. Shapiro, A.L., Viñuela, E. and Maizel, J.V. (1967) Biochem. Biophys. Res. Commun. 28, 815-820.
3. Maizel, J.V. (1971) in Methods of Virology, Maramorosch, K. and Koprowski, H. eds., Acad. Press, New York, Vol. 5, pp. 179-246.
4. Raymond, S. and Weintraub, L. (1959) Science 130, 711.
5. Weber, K., Pringle, J.R. and Osborn, M. (1972) in Methods in Enzymology, Colowick, S.P., Kaplan, N.O. and Hirs, C.H.W. eds., Acad.Press, New York,Vol.26/C, pp.3-27.- Weber,K. and Osborn,M. in The Proteins, Neurath, H. and Hill, R.L. eds.,same publ.,3rd ed.,Vol.1.,pp.179-223.
6. Chrambach, A., Jovin, T.M., Svendsen, P.J. and Rodbard, D. (1976)in Methods of Protein Separation, Catsimpoolas, N. ed., Plenum Press, New York, Vol.2, pp. 27-144; AA purification: p. 52 and 138.
7. Segrest, J.P. and Jackson, R.L. (1972) in Methods in Enzymology, Colowick, S.P., Kaplan, N.O. and Ginsburg, V. eds., Acad. Press, New York, pp. 54-63.
8. Panyim, S., Thitipongpanich, R. and Supatimusro, D. (1977) Anal. Biochem. 81, 320-327.
9. Rodbard, D. (1976) in Methods of Protein Separation, Catsimpoolas, N. ed., Plenum Press, New York, Vol. 2, pp. 145-217.
10. Maurer, H.R. (1971) Disc Electrophoresis, de Gruyter, Berlin, pp.222.
11. Cox, H.C. and Teven, J.M.G. (1976) J. Chromatog. 123, 261-270.
12. Reynolds, J.A. and Tanford, Ch. (1970) J. Biol. Chem. 245, 5161-5165 and Proc. Nat. Acad. Sci, US. 66, 1002-1003.
13. Collins, R.C. and Haller, W. (1973) Anal. Biochem. 54, 47-53.
14. Mattice, W.L., Riser, J.M. and Clark, D.S. (1976) Biochemistry 15, 4264-4272.
15. Wyckoff, M., Rodbard, D. and Chrambach, A. (1977) Anal. Biochem. 78, 459-482.
16. Stegemann, H., Hillebrecht, R. and Rien, W. (1965) H.-S. Z. Physiol. Chem. 340, 11-17.
17. Raymond, S. and Wang, Yi-Ju. (1960) Anal. Biochem. 1, 391-396 and manual of the E.-C. Apparatus Comp., Philadelphia.
18. Steinhardt, J., Stocker, N., Carroll, D. and Birdi, K.S. (1974) Biochemistry 13, 4461-4468.
19. Birdi, K.S. (1976) Anal. Biochem. 74, 620-622.
20. Steinhardt, J., Scott, J.R. and Birdi, K.S. (1977) Biochemistry 16, 718-725.
21. Matheka, H.D., Enzmann, P.-J., Bachrach, H.L. and Migl, B. (1977) Anal. Biochem. 81, 9-17.
22. Rosenthal, A.L. and Lacks, S.A. (1977) Anal. Biochem. 80, 76-90.
23. Koenig, R., Stegemann, H., Francksen, H. and Paul, H.L. (1970) Biochim. Biophys. Acta 207, 184-189.
24. Stegemann, H., Francksen, H. and Macko, V. (1973) Z. Naturforsch. 28 C, 722-732.
25. Grossfeld, R.M. (1968) Ph.D. Thesis, Stanford Univ., Palo Alto, Cal.
26. Laemmli, U.K. (1970) Nature 227, 680-685.

27. Studier, F.W. (1973) J. Mol. Biol. 79, 237-248.

28. Stegemann, H. (1959) H.-S. Z. Physiol. Chem. 315, 137-140.

29. Stegemann, H. (1977) Z. Acker- u.Pflanzenbau (J.Agron. Crop Science) 144, 157-161.

30. Sato, T., Ito, K. and Yura, T. (1977) Europ. J. Biochem. 78, 557-567.

31. Wilson, D.L., Hall, M.E., Stone, G.C. and Rubin, R.W. (1977) Anal. Biochem. 83, 33-44.

32. Hegg, P-O. and Löfqvist, B. (1977) J. Sci. Food Agri. 28, 103-112.

33. O'Farrell, P.H. (1975) J. Biol. Chem. 250, 4007-4021, Methods Cell Biol. (1977) 16, 407-420.

34. Garten, W., Hindennach, I. and Henning, U. (1975) Europ. J. Biochem. 59, 215-221.

35. Flanagan, P.R. and Forstner, G.G. (1977) Biochem. Soc. Trans. 5, 1542-1543.

36. Hundt, E. and Kadenbach, B. (1977) H.-S. Z. Physiol. Chem. 358, 1309-1314.

37. Bloemendahl, H., Zweers, A., Koopmans, M. and van den Broek, W. (1977) Biochem. Biophys. Res. Commun. 77, 416-425.

38. Morell, P., Wiggins, R.C. and Gray, M.J. (1975) Anal. Biochem. 68, 148-154.

39. Toennies, G. and Homiller, R.P. (1942) J. Am. Chem. Soc. 64, 3054-3056.

40. Stegemann, H. (1958) H.-S. Z. Physiol. Chem. 312, 255-263.

41. Francksen, H. and Garadi, R. (1974) Z. Anal. Chem. 271, 340-344.

42. Loeschcke, V. and Stegemann, H. (1966) Z. Naturforsch. 21 B, 879-888.

43. Bell, D.Y. and Hodgson, E. (1977) Gen. Pharmac. 8, 113-119 and 121-127.

44. Gofman, Yu.Ya. (1965) Biochemistry (engl. transl. of Biokhimiya) 30, 996-999.

45. Stegemann, H. (1967) H.-S. Z. Physiol. Chem. 348, 951-952.

46. Bryan, J.K. (1977) Anal. Biochem. 78, 513-519.

47. Lambin, P., Rochu, D. and Fine, J.M. (1976) Anal. Biochem. 74, 567-575.

48. Stegemann, H. (1968) Z. Anal. Chem. 243, 573-578.

49. Fazekas de St. Groth, S., Webster, R.G. and Datyner, A. (1963) Biochim. Biophys. Acta 71, 377-391.

50. Takagi, T., Kubo, K. and Isemura, T. (1977) Anal. Biochem. 79, 104-109.

51. Nelles, L.P. and Bamburg, J.R. (1976) Anal. Biochem. 73, 522-531.

52. Condeelis, J.S. (1977) Anal. Biochem. 77, 195-207.

53. Scott, P.G., Telser, A.G. and Veis, A. (1976) Anal. Biochem. 70, 251-257.

54. Tuszynski, G.P., Damsky, C.H., Fuhrer, J.P. and Warren, L. (1977) Anal. Biochem. 83, 119-129.

55. Sorimachi, K. and Condliffe,P.G. (1977) Biochem. Biophys. Res. Commun. 77, 526-532.

56. Mayer, J.W. (1976) Anal. Biochem. 76, 369-373.

57. Schmidt-Ullrich, R., Knüfermann, H. and Wallach, D.F.H. (1973) Biochim. Biophys. Acta 307, 353-365.

58. Waite, J.H. and Wang, C.-Y. (1976) Anal. Biochem. 70, 279-280.

59. Stegemann, H. (1972) Z. Anal. Chem. 261, 388-391 and Glas- u. Instr. technik 16, 827-830 and 925-928.

60. Asao, T. (1977) Anal. Biochem. 77, 321-331.

61. Rüchel, R., Mesecke, S., Wolfrum, D.-I. and Neuhoff, V. (1973) H.-S. Z. Physiol. Chem. 354, 1351-1368.

62. Rüchel, R., Mesecke, S., Wolfrum, D.-I. and Neuhoff, V. (1974) H.-S. Z. Physiol. Chem. 355, 997-1020.

63. Rüchel, R., Richter-Landsberg, Ch. and Neuhoff, V. (1975) H.-S. Z. Physiol. Chem. 356, 1283-1288.

64. Klose, J. (1978) Workshop on Methods in Mammalian Genetics, Oct. 1976, Genetics, in press.

65. Poehling, H.M., Wolfrum, D.-I. and Neuhoff, V. (1976) Entomol. Exp. Appl. 19, 271-286.

66. Bhakdi, S., Bjerrum, O.J. and Knüfermann, H. (1976) Biochim. Biophys. Acta 446, 419-431.

67. Schäfer Nielsen, C. and Bjerrum, O.J. (1975) Scand. J. Immunology 4 (Suppl. 2), 73-80.

68. Neuhoff, V. and Mesecke, S. (1977) H.-S. Z. Physiol. Chem. 358, 1623-1637.

69. Stegemann, H. and Toennies, G. (1956) Nature 177, 440.

70. Tuszynski, G.P. and Warren, L. (1975) Anal. Biochem. 67, 55-65.

71. Danno, G.-I. (1977) Anal. Biochem. 83, 189-193.

72. Ames, G.F.-L. and Nikaido, K. (1976) Biochemistry 15, 616-623.

73. Macko, V. and Stegemann, H. (1970) Anal. Biochem. 37, 186-190.

74. Zechel, K. (1977) Anal. Biochem. 83, 240-251.

75. Schick, M.J. and Manning, D.J. (1966) J. Am. Oil Chemists' Soc. 43, 133-136.

76. Makino, S., Reynolds, J.A. and Tanford, C. (1973) J. Biol. Chem. 248, 4926-4932.

77. Helenius, A. and Simons, K. (1975) Biochim. Biophys. Acta 415, 29-79.

78. Clarke, S. (1975) J. Biol. Chem. 250, 5459-5469.

79. Kaltschmidt, E. and Wittmann, H.G. (1970) Anal. Biochem. 36, 401-412.

80. Dale, G. and Latner, A.L. (1969) Clin. Chim. Acta 24, 61-68.

81. Macko, V. and Stegemann, H. (1969) H. S. Z. Physiol. Chem. 350, 917-919.

82. Kenrick, K.G. and Margolis, J. (1970) Anal. Biochem. 33, 204-207.

83. Rüchel, R. (1977) J. Chromatog. 132, 451-468.

84. Stegemann, H. (1975) in The Chemistry and Biochemistry of Plant Proteins, Ghent, Sept. 73, Harborne, J.B. and Van Sumere, C.F. eds., Academic Press, London, pp. 71-88.

85. Stegemann, H. (1977) in Electrofocusing and Isotachophoresis 1976, Graesslin, D. and Radola, B.J. eds.,de Gruyter,Berlin, pp.385-394.

86. Klose, J. (1975) Humangenetik 26, 231-243.

87. Klose, J. (1975) in New Approaches to the Evaluation of Abnormal Embryonic Development, Neubert, D. and Merker, H.J. eds., Thieme Publ., Stuttgart, pp. 375-387.

88. Klose, J., Blohm, J. and Gerner, I. (1977) in Methods in Prenatal Toxicology, Neubert, D., Merker, H.J. and Kwasigroch, T.E. eds., Thieme Publ., Stuttgart, pp. 303-313.

89. Barrett, T. and Gould, H.J. (1973) Biochim. Biophys. Acta 294, 165-170.

90. Bhakdi, S., Knüfermann, H. and Wallach, D.F.H. (1974) Biochim. Biophys. Acta 345, 448-457.

91. MacGillivray, A.J. and Rickwood, D. (1974) Europ. J. Biochem. 41, 181-191.

92. Righetti, P.G. and Drysdale, J.W. (1974) J.Chromatog. 98, 271-321.

93. Iborra, F. and Buhler, J.-M. (1976) Anal. Biochem. 74, 503-511.

94. Conrad, M.J. and Penniston, J.T. (1976) J.Biol.Chem. 251, 253-255.

95. Booth, A.G. (1977) Biochem. J. 163, 165-168.

96. Neuhoff, V. and Lezius, A. (1968) Z. Naturforsch. 23 B, 812-819.

97. Liefländer, M. and Stegemann, H. (1968) H.-S. Z. Physiol. Chem. 349, 157-160.

98. Cleveland, D.W., Fischer, S.G., Kirschner, M.W. and Laemmli, U.K. (1977) J. Biol. Chem. 252, 1102-1106.

99. Marshall, D.R. and Brown, A.H.D. (1975) J.Mol.Evolut. 6, 149-163.

100. Lehtovaara, P. (1978) Biochem. J. 169, 251-253.

101. Williams, J.G. and Gratzer, W.B. (1971) J. Chromatog. 57, 121-125.

102. Wittmann-Liebold, B. and Wittmann, H.G. (1978) personal communication and (1974) in Ribosomes (Cold Spring Harbor Lab) pp. 93-114.

103. Stegemann, H. (1978) Proc. Symposium Seed Proteins, Gatersleben, DDR, 1977, in press.

104. Khan, K. and Bushuk, W. (1977) Cereal Chem. 54, 588-596.

105. Abraham, G. and Cooper, P.D. (1976) Anal. Biochem. 73, 439-446.

106. Makino, S. and Niki, R. (1977) Biochim. Biophys. Acta 495, 99-109.

107. Koenig, R. (1972) Virology 50, 263-266.

108. Koenig, R., Lesemann, D., Brunt, A.A. and Kühne, H. (1973) Intervirology 1, 348-353.

Electrokinetic Separation Methods,
P.G. Righetti, C.J. van Oss and J.W. Vanderhoff

PREPARATIVE ELECTROPHORESIS, ISOTACHOPHORESIS AND ELECTROFOCUSING ON POLYACRYLAMIDE GEL

ANDREAS CHRAMBACH and NGA Y. NGUYEN
Endocrinology and Reproduction Research Branch, National Institute of Child Health and Human Development, National Institutes of Health, Bethesda, Maryland 20014 (USA)

PREPARATIVE POLYACRYLAMIDE GEL ELECTROPHORESIS

Over the last 20 years, the resolving power of electrophoretic methods of protein separation has increased dramatically twice: First, when Smithies on starch gels (1) Ornstein (2) and Raymond (3) on polyacrylamide gels, extended the serum protein pattern from 5 zones to 30. Secondly, when O'Farrell by use of 2-dimensional techniques increased the resolution of an E. coli extract from 30-50 zones to more than 1,000 (4).

Each of these 2 quantum jumps in separation science started by providing zone patterns. All one could do with these patterns was to inspect them subjectively, and to make subjective pattern comparisons. Even to the present day the vast majority of biochemists continue to make use of these separation techniques in that same qualitative and subjective fashion.

It seems more useful to define and identify protein components by specific, characteristic numbers, which allow one to derive physical-chemical information as to the nature of each species, and thereby provide an objective basis for distinguishing among them. Such objective characterization necessitates the control of fractionation conditions to a degree sufficient to make the numbers reproducible. Sufficient accuracy of the numbers is obtained by mechanizing and automating the measurements. Automated data processing eliminates calculation error and allows for an exhaustive statistical treatment of the characteristic numbers. This "quantitative" approach to analytical gel electrophoresis (5,6) has extended the usefulness of the Ornstein-Raymond type analysis of protein systems, but remains burdened with the basic inadequacy of all purely analytical resolution: A component, no matter how well characterized with regard to molecular size and net charge, remains essentially undefined with regard to chemical composition, conformational parameters and, most important, its specific activity or activities (i.e. activity per mole, or per mass of protein).

It is not surprising, therefore, that in more than fifteen years of polyacrylamide gel electrophoresis (PAGE) a vast number of attempts have been made to translate the newly found resolution to a preparative level and to develop suitable apparatus and procedures (7 and Table 1). In the perspective of the ultimate goal of biochemistry, i.e. elaboration of a chemistry of living matter, with thousands of chemically homogeneous, isolated macromolecules, this attempt to generate these species by electrophoretic (or other) methods has been a dismal failure to date. Not only have most macromolecular species eluded purification to homogeneity, but no general procedures have been developed by which the average biochemist could possibly arrive at an either adequate or efficient isolation of any molecule of interest.

This review is concerned with milligram-preparative methods since chemical analysis, and physical-chemical methods such as spectroscopy, still depend on the availability of milligram amounts of protein. The gram-preparative level, although industrially important, has barely been attained to date since it is burdened with unsolved problems not present in milligram-preparative methods.

Since a preparative PAGE does not exist as yet in a meaningful, widely applicable and efficient form, we will not provide an exhaustive review of the literature on preparative gadgetry but rather focus attention on the various developments in apparatus and procedural designs which seem capable of being developed into generally applicable isolation tools.

Reviewing the substantial number of preparative PAGE apparatus types in the literature (Table 1) one is amazed by the fact that most have never been used on any serious separation problems, or have been applied only once to the single problem for which they were developed (7). If apparatus and procedures were as easy to use and as well investigated as claimed by many of the inventors, one would have expected waves of useful applications in the wake of the publication of each new apparatus type. Applications are, however, conspicuously absent, and thus one must doubt the adequacy of most apparatus and procedures in practical, routine application.

The apparatus of Brownstone (8) is a good example of an electronically advanced apparatus design of which the potential usefulness has been greatly reduced due to failure to provide a sufficiently simple procedure and due to failure to provide the user with the means to arrive at an optimized set of conditions for the particular fractionation. It seems notable that after 6 years of uninterrupted operation of this apparatus not more than a handful of published isolations have resulted (8). Needless to say, the record for other apparatus is worse. A relatively mature and sophisticated design like that of Brownstone is therefore obviously not enough, unless it is matched by equal efforts toward procedural simplification and the provision of optimized conditions.

In general, it is a widespread misconception that preparative PAGE has failed to be useful in biochemistry solely for lack of a "good apparatus". In fact, many useful types of preparative PAGE apparatus are available both in the literature (Table 1) and commercially. Most of these apparatus, although non-optimal, are useful if higher than average skill and some knowledge of principles are applied. It is also important that the conditions of preparative PAGE have first been carefully optimized at the analytical scale. It is usually for these reasons rather than apparatus design that preparative PAGE has failed. Hundreds of unused pieces of preparative apparatus relegated to the museum corner of laboratories testify to this fact.

To date, procedures of preparative PAGE vary widely and no consensus on optimal steps has been reached, leaving the amateur biochemist free to devise his own - a freedom for which he is usually ill-equipped. Therefore, to be of practical help, we will deal extensively with ways of optimizing conditions of preparative PAGE procedure in application to a presently common preparative apparatus design type, originally introduced by Jovin et al (9), without any claim or prejudice in favor of the superiority of this particular design over other types (Fig.1).

In general terms, preparative PAGE will only be useful to the majority of biochemists if it can be reduced to the degree of operational simplicity that has made gel filtration such an overwhelming success among fractionation methods. Since effective preparative gel electrophoresis requires an optimization of pH, pore size, voltage, length of gel, and a number of other conditions (10), and is inherently a multistep procedure, such operational simplification can be achieved only with far more complicated and sophisticated apparatus design, electronic control features and ready-to-use computer programs than available today. Presently the determination of optimal pH and pore size conditions at the analytical level for application to preparative PAGE can already be carried out in such computerized, easy-to-use fashion, under elimination of guesswork and elements of art (6). However, as will be shown below, the particularly preparative parameters which govern resolution as a function of the diffusion coefficients of proteins in gels, are not equally defined and available as yet. Finally, apparatus of the necessary degree of electronic sophistication required for extreme operational simplicity - the push-button "fractionation machine" - does not exist as yet.

A) Optimization of preparative PAGE conditions

The theoretical approaches to optimizing resolution in preparative PAGE have recently been treated by Rodbard et al (10) a practical guide towards optimization has been reported by Marker et al (11). These treatments deal with the question of resolving two zone distributions from another, where each is defined by a particular migration distance in the gel and by a band width which increases as a function of time. It is this time dependent zone spreading, or, in physical-chemical terms, the problem of determining the diffusion coefficient in polyacrylamide gel for each component of the pair to be separated, which is at the heart of the problem of

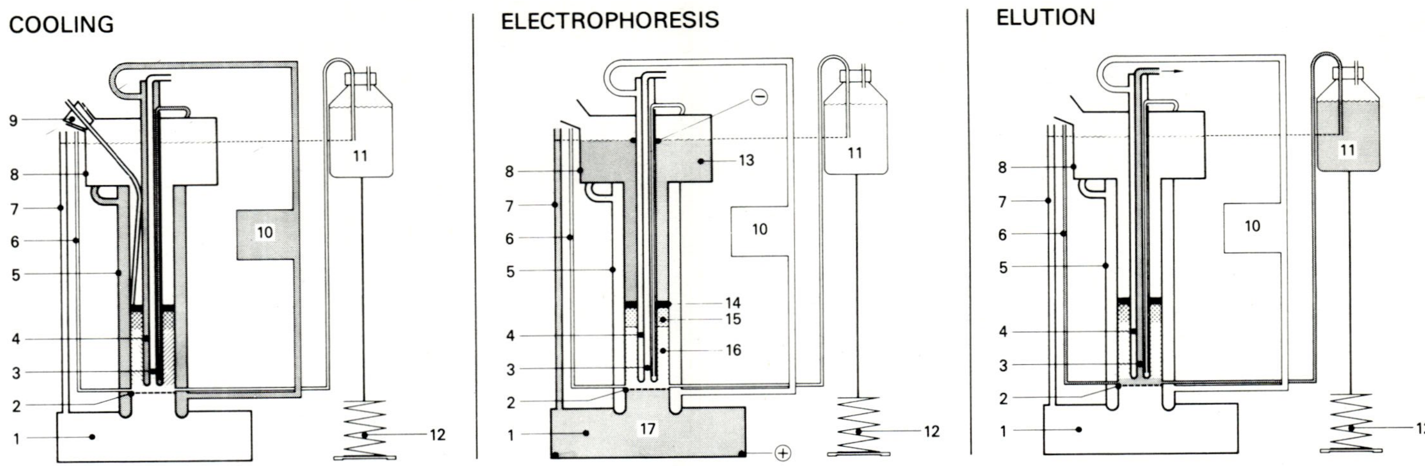

1. Lower Buffer Reservoir
2. Semipermeable Membrane
3. Capillary Elution Tube
4. Inner Cooling Jacket
5. Outer Cooling Jacket
6. Manometer Tube
7. Hydrostatic Balancing Tube
8. Upper Buffer Reservoir
9. Sample Applicator
10. Thermostated Water Bath
11. Elution Buffer Reservoir
12. Lab-Jack

Fig. 1.

optimizing conditions for preparative PAGE. Thus, in one approach one can determine the free diffusion coefficient by measuring zone width as a function of time in gels, at given protein load, pH, ionic strength, current density and temperature (12) and compute resolution for particular values of these parameters, taking into account the general functions relating them to protein zone width. In conjunction with Ferguson plot parameters experimentally obtained by PAGE, this approach allows one to predict elution times for a gel of given length and conduct computer simulations for gel fractionations at various lengths and operative under various voltages by use of a program of Rodbard (Fig.5 of (15)). Alternatively, one can, for an optimization of preparative conditions, measure, for each unknown species and its closest contaminant, from which it is to be separated, the actual band width under defined conditions of pore size, pH, ionic strength, protein load, temperature and current density of PAGE, using the identical sample which is to be separated preparatively - and by taking into account the ratio of protein loads (peak areas) between the unknown species and its contaminant. This band width determination for protein and closest contaminant in the particular gel of interest depends on a mathematical model when peaks are asymmetric, non-Gaussian, partially overlapping or located on a slanting baseline.

Clearly, to carry out all these steps prior to any preparative experiment would be painful even if the tools for automated measurement and computation were widely available. Also, the present procedures using UV-optics consume undue amounts of protein which, although not pure, at least must be sufficiently purified prior to preparative PAGE (sic) to present a single protein distribution within a complex pattern, or a distribution overlapping with a single neighboring contaminant. However, at least apparatus to measure band width in gels during electrophoresis in a continuous fashion and computer interfacing of zone width data have been developed and are operative (13), and the first free diffusion coefficient of a protein has just been determined by gel electrophoresis (14). Therefore, in present practice, it seems hardly any more reasonable to assume that the diffusion coefficient of a protein in a substantially restrictive gel is one order of magnitude higher than the free diffusion coefficient (12). Both the procedures and commercially available apparatus (Bascom-Turner Instr. Co., Newton, MA) exist for determining it on any protein.

A reasonable choice for gel concentration, as a first approximation, is the "gel concentration for maximal separation", T_{max}, obtained from the Ferguson plots of the protein and the contaminant by calculating the value for $0.368 \times \mu$, where μ is the antilog R_f at the crossing point of the 2 Ferguson plots (16) or by use of program T-OPT (6,16). Voltage gradient and gel length can be optimized by systematic computer simulation of the efficiency of fractionation, expressed as the number of theoretical plates, N, (17, Fig.5 of (15))[1] or by experiment (11). Since protein zones in preparative PAGE with a conventional elution mechanism (Fig. 1) migrate through the entire gel length, gel height should be equal to or less than that adopted at the analytical scale. Voltage of course is limited by the Joule heat dissipation characteristics of the apparatus used. In conventionally designed apparatus (Fig.1), at least 1 watt per vertical centimeter may be tolerated without a significant radial temperature gradient across the gel. The radial temperature differential between the center of the gel and the cooled periphery should not exceed 0.1°C. At such power, temperature differentials of 2-6°C between thermostated bath and the center of gels of 18 cm diameter have been measured with conventional types of apparatus (20).

The choice of buffer system and pH for preparative PAGE has to be made at the analytical scale, with the purpose of achieving optimal "charge separation" as close to the pI's of the species to be separated as compatible with adequate mobility (6,10).

In the early days of preparative PAGE development, it was considered important to select volatile buffers for preparative PAGE, in order to be able to lyophilize eluate fractions directly. Since that time, however, ample evidence has accrued showing that eluates from preparative PAGE contain polyacrylamide-like impurities which must be removed chromatographically before the eluate can be lyophilized (6,21). Since this chromatographic purification can and should be conducted in volatile buffers, it is unnecessary to be concerned with the volatility of preparative PAGE buffers.

Another concept prevalent in the early days of preparative PAGE was that of sequential fractionation of multicomponent systems by successive elution of components in an elution apparatus (Fig.1). Although the goal remains desirable, it has met with no success whatsoever. It is apparently not possible to optimize pH, pore size or, in isotachophoresis, "stacking limits" (6) for more than two components at once, so that when one attempts to separate many components simultaneously from another, separation becomes relatively inefficient for all.

In summary, before ever attempting a preparative separation, it is essential to do a sufficient amount of "legwork" at the analytical scale and calculation, preferably by computer, leading to the choice of optimal conditions for the particular separation problem of interest by application of the general strategy of PAGE previously oulined (Fig.1 of (6)).

B) Apparatus

1) Design types. Conventional preparative PAGE apparatus has been designed to allow zones to migrate successively through the entire gel length, with elution at

the bottom of the gel into either a buffer chamber wholly or partially closed by a semipermeable membrane, or into a chromatographic medium from which zones can be eluted. This type of device is designated as "successive zone elution PAGE" apparatus (Fig.1). A second type of apparatus allows one to elute the zones at any position along the gel, through electrophoresis into buffer at a right angle to the first direction of migration. Alternatively, a compartmentalized apparatus suitable for the selective removal of any particular section, or sectioning of the gel followed by diffusion or electrophoresis of protein from the gel slices may be used. This type of device may be designated as "simultaneous zone elution PAGE" apparatus. Table 1 lists some examples from the literature for apparatus of both types.

A consensus, or even a comparative study, with regard to the choice between "consecutive" and "simultaneous" zone elution apparatus and procedure does not exist. Nor can one pinpoint the optimal apparatus type within each of the two categories on the basis of objective criteria. In fact, there is no agreement of what these criteria might be.

It is clearly advantageous to isolate all components simultaneously. The one important, and possibly overriding advantage of most simultaneous zone elution techniques is that they seem relatively safe. It is usually not possible to lose substantial amounts of protein during gel electrophoresis, sectioning and recovery, by diffusion or electrophoresis, of the protein. The second advantage is that the method is equally compatible with any value of mobility. Thus, slowly migrating proteins can be processed. High gel concentrations, frequently "optimal" (see above) are applicable. Only laboriousness, and the absence of integrated, modular apparatus capable of the entire operation are present drawbacks of "simultaneous zone elution apparatus".

The closest approximation to an integrated "simultaneous elution" apparatus has been the "elution-convection" design of Raymond and Jordan (27). But in the practice of this barely mg-preparative, 6 mm gel slab apparatus, (22b), the 30 serum components resolved by staining methods are not resolved, but only 5-7 are. It is clearly difficult, and conceivably impossible, to electrophorese a long zone at a right angle without enormous zone spreading and loss of resolution; this is presumably the reason why the collecting system did not allow for resolution of zones closer adjacent than 6 mm.

A second design type potentially applicable to the collection of both slow and fast moving zones simultaneously is the vertebral, segmented column design of Stathakos (23). However, since segments of such a column no matter how much reduced in size still must have substantial height, the device seems suited for fractionation into subgroups, not into single homogeneous proteins.

The third type of "simultaneous" zone elution apparatus is the scaled up analytical gel with provision for sectioning of the gel and elution of the protein from gel slices by diffusion or electrophoresis (24,25). A potentially interesting approach to recovery of protein from gel slices would be to solubilize the gel by selective cleavage of crosslinks (79,80,81) with subsequent separation of linear polymers from protein. In the past, gel slice elution has been frequently dismissed as being a primitive, rudimentary and excessively laborious way of doing preparative PAGE. But since it is also an approach with minimal risks of losing the protein and the least apparatus requirements, it may very well turn out the ultimately preferable way, in particular also in view of the possibility to isolate a number of proteins at once and independently of the magnitude of their mobilities. It will be necessary, however, once this approach is accepted as being promising, to assemble suitable, non-laborious procedures and apparatus modules for sectioning, protein extraction and recovery as a functional unit.

All other major design types of preparative PAGE apparatus provide for consecutive elution. The type which resembles Raymond's design superficially by being a slab apparatus with provision for electrophoretic elution is that of Hodson and Latner (26) or the flat bed slab apparatus of Gordon (27) and of Polsky _et al_ (28) which are potentially applicable to polyacrylamide. In this type, zones on a slab gel are allowed to migrate into an elution chamber at the end of the gel (26,27) or cut into the gel (28); the chamber is emptied and refilled in timed or programmed fashion. This is clearly "successive zone elution" apparatus which, although it has the appearance of a gel slab, is essentially very similar to the more familiar "folded slab" type of preparative PAGE apparatus (Fig.1) which externally gives the appearance of a column. Assuming that Schenkein's _et al_ (29) design of a vertical gel column with discontinuous elution, or Brownstone's programmed elution (8) were combined with a vertical folded slab design (Fig.1) - as done by Jaarsveld _et al_ (30), it should function pretty much like a Hodson-Latner or a Polsky apparatus. If one then further discounts the intuitive notion that discontinuous elution WITHOUT A PROTEIN DETECTION MECHANISM in the elution chamber would give rise to less eluate dilution than continuous elution (20), the 2 slab designs appear more and more indistinguishable from not only Nees' slab design (31) but even the original Jovin model (Fig.1). Thus, although there is no consensus on the detailed design of the elution mechanism, the independently developed apparatus types listed in Table 1 do not seem profoundly different from each other. This view lends support to the notion that with sufficient ingenuity and practice of model fractionations any one of the major models of preparative PAGE apparatus may be workable. In other words, the problems involved in using preparative PAGE as an ordinary tool in biochemistry lie in procedure employed,

TABLE I

PREPARATIVE POLYACRYLAMIDE GEL ELECTROPHORESIS APPARATUS

Zone elution	Principle of Elution	Authors (Ref.)
successive	concentric upward buffer flow	Jovin *et al* (9)
		Gordon and Louis (36)
		Vogel and Reinmuth (37)
		Duesberg and Rueckert (41)
		Malacinski (40)
		Coy and Woo (48)
	concentric downward buffer flow	Hjerten *et al* (33)
		van Welzen and Zuidweg (34)
		Svendsen (39,43)
		Radhakrishnamurthy *et al* (46)
		Bergrahm and Harlestam (49)
		Smith and Moss (65)
	crossflow	Nees (31)
		Racusen and Calvanico (47)
		Maizel (63)
	serpentine groove flow	Hochstraesser (42a)
		Hollmen and Kulonen (42b)
	discontinuous radial upward buffer displacement	Brownstone (8)
		Jaarsveld *et al* (30)
	discontinuous downward vertical buffer displacement	Hodson and Latner (26)
		Polsky *et al* (28)
	discontinuous migration into a bag	Reznick *et al* (45)
		Kawata *et al* (50)
		Suzuki *et al* (51)
		Wachslicht and Chrambach (52)
		Sun and Hall (66)
simultaneous	elution-convection	Raymond and Jordan (22)
	electrodecantation	Posner (67)
	segmented gel column	Stathakos (23)
	slice elution	Dudman and Zerner (24)
		Sulitzeanu *et al* (25)
		Lewis and Clark (68)
		Karsnas and Roos (69)
		Bont *et al* (70)
		Weliky *et al* (71)
continuous		Boyde and Remtulla (32)
microgram-preparative		Voytek (72)
		Houghten and Chrambach (73)
		Popescu *et al* (74)
		Strauch (75)
		Ray *et al* (76)
		Mann and Huang (77)
		Pajdak (78)

rather than in apparatus design, or at least not as predominantly in apparatus design as is commonly thought.

If we tentatively accept such a unified viewpoint, what are the key differences in apparatus design within the general model? We have already mentioned continuous *vs.* discontinuous elution of zones from the elution chamber. Clearly, discontinuous elution coupled to an optical detection system in the elution chamber and programmed by such detection system is preferable but doesn't exist as yet with a sufficient degree of sensitivity (i.e. sensitivity higher than that of chromatographic UV-monitors). Timed discontinuous elution in the absence of continuous monitoring does not seem to provide much advantage in preventing undue dilution (20).

Another relatively trivial distinction is between elution systems providing concentric upward *vs*. downward flow, cross-flow or flow through a chamber as opposed to periodic change of a dialysis bag or cup at the bottom of the gel column. Again, when this is done periodically in the elution of non-colored proteins, the degree of dilution should be comparable to that in continuous elution. Thus, most of the designs of preparative PAGE apparatus may be grouped into a central model, making it possible to talk in exclusion of the minority of peripheral designs about procedure in terms that will benefit the majority of users independently of the particular apparatus type used.

In considering "simultaneous" *vs*. "successive" elution devices, we have already considered a distinct class of preparative devices introduced by what we may call the "conservative" designers (24,25) which are scaled up analytical gel slabs or cylinders, with component detection by staining of a guidestrip, slicing and elution from gel slices. These designs obviously excel in simplicity of construction and operation and should be taken seriously in as far as the burden of proof for greater resolving or load capacity resides with the more complicated, expensive and difficult to operate devices.

The corresponding "radical" preparative PAGE apparatus designers are clearly headed by Boyde and Remtulla (32) with their concept of a rotating gel cylinder of annular cross-section for continuous loading and elution. Once a consensus on the optimal stationary gel slabs of the same type is reached, the transformation to a model providing continuous operation certainly would appear promising.

2) *Load capacity*. Load capacity in PAGE is proportional to gel surface area. To know the required area, one needs a realistic estimate of resolution between the zone of interest and its nearest neighboring contaminant as a function of load. In the absence of plots of resolution between proteins on gels as a function of protein load (see section A), the best guess, at present, based on experience with successive zone elution, seems to be that two narrowly adjacent, equally concentrated zones of R_f 0.3-0.4 cannot be resolved by preparative PAGE when their

concentrations exceed 0.1 mg/component/cm^2 of gel. The estimate appears very low compared with some claims in the literature, but this is probably due to the widespread practice of referring load capacity to multicomponent mixtures rather than to the single component of interest migrating within approximately one band width from its closest contaminant, as was done in our estimate. The value does not seem inconsistent with the widely appreciated fact that 30-50 component mixtures may be resolved by analytical PAGE at load capacities of not more than 800 µg of the mixture/cm^2 of gel. Assuming our estimate is correct, preparative PAGE could not become mg-preparative unless one used at least 10-20 cm^2 of gel surface area. Preferably, the gel surface should be much larger. Among the apparatus types listed in Table 1 only very few attain such gel dimensions. Of these, the majority are in the 10-20 cm^2 range of area, including some commercially available apparatus.

The reason for the frequently minimal or inadequate size of gel surface area used in preparative apparatus is threefold: first, limitations in the efficiency of Joule heat dissipation and dissipation of the heat of polymerization. Beyond this limit, temperature dependent gel swelling and shrinking would endanger its mechanical stability, and uneven mobilities across the zone would result, with consequent band distortion and failure of protein along the zone diameter to enter into the elution chamber (Fig.1) at the same time. Second, the larger the gel, the more difficult is its mechanical stabilization in vertical (open or folded) gel slabs, and, thirdly, eluate collection without undue dilution becomes increasingly difficult with increasing gel size. (The first two classes of considerations regarding the limitation of gel thickness to 15 mm will be discussed below.) The limitation to a gel thickness of 15 mm translates into a maximal gel surface area of 20 cm^2 for a slab of 13.3 cm length. Since elution along a slab that long seems difficult, the gel sandwich may be conveniently folded into a cylinder with annular cross-section cooled internally and externally (9). Unfortunately, the load capacity even of a folded slab of this type (Fig.1) cannot be increased much beyond 2 mg/zone of 20 cm^2, in view of the fact that the volume of the elution chamber increases faster than the load capacity (surface area), and in view of difficulties in mechanical stabilization which rapidly increase with expanding diameter of the gel cylinder. Several ways have been proposed to free preparative elution PAGE from this limitation to 20 cm^2. The first, by Brownstone, achieves a gel surface area up to 150 cm^2 by cooling across the flat dimension of a gel donut, reduction in the size of the elution chamber, multiple points of collection of eluted zones and mechanical support (8). The second by Hjerten et al (33), introduces multiple cooling coils along both vertical faces of a gel block of 84 cm^2 surface area and avoids excessive eluate dilution by passage of the eluate into an elution chamber filled with agarose. A third approach, by Hodson and Latner (26), Gordon (27) and recently by Polsky et al (28), provides troughs in or at the ends of a horizontal gel slab filled with buffer and periodically emptied after migration of the zones into the troughs. Assuming that this principle could be applied to a gel slab of 1 cm thickness and 50 cm length, collection from a trough of 5 mm thickness would allow the collection of 5 mg protein per zone into a volume of 25 ml, which is not a significant improvement. In principle, electrophoretic collection of a very wide protein zone at a 90° angle to the direction of migration proposed by Raymond and Jordan (22) could also provide a similar load capacity. Another potential way to provide a large gel surface area utilizes a pyramidal gel shape to achieve a large surface area and small elution chamber (34). The problem associated with this design idea is that the current density rises sharply in proportion to the degree of constriction in the apex of the pyramid, requiring proportionately improved heat dissipation in the constricted parts of the gel. This is difficult to provide without considerably slowing down the rate of electrophoresis, or without a mode of continuous power decrease.

Finally, preparative PAGE could be scaled up by making it continuous. Reference is made to the rotating gel slab of Boyde and Remtulla (32) with continuous sample application and elution into agarose. Although it is a conceptually intriguing design and may very well turn out to be useful, the model has not matured as yet beyond a prototype and a single set of model separations.

In summary, a number of designs for gel surface areas of 15-150 cm^2 of gel are feasible which would provide one with up to several milligrams of protein per zone. All smaller apparatus - and that is the majority of proposed designs in the literature - are not milligram preparative in PAGE (6). Microgram-preparative designs listed in Table 1 are important for "final step PAGE" and may become important for preparative isotachophoresis (35)[2]. "Final step PAGE" refers to the situation where, after a series of first preparative PAGE fractionations, a pool of nearly pure product has been prepared in amounts of a few milligrams. If one were to re-apply these few milligrams on a 20 cm^2 gel, nearly all would be lost through adsorption which at 0.1 mg/cm^2 can be as much as 70% (20). Thus it is necessary in such a case to apply a small volume preparative PAGE apparatus holding a few milliliter of gel (20).

3) Temperature control. Efficient dissipation of the heat of polymerization and of Joule heat are needed a) to maintain zone symmetry allowing each zone to enter at the same time into the elution chamber or collecting system, and b) to minimize temperature dependent gel swelling or shrinkage which would lead to mechanical instability and to current leakage along the apparatus walls. c) Another reason for strict temperature control of preparative PAGE is that the elution times for the band(s) of interest can only be predicted on the basis of the characteristic R$f values for each of the zones. Analytical and preparative values will only agree if both have been conducted at identical temperatures. For the same reason, thermal

changes between polymerization and electrophoresis should be avoided by conducting both in the same vessel. Thus, gel transfer from a polymerization mold to an electrophoresis vessel (8,30) is undesirable.

Geometrically, gel cylinders are least suited to provide efficient heat transfer. Gel slabs excel in surface to mass ratio in proportion to their thinness. This holds true whether or not slabs are folded into cylinders of annular cross-section. The latter have certain advantages over slabs cooled from both surfaces with regard to the ease of sample application and of eluate collection but in view of the discrepancy between inner and outer gel surface areas exposed to coolant unfolded slabs only can be uniformly cooled from both sides (31). Equal cooling on both faces also depends on coolant flow to both in parallel, rather than in series (33).

Gel thickness compatible with maintenance of uniform temperature (radial temperature gradients of less than 0.1°C) at reasonably rapid rates of migration (measured in hours, not in days) is restricted to approximately 15 mm, if the gel is surrounded by, and in contact with the coolant across thin glass walls. This value of 15 mm is again an estimate which should be redetermined for any given apparatus by radial temperature measurements, but it seems a widely accepted result among the design types (Table 1). Apparatus with 25 mm or more gel thickness is available but does not fulfill the above-stated criteria for adequate performance. However, a substantial improvement of heat dissipation techniques over the best present methods (liquid coolant flow at or above 1.5 l/min) could change that limitation drastically.

Vertical slabs are compatible with batteries of multiple cooling coils (33) and can therefore be extended in size without creation of increased dead space in the center of the elution chamber. Sandwich-type horizontal preparative slab apparatus with two cooled surfaces may present sample application problems, while gel surfaces cooled from just a single face (e.g. 26,27) are burdened with having to maintain a saturated atmosphere above the open face of the gel to prevent hydration changes at the surface. An open slab apparatus also necessarily gives rise to a temperature gradient between the cooled lower face and the open upper face and produces some drying of the gel, gel shrinkage at the open surface depending on humidity levels above and moisture values inside the gel. It remains unclear to what degree these difficulties outweigh the apparent advantages of horizontal slab design, e.g. suitability for mechanically labile gel types and simplicity of construction.

From the viewpoint of heat dissipation efficiency, thin glass remains the preferred material over plastics for the walls of the gel container. Also, designs which feature a direct contact of the coolant jacket with the gel are preferable to designs featuring an intermediate channel of elution buffer between coolant and gel (e.g. (36)), or interposing insulating materials between coolant and gel as in many of the present horizontal slab designs. In particular, Vogel and Reinmuth's idea of eliminating such "intermediate channels" by making the elution buffer the coolant seems valuable (37), although it poses a flow regulation problem since a reasonable elution buffer flow rate is 1 elution chamber volume per minute (frequently about 1 ml per minute) while an effective coolant flow rate is 1.5 liter/min.

Optimally, one would like to disregard the problems of cooling through liquid flow altogether, and replace it by electronic methods of heat transport, as pointed out above. However, no such design features have been incorporated into any of the preparative PAGE apparatus types, since the technology for electronic cooling remains costly as compared to coolant flow systems.

It remains possible, of course, to regulate the temperature of preparative PAGE by conducting it in a cold room, but this approach seems objectionable in view of the relative inefficiency in heat transfer by gases as compared to liquids.

4) Devices for the control of hydrostatic and mechanical pressure. Preparative PAGE appears to depend on firm and stable adherence of the gel to the glass walls of the apparatus (6) although even this point has been questioned (8,30). While gels resting on horizontal plates are not mechanically stressed, this being the foremost advantage of horizontal apparatus designs (e.g. 26-28), mechanical instability is an inherent problem for vertical gels. As pointed out above, the problem of wall adherence is aggravated by the requirement for large gel surface areas, i.e. for large gel volumes which are heavy enough to severely stress the wall adherence, particularly at low gel concentrations and gel height. This is especially painful in successive zone elution methods (Table 1) where the need for a zone to migrate through the entire gel length makes it imperative to work with relatively short gel columns to maintain reasonable fractionation times. In vertical designs, stress on the gel results from the weight of the Upper Buffer above it, requiring either a mechanical gel support (see section a) below) or adequate hydrostatic equilibration through a U-tube arrangement by which this weight is counterbalanced with a liquid column of equal height (Fig.1). Additionally the elution buffer may exert hydrostatic pressure on the gel unless it is balanced by appropriate height adjustment of the reservoir and is flowing under pump pressure alone. Finally, any pulsation of the elution pump exerts some mechanical pull on the gel. In addition to these purely physical stresses on wall adherence there are chemical stresses as well. Gel swelling or shrinkage must necessarily accompany the pH changes concomitant with the transition from the setting phase (phases BETA or GAMMA) to the operative phases (PHASES ZETA or PI) in discontinuous (multiphasic) buffer systems (6,38). Even more dramatically, osmotic perturbations of gel stability due to pH changes affect gel electrofocusing. In

gel isotachophoresis the high protein concentrations in the stack appear to produce substantial localized gel swelling; these hydrated zones traversing the gel weaken its wall adherence (see section II).

a) Mechanical supports. One way to counteract the above-mentioned causes of gel instability in vertical preparative PAGE apparatus is to provide mechanical support for the gel. Porous polyethylene or polypropylene saturated with polyacrylamide gel, porous glass (39-43), and agarose or Pevikon (33) have been used for this purpose. Qualitatively, insertion of resistive materials into the electric field seems undesirable, since it must give rise to enhanced voltage gradients and Joule heating inside the supporting medium even if one succeeds in removing airbubbles quantitatively from the porous medium and preventing occlusion of the elution line (31,33). Certainly, stretched Nylon mesh membranes, or rigid grids seem preferable as supports to porous plastic or glass materials. However, Nylon mesh supports are presently not rigid and a mature design for their insertion has not been developed. The use of agarose or Pevikon to fill the elution chamber and support the gel (33) may raise problems in obstructing elution flow and of clogging.

Avoidance of pressure against the gel during the preparative gel assembly is a central procedural requirement which will be discussed below. Preventing stress on the gel also necessitates the choice of elution flow pumps with minimal pulsation: any pulsating pressure applied on the gel by the pump over many hours of elution must weaken its wall adherence.

b) Chemical modifications of wall properties. Adherence of polyacrylamide gels to glass walls is clearly better than to plastics. It can be improved by cleaning the glass with methanolic KOH and coating with linear polyacrylamide in addition to mechanical support (44). At least in gel electrofocusing, these measures have been noted to counteract osmotic and/or capillary effects along the walls. Presumably, perfect wall adherence is equally important in PAGE. This seems doubly true for relatively large and heavy preparative gels which are, at least momentarily, stressed mechanically during handling in polymerization and setup of the electrophoretic system. Etching of glass by hydrofluoric acid, mechanical grinding of glass surfaces, notching, or serration of glass surfaces should be beneficial but have been seldom applied to preparative PAGE apparatus, presumably in view of field distortions arising from such measures. There also remains the possibility that plastic gel columns which are normally unsuitable for adherence to polyacrylamide due to their hydrophobicity, could be rendered suitable by sulfuric acid treatment, electric discharge (82) or hydrophilic coating (although this would not alleviate problems arising from their poor heat transfer characteristics).

c) Hydrostatic equilibration. The weight of the Upper Buffer on vertical gels and the hydrostatic pressure of the elution buffer against closed elution chambers can be balanced against another by providing a U-tube arrangement ("hydrostatic balancing tube") (Fig.1). Zero net pressure in the elution chamber should be continuously indicated by equal fluid levels in the manometer tube connected to the elution chamber, the air inlet tube to the elution buffer reservoir and the Upper Buffer (Fig.1). The original "successive elution" apparatus of Jovin et al (9) was fully suspended in Lower Buffer (about 12 liters) and thereby hydrostatically equilibrated, although in a rather inconvenient fashion. Most subsequent "successive elution" apparatus other than the early versions of the Buchler "Polyprep" design have neglected hydrostatic equilibration, and it is doubtful that even these particular designs provided an effective "U-tube" since they were blocked in the elution chamber by a disc of Corning No.7930 glass (Fig. 2 of Appendix II to (20)). Most designs could tolerate that omission because they utilized relatively small gels, provided mechanical support for the gel, or balanced the weight of the Upper Buffer by height adjustment of a Mariotte bottle containing the elution buffer (Fig.1). Nonetheless, an optimized design of vertical preparative apparatus should attempt to utilize hydrostatic equilibration by a simple immersion technique or "U-tube arrangement" to aid in the mechanical stabilization of the gel.

d) Polymerization molds. Since mechanical stress on the gel has to be avoided, polymerization in situ must be preferred to transfer of an externally polymerized gel into the electrophoresis apparatus (8,30). Even if at some particular values of %T and %C a mechanical press-fit and gel swelling should, to some degree, restore wall-adherence after a gel transfer, this can only hold for a few selective conditions to which one would not want to limit the performance of preparative PAGE.

5) Electric field symmetry and unobstructedness. Preparative PAGE depends on simultaneous entrance of each zone into the elution chamber (Fig.1). This simultaneity is achieved only if the electrical field is symmetrical with respect to the entire zone, and if obstructions in the field such as air bubbles, mechanical gel supports (39-43), cooling tubes (33) or elution tubes (8,30) do not cross the zone, or impede the field uniformly across the zone. This consideration prohibits a great many design features of the various preparative PAGE apparatus types such as conical elution chambers with permeable membranes smaller than the gel surface area (32), stirrers below the membrane (36,45) if these exceed the diameter of the central cooling finger, mechanical support fingers emanating from the wall in which the gel is held, etc. It is also important to realize that uneven Joule heat distribution across the diameter of the gel must result from a variable coolant flow rate or a variable ratio of coolant surface area to gel surface area between the inner and outer walls of the gel. Any Joule heat differences along the zone width must lead to non-uniformity in mobility of the zone.

6) Elution chamber. Much thought has been given to the design of the elution

chamber in the literature of preparative PAGE.

A) "Jovin-type" elution chamber. Some of the features of upward (9,36,37) vs. downward elution (33,34,43,46) in vertical apparatus have recently been reviewed and will therefore only be referenced (6). The only additional (although obvious) point to be made is that upward elution removes air bubbles from the elution chamber, while downward elution does not. This is particularly the case if the bottom of the central elution tube is conically shaped, as in the Buchler "Polyprep" design. The argument that it is difficult to sweep the dense, sedimenting protein eluate upward is valid only at low elution flow rates which are, regrettably, ruled out for reasons stated below.

The elution chamber should be of minimal height to minimize eluate dilution. Frequently, the height is taken as 1 mm, since greater chamber heights lead to an unnecessarily high eluate dilution and shallower chambers to occlusion and capillary flow problems. Alternatively, the volume of the chamber can be reduced by filling it with an anticonvectant medium (33), or closing it with provision of serpentine grooves under the gel (42), assuming that the materials used do not obstruct the field appreciably. Elution chambers without a bottom membrane have been described (34) (see below) but most elution chambers feature such a membrane. A stretched membrane pulsating with the pump pressure against a 1 mm shallow chamber occludes it very easily through attachment of the membrane to the gel. Therefore, the rigid Corning 7930 glass membrane (introduced by Buchler Instruments Co. in 1964) proved very useful although its permeability characteristics seem ion-selective and therefore imperfect. It is also possible to use polyacrylamide at high gel concentration as a rigid membrane, but this may occlude the elution chamber by swelling or by giving off small gel particles during prolonged elution.

Common defects of all present membranes are the ease with which they adsorb proteins irreversibly and the pH anomalies they introduce in the elution chamber through selective ion permeability (the Bethe-Toropoff effect). These have been considered by the designs of Hjerten et al (33) and of Nees (31). The procedural and constructional burden of either counteracting the Bethe-Toropoff effect by a double elution chamber with counterflow in the two chambers (31), by a serpentine elution buffer flow (42) or by elution of zones into a medium of similar density to the polyacrylamide (at 5-40%T) from which the zone emerges i.e. into Sephadex, Pevikon or agarose at suitable concentrations (33), seems considerable. It does not seem clear, however, to what degree the Bethe-Toropoff effect perturbs the elution from the elution chamber of protein under "standard conditions" (6), i.e. by a several-fold concentrated buffer of the same composition as the operative gel phase and viscosity similar to the gel, at the rate of 1 chamber volume/minute (6).

Conceivably, membranes could be charged (e.g. using the method of neutralizing a net negative charge recently applied by Brownstone (8)) so as to repel the protein electrostatically; but stationary charges in the field would presumably give rise to electro-osmotic problems, change in the conductance of the elution buffer, etc. Alternatively, the membrane could be eliminated if the protein emerging from the gel could be cleared from the elution chamber quantitatively by a design of the elution chamber producing very effective buffer sweeping (34). But this has not been realized as yet without raising new problems, such as increased Joule heating and band distortion in an elution chamber smaller than the gel or vastly excessive flow rate when the elution chamber has the same dimensions as the gel. The effectiveness of polyacrylamide gel as a membrane (24,47), like that of the "no-membrane", depends on a rapid eluion buffer flow rate.

The original elution chamber of Jovin et al was fully open to the concentric inflow of elution buffer; the price to be paid for this clear advantage was that the gel was separated from the coolant by an intermediate layer of elution buffer. While this did not matter for a column totally suspended in cooled Lower Buffer, it seemed an intolerable sacrifice of the efficiency of heat dissipation in later commercial models which, for reasons of operational convenience, removed the gel column from immersion in buffer. For many years gel columns were constructed which sacrificed unhindered buffer inflow into the elution chamber to heat dissipation efficiency. In these designs, originated by Gordon (36), a coolant jacket was brought into direct contact with the gel, but eluent consequently had to be introduced into the elution chamber by a finite, relatively small number of openings in the wall surrounding the elution chamber. Resistance to flow through these openings is large compared to the circumferentially free elution chamber of Jovin et al (9), giving rise to inefficiency in the sweeping of the elution chamber and frequently - through occlusion of holes by particles of polyacrylamide gel or airbubbles - to asymmetry of eluate flow. It was only the more recent design by Vogel and Reinmuth (37) which returned to the fully open elution chamber without sacrificing the convenience of apparatus not suspended in liquid. The crucial realization (37,48) was to use the Lower Buffer as the coolant, just as in Jovin's earlier design. It simplifies the design of preparative PAGE apparatus greatly, without decreasing efficiency of resolution or procedural simplicity with the sole possible exception that it may be necessary to exchange the Lower Buffer made as a GAMMA phase to one made as a PI phase (6) after entrance of the front moving boundary into the elution chamber. Svendsen's recent elution chamber design for downward eluate flow is also able to provide an unrestricted elution buffer flow along the entire circumference of the elution chamber (43,49).

Whether or not the direction of elution is vertical or horizontal, an essential element of elution chamber design is the provision of an outlet to a "manometer tube", i.e. a transparent tube attached along the vertical axis of the apparatus to indicate the meniscus at a level identical with that of the Upper Buffer and the air inlet into the Elution Buffer Reservoir (Fig.1).

Variable size elution chambers seem unncessary. Drastic pH changes between setting and operative phases of multiphasic buffer systems giving rise to a severe hydration change can be prevented by the appropriate input restriction in the computer generation of these systems (18,38). Gel swelling can also be minimized by the appropriate choice of crosslinking agents and crosslinker concentration at the analytical level, as part of the optimization of preparative PAGE conditions. It therefore does not appear necessary to design an apparatus with provisions for an elution chamber of variable size for preparative PAGE, although this may not hold for gel electrofocusing.

B) _Other elution chamber designs_. The elution chamber designs for successive zone elution apparatus other than the Jovin-type (Fig.1) are: a) the impermanent, pressure-induced elution space of Brownstone (8) and Jaarsveld _et al_ (30); b) the chambers with discontinuous buffer displacement of w?v&!c&mn _et al_ (29), Hodson-Latner (26), Gordon (27) and Polsky _et al_ (28); c) the elution bags or cups of Reznick _et al_ (45), Kawata _et al_ (50), Suzuki _et al_ (51) and Wachslicht-Chrambach (52).

a) The periodically (30) or programmably (8) pressure-induced elution space at the bottom of the elution chamber disregards the need for firm wall adherence, prevention of current flow along the lateral surface of the gel or along the elution veins. Possibly, these effects of wall separation are negligible with a gel as large as 150 cm^2; evidently the preparation of the gel outside of the apparatus shows that these designers think that it is. Also, this design depends on the elasticity of the gel and must be limited, therefore, to a relatively narrow range of %T, %C and choice of crosslinking agents. At present, this range is unknown; even if it were known, it would represent an unnecessary limitation for the design of preparative PAGE apparatus.

b) The elution chambers or cavities periodically or programmably emptied and refilled with elution buffer are designed to yield a minimum in eluate dilution, but probably do not achieve any substantial eluate concentration compared to continuous elution devices if they are used blindly, i.e. without protein monitoring and coupling of the elution mechanism to the appearance of protein in the elution chamber. Once this is achieved, continuous elution obviously would have become obsolete. Before this is achieved, elution chamber type b) probably does not represent an advance over continuous elution devices. Like elution chamber type a), type b) seems limited with regard to gel concentration. Here, the mechanical stability of the gel must be sufficient to maintain the elution cavity during the entire course of the fractionation. Presumably for this reason, the reported elution cavities are relatively large compared to the 1 mm high chambers of the Jovin type. This fact means an increase in eluate dilution and it further weakens the claim that concentration of eluate is maintained. This is particularly true if one makes these apparatus types mg-preparative through increases in the gel surface area to 20 cm^2 or more. Discontinuous buffer changes in the elution chamber have been reported using periodic activation of a syringe mechanism (20,29). This must endanger wall adherence of the gel by exertion of pressure against it. Open elution chambers (26-28) avoid this problem.

c) Elution bags or cups have the obvious disadvantage that they need to be emptied and replaced periodically - in addition to excessive size of the elution chamber. Compared with discontinuously filled and emptied elution chambers, elution bags or cups seem to have no advantage other than visibility of the elution chamber in those applications to colored proteins where only one fraction (e.g. a stack marked by tracking dye) or a few are being collected (51-54). Their sole *raison d'être* is simplicity of construction. The claim that large elution chambers of type c) would not give rise to losses of protein on the membrane or give rise to less loss than a well-flushed narrow elution chamber remains doubtful for microgram loads/cm^2 of gel but holds up for milligram loads per zone (52).

7) *Electrophoresis buffers*.

a) *Buffer vessels*. The preparative apparatus design of Jovin *et al* (9) provided 2 liter Upper Buffer reservoirs. This value was based on an estimate for the electrophoretic mass transport of the trailing constituent (terminator) during 24 hours of PAGE under representative conditions and assuming that the composition of the Upper Buffer was that of an operative stacking phase (phase ZETA) (see 6). Later modification of the apparatus by Gordon and Louis (36) eliminated the large buffer chambers for reasons of convenience in handling of the apparatus and replaced them by small reservoirs continuously equilibrated through continuous pumping with large outside reservoirs. Even that may be unnecessary. The approximately 20 mmoles of trailing constituent consumed per day can easily be supplied in Buffer chambers of 100 ml (0.2 M), eliminating the need for pumping entirely when, as in the application of multiphasic buffer systems, the ionic strengths of anolyte and catholyte become irrelevant to the operative concentrations inside the gel. Since preparative experiments rarely last longer than 5 hours, and for 1 day only under rather unusual conditions, such a small volume for the Upper and Lower buffer reservoirs seems quite sufficient.

b) *Electrode compartments*. In the design of PAGE apparatus little attention has, in general, been paid to electrode chambers. Since the electrodes are contained in open vessels, gas evolution seems inconsequential, unless electrolysis produces chemically reactive agents that affect the pH and ionic strength of the buffer phases. Whether or not electrolytic products at concentrations noxious to the proteins or modifying the electrophoretic system effectively are generated with daily changes or buffer has been a matter of disagreement unsupported by facts on either side of the argument. The protagonists of damage through electrolytic products have introduced various means of separating the electrodes from the buffer reservoirs by means of baffle plates, conductive bridges, wicks etc. Evidently these devices may only retard but cannot prevent the migration of the putative noxious agents. Thus they appear ineffectual unless accompanied by rapid equilibration of electrode and buffer compartments with large reservoirs (33,42a).

c) Catholyte-anolyte mixing. A related measure aimed at avoiding electrode polarization is the continuous mixing of Upper and Lower buffers. However, buffer mixing is evidently inadmissible in application to discontinuous (multiphasic) buffer systems where cathodic and anodic buffers serve the disparate functions of reservoirs for the trailing constituent or the common buffer constituent respectively. Since one would not want to design preparative PAGE apparatus inapplicable to multiphasic buffer systems, buffer mixing appears generally counterindicated.

8) Elution buffer reservoirs. Reported or commercially available preparative PAGE apparatus may employ open elution buffer reservoir bottles or sealed Mariotte bottles; single or double reservoirs; cooled or room temperature reservoirs, with or without cooled elution lines; volumes of 2 liter or less; positioning at bench level or above it, constant or readily variable positioning. Limiting ourselves to the basic Jovin-type designs (Fig.1) it is clear that a variable positioning of the air-buffer interface level in a sealed elution buffer reservoir is essential to provide hydrostatic equilibration of the gel. Easy and continuous vertical adjustments can be provided by mounting the elution bottles onto a "lab-jack" (e.g. A. H. Thomas No. 8554-J10) (6); electronic "elevators" would be costly but even more convenient.

Present 20 cm^2 apparatus requires an elution buffer flow rate of approximately 1.5 ml/min. A 2 liter capacity would therefore correspond to 24 hours operation for each reservoir. That is clearly conservative but does not seem to be excessive, allowing a safety margin by a factor of 2. Two bottles, one suitable for maintenance of a GAMMA phase at the bottom of the gel, the other adjusted to the PI phase, are required in application to multiphasic buffer systems. Cooling of the buffer bottles increases buffer stability with time of storage in the bottles. Since the apparatus is, in general, cooled by flow pumps circulating liquid coolant, cooling of the bottles just requires another set of pumps. It can be achieved by cold fingers in the bottles (6), although water jackets would be far more efficient.

Jacketed connecting lines. When the connecting line to the elution chamber is cooled, a thermostated elution buffer also helps in maintaining an equal temperature of buffer in the elution chamber and the lower gel surface. Without a cooled inlet line, the advantage is lost, and gas evolution upon warming in the line endangers even buffer flow and may introduce air bubbles into the elution chamber. It can therefore only be advantageous to thermostat the buffer line connecting the elution buffer reservoir with the elution chamber.

C) Procedures

It was already pointed out above that even an optimally constructed apparatus for preparative PAGE will not make PAGE a generally useful, easy-to-use preparative tool in biochemistry unless it is accompanied by computer programs providing optimal conditions for the specific fractionation problem of interest, and unless non-laborious and fool-proof procedures are elaborated. Ultimately such a procedure should be fully mechanized and automated so as to become feasible in one or a very few experimental steps. Therefore, in the selection of procedures attention should be paid to the potential for automation at each step.

Nonetheless, it would not be realistic to postpone the application of preparative PAGE until fully optimized conditions can be provided. Most presently available preparative PAGE apparatus can be made to work. It is the widespread unawareness of already available principles and procedures developed over the years which, more than any other factor, has been responsible for the disappointment of users who after a few first catastrophic trials decided to scrap their preparative PAGE appratus. We will attempt, therefore, in the following section, to demonstrate generally applicable step-by-step procedural rationales that will allow the reader, for many apparatus types, to develop a concise and easy-to-follow procedure. At present, this will still be a multistep procedure, in which the correct execution of each step is necessary before it is possible to proceed to the next, and the success of which therefore depends on some practice fractionations with colored proteins.

1) Polymerization. The gel concentration for preparative PAGE needs to be rationally optimized on the basis of prior analytical PAGE establishing, first, the optimal pH, and then, the optimal gel concentration for the separation of the protein of interest and its closest contaminants. The rationales, procedures and computer programs used in these optimizations have been previously described (6,10). Thus, for the time being, it appears reasonable to approximate the optimal gel concentration for preparative "elution-PAGE" (in contrast to analytical separations) as the "gel concentration for optimal separation, T_{max}" (10). As pointed out above, an analytical solution for any %T and the optimized gel volume at a particular current density in a particular buffer system can be approximated experimentally (11) or by computer simulation (10) after measurement of the zone spreading (diffusion coefficient) of the protein in polyacrylamide gel.

Preparative gels of 2 $\%C_{Bis}$ appear to give in general the best wall adherence above 4 %T. For open-pore ("non-restrictive") gels, the sticky and elastic N,N'-diallyltartardiamide (DATD) crosslinked gels appear preferable to gels crosslinked with Bis (55).

To obtain chracteristic R_f values for the eluting proteins, identical between preparative and analytical PAGE, it is important to conduct analytical and preparative PAGE under identical conditions; the only exception is that it appears necessary to double polymerization times for gels of 15-20 mm cross-section as compared to the 5-6 mm diameter gels used analytically in order to achieve an adequate polymerization efficiency. (This is a subjective estimate; measurements of polymerization efficiency (56) on wide as opposed to narrow diameter gels remain to be carried out.) Attention has to be paid to providing sufficient concentrations of the polymerization catalysts to achieve equal polymerization efficiency on top of the gel (where there is relatively much oxygen inhibition) and at the bottom (where there is less), in order to avoid the appearance of lowered mobilities at the gel bottom as noted by Schenkein *et al* (29).

Since preparative gels[3], compared to analytical ones, are heavy, close attention has to be paid to preparing immaculately clean, and possibly linear polyacrylamide coated (44) polar glass (Pyrex) walls for optimal wall adherence of the gel. Although it is not required, a mechanical support for the gel is highly desirable; a number of designs have been discussed above. The chemical nature of the polymerization support suitable for providing a mechanically smooth and chemically homogeneous Lower Gel surface has been previously discussed (6). It is more difficult but equally important than in analytical technique to overlayer the gel without disturbance of the gel surface. This is most easily done by proportioning pump (Appendix II and III of (20)).

Procedurally, the most important consideration in the formation of preparative PAGE gels is to avoid exertion of mechanical and hydrostatic pressure on the gel. One critical operation is the removal of the gel support without stressing the gel. This can be done by very slow and gentle separation of the gel from its lower support plate, trying to allow for a gradual air inlet into the space between gel and support surface.

Wall adherence is most precarious on the inner surface of gel cylinders with annular cross-section (Fig.1). Any movement of the center capillary and cooling tube must be avoided once the gel has formed.

Once the gel has been separated from its support surface, it should be hydrostatically balanced as soon as possible by filling the elution chamber with elution buffer and by proper height adjustment of the elution buffer reservoir so that the manometer tube connected to the elution chamber is at the same level as the gel surface (Fig.1). When the elution chamber is formed, it should be pre-filled with buffer, and the gel should be placed on top of the chamber very gently, with care not to press it abruptly into the buffer and against the O-ring seal of the elution chamber, or to tighten the wing screws connecting gel column and Lower Buffer reservoir in an abrupt or uneven manner. When the elution chamber is filled with elution buffer, the buffer is usually allowed to enter by gravity to avoid undue delay in filling. Also, the apparatus has to be tilted at this step to displace airbubbles from the elution chamber. Again, this needs to be done as gently as possible and with minimal delay of the hydrostatic equilibration of the gel.

Clearly, the elaborate precautions necessary to avoid pressure on the gel are incompatible with the ultimately necessary degree of procedural ease. Possibly, one could polymerize on top of a 50-80% sucrose solution in the preformed elution chamber with immediate hydrostatic equilibration of the gel after pouring of the polymerization mixture. This would avoid all the stresses on the gel during formation of the elution chamber after polymerization. Also, as pointed out above, wall adherence could be greatly strengthened through appropriate Nylon-mesh or other "non-obstructive" supports, and through notching or chemical treatment of the glass walls.

2) Electrophoresis. Electrophoresis at the preparative scale differs from the analytical scale only in two ways: Intermittent adjustment of the height of the elution buffer reservoirs has to be made to keep the gel hydrostatically equilibrated throughout electrophoresis, and wattage has to be reduced to allow for electrophoresis at maximal voltage without an appreciable cross-sectional temperature gradient in the gel. Usually, it is assumed, that deviations in temperature along the cross-section of the gel of more than $0.1^{o}C$ are intolerable for the purposes of maintaining zone symmetry and simultaneous entrance of the zone into the elution chamber. Since resolution improves with increasing voltage, maximal rates of heat dissipation from the gel are desirable. These can be obtained by maximal rates of coolant flow, preferably of more than 1.5 liter/min, across thin glass walls. Potentially more potent electronic methods of heat dissipation have been mentioned above. Thus, electrophoresis does not present any procedural difficulties, once the maximally tolerable wattage is determined and suitable conditions of cooling have been defined for any given apparatus and gel.[4] We have pointed out above that load should not exceed 0.1 mg/cm^2 of gel of the component of interest. However, total protein load should be of the order of 1-3 mg/cm^2 to saturate the polyacrylamide adsorption sites (20). Thus, when the sample contains few impurities, very small gels are required to minimize losses on adsorption, or addition of an easily removable "extraneous protein" may be necessary (20).

3) Elution. Elution of protein zones from the gel and their collection is the preparative PAGE problem *par excellence*, both in design and in procedure.

The elution buffer has to be made in such a way so as to induce the protein zones to enter into it from the gel. However, mobility in the buffer should be slow enough to prevent passage across the mere 1 mm height of the elution chamber in the time allotted for the flushing of the elution chamber, to prevent adsorption of the protein on the semipermeable membrane at the bottom of the elution chamber or electrophoretic migration out of the elution chamber in a membrane-less device. Also, the viscosity of the buffer should be high enough to simulate that of the gel and its pH and ionic strength should be similar enough to the gel so as to prevent formation of a bottom layer in the gel with drastically altered mobility characteristics for the protein. Evidently, these are in part contradictory requirements. Equally contradictory to the need for efficient flushing of the elution chamber is the desirability for minimal dilution of the eluate.

In practice, we have used the following compromise solution to the make-up and flow rate of the elution buffer. The pH of the elution buffer has been made identical to that of the gel. This means, in application to multiphasic buffer systems, that prior to the entrance of the front moving boundary into the elution chamber the elution buffer has the composition of the GAMMA phase, thereafter that of the PI phase. The necessary degree of protein retardation in the PI phase, preventing adsorption of the protein on the membrane, is provided through an increase in ionic strength (Appendix II and III of (20)). It should, however, be noted that this may lead to an undesirable retardation of the protein at the gel end, osmotic dehydration of the gel end, and consequent deterioration of resolution as noted by Nees (31). The density increase is brought about by the addition of sucrose to the PI phase buffer, i.e. at the time of protein elution. The degree of dilution can be reduced by a continuous flow rate decelerator (57), and potentially by discontinuous elution governed automatically by an optical sensing device in the elution chamber.

Thus, in spite of conflicting requirements for the design of elution buffers, generally applicable well-defined compromise solutions may be defined which make it unnecessary for the biochemical user of preparative PAGE to engage in methodological investigations on that point.

4) Eluate analysis, pooling of fractions and eluate concentration. In view of the limited load capacity of preparative PAGE and the substantial dilution of the eluate, the continuous optical detection devices developed for chromatography are inapplicable to preparative PAGE (6), both in continuous, discontinuous (29), decelerated continuous (57) or programmed discontinuous (8) elution. An exceptional case was reported by Reznick *et al* (45) in which widely divergent R_f values of the components allowed for enormous loads which could be visualized by continuous spectrophotometry. Potentially, it will be possible to develop a meaningful optical monitoring system prior to entrance of the protein into the elution flow system, i.e. at the time of its entrance into the elution chamber. At

the present, only 2 methods are available for a meaningful eluate analysis in preparative PAGE, i.e. one capable of determining which fraction contains the uncontaminated protein of interest. The first is rerun of fractions on analytical PAGE. Since the eluate is necessarily dilute, Stacking Gels of at least twice the volume of the eluate fraction are needed, requiring slab gels of at least 3 mm cross-section and slot formers of greater length than customarily used for PAGE on slabs. Since Stacking Gels are required, horizontal slab apparatus is not conveniently applicable. If dilution is great, it may even be necessary to employ large numbers of cylindrical gels to be able to allocate sample volumes of one or several milliliter containing the requisite microgram amounts of protein for staining. Evidently, eluate analysis should employ the gel concentrations for optimal resolution (T_{opt}) between the protein of interest and its closest contaminant (on either side). The second eluate analysis method is immunological. It excels in sensitivity but depends on the preparation of antibodies directed against the protein mixture loaded into preparative PAGE. This method has been described previously in detail (6,43).

Frequently, a first preparative PAGE step will fail to resolve the overlapping distributions of the protein of interest and its nearest contaminants. As in chromatographic fractionations, homogeneous protein is obtained as a rule upon pooling of ascending, center and descending cuts of the peak of interest from several runs and re-submission to preparative PAGE under the optimal conditions of pH and %T specific for each of the adjacent contaminants. In exceptional cases, the protein of interest, or its contaminants, can be specifically stacked and eluted in highly concentrated form as a stack; if no contaminants more slowly migrating than the unstacked protein of interest occur in the system, specific restacking of that protein is possible (Kapadia, G. and Chrambach, A., in preparation).

Eluate concentration can employ a number of methods, but by far the easiest is to place the eluate directly onto a large stacking gel and to collect the protein visualized by tracking dye into a collection cup of 1 ml volume (52). This method is maximally non-laborious and seems to provide high recovery.

5) _Purification, lyophilization, and storage of the eluted protein_. Eluates from polyacrylamide are contaminated by polyacrylamide-like material (6). This can be removed by gel filtration on purified Sephadex G-50 which at the same time serves to transfer the protein into volatile (home-made, residue free) ammonium bicarbonate buffer (21). The 1 ml concentrate is again diluted by the chromatographic step, but this is of no consequence since the step is followed by lyophilization or protein precipitation by alcohol, acetone, etc. and drying. Storage in liquid nitrogen seems singly suited for the maintenance of protein homogeneity over prolonged periods of time (6,58).

It can be concluded that generally applicable, not excessively laborious and high-recovery methods and procedures for preparative PAGE already exist. What does not exist, and is clearly desirable, is a modular apparatus assembly for separation, elution, eluate analysis, concentration, fraction purification, lyophilization and storage in an organized, assembly-line fashion. It would represent a first step toward the full automation of the process of protein isolation by preparative PAGE, and, at this time, a ready-made tool for the biochemical consumer.

PREPARATIVE GEL ISOTACHOPHORESIS AND ELECTROFOCUSING

When facing a pure "charge fractionation" problem, i.e. when the species to be separated have similar molecular sizes but differ in net charge, or when the analytical solution to pore size optimization based on the Ferguson plots of the protein of interest and its closest contaminants indicates an optimum at "0 %T" (6), the most efficient preparative isolations are those by isotachophoresis (ITP) or electrofocusing (EF). These two methods, and in particular the former, excel in load capacity over PAGE by a factor as much as 3 orders of magnitude, and are preferred over preparative PAGE.

In view of the relative advantage of charge fractionation methods with regard to load capacity, it appears advantageous to utilize size fractionation (gel filtration) for gross fractionation of multicomponent systems prior to preparative electrophoresis. The more effectively this is done, e.g. by use of very long gel filtration columns and/or recycling, the more effective is a subsequent high load capacity charge fractionation method in generating a homogeneous product (58). Preparative ITP and EF have been recently reviewed (35) and will therefore not be discussed explicitly. However, a complete, balanced and realistic view of present-day preparative gel methods does depend on their consideration, and reprints of ref. (35) will be provided upon request to the interested reader to allow him to take such a view of presently available macromolecular isolation methods. The review of these two methods by ref. (35) needs to be supplemented and updated by some considerations and references:

1) Isotachophoresis (ITPPA) and isoelectric focusing (IFPA) on polyacrylamide gel, in view of their load capacities 2 or 3 orders of magnitude higher than that of PAGE appear as the preparative electrophoretic methods *par excellence*.

With the sole exception of ITPPA separations (61) in the buffer systems of the multiphasic zone electrophoresis buffer output of Jovin *et al* (18), preparative ITPPA and IFPA do not exist as yet in a meaningful sense, having been applied to date to model systems only. No applications of either of the two families of preparative methods have been reported which would have achieved a biochemical isolation not previously solved or potentially solvable by other fractionation methods. In no case reported to date has the resolving power exceeded that of more conventional methods. Although some new ITPPA and IFPA methods have been shown (59,60) to work "in principle", no widely useful preparative procedures have as yet been developed. This is not being said with the purpose of discrediting the two methods (with the development of which we are intimately involved) but rather to discourage premature and unrealistic expectations.

2) In order to increase resolving power of ITPPA and IFPA, various measures have been applied which resulted in flattened pH gradients in both methods (60). It appears now that such pH gradient flattening and preparative applicability by "successive" zone elution techniques of ITPPA or "cascade stacking" are antagonistic. "Simultaneous" zone elution techniques appear suitable but have not been applied to this problem as yet.

3) Recent work by Kopwillem _et al_ (62) and by Svendsen (43) support the notion that preparative gel ITP with mixtures of aminoacid spacers (62) or Ampholine spacers (43) is presently available as a ready-to-use, powerful, fractionation method. This contrasting interpretation of the data is clearly due to differences in the applied criteria for evaluation. By our own criteria, ITP with mixtures of spacers fails to provide:

a) Evidence that separation is isotachophoretic by pinpointing the positions of the leading and trailing moving boundaries relative to the protein zones;

b) evidence of spacer specificity for a particular protein pair, i.e. demonstration of intermediate constituent mobility based on either the use of labeled spacers (Nguyen and Chrambach, in prepration) or determination, experimentally or by computation, of the constituent mobilities of the proteins to be separated and of the spacers;

c) evidence that the particular spacer load, and spacer multiplicity applied to the separation is compatible with attainment of the steady-state of ITP within practical limits of time and gel length;

d) resolving power and load capacity equal to or better than other presently available methods.

4) Preparative ITPPA and ITPA will require new apparatus. This apparatus will have to be small ("microgram preparative" by the standards of preparative PAGE) to accomodate milligram loads. This is particularly true for ITPPA without spacers where resolution is proportional to load (61,72). Furthermore, the swelling of protein zones in ITPPA, and the pH dependent differential swelling of the gel in IFPA raise new mechanical stability problems that need to be overcome. Finally, the disproportionate development of Joule heat within the protein zones of ITPPA raises some additional heat dissipation problems. In addition, the use of strongly acidic and basic anolyte and catholyte in IFPA, though avoidable (79), bars the use of plastic apparatus materials like Plexiglas and of the conventional glass-to-plastic bonding agents.

FOOTNOTES

[1]Systematic variation in the input can be evaluated against the predicted efficiency of the separation (resolving power in terms of the number of theoretical plates, N (17)). The input parameter "mobility of the front" is the value of RM(1,PI) of the Multiphasic Buffer Systems Output (18). "Velocity of the front" is given in the same output as "boundary displacement", ν (ml/Coulomb). Viscosity refers to the gel and is an estimate. All remaining input parameters, except D, are experimental and derived from the Ferguson plots. These are the retardation coefficients, K_R, and the y-intercepts on the Ferguson plots, Y_o, and the molecular radii obtained from K_R by means of a standard curve (16,19). Given these input parameters, the output yields, both numerically and graphically, the elution profiles and elution times of protein and contaminant.

[2]Provided upon request

[3]Procedural suggestions refer to the Jovin-type design (Fig.1).

[4]The producers of the preparative PAGE apparatus should provide graphs defining temperature inside the gel as a function of ionic strength and pH, thereby defining the heat dissipation characteristics of the apparatus.

ACKNOWLEDGEMENTS

We wish to thank Drs. Mike Conn, Larry Hjelmeland and Bob Mandle for a critical review of the manuscript.

REFERENCES

1. Smithies, O. a) (1955) Biochem. J., 61, 629-641; b) (1959) Adv. Prot. Chem., 14, 65-113.

2. Ornstein, L. and Davis, B. J. a) (1962) in Disc Electrophoresis, Preprinted by Distillation Products Industries, Rochester, NY; b) Ornstein, L. (1964) Ann. N. Y. Acad. Sci., 121, 321-349.

3. Raymond, S. and Weintraub, L. (1959) Science, 130, 711.

4. O'Farrell, P. H. (1975) J. Biol. Chem., 250, 4007-4021.

5. Chrambach, A. and Rodbard, D. (1971) Science, 1972, 440-451.

6. Chrambach, A., Jovin, T. M., Svendsen, P. J. and Rodbard, D. (1976) in Methods of Protein Separation, Catsimpoolas, N., ed., vol.2, Plenum Press, New York, NY, pp. 27-144.

7. Shuster, L. 91969) in Methods in Enzymology, Jakoby, W. B., ed., vol. 22, Academic Press, New York, NY, pp. 412-433.

8. Brownstone, A. D. a) (1976) Anal. Biochem., 70, 572-591; b) (1969) Anal. Biochem., 27, 25-46.

9. Jovin, T., Chrambach, A., and Naughton, M. A. (1964) Anal. Biochem., 9, 351-369.

10. Rodbard, D., Chrambach, A., and Weiss, G. (1974) in Electrophoresis and Isoelectric Focusing on Polyacrylamide Gel, Allen, R. C. and Maurer, H. R., eds., Walter de Gruyter, Berlin, pp. 62-105.

11. Marker, S. C., Melby, P. and Jahrling, P. B. (1977) Anal. Biochem., 82, 423-434.

12. Lunney, J., Chrambach, A., and Rodbard, D. (1971) Anal. Biochem., 40, 158-173.

13. Catsimpoolas, N., Griffith, A. 1., Williams, J. M., Chrambach, A., and Rodbard, D. (1975) Anal. Biochem., 69, 372-384.

14. Chen, B. (1978) Fed. Proc., in press.

15. Ben-David, M., Becker, R., Rodbard, D., and Chrambach, A. (1974) Endocrinol. Res. Commun., 1, 211-228.

16. Rodbard, D., and Chrambach, A. (1974) in Electrophoresis and Isoelectric Focusing on Polyacrylamide Gel, Allen, R. C. and Maurer, H. R., eds., Walter de Gruyter, Berlin, pp. 28-61.

17. Rodbard, D., and Chrambach, A. (1970) Proc. Nat. Acad. Sci. USA, 65, 970-977.

18. Jovin, T. M., Dante, M. L., and Chrambach, A. (1970) Multiphasic Buffer Systems Output, National Technical Information Service, Springfield, VA. 22151, PB NO. 196085-196091, 203016, 259309-259312.

19. Rodbard, D., and Chrambach, A. (1971) Anal. Biochem., 40, 95-134.

20. Kapadia, G., and Chrambach, A. (1972) Anal. Biochem., 48, 90-102 (Appendix II: Preparative PAGE Procedure: Apparatus B. Appendix III: Preparative PAGE Procedure: Apparatus C and D).

21. Yadley, R. A., Rodbard, D., and Chrambach, A. (1973) Endocrinology 93, 866-873 (Appendix I).

22. a) Raymond, S. and Jordan, E. M. (1966) Sep. Sci., 1, 95-111; b) Broome, J. and Jordan, E. M. (1966) Sep. Sci., 1, 319-328.

23. Stathakos, D. (1975) in Progress in Isoelectric Focusing and Isotachophoresis, Righetti, P. G., ed., Associated Scientific Publishers, Amsterdam, pp. 65-76.

24. Dudman, N.P.R. and Zerner, B. (1973) Anal. Biochem., 57, 14-26.

25. Sulitzeanu, D., Slavin, M. and Yecheskeli, E. (1967) Anal. Biochem., 21, 57-67.

26. Hodson, A. W. and Latner, A. L. (1971) Anal. Biochem., 41, 522-531.

27. Gordon, A. H. (1962) Biochem. J., 82, 531-540.

28. Polsky, F., Edgell, M. H., Seidman, J. G. and Leder, P. (1978), Anal. Biochem., in press.

29. Schenkein, I., Levy, M. and Weis, P. (1968) Anal. Biochem., 25, 387-395.

30. van Jaarsveld, P. P., van der Walt, B. J. and Le Roux, C. H. (1976) Anal. Biochem., 75, 363-373.

31. Nees, S. (1974) in Electrophoresis and Isoelectric Focusing on Polyacrylamide Gel, Allen, R. C. and Maurer, H. R., eds., Walter de Gruyter, Berlin, pp. 189-198.

32. Boyde, T.R.C. and Remtulla, M. A. (1973) Anal. Biochem., 55, 492-508.

33. Hjerten, S., Jerstedt, S., and Tiselius, A. a) (1969) Anal. Biochem., 27 108-129; b) (1965) Anal. Biochem., 11, 211-218.

34. van Welzen, H., and Zuidweg, M.H.J., (1974) Anal. Biochem., 59, 306-315.

35. Chrambach, A. and Nguyen, N. Y. (1977) in Proc. Internatl. Workshop on Technology for Protein Separation and Improvement of Blood Plasma Fractionation, Sept. 7-9, Reston VA, pp. XXX-XXX.

36. Gordon, A. H. and Louis, L. N. (1967) Anal. Biochem., 21, 190-200.

37. Vogel, O., and Reinmuth, H. (1975) Anal. Biochem. 62, 461-475.

38. Jovin, T. M. (1973) Biochemistry, 12, 871-898.

39. Svendsen, P. J. (1973) Sci. Tools, 20, 1-4.

40. Malacinski, G. M. (1970) Anal. Biochem., 38, 288-291.

41. Duesberg, P. m. and Rueckert, R. R. (1965) Anal. Biochem., 11, 342-361.

42. a) Hochstraesser, H., Skeggs, O. T., Jr., Lentz, K. E. and Kahn, J. R. (1963) Anal. Biochem., 6, 13-26. b) Hollmen, T. and Kulonen, E. (1966) Anal. Biochem., 14, 455-466.

43. Svendsen, P. J. (1978) Biochim. Biophys. Acta, in press.

44. Nguyen, N. Y., and Chrambach, A. (1978) Anal. Biochem., in press.

45. Reznick, A. Z., Allen, H. J. and Winzler, R. J. (1973) Anal. Biochem. 52, 395-401.

46. Radhakrishnamurthy, B., Dalferes, E. R. Jr. and Berenson, G. E. (1965) Biochim. Biophys. Acta, 107, 380-383.

47. Racusen, D. and Calvanico, N. (1964) Anal. Biochem., 7, 62-66.

48. Coy, D. H. and Wuu, T.-C. (1971) Anal. Biochem., 44, 174-181.

49. Bergrahm, B. and Harlestam, R. (1968) Sci. Tools, 15, 26-28.

50. Kawata, H., Chase, M. W., Elyjiw, R. and Machek, E. (1971) Anal. Biochem., 39, 93-112.

51. Suzuki, T., Benesch, R. E., Yung, S. and Benesch, R. (1973) Anal. Biochem. 55, 249-254.

52. Wachslicht, H. and Chrambach, A. (1978) Anal. Biochem., 84, 533-538.

53. McCormick, A., Miles, L.E.M. and Chrambach, A. (1976) Anal. Biochem., 75, 314-324.

54. McCormick, A. G., Wachslicht, H. and Chrambach, A. (1978) Anal. Biochem., 85, XXX-XXX.

55. Baumann, G. and Chrambach, A. (1976) Anal. Biochem., 70, 32-38.

56. Chrambach, A. and Rodbard, D. (1972) Sep. Sci., 7, 663-704.

57. Ellwein, L. B., Huff, R. W. and Chrambach, A. (1977) Anal. Biochem., 82, 46-53.

58. Ben-David, M. and Chrambach, A. (1977) Endocrinology 101, 250-261.

59. Nguyen, N. Y., Rodbard, D., Svendsen, P. J. and Chrambach, A. (1977) Anal. Biochem., 77, 39-55.

60. Nguyen, N. Y. and Chrambach, A. (1978) Fed. Proc., in press.

61. Baumann, G. and Chrambach, A. (1976) Proc. Nat. Acad. Sci. USA, 73, 732-736.

62. Kopwillem, A., Merriman, W. G., Cuddeback, R. M., Smolka, A.K.J., and Bier, M. (1976) J. Chromatography, 118, 35-46.

63. Maizel, J. V. Jr. (1964) Ann. N. Y. Acad. Sci., 121, 382-390.

64. Altschul, A. M., Evans, W. J., Carney, W. B., McCourtney, E. J. and Brown, H. D. (1964) Life Sci., 3, 611-615.

65. Smith, J. K. and Moss, D. W. (1968) Anal. Biochem., 25, 500-509.

66. Sun, S. M. and Hall, T. C. (1974) Anal. Biochem., 61, 237-242.

67. Posner, I. (1976) Anal. Biochem., 72, 491-501.

68. Lewis, U. J. and Clark, M. O. (1963) Anal. Biochem., 6, 303-315.

69. Karsnas, P. and Roos, P. (1977) Anal. Biochem., 77, 168-175.

70. Bont, W. S., Geels, J. and Rezelman, G. (1969) Anal. Biochem., 27, 99-107.

71. Weliky, N., Leaman, D. H. and Kallman, B. J. (1975) Anal. Biochem., 67, 507-514.

72. Voytek, P. (1972) Anal. Biochem., 47, 629-640.

73. Houghten, R. and Chrambach, A. (1977) Anal. Biochem., 77, 303-309.

74. Popescu, M., Lazarus, L. H. and Goldblum, N. (1971) Anal. Biochem., 40, 247-253.

75. Strauch, L. (1967) in Prot. Biol. Fluids Proc. Colloq. Bruges, Peters, H., ed., vol. 15, Pergamon Press, London, pp. 535-539.

76. Ray, D. K., Troisi, R. M. and Rapoport, H. P. (1969) Anal. Biochem., 32, 322-330.

77. Mann, M. B. and Huang, P. C. (1969) Anal. Biochem., 32, 138-144.

78. Pajdak, W. (1973) Clin. Chim. Acta, 48, 113-115.

79. Nguyen, N. Y. and Chrambach, A. (1977) Anal. Biochem., 79, 462-469.

80. O'Connell, P.B.H. and Brady, C. J. (1976) Anal. Biochem. 76, 63-73.

81. Anker, H. S. (1970) FEBS Lett. 7, 293.

82. Hansen, J. N. (1976) Anal. Biochem. 76, 37-44.

83. White, F. K. (1976) US Patent No. 3,960,499.

Electrokinetic Separation Methods,
P.G. Righetti, C.J. van Oss and J.W. Vanderhoff

ELECTROPHORESIS AND ISOELECTRIC FOCUSING OF INTERACTING SYSTEMS

JOHN R. CANN
Department of Biophysics and Genetics, University of Colorado Medical Center, Denver, Colorado 80262 (U.S.A.)

INTRODUCTION

Electrophoresis and isoelectric focusing are powerful methods for the separation and characterization of proteins and other macromolecules with respect to electrophoretic mobility and isoelectric point, respectively. In principle, a homogeneous protein will give electrophoretic and isoelectric focusing patterns showing a single peak, while a heterogeneous protein will show multiple peaks corresponding to separated components of the mixture. However, this behavior pertains to ideal situations uncomplicated by macromolecular interactions, and it is well documented in both theory and practice that interactions can give rise to patterns exhibiting multiple peaks which do not faithfully reflect the inherent state of homogeneity of the macromolecule [1-9]. Several classes of interaction can give multiple peaks. These include macromolecular association-dissociation; kinetically controlled isomerization; complex formation between different macromolecules; and macromolecular interactions induced by the binding of constituents of the supporting medium such as undissociated buffer acid or carrier ampholyte. It is the latter class of interactions which will be considered below with respect to detection, characterization, and implications for conventional applications of electrophoresis and isoelectric focusing.

ELECTROPHORESIS

Longsworth and Jacobsen [10] were the first to recognize that a homogeneous protein can give moving-boundary electrophoretic patterns showing multiple peaks due to interaction, but the nature of the interaction was not understood at that time. Some years later we [11] observed that a variety of highly purified proteins display bimodal, moving-boundary patterns in acidic media containing acetate or other carboxylic acid buffers. Evidence was advanced to support interpretation of the patterns in terms of rapid and reversible binding of undissociated buffer acid by the protein, with concomitant increase in the net positive charge on the protein molecule (decrease in net negative charge at pH values alkaline to the isoelectric point) [12] without significant change in its frictional coefficient. It was shown that the change in net charge is most probably due to binding of the acid to side-chain carboxyl groups with accompanying increase in their pK [13]. At the buffer concentrations used for electrophoresis binding induces subtle macromolecular structural changes as well [11, 13, 14], and higher concentrations cause major changes in the tertiary structure of ribonuclease A without substantial alteration in secondary structure [14]. The latter interaction was shown to be a cooperative phenomenon, but binding of undissociated acid to a protein need not necessarily be cooperative for resolution of bimodal reaction boundaries [13, 15].

A theoretical basis for interpretation of the electrophoretic patterns was provided [16] by a theory of electrophoresis of reversibly interacting systems of the type

$$P + nHA \rightleftarrows P(HA)_n \qquad (1)$$

where P represents a protein molecule or other macromolecular ion in solution and $P(HA)_n$ its complex formed by binding n moles of a small, uncharged constituent, HA, of the solvent medium, *e.g.*, undissociated buffer acid. It is assumed that P and $P(HA)_n$ possess different electrophoretic mobilities and that equilibrium is established instantaneously. The calculations account for the essential features of the electrophoretic behavior of proteins in acidic media containing varying concentrations of carboxylic acid buffers. In particular, they predict the experimental observation that a single macromolecule can give moving-boundary patterns showing two well-resolved peaks due to reversible and rapid interaction with the buffer acid.

Over the years moving-boundary electrophoresis has been superseded by a variety of zone electrophoretic techniques which enjoy myriad applications to biologic problems. Accordingly, our calculations on interacting systems were extended [17] to include zone electrophoresis. As evidenced by the theoretical patterns presented in Fig. 1, zone electrophoretic patterns can also show two well-resolved peaks due to interaction of a single macromolecule with a small neutral molecule in accordance with Reaction 1. This is so despite rapid re-equilibration during differential transport of P and $P(HA)_n$. Resolution is dependent upon changes in the concentration of HA accompanying re-equilibration and maintenance of the resulting concentration gradients of the electrically neutral molecule along the electrophoresis column (dashed curves in Fig. 1), the peaks in the pattern corresponding to different equilibrium compositions and not simply to P and $P(HA)_n$. Even though the protein concentration eventually becomes very small between the peaks, it never vanishes. In other words, the two peaks constitute a single reaction zone, but in practical terms it can be said that a single macromolecule can give two zones due to reversible and rapid interaction with the buffer medium. An experimental verification of this important prediction is presented in Fig. 2.

It now appears that multiplicity of zones due to interaction is of rather frequent occurrence. Moreover, the zones may be quite well resolved as in Fig. 2b and could easily be misinterpreted as indicating inherent heterogeneity. Clearly, unambiguous interpretation of zone patterns depends upon fractionation experiments. The protein eluted from each unstained zone is resubjected to zone electrophoresis

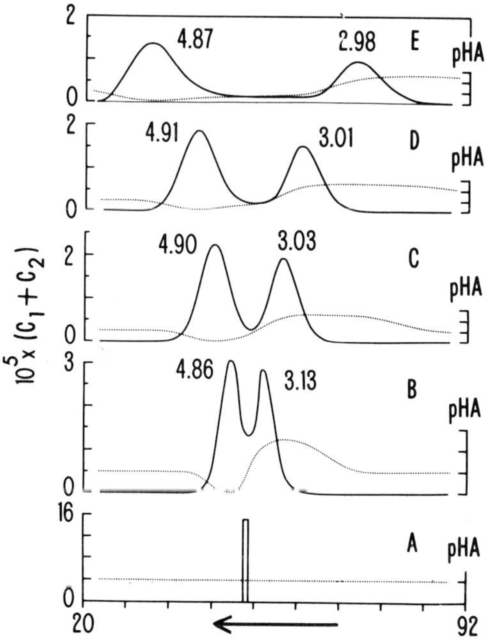

Fig. 1. Time course of development of the theoretical zone electrophoretic pattern of a single macromolecule interacting rapidly and reversibly with a small neutral molecule (*e.g.*, undissociated buffer acid) in accordance with Reaction 1. ——, electrophoretic pattern; – – – –, p*HA*. Time of electrophoresis: A, t=0; B, 3.46×10^3 sec; C, 6.92×10^3 sec; D, 1.04×10^4 sec; E, 2.01×10^4 sec. Unless stated otherwise, all theoretical transport patterns presented herein are plots of constituent concentration of macromolecule against position. Adapted from Cann and Goad [17].

in the same buffer used for the original separation. Since the production of multiple zones due to interaction will, in general, be sensitive to protein concentration [19], ideally the fractions should be reconcentrated by miniaturized ultrafiltration for example. Alternatively, the test can be calibrated by analyzing the unfractionated material at several concentrations. Fractionation is then carried out at a concentration sufficiently high that the concentration of the fractions falls within the range over which the original material exhibits multizoning. For interactions, the fractions will behave electrophoretically like the unfractionated material and show multiple zones (Figs. 2c and 2d), while for heterogeneity a single zone will be obtained. In practice this interpretation of the fractionation test is subject to the proviso that the reacting protein system is not microheterogeneous with respect to the conditions required for the interaction [20]. If this should obtain, even a rapidly interacting system may behave deceptively like a noninteracting, heterogeneous system in the sense that each fraction will show a single zone as in the case of the N-F transformation of bovine serum albumin [21].

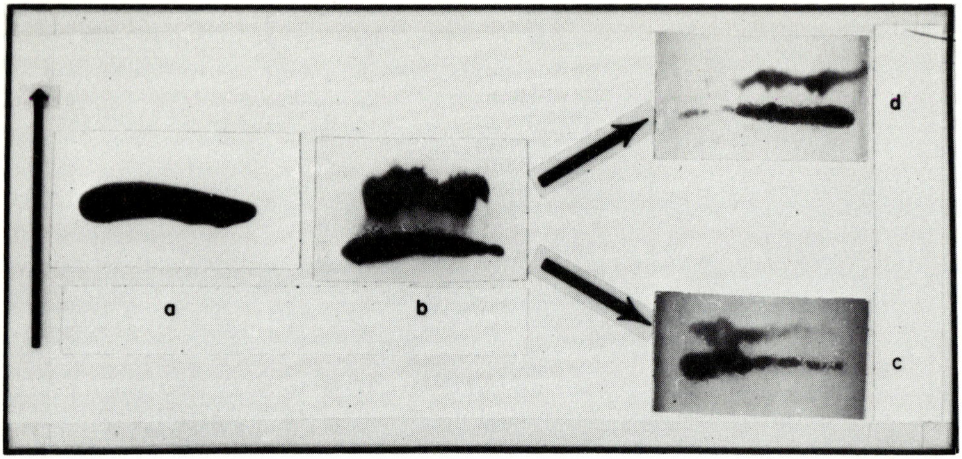

Fig. 2. Demonstration that the two electrophoretic zones shown by bovine serum albumin (BSA) in phosphate-borate buffer on cellulose acetate arise from reversible interaction of the protein with boric acid or borate anion and are not due to an inherent heterogeneity: a, zone pattern of BSA in 0.0043M Na_2HPO_4 -0.003M NaH_2PO_4, pH 6.07; b, pattern in 0.012M Na_2HPO_4 -0.0081M NaH_2PO_4 -0.4M H_3BO_3, pH 6.15; c, zone pattern in phosphate-borate buffer of material eluted from the slower migrating zone of pattern b; d, pattern of material eluted from faster zone. Vertical arrow indicates direction of migration. From Cann[19].

In conventional applications of electrophoresis, it is obviously desirable to avoid conditions conducive to the production of multizoned patterns which do not reflect the actual state of homogeneity of the protein. This can often be achieved by judicious choice of pH, ionic strength and buffer: Multiple zones due to protein-buffer interactions have been encountered with carboxylate, glycine (at acid pH, extrapolation from moving-boundary), borate, Tris-borate, phosphate-borate and even phosphate buffers; while barbital and Tris buffers appear to be free of these complications. If strong interaction with the buffer is unavoidable, an attempt can be made to minimize interpretive difficulties by lowering the protein concentration and/or increasing the buffer concentration. At a sufficiently low protein to buffer concentration ratio an inherently homogeneous protein will, in principle, show a single zone. However, such conditions may be difficult to realize in practice, and failure to do so must not be accepted in lieu of fractionation as indicative of inherent heterogeneity.

ISOELECTRIC FOCUSING

Above we have seen that the resolution of bimodal reaction boundaries and zones due to small molecule-macromolecule interaction depends upon generation of stable concentration gradients of unbound small molecule along the electrophoresis column by re-equilibration during differential transport of macromolecular reactant and product. In contrast to electrophoresis, steady-state gradients of carrier ampholytes and pH are inherent to the method of isoelectric focusing, thereby predisposing it to complications arising from macromolecular interactions induced by binding of carrier ampholyte and from pH-dependent conformational transitions. There are several examples in the literature of isoelectric focusing patterns exhibiting multiple peaks due to reversible interaction as ascertained by fractionation experiments. These include an acidic protein from wool [22], some basic microbial proteases [23], bovine serum albumin [24,25], and t-RNA [26,27]. In each case, binding of carrier ampholytes ("Ampholine") has been implicated, although in no case is the mechanism of interaction understood. In the other vein Talbot [28] has reported an instructive study of the 12S subunit protein of the capsid of foot-and-mouth disease virus, which demonstrates that pH-dependent conformational transitions can give isoelectric focusing patterns showing two peaks. Recently, we [5-8] have formulated a phenomenological theory of isoelectric focusing which predicts multiple peaks due to any of several types of rapidly reversible interaction. The results of these calculations, which are summarized below, have practical implications for unambiguous interpretation of isoelectric focusing patterns and furnish explanations for hitherto puzzling experimental observations.

Carrier ampholyte-induced interactions

Theoretical transient and equilibrium isoelectric focusing patterns have been computed numerically for both macromolecular isomerization and association-dissociation reactions induced by binding of carrier ampholyte molecules. The steady-state parameters of the isoelectric focusing column and the electrophoretic velocity gradients of the macromolecular species are presented in Fig. 3. Most of the calculations assume specific binding by an amphoteric macromolecule of a single species of carrier ampholyte (or a closely spaced family with Gaussian envelope) position in the region of its isoelectric point (pI). Both broad and sharp distributions of ampholyte (Fig. 3A), roughly corresponding to the extreme widths of the discrete, steady-state zones of "Ampholine" visualized experimentally, have been examined.

Many of the calculations simulate the experimental procedure in which a narrow zone of macromolecule is inserted into the pH gradient at some chosen position. Those for a broad initial zone can be placed into correspondence with the alternative procedure in which both the macromolecule and the carrier ampholytes are distributed uniformly throughout the column initially [5,7]. Of course, the equilibrium isoelectric focusing pattern is the same for all arbitrary initial distributions of a given amount of macromolecule, although the paths to equilibrium may differ.

Isomerization. Let us first consider the two-state, carrier ampholyte-induced macromolecular isomerization reaction

$$M + nA \rightleftarrows NA_n \qquad (2)$$

where M is an amphoteric macromolecule, and A is a carrier ampholyte of which a fixed number, n, are bound into the complex NA_n, in which N is an isomer with different electrophoretic mobilities at the same pH-values and a different pI than M. Unless stated otherwise, the calculations considered here are for $n=4$, and the pI of N lower than M. (The locations of the pI's in the isoelectric focusing column are indicated by vertical arrows in the several figures.) The calculations are for the limit in which the rates of reaction are so fast that, in effect, chemical equilibrium is established instantaneously.

Computed equilibrium isoelectric focusing patterns are displayed in Figs. 4 (curve g), 5A (curve d) and 5B (curve b). It is immediately apparent that carrier ampholyte-induced isomerization reactions can give patterns showing two peaks for either broad or sharp distribution or ampholyte, although the peaks are more highly resolved in the latter case. Resolution of the peaks can be understood in terms of a dynamic equilibrium situation, in which the sense of the steady-state distribution of ampholyte governs the local equilibrium composition of macromolecule along the isoelectric focusing column. Thus, mass action favors NA_n in the region of its pI where the ampholyte concentration is high, while M is favored in the region of its pI where the ampholyte concentration is relatively low. This principle acting in concert with transport of individual macromolecular species, toward their respective pI's for electrophoretic transport (net diffusional and electrophoretic transport of constituent macromolecule being zero), prescribes a higher concentration near the ends of the pattern than in the middle. In other words, the two peaks correspond to different equilibrium compositions each enriched in the isomer of proximal pI (Fig. 4B), and not to separated isomers. That is to say, the two peaks constitute a focused bimodal reaction zone. In practice, however, a pattern such as curve d in Fig. 5A could easily be mistaken for inherent heterogeneity. (See Cann and Stimpson[5] for conditions which can give a unimodal reaction zone focused midway between the pI's of M and NA_n.)

The time-course of approach to the bimodal equilibrium pattern follows biphasic kinetics (Figs. 4A and 5). During the first phase, the transient isoelectric focusing pattern resolves into two peaks of about the same size at a relatively rapid rate controlled by electromigration of the two isomers. During the slower second phase, which is diffusion dominated, one peak grows at the expense of the other. Briefly, the biphasic nature of the kinetics is due to the fact that the constituent velocity of the macromolecule vanishes at a point (x_i) in the column midway between the pI's so that material can be transported across x_i only by diffusion. The better resolved the peaks are, the smaller at x_i is the concentration gradient of macromolecule upon which the diffusional flow depends; and, thus, the slower the rate of the second kinetic phase. Indeed, comparison of Figs. 4A, 5A, and 5B for the effect of ampholyte distribution and electric field strength upon these kinetic parameters

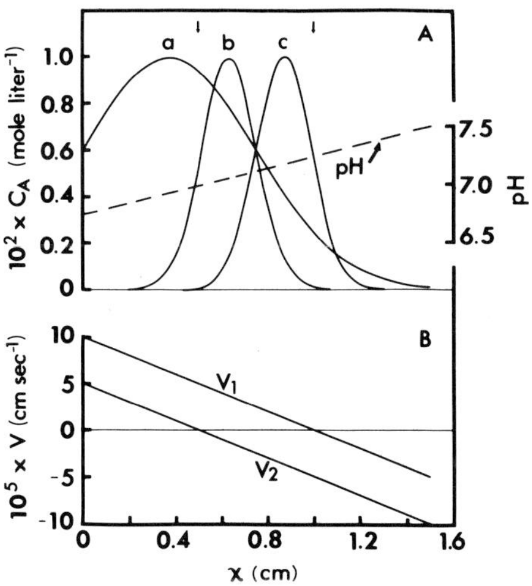

Fig. 3. Parameters of the isoelectric focusing calculations. A - Environment created by the steady-state distribution of carrier ampholytes: a, broad Gaussian distribution of carrier ampholyte used to calculate the isoelectric focusing patterns presented in Figs. 4, 9B, 10 and 11; b, sharp distribution of ampholyte pertaining to Figs. 5, 6 and 9A; c, sharp distribution pertaining to Fig. 12. B - Electrophoretic velocity gradients of macromolecular species: V_1, *M* in Figs. 4-6A and 9-12; V_2, *N* in Figs. 4-6A and M_2 in Figs. 9-12. Adapted from Cann, Stimpson and Cox [7].

indicates that, in certain experimental situations, it may not be possible to realize the equilibrium isoelectric focusing pattern in a practical length of time. In addition to the foregoing considerations, the time-course of approach to equilibrium is dependent upon the initial distribution of macromolecule in the isoelectric focusing column. Whereas the kinetics for a broad initial zone encompassing both pI's are essentially the same as those just described for a narrow zone centered between the pI's, a rather different picture is seen when a narrow initial zone is inserted into the pH gradient either anodically or cathodically to both pI's. As illustrated in Fig. 6A, a major peak is stationed just internally to the proximal pI during the first kinetic phase, after which a second peak grows very slowly in the vicinity of the distal pI. Since this slow growth is due to the almost vanishing gradient of macromolecule at x_i, it can be speeded up to some extent by decreasing the electric field strength thereby increasing the breadth of the peaks, which, in turn, increases the gradient at x_i (Fig. 6B). Nor is it a foregone conclusion that in practice the rate of approach to equilibrium will necessarily be so slow. As already stated, the calculations are for the limit of instantaneous establishment of chemical equilibrium, but actually the rates of reaction need not be rapid. For

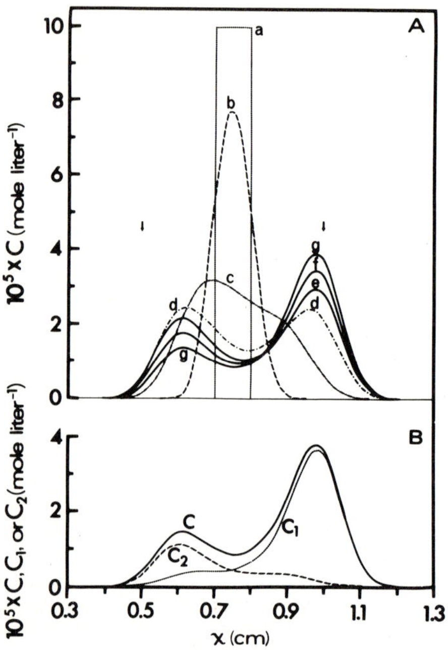

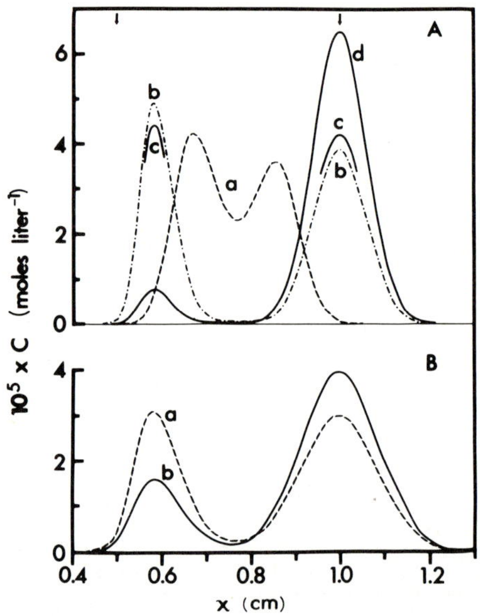

Fig. 4. (left). Theoretical isoelectric focusing patterns for two-state, carrier ampholyte-induced macromolecular isomerization (Reaction 2); broad distribution of ampholyte, curve a in Fig. 3A. A - Time-course of approach to the equilibrium pattern: a, t=0; b transient pattern at $t=2.5\times10^3$ sec; c, 1.75×10^4 sec; d, 5×10^4 sec; e, 1×10^5 sec; f, 2×10^5 sec; g, equilibrium pattern. B - Transient pattern at 4×10^5 sec showing concentration profiles of constituent macromolecule, C; species *M*, C_1; and species NA_4, C_2. From Cann and Stimpson [5].

Fig. 5. (right). Isoelectric focusing patterns calculated for two-state, ampholyte-induced isomerization (Reaction 2); sharp distribution of ampholyte, curve b in Fig. 3A; t=0 as in Fig. 4. A - Time-course of approach to the equilibrium pattern: a, $t=5\times10^3$ sec; b, 2×10^5 sec; c, 4×10^5 sec; d, equilibrium pattern; similar results for $n=2$. B - Increased rate of approach to equilibrium at halved electric field strength: a, transient pattern at 2×10^5 sec; b, equilibrium pattern. From Cann and Stimpson [5].

sufficiently slow rates of reaction the constituent velocity of the macromolecule will not vanish at x_i during the approach to the equilibrium pattern (although it will do so once equilibrium is attained). In that case, the rate of approach will be faster than predicted by the calculations in hand.

Having considered the region of the isoelectric focusing column in the immediate vicinity of the pI's, we turn to another aspect of isoelectric focusing; namely, the migration of the zone of macromolecule down the column enroute to its isoelectric region. Of interest here is possible isomerization of the macromolecule induced by binding of carrier ampholytes positioned along the route. Conceivably,

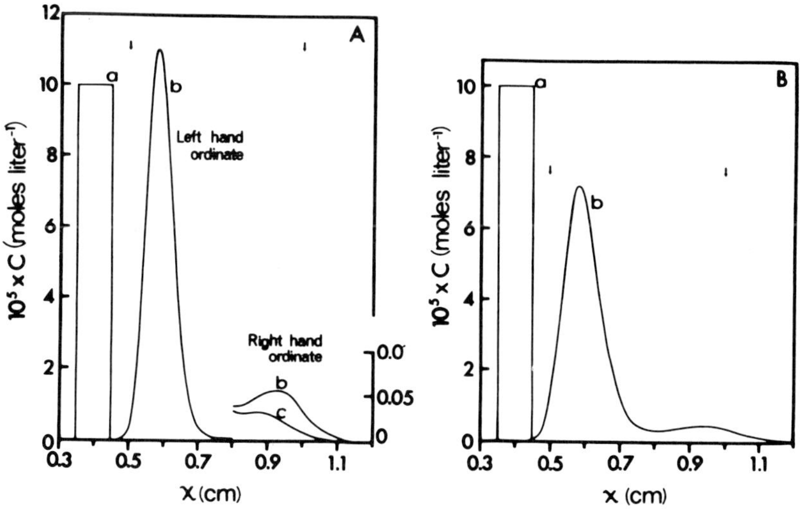

Fig. 6. Time-course of approach to isoelectric focusing equilibrium from an initially narrow zone of macromolecule positioned anodically to both pI's; two-state ampholyte-induced isomerization (Reaction 2). A - Distribution of ampholyte and other parameters as in Fig. 5A: a, t=0; b, 4×10^4 sec; c, 3×10^4 sec, equilibrium pattern given by pattern d in Fig. 5A. B - Effect of halving the electric field strength: a, t=0; b, 8×10^4 sec, which is the field-equivalent time of pattern b in Fig. 6A; equilibrium pattern given by pattern b in Fig. 5B. From Cann and Stimpson [5].

such interaction could involve either a specific ampholyte or all of the ampholytes in accordance with the reversible, two-state Reaction 2. The continuously changing profile of the zone as affected by interaction with all of the ampholytes positioned along the route of migration is shown in Fig. 7. The zone of macromolecule successively hypersharpens and resolves into two peaks, which then coalesce into one, each time it crosses one of the peaks in the distribution of carrier ampholytes. Interaction with a specific ampholyte gives only a single such cycle of resolution followed by coalescence during migration down the column. Recurrent cycles of resolution-coalescence of a zone of DNT insulin moving down a gel column has been observed by Mayers [29], while a single cycle has been observed by Catsimpoolas [30] for a protein in a density-gradient column. Our calculations furnish an explanation for these observations in terms of interaction.

Finally, a macromolecule, which complexes reversibly with several species of carrier ampholyte located at different positions along the isoelectric focusing column, can give an equilibrium isoelectric focusing pattern showing multiple peaks provided that each species of ampholyte induces a different isomer with distinctive pI. The number of peaks in the patterns depends upon the number of carrier ampholytes with which the macromolecule can form complexes, the strength of complex formation and the concentrations of the several carrier ampholytes. An illustrative example is the multi-state isomerization reaction set

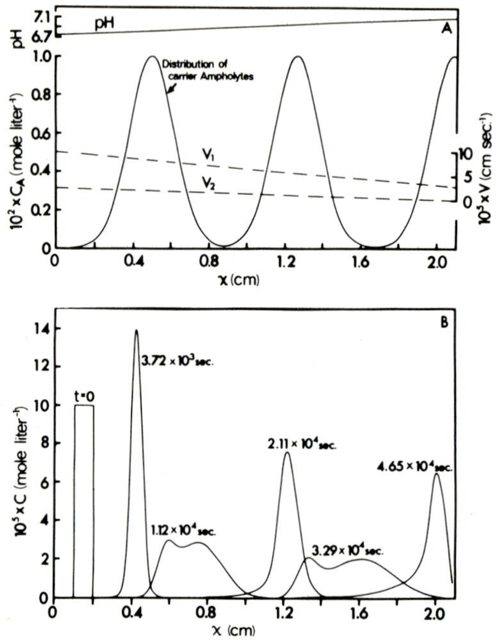

Fig. 7. Effect of interaction with all carrier ampholytes (in accordance with Reaction 2) upon the shape of a zone of macromolecule migrating down an isoelectric focusing column enroute to its isoelectric region. A - Environment created by the steady-state distribution of carrier ampholytes and electrophoretic velocities: V_1, velocity of M; V_2, NA_4. B - Transient profiles of macromolecule concentration. From Cann and Stimpson [5].

$$\begin{array}{lll} M+4A_1 \rightleftarrows N(A_1)_4 & ,K_1 , & \\ M+4A_2 \rightleftarrows P(A_2)_4 & ,K_2 , & \\ P(A_2)_4+4A_3 \rightleftarrows Q(A_3)_4+4A_2 & ,K_3 , & (3) \\ N(A_1)_4+4A_4 \rightleftarrows R(A_4)_4+4A_1 & ,K_4 , & \end{array}$$

in which M is an amphoteric macromolecule; A_1, A_2, A_3 and A_4 are four species of ampholyte; N, P, Q and R in the four macromolecule-carrier ampholyte complexes, *e.g.*, $N(A_1)_4$, are isomers; and the K_i's are equilibrium constants. The idealized distribution of carrier ampholytes is given by the dashed curve in Fig. 8, and the computed equilibrium isoelectric focusing pattern by the solid curve. The pattern exhibits five well-resolved peaks, one corresponding to the uncomplexed macromolecule and four due to the four macromolecule-carrier ampholyte complexes. Moreover, inductive reasoning leads to the generalization that, for interaction with p species of ampholyte, the pattern could show p+1 peaks under appropriate conditions.

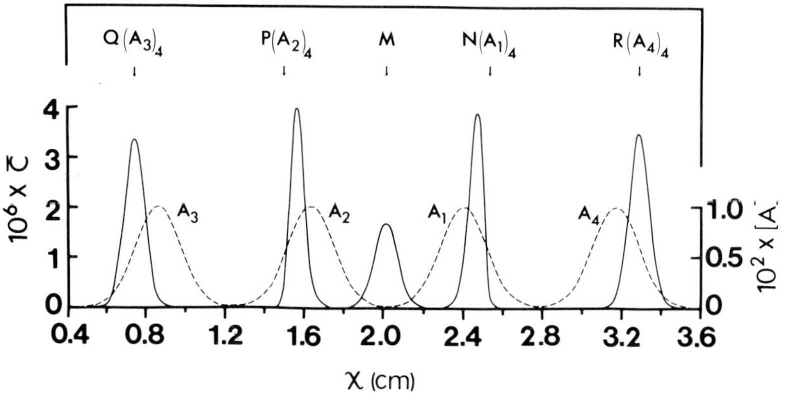

Fig. 8. Equilibrium isoelectric focusing pattern (——) calculated for multistate, ampholyte-induced isomerization (Reaction Set 3); positions of pI's of indicated macromolecular species designated by vertical arrows. $K_1=K_2=1.48\times10^9M^{-4}$ and $K_3=K_4=0.25$. From Hare, Stimpson and Cann [8].

Association-dissociation. Although it is not generally possible to predict *a priori* the relationship of the electrophoretic properties of an aggregated protein to those of the monomer, it is clear that changes in state of association sometimes do change the electrophoretic mobility in free solution and the pI. Accordingly, we have extended our theoretical calculations to include the isoelectric focusing behavior of the rapidly reversible, ampholyte-induced macromolecular dimerization reaction

$$2M+4A \rightleftarrows M_2A_4 , \qquad (4)$$

for the case where the pI of M_2A_4 is lower than for M. Because of the non-linearity of the problem direct calculation of equilibrium isoelectric focusing patterns is a formidable mathematical problem. The tack was to compute transient isoelectric focusing patterns numerically as a function of time thereby constructing the path to equilibrium. Convergence to the equilibrium pattern was tested by comparison of the amplitudes and positions of the extrema in the transient patterns with those predicted for the equilibrium pattern by an analytical relationship. In all cases considered here equilibrium had been virtually attained in the longest time of focusing shown in the figures.

The most provocative result is that during the approach to equilibrium one of the two peaks in the transient focusing pattern grows at the expense of the other, which under certain conditions may disappear completely as illustrated in Fig. 9. The latter behavior can be understood in terms of the coupling between transport and the principle of mass action, which stipulates that, given the assigned value of the equilibrium constant, relatively high concentrations of both macromolecule and carrier ampholyte are required for stability of the dimer. At the low concentration of macromolecule applied to the column the bimodality of the transient pattern cannot be sustained against: (a) the tendency of the dimer to dissociate in the decreasing concentration gradient of ampholyte which obtains between the pI's of dimer and monomer and (b) the removal of monomer from the local reaction arena by transport towards its pI. Accordingly, the transient pattern converges upon a symmetrical unimodal reaction zone focused close to the pI of monomer. When the concentration of macromolecule is increased, other things held constant, the bimodality of the transient pattern is preserved in the equilibrium pattern (Fig. 10). With further increase in concentration, the peak proximal to the pI of dimer grows in size (Fig. 11, which also illustrates that the pattern is a focused, bimodal reaction zone) until at sufficiently high concentration the equilibrium pattern would show a single peak focused close to the pI of dimer.

The results presented in Figs. 10 and 11 are for a broad distribution of carrier ampholyte; had the ampholyte distribution been sharp, the peaks would have been more highly resolved. This effect is illustrated in Fig. 12 for the ampholyte-induced dissociation of a subunit macromolecule into its isoelectrically identical half-mers

$$M_2+4A \rightleftarrows 2MA_2 \quad . \tag{5}$$

<u>pH-Dependent conformational transitions</u>

As described earlier, macromolecular isomerization (Reaction 2) can give well-resolved, bimodal focusing patterns when the sense of the steady-state gradient of ampholyte to which the interaction is coupled, dictates stabilization of the isomers in the vicinity of their respective pI's. An analogous behavior is to be expected for the rapidly equilibrating, pH-dependent conformational transition

$$M+nH^+ \rightleftarrows NH_n^+ \quad , \tag{6}$$

where once again M is an amphoteric macromolecule, while N is its conformer of lower pI formed *via* cooperative binding of n hydrogen ions. For such a transition, however, the value of n would have to be much larger than for ampholyte-induced isomerization, because the gradient of H^+ along the isoelectric focusing column is shallow compared to the gradients of ampholyte. These expectations are borne out by the theoretical isoelectric focusing patterns displayed in Figs. 13 and 14: Note especially the biphasic time-course of approach of the transient patterns to the bimodal equilibrium pattern and the strong dependence of the rate of the second phase of approach upon the position of the narrow initial zone in the pH gradient. (Here too the rate of the second phase can be speeded up by decreasing the electric field strength.) It is gratifying that these calculations accurately simulate the experimentally observed isoelectric focusing behavior of the 12S protein subunit of

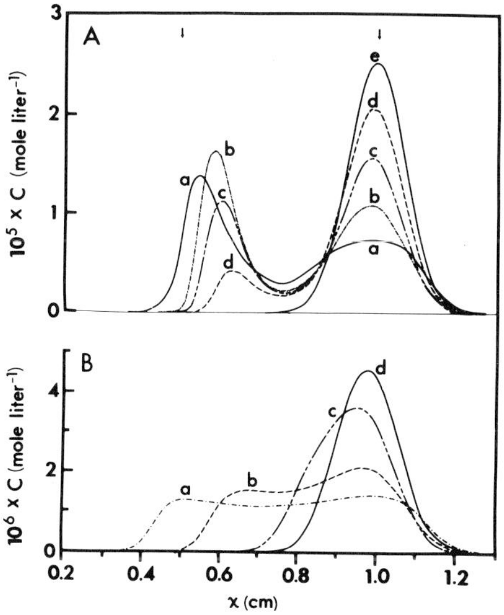

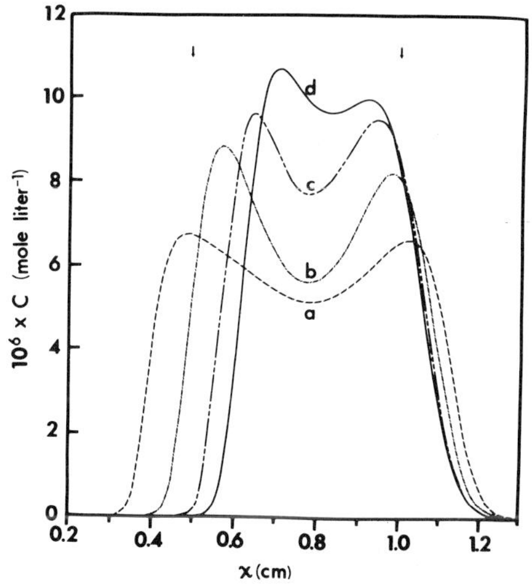

Fig. 9. (left). Isoelectric focusing patterns calculated for carrier ampholyte-induced macromolecular dimerization (Reaction 4). A - Sharp distribution of ampholyte (curve b in Fig. 3A); value of apparent dimerization constant at inflection point of ampholyte distribution, $K'=1\times10^5 M^{-1}$; concentration of macromolecule applied to column in a broad initial zone encompassing both pI's, $C^o=5\times10^6 M$: a, 3.75×10^3 sec; b, 1×10^4 sec; c, 3×10^4 sec; d, 6×10^4 sec; e, 1.8×10^5 sec. B - Broad distribution of ampholyte (curve a in Fig. 3A); $K'=1\times10^5 M^{-1}$; $C^o=1\times10^{-6} M$: a, 3.75×10^3 sec; b, 1×10^4 sec; c, 2.5×10^4 sec; d, 1×10^5 sec. From Cann, Stimpson and Cox [7].

Fig. 10. (right). Isoelectric focusing patterns calculated for carrier ampholyte-induced macromolecular dimerization (Reaction 4). Same conditions as in Fig. 9B except that $C^o=5\times10^{-6} M$: a, 3.75×10^3 sec; b, 1×10^4 sec; c, 2.5×10^4 sec; d, 1×10^5 sec. From Cann, Stimpson and Cox [7].

the capsid of foot-and-mouth disease virus, which undergoes a pH-dependent conformational change below pH 7: Compare pattern e in Fig. 13B and the patterns in Fig. 14 with the experimental ones reported by Talbot [28].

Theoretical isoelectric focusing patterns have also been computed for the set of two sequential transitions

$$\begin{aligned} M+nH^+ &\rightleftarrows NH_n^+ \quad , \\ NH_n^+ + mH^+ &\rightleftarrows PH_{n+m}^+ \quad , \end{aligned} \qquad (7)$$

in which M, N and P are conformers with different pI's, which decrease in the same order. As shown in Fig. 15, systems of this type can give equilibrium patterns exhibiting three well-resolved peaks. Furthermore, inductive reasoning leads to the conclusion that, for a sequence of q transition, the pattern could show q+1 peaks under appropriate parametric conditions.

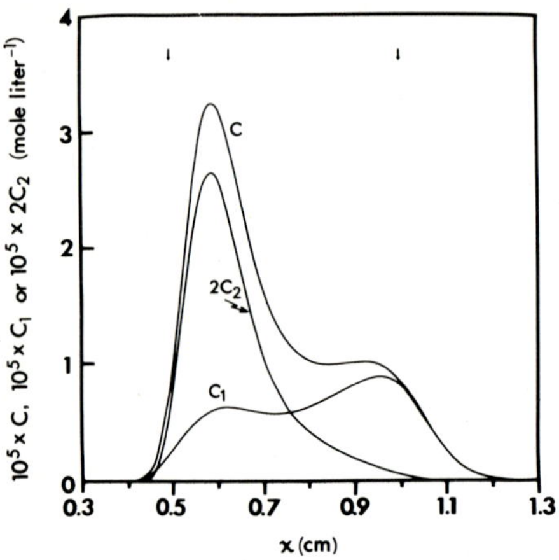

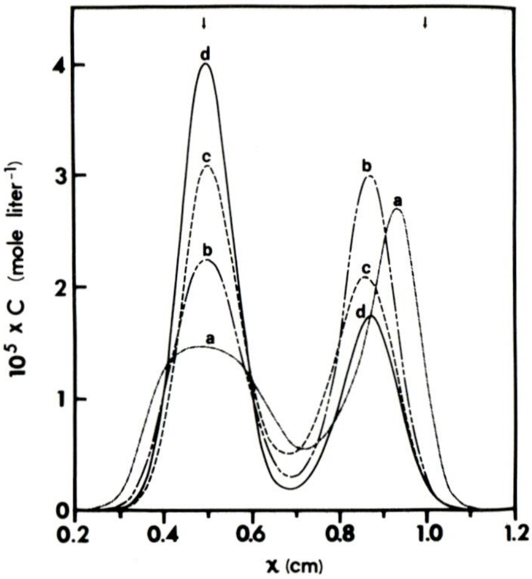

Fig. 11. (left). Isoelectric focusing pattern calculated for carrier ampholyte-induced macromolecular dimerization (Reaction 4) showing concentration profiles of constituent macromolecule, C; species M, C_1; and species M_2A_4, $2C_2$. Same conditions as in Fig. 9B except $C^o=1\times10^{-5}M$; $t=2\times10^5$ sec. From Cann, Stimpson and Cox [7].

Fig. 12. (right). Isoelectric focusing patterns calculated for carrier ampholyte-induced dissociation of a subunit macromolecule into its isoelectrically identical half-mers (Reaction 5). Sharp distribution of ampholyte (curve c in Fig. 3A): a, 3.75×10^3 sec; b, 1×10^4 sec; c, 3×10^4 sec; d, 1×10^5 sec. From Cann, Stimpson and Cox [7].

Practical implications

The new insights provided by these calculations have important implications for the many analytical and preparative applications of isoelectric focusing to biochemistry and molecular biology. In particular, a single protein or other macromolecule can give both transient and equilibrium isoelectric focusing patterns showing two or more peaks due to rapidly reversible interactions induced by binding of either carrier ampholyte or hydrogen ions. Moreover, their profiles ape those for a mixture of noninteracting proteins with different pI's and could easily be misinterpreted as indicative of inherent heterogeneity. Unambiguous interpretation depends upon fractionation experiments in which protein is isolated from each peak. The resulting fractions are reconcentrated by either precipitation-dissolution or miniaturized ultrafiltration, a measure which takes cognizance of possible macromolecular association-dissociation, and resubjected to isoelectric focusing taking care to position the initial zone as in the original separation. For interaction, each of the fractions will behave like the starting material and show two or more peaks; while for heterogeneity a single peak focused at a reproducible pI

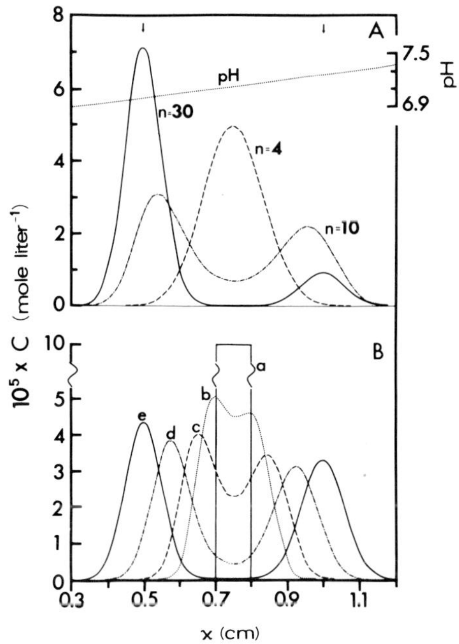

Fig. 13. Isoelectric focusing patterns calculated for the pH-dependent conformational transition given by Reaction 6. A - Equilibrium patterns for indicated values of n. B - Time-course of approach to the equilibrium pattern shown for $n=30$ in Fig. 13A: a, t=0; b, 1.25×10^3 sec; c, 5×10^3 sec; d, 1.25×10^4 sec; e, 1×10^5 sec; the amount of material in each peak changed by only ∿1.5% between 1×10^5 and 4×10^5 sec. From Stimpson and Cann [6].

will be obtained. Both theory and experiment [28] prescribe that the starting material and its fractions be focused from initial zones located at different positions along the isoelectric focusing column. The investigation of Talbot [28] on the pH-dependent conformational transition of the 12S subunit protein of foot-and-mouth disease virus is exemplary in the way these procedural variations were combined systematically with fractionation and ultracentrifugation to leave no doubt as to the interacting nature of the system.

A corollary pertaining to the routine determination of pI's is that the sample should be focused from different positions along the column. If, for example, an interacting protein were to be focused only from a point cathodic (or anodic) to its isoelectric region, the interaction might well escape attention (Figs. 6 and 14), in which case the apparent pI of the presumed noninteracting protein could be in error, perhaps, by as much as ± 1.5 pH units (See Fig. 21.1 of Talbot [28].).

Throughout, it has been emphasized that interacting systems can give rise to multimodal transient and equilibrium focusing patterns, but it should be noted that

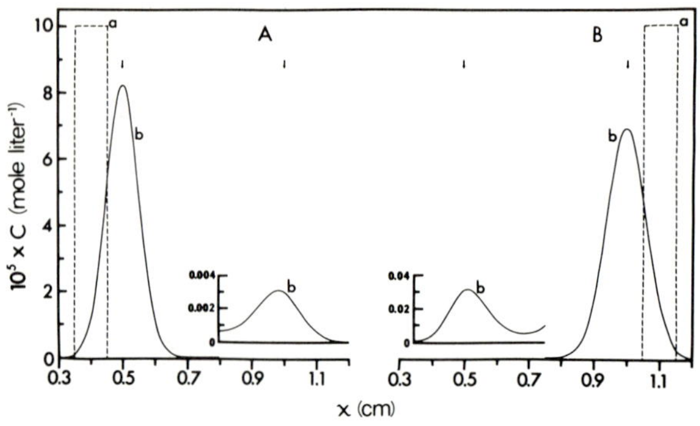

Fig. 14. Transient isoelectric focusing patterns for approach to equilibrium from a narrow initial zone positioned either anodically, A, or cathodically, B, to both pI's; calculated for Reaction 6 with *n*=30: a, t=0; b, 1×10^5 sec. From Stimpson and Cann [6].

under nonideal conditions a noninteracting macromolecule can also give transient, but not equilibrium, patterns exhibiting two peaks. The conditions have been derived by Dishon and Weiss [31] in a theoretical study of isoelectric focusing in a nonlinear pH gradient. Their calculations, which are for a broad initial zone of material, predict two transient peaks arising at the ends of the initial zone, migrating towards each other as time goes on and finally merging at the pI. Certain features of this behavior are similar to those which may be shown by associating-dissociating systems (Fig. 9), and it is conceivable that in practice the two could be confused. It should be possible, however, to distinguish the effect of a nonlinear pH gradient from interactions by comparison of transient and equilibrium patterns given by a broad initial zone with those from a narrow zone located at different positions along the isoelectric focusing column.

PERSPECTIVES

The foregoing discussions have dwelt on the implications of small molecule-macromolecule interactions for conventional applications of electrophoresis and isoelectric focusing, but it is anticipated that the same concepts will also find application in fundamental studies on a variety of biochemical reactions. These include protein-drug interactions, the interaction of enzymes with cofactors and allosteric affectors, and the interaction of macromolecules with each other. An example [32] is the determination from acrylamide gel electrophoresis of the equilibrium constants for the binding of proflavin to chymotrypsin, the binding of complementary tetranucleotide to t-RNA, and complex formation between pairs of t-RNA molecules whose anticodon triplets are complementary to each other. Preliminary

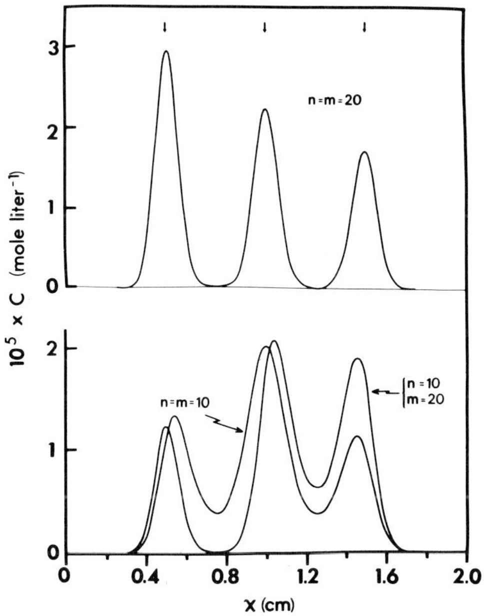

Fig. 15. Equilibrium isoelectric focusing patterns calculated for a set of two sequential, pH-dependent conformational transitions (Reaction Set 7). From Stimpson and Cann [6].

findings indicate that isoelectric focusing not only holds promise but offers unique advantages for the study of these several classes of interaction [33]. Offhand, one expects that the peculiar sensitivity of isoelectric focusing might permit detection of subtle, ligand-induced conformational changes which are beyond the ready reach of other methods. It has already been demonstrated that the method is useful for detecting and characterizing pH-dependent conformational transitions [28].

Finally, it must be emphasized that the production of multiple electrophoretic and isoelectric focusing peaks due to interaction is not unique for any one reaction mechanism [1,3,5-8,18,34]. While fractionation experiments will, in general, yield information on reversibility and, possibly, rates of reaction, elucidation of the mechanism depends upon the combined application of electrophoresis or isoelectric focusing with one or more independent physical methods such as ultracentrifugation, circular dichroism, ligand-binding measurements and relaxation kinetics. The greater the diversity of methods brought to bear on the problem, the more precisely can the interaction be described.

ACKNOWLEDGEMENTS

Supported in part by Research Grant 5R01 HL 13909-26 from the National Heart, Lung and Blood Institute, National Institutes of Health, United States Public

Health Service. This is publication No. 709 from the Department of Biophysics and Genetics, University of Colorado Medical Center, Denver, Colorado 80262 (U.S.A.).

REFERENCES

1. Cann, J.R. (1970) Interacting Macromolecules The Theory and Practice of Their Electrophoresis, Ultracentrifugation, and Chromatography, Academic Press, New York.

2. Cann, J.R. (1972) in Methods in Enzymology (Hirs, C.H.W. and Timasheff, S.N., eds.), Vol. 25, pp. 157-178, Academic Press, New York.

3. Cann, J.R., Bethune, J.L. and Kageles, G. (1973) in Subunits in Biological Systems (Fasman, G.D. and Timasheff, S.N., eds.), Part B, pp. 253-302, Marcel Dekker, New York.

4. Wrigley, C.W. (1976) in Isoelectric Focusing (Catsimpoolas, N., ed.), pp. 93-117, Academic Press, New York.

5. Cann, J.R. and Stimpson, D.I. (1977) Biophys. Chem. 7, 103-114.

6. Stimpson, D.I. and Cann, J.R. (1977) Biophys. Chem. 7, 115-119.

7. Cann, J.R., Stimpson, D.I. and Cox, D.J. Anal. Biochem., in press.

8. Hare, D.L., Stimpson, D.I. and Cann, J.R. Arch. Biochem. Biophys., in press.

9. Righetti, P.G. (1978) in Electrokinetic Separation Methods (Righetti, P.G., van Oss, C.J. and Vanderhoff, J.W., eds.), Chapter 21, Elsevier/North Holland, Amsterdam.

10. Longsworth, L.G. and Jacobsen, C.F. (1949) J. Phys. Colloid Chem. 53, 126-135.

11. Literature cited in Cann, J.R. (1970) Interacting Macromolecules The Theory and Practice of Their Electrophoresis, Ultracentrifugation, and Chromatography, pp. 178-179, Academic Press, New York.

12. Cann, J.R. and Goad, W.B. (1968) Advan. Enzymol. 30, 139-177.

13. Cann, J.R. (1971) Biochemistry 10, 3707-3712.

14. Cann, J.R. (1971) Biochemistry 10, 3713-3722.

15. Cann, J.R. and Kegeles, G. (1974) Biochemistry 13, 1868-1874.

16. Cann, J.R. and Goad, W.B. (1965) J. Biol. Chem. 240, 148-155.

17. Cann, J.R. and Goad, W.B. (1965) J. Biol. Chem. 240, 1162-1164.

18. Cann, J.R. (1973) Biophys. Chem. 1, 1-10.

19. Cann, J.R. (1966) Biochemistry 5, 1108-1112.

20. Kegeles, G. (1977) Arch. Biochem. Biophys. 180, 530-536.

21. Foster, J.F., Sogami, M., Petersen, H.A. and Leonard, W.J., Jr. (1965) J. Biol. Chem. 240, 2495-2502.

22. Frater, R. (1970) J. Chromatogr. 50, 469-474.

23. Yates, J.R. quoted by Frater, R. (1970) J. Chromatogr. 50, 469-474.

24. Kaplan, L.J. and Foster, J.F. (1971) Biochemistry 10, 630-636.

25. Wallevik, K. (1973) Biochim. Biophys. Acta 322, 75-87.

26. Drysdale, J.W. and Righetti, P. (1972) Biochemistry 11, 4044-4052.

27. Galante, E., Carvaggio, T. and Righetti, P.G. (1976) Biochim. Biophys. Acta 442, 309-315.

28. Talbot, P. (1975) in Isoelectric Focusing (Arbuthnott, J.P. and Beeley, J.A., eds.), pp. 270-274, Butterworths, London.

29. Mayers (1973) Ann. N.Y. Acad. Sci. 209, Discussion p. 63.

30. Catsimpoolas, N. (1973) Ann. N.Y. Acad. Sci. 209, Discussion p. 63.

31. Dishon, M. and Weiss, G.H. (1977) Anal. Biochem. 81, 1-8.

32. Eisinger, J. and Blumberg, W.E. (1973) Biochemistry 12, 3648-3662.

33. Drysdale, J.W. (1975) in Methods of Protein Separation (Catsimpoolas, N., ed.), pp. 93-126, Plenum Press, New York.

34. Cann, J.R. and Oates, D.C. (1973) Biochemistry 12, 1112-1119.

© 1979 Elsevier/North-Holland Biomedical Press
Electrokinetic Separation Methods,
P.G. Righetti, C.J. van Oss and J.W. Vanderhoff

ISOELECTRIC FOCUSING

PIER GIORGIO RIGHETTI
Department of Biochemistry, University of Milano, Via Celoria 2, Milano 20133, Italy

INTRODUCTION

Twelve years after its official introduction in biochemistry, isoelectric focusing (IEF) is still alive and well, as documented by the more than 3000 articles dealing with its use or development, and by the five international symposia held on this subject (in 1972 in New York[1] and Tübingen[2], in 1973 in Glasgow[3], in 1974 in Milano[4] and in 1976 in Hamburg[5]). The principle of IEF, in comparison with conventional or disc-electrophoresis, is depicted in Fig. 1 (taken from a japanese review article[6], remarkable because at that time the father founders of IEF, Svensson[7] and Kolin[8] had not even dreamed of reviewing the newly introduced technique). Amphoteric species, applied at pH 7 in a narrow zone in electrophoresis, keep migrating and progressively diffusing as long as the voltage is applied. On the other hand, in IEF, even if the sample is applied as a broad zone, all the species "condense" or "focus" at their iso electric point (pI) where they are packed in narrow zones of zero mobility and zero net charge. This is because IEF is an equilibrium method in which amphoteric compounds are segregated according to their pI's in a pH gradient, formed by electrolysis of amphoteric buffer substances known as carrier ampholytes. In this system, an electrical current directed toward increasing pH sweeps a protein, which is positively charged in regions where pH<pI, toward the isoelectric zone, where it is arrested due to loss of charge. Similarly, proteins which are negatively charged in regions where pH>pI, are swept in the direction opposite to the current toward the isoelectric zone, where they come to a stop. Thus, eventually all of the amphoteric species are packed in sharp isoelectric zones, which represent an equilibrium between electrical mass transport and back diffusion. This condensation or focusing effect is depicted in Fig. 2, which shows how the same protein ions, applied as cationic and anionic species to a preformed pH gradient, converge and merge into the same sharp isoelectric zone at equilibrium[9].

IEF has now been refined to a level that permits the resolution of ampholytes whose pI's differ by as little as 0.02 pH units or less[10] (indeed, in 25 cm long polyacrylamide gels, at high voltages, and narrow pH gradients, Allen *et al.*[11] have claimed a resolution of 0.005 pH units in the typing of α_1-proteinase inhibitor phenotypes). This resolving power cannot be achieved by conventional electrophoretic or chromatographic procedures. With these latter methods, specially adjusted conditions have to be devised for particular separations, while IEF, by virtue of being an equilibrium process, has a built in resolution which allows one to separate in the same experiments all components with measurably different pI's. IEF is particularly suitable

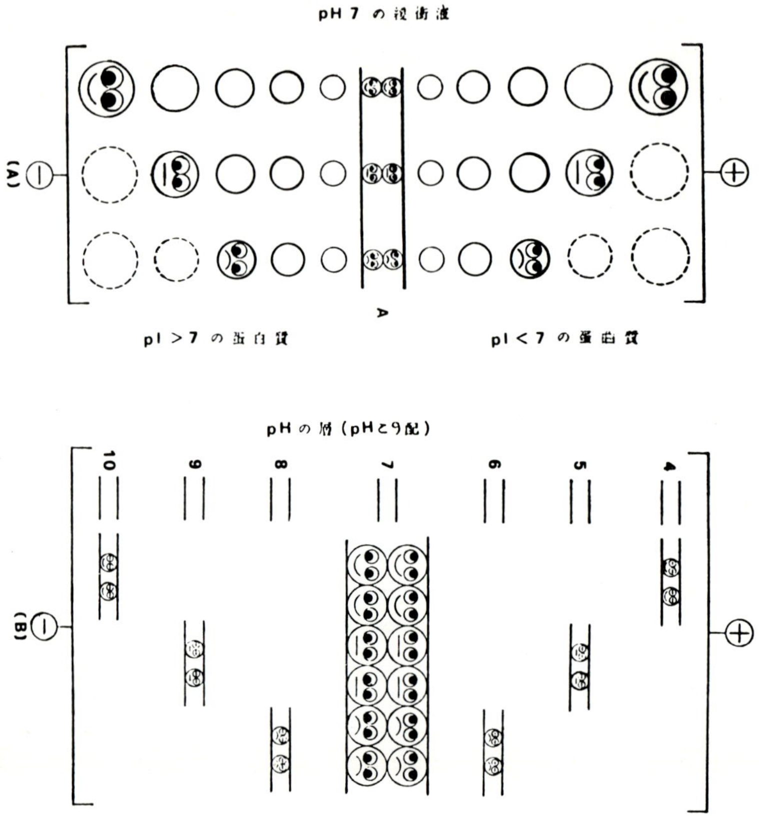

図1. ⊕：陽極, ⊖：陰極。目玉の向き（上, 下）はその物質の移動の方向を意味しており, 閉じた目はその物質が移動せず停止状態にあることを意味している。顔(円)の大きさは拡散（大きな顔）と，濃縮（小さな顔）を意味する。

Fig. 1

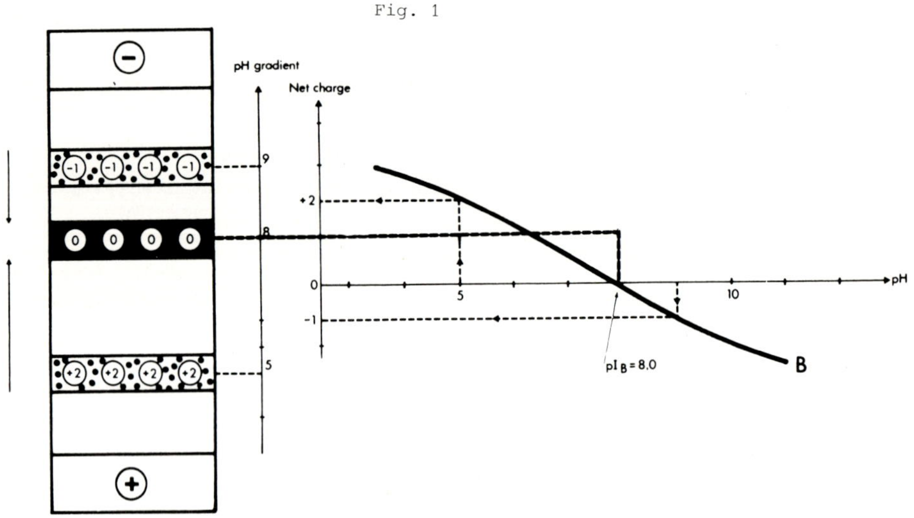

Fig. 2.

for differentiating closely related ampholytes and provides a valuable criterion of homogeneity.

HISTORICAL

It has been customary to attribute early achievements on IEF to rough separations of ampholytes described by Ikeda and Suzuki[12], Williams and Waterman[13] and Tiselius[14]. This seems to be due to a phenomenon of "defocusing of the origins of ideas" which, according to Kolin[15], "consists of blurring of historical perspectives and usually takes the form of shifting the responsability for discoveries and ideas to multiple sources in the remote past". Indeed, in 1954-56, Kolin[16-18] conceived the idea of "focusing ions in a continuous pH gradient", stabilized by a sucrose density gradient. He used the term "isoelectric spectrum" to denote a distribution pattern established by a sorting process. The pH gradient was generated by placing the substance to be separated at the interface between an acidic and a basic buffer, in a Tiselius-like apparatus, and allowing diffusion to proceed under an electrical field. In these "artificial" pH gradients, Kolin was able to obtain "isoelectric line spectra" of dyes, proteins, cells, microorganisms and viruses on a time scale ranging from 40 sec up to a few minutes, a rapidity still unmatched in the field of electrophoresis. Concomitant with the pH gradient, a density gradient, an electrical conductivity gradient and a vertical temperature gradient were acting upon the separation cell. The main drawback was the instability of these gradients, due to rapid migration of the buffering electrolytes during electrophoresis.

In the 1960's, Svensson (alias Rilbe)[19-21] laid the foundations of IEF in its present form with a series of theoretical articles entitled "isoelectric fractionation, analysis and characterization of ampholytes in natural pH gradients". He introduced the law of pH monotony, and studied the protolytic properties of carrier ampholytes, their conductivity, their buffering capacity, their titration curves and the resolving power of the technique in presence of suitable ampholytes, capable of generating stable pH gradients. These concepts were achieved in practice with the synthesis by Vesterberg[22] of carrier ampholytes with many of the properties described by Svensson.

Fig. 1. Principles of conventional electrophoresis (left) and IEF (right). In the first case, a narrow zone of sample ions has to be applied to the support and the anionic (upward looking) and cationic (downward looking) species keep migrating and progressively diffusing (enlarged circles) as long as the current is applied. In IEF, even if the sample is applied as a broad zone (here at pH 7) each species condenses in an isoelectric, narrow zone, where it comes to a rest (closed eyes) (from Matuo and Horio[6]; freely interpreted from japanese).

Fig. 2. Condensation or focusing effect in IEF. A protein with pI = 8 is applied to the IEF column, in a preformed pH gradient, as an anionic (negatively charged, pH 9) as well as a cationic (positively charged, pH 5) species. Both forms converge and merge in the same isoelectric zone (net charge zero) at pH 8. On the right, the protein titration curve is depicted (Courtesy of LKB Produkter AB. See ref. 9).

In a recent historical article[15] Kolin has pointed out that, in a sense, IEF can be considered as a null method in which particles reach equilibrium in a focal zone of vanishing force. He has generalized the idea of IEF and shown that it may be considered as a member of a family of phenomena - isoperichoric (isos = equal and perichoron = environment) focusing effects - "in which particles migrating in a force field along a chemical concentration gradient are swept into sharp, stable stationary zones of vanishing force in which a physical parameter of the suspension medium matches that of the particle. Examples of such condensation phenomena are isopyknic-, isoconductivity-, isodielectric-, isomagnetic-, isoparamagnetic- and isodiamagnetic-focusing". This article[15] as well as a similar historical article by Rilbe[23], on the "isoelectric" research group in Stockholm, make a fascinating reading.

Several review articles have appeared covering general aspects of IEF[24-30], as well as specialized topics: preparative IEF[31,32], gel isoelectric focusing[33-35], zymogram techniques[36], use of pH markers[37], applications to immunoglobulins[38] and to clinical chemistry[9,39] and a table of pI's and MW's of proteins[40]. Recently, a general book on IEF[41] and a laboratory manual, a monograph[42], covering all the theoretical and practical aspects of IEF, have appeared. A useful literature reference list, called Acta Ampholinae, is published by LKB Produkter AB. A volume covering the years 1960-1974 is supplemented yearly with references on both, IEF and isotachophoresis (ITP).

THEORY AND FUNDAMENTAL ASPECTS OF IEF

This chapter can only be a brief survey of the fundamental equations pertaining to IEF. For a detailed treatise, the reader is referred to the early articles by Svensson[19-21], as well as to more recent surveys and extensions of his theory[44-47]. It must be emphasized that, even though Rilbe's theories have been challanged (especially on account of steady-state decay in gel IEF, as well as lack of uniform conductivity, buffering capacity and pH gradients in present-state IEF) no alternative theory has been presented to date, and Rilbe's basic concepts remain valid.

The differential equation of IEF

Under steady-state conditions, as in ideal IEF, the following differential equation (which describes the equilibrium condition between simultaneous electrophoretical and diffusional mass transport) can be derived:

$$d(CuE) = \frac{d}{dx} D \frac{dc}{dx} \qquad (1)$$

where C is the protein concentration at the level x in the separation column, u is its mobility at that point, E is the field strength and D is its diffusion coefficient. Equation (1), upon integration, gives:

$$CuE = D \frac{dc}{dx} + \text{integration constant} \qquad (2)$$

This equation must be valid everywhere in the column, even outside the focused protein zone, where the protein under consideration is absent. Thus the integration constant is zero, and we obtain:

$$CuE = (dC/dx)\ D \qquad (3)$$

which is the general differential equation for the final steady state in IEF. Eq. (3) is not applicable to components which cannot reach a steady state (such as salts, which are split up in their ionic constituents) nor to non electrolytes (such as sucrose, sorbitol, glycerol, often used in IEF as anticonvective media). However, a constant concentration gradient of a non electrolyte in IEF does not violate eq. (3) if it can be kept reasonably constant with time. In theory, equation (3) can be applied also to acids and bases which migrate to the anode and cathode, respectively, and are concentrated there. Therefore, one free acid at the anode and one free base at the cathode are compatible with steady state conditions. However, if mixtures of acids and/or bases are used, they will not necessarily come to a steady-state. For instance, when one strong and one weak acid simultaneously migrate to the anode, the strong one collects there more rapidly than the weak one since the former suppressed the dissociation of the latter. Thus, the electrophoretic migration of the weak acid is stopped, but its diffusion proceeds through the strong acid layer toward the anode. Since no electric mass transport can balance this diffusional mass transport, a steady-state cannot be reached (see also the section "pH gradients obtained with mixtures of acids, bases and ampholytes"). Components unable to reach the steady-state may slowly alter the conditions underlying the steady-state of other components. The latter states then become only quasi-stationary, subjected to slow drift.

Buffering capacity in the isoprotic state

The molar buffer capacity (B_i) of an ampholyte in the isoprotic state (isoprotic and isoelectric states coincide when the absolute mobilities of the cationic and anionic species of an amphoteric substance are equal) is given by:

$$B_i = (\ln 10)\ /\ (1 + s) \qquad (4)$$

where the parameter s is defined as:

$$s = f\ (K_1\ /\ 4K_2)^{1/2}, \qquad \log s = 1/2\ \Delta\ pK + \log\ (f/2) \qquad (5)$$

i.e. it is a function of the activity factor f and of the two dissociation constants, K_1 and K_2, of a bivalent ampholyte. The maximum buffer capacity of a monovalent weak acid or base, *i.e.* its buffer capacity at pH = pK, is:

$$B_i = (\ln 10)/4 \qquad (6)$$

The relative buffer capacity (b_i), defined as the ratio between (5) and (6), is thus:

$$b_i = 4/(1 + s) \qquad (7)$$

Since this cannot be possibly more than 2, it follows that s cannot possibly be smaller than unity ($s \geq 1$). By virtue of eq. (5) this is equivalent to the conditions for bivalent protolytes:

$$\Delta pK' = pK'_2 - pK'_1 > \log 4 \qquad (8)$$

where $K'_1 = fK_1$ and $K'_2 = K_2/f$ are the stoichiometric dissociation constants. Eqs. (7)

and (8) reveal that the buffer capacity in the isoprotic state decreases with increasing s, *i.e.* with increasing Δ pK across the isoprotic point (see Fig. 3).

Conductivity in the isoprotic state

The conductivity is proportional to the degree of ionization α, defined by the equation:

$$\alpha = (C_+ + C_-)/C \qquad (9)$$

where C_+ and C- are the concentrations of cationic and anionic forms, respectively, of a zwitterion, while C is the sum of C_+, C_- and C_o (concentration of the undissociated or zwitterionic, or both, forms of the ampholyte). By the use of eqs. (4) to (8), we find:

$$\alpha_i = 1\ (1 + s) \qquad (10)$$

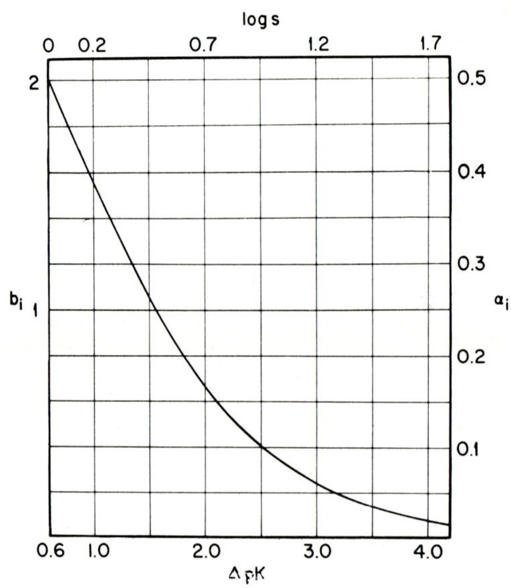

Fig. 3. Relative buffer capacity (b_i) and degree of ionization (α_i) in the isoprotic state as a function of ΔpK and the parameter s (see eq.5. From Rilbe[45]. Source: Academic Press).

which, by comparison with (7), gives the relation:

$$b_i = 4\alpha_i \qquad (11)$$

Thus, buffering capacity and conductivity in the isoprotic state are always complementary. The requirement of good values for these two properties has been found to be equivalent to the requirement of a small pK difference across the isoprotic point (see Fig. 3). Svensson[20] has defined as *carrier ampholytes* compounds which exhibit good conductivity aed buffering capacity in the isoelectric state.

Law of pH monotony

Formulated by Svensson,[48] it states that a natural pH gradient, developed by the current itself, is positive throughout as the pH increases steadily and monotonically from anode to cathode. As a corollary, two ampholytes can never be completely separated from one another unless a third ampholyte, with intermediate pI, is present in the system. This means that, at any pH removed from pH 7, no zone of pure water can develop between two adjacent ampholytes, since H_2O can be considered as an ampholyte with pI 7. Even if theoretically compatible at pH 7, a zone of pure water developing in IEF would be highly deleterious since it will develop considerable joule heat and absorb almost all of the available voltage. Even with a current density (I) as low as 10^{-4} A/cm^2, the field strength (E = I/x) in pure water would be 10 000 V/cm and the joule heat ($q = I^2/x$, where x is the distance along the column) 1W/cm^2.

Resolving power

In general, focused proteins zones are usually very narrow, *i.e.* are confined within a narrow pH interval, where the pH gradient, d(pH)/dx, and the mobility slope, du/d(pH), can be regarded as constant. With the x axis pointing in the direction of the current, the pH gradient becomes a positive quantity, but the mobility slope is always negative. Thus p, defined as:

$$p = -\frac{du}{dx} = -\frac{du}{d(pH)}\,\frac{d(pH)}{dx} \qquad (12)$$

is a positive quantity and, since u=0 for x=0, the relation between u and x becomes: u = -px. Inserted into eq. (3), this gives:

$$dC/C = -\ (pE/D)\ xdx \qquad (13)$$

If also E and D are considered constant within the focused zone, this equation has the solution:

$$C = C_{(o)}\ \exp(-pEx^2/2D) \qquad (14)$$

where $C_{(o)}$, the integration constant, is the concentration maximum at the level x=0. Eq. (14) expresses a gaussian concentration distribution with inflection points lying at:

$$\sigma = \pm\ (D/pE)^{1/2} \qquad (15)$$

Narrow zones are thus favored by a low diffusion coefficient, a high field strength, a large mobility slope and a steep pH gradient. If two adjacent zones of equal mass have a peak to peak distance three times larger than the distance from peak to inflection point, there will be a concentration maximum approximating the two outer inflection points. Taking this criterion for resolved zones, Rilbe has derived the following equation for minimally but definitely resolved zones:

$$\Delta pH = 3\sigma\ (d(pH)/dx) \qquad (16)$$

By changing Δ(pH) to Δ(pI) and by insertion of σ from (15), with the aid of (12), we obtain:

$$\Delta(\mathrm{pI}) = 3\left(\frac{D(d(\mathrm{pH})/dx)}{E(-du/d(\mathrm{pH}))}\right)^{1/2} \qquad (17)$$

It can thus be seen that a shallow pH gradient favors a high resolution, whereas a steep pH gradient favors narrow zones. This equation does not apply to the focusing of oligopeptides, which have a D/u ratio (diffusion *vs* mobility) different from proteins[45].

Peak capacity

It has been defined as the maximum number of components resolvable by a given technique under specified conditions[47]. In IEF, the peak capacity n is given by:

$$n = \left(\frac{-FE\ (dq/d(\mathrm{pH})(d(\mathrm{pH})/dx)}{L6RT}\right)^{1/2} \qquad (18)$$

where F is the faraday (96 500 coulombs), R the gas constant, T the absolute temperature, q the effective charge of the particle, x the distance moved and L the total path length. This equation shows that peak capacity increases in proportion to total path length and is a function of the square rooth of both, the slope of the pH-gradient and protein mobility slope.

Mass content of a protein zone

Rilbe and Pettersson[49] have related the mass load (m) of a protein zone in liquid-phase IEF to the resolving power, in the following inequality:

$$m < \frac{45qD^2(dC/dx)}{E^2(\mathrm{pI})^2\ (du/d(\mathrm{pH}))^2} \qquad (19)$$

where dC/dx is the sucrose density gradient along the IEF column and q its cross sectional area. Thus, the capacity of a density gradient column rises with the square of the resolving power with which it operates for a given protein system. This means that an experiment with a pH gradient favoring the purity of protein fractions (*i.e.* a shallow pH gradient) also favors the quantity of pure protein that can be separated. For the load capacity of IEF in gel media, Radola has found experimentally an upper load limit of about 10-12 mg/ml gel suspension (Sephadex G-75 superfine)[50].

Conditions for a stepless pH course

If too few carrier ampholytes are present in a pH gradient in IEF (5 to 10 species per pH unit), with a high ΔpK value, the pH course, rather than smooth, will be more like a staircase. The question of how many individual carrier ampholytes are really necessary in order to obtain a stepless pH course can be answered by formulating a condition for completely unresolved double zones of adjacent carrier ampholyte peaks. As demonstrated by Svensson[51], the sum of two similar gaussian curves has one flat maximum and no minimum at all for a peak to peak distance equal or less than two standard deviations. If all individual carrier ampholytes lie as closely spaced as that, all of them are completely unresolved from their neighbors. No plateaus in the pH course can then be expected. Therefore, the greatest allowable pI difference between adjacent carrier ampholytes compatible with a smooth pH course is given by the inequality (partly derived from eq. (17)):

$$\Delta(\mathrm{pI}) < 2\left(\frac{D(d(\mathrm{pH})/dx)(1+s)}{Eu \ln 10}\right)^{1/2} \qquad (20)$$

This inequality is in full agreement with Almgren's conclusions. The parameter s is the quality mark of ampholytes: the lower the s value, the better are the conductivity and buffering capacity. Condition (20) shows that ampholytes of high quality must lie more closely spaced on the pH axis than ampholytes of a lower quality. Ampholytes with low s values are especially needed in the pH region around the neutral point in order to compensate for the very low conductivity of water ions in that region. According to Rilbe[45], high quality as well as low-quality carrier ampholytes, with a preponderance for the high quality in the neutral range, are essential for a successful isoelectric focusing. According to Almren[46], the use of ampholytes with widely separated pK values reduces the number of ampholytes required to give a good pH gradient. However, if ΔpK is too large, more ampholyte must be added to give a certain conductivity and buffering capacity and the system will approach the steady-state more slowly.

SYNTHESIS, FRACTIONATION AND GENERAL PROPERTIES OF CARRIER AMPHOLYTES

Rilbe's theory was difficult to be put to practical use, due to the lack of suitable commercial carrier ampholytes, especially in the pH range 5-7, where only very few amphoteric species are present. Thus, at the beginning, the "focusing group" in Stockholm used protein hydrolysates as buffering substances in IEF, notwithstanding the drawbacks associated with them (absorption at 280 nm, isoelectric precipitation, difficulties in separating large peptides from proteins, etc.).

A real breakthrough came in august-26-1964, when Olof Alfred Yngve Vesterberg filed a swedish patent on the synthesis of carrier ampholytes suitable for IEF (a more comfortable english version can be read as the USA extension granted in dec-23-1969 as patent No. 3,485,736) by a remarkably elegant and simple process. He synthesized a mixture of a large number of homologues and isomers of aliphatic polyamino polycarboxylic acids with different pK's and pI's closely spaced in the pH 3.5-10 range. The synthesis involves the coupling of propanoic acid residues to polyethylene polyamines. Five years later the scientific community was informed about it in a scientific memory[22]. Subsequently, Vesterberg synthesized also more acidic and more basic carrier ampholytes, thus extending the useful fractionation range in IEF from pH 2.5 to pH 11. Their synthesis and their properties, though, have never been really described. Thus, even though Vesterberg should be given full credit for having made possible present-day IEF, he should also be given a prize as a "suspense" writer, for having kept so hermetic and tight-lipped about his carrier ampholytes.

Synthesis and fractionation of carrier ampholytes (CA)

Vinogradov *et al.*[54] and Righetti *et al.*[55] spurred a wave of interest in "home-made" CA by describing in more details the synthetic process, which can be easily and inexpensively adopted in any laboratory. Carrier ampholytes of suitable properties are formed by reacting appropriate amounts of acrylic acid with different polyethylene polyamines in H_2O at 70°C until all the acid has been consumed. The synthesis goes via an anti-markovnikov addition after the mechanism of the Michael reaction and therefore β-amino acids are obtained. All polyamines and acrylic acid are distilled under nitrogen and reduced pressure immediately before use. To a mixture of polyamines in a two-necked flask (we use pentaethylenehexamine (PEHA), tetraethylenepentamine (TEPA), triethylenetetramine (TETA) and hexamethylenetetramine (HMTA) in ratios 60:25:10:5 %) acrylic acid is added dropwise, under nitrogen and continuous stirring, during 1 hr, to a final N/-COOH ratio of 2:1. The flask is then stoppered and held for *ca* 20 hrs at 70°C in a Dubnoff shaker. The reaction can be summarized as follows:

$$\text{PEHA + TEPA + TETA + HMTA + } CH_2\text{=CH-COOH} \xrightarrow[\text{20 hrs}]{H_2O,70°C} \text{carrier ampholytes}$$

The product is stored frozen in dark bottles. Yellow compounds formed during the synthesis can be removed by repeated charcoal treatment[54]. The distillation set-up and the reaction chamber are depicted in Fig. 4.

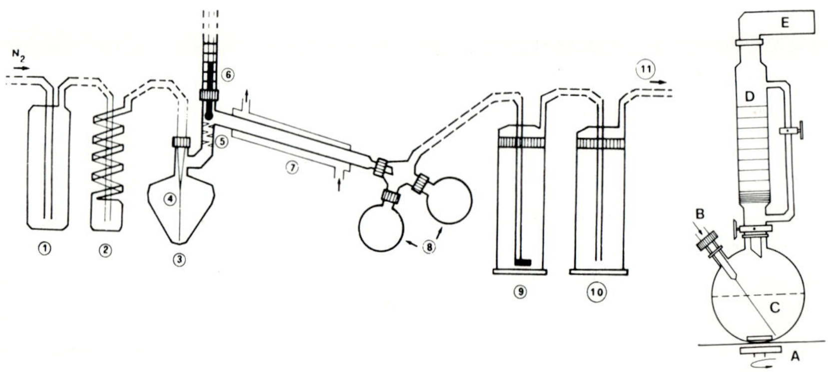

Fig. 4. Distillation set-up (left) and reaction chamber (right) for production of carrier ampholytes. Scheme of the distillation apparatus: 1) copper wire catalyst, kept in a oven at 450°C; 2) cooling serpentine; 3) distillation flask; 4) capillary for nitrogen flushing; 5) Vigreux column; 6) thermometer; 7) condenser; 8) collection flask; 9) Fresenius tower with conc. H_2SO_4; 10) $CaCl_2$ trap; 11) connection pump. The shaded regions represent ground-glass joints. (From Righetti *et al.*[55]. Source: Elsevier). Reaction chamber: A) magnetic stirrer; B) capillary for nitrogen flushing; C) polyamine solution; D) acrylic acid in a burette with a side arm; E) $CaCl_2$ trap for gas escape. (Gianazza, E. and Righetti, P.G., unpublished).

Later on, Grubhofer and Borja[56] described new types of CA's. Different polyamines were obtained by condensing ethylene imine with propylene diamine and distilling the product in order to eliminate compounds with MW's greater than 400. Alkylation of such amines is achieved with propane sultone and chloromethylphosphonic acid. These CA's, available commercially as Servalyt or Biolyte, have to be supplemented also with "carboxyl ampholytes" since sulfonic and phosphonic derivatives do not buffer in the pH region 3.5-6. All these compounds exhibit a minimum of conductivity and buffering capacity centered around pH 6.2[57-61] (Servalyt having the deepest minimum of buffer capacity, of only 1.8 μequiv./pH.ml at pH 6.2[59]). Probably the minimum is due to the lack of appropriate pK values of the polyamines in the pH region 5.5-6.5. By replacing acrylic with itaconic acid, it has been possible to effect substantial improvements in the pH range 5.5-6.5, as itaconic acid has a pK_2 value of 5.45. Upon coupling, the pK of the γ-carboxyl group remains virtually unaltered. With itaconic acid CA's it has been possible to improve the conductivity in the pH range 5.5-6.5 by as much as 400% compared with conventional CA's.

The actual number of CA's in the wide pH range mixture (pH 2.5-11) has been the object of debate for several years. By the glucose-caramelization technique[62] about 40 bands were detected, a few dozens could be distinguished by precipitation with Coomassie Violet in half-saturated picric acid and 5% acetic acid[63] and 60 to 70 peaks could be separated by ion-exchange chromatography[64]. CA's can also be visualized on the basis of their different refractive indexes, both in liquid support media[65] as well as in gels[55]. Based on theoretical considerations, Vesterberg had calculated that more than 360 different amphoteric species should be produced by his synthetic process. Actually, in a recent investigation, Radola *et al.*[66] have substantiated this claim and demonstrated that probably 500 to 600 different amphoteric species are present in the pH 2.5-11 mixture, since about 150 species can be resolved within any narrow range of 2 pH units.

For production of narrow pH range cuts, the synthetic mixture can be fractionated by exploiting the IEF technique itself. Commercially, this is done by using multi-compartment electrolyzers[22,56] (Grubhofer uses a 60-liter chamber, divided into 21 compartments by porous ceramic diaphragms into which a 2% ampholyte solution is focused in *ca* 20 hours). In the laboratory scale, this can be done by using the zone convection apparatus of Valmet[67] or by a continuous flow IEF technique[68], on Sephadex anticonvective media, after the principle of Fawcett[69]. There is also the possibility of preparing narrow-pH ranges by free-flow IEF, in a modified Hannig-apparatus, as described by Just and Werner in this book (Ch.9).

Molecular size

The MW of CA's was reported to be in the range 300-1000 daltons, with the main peak centered at 600 d[24]. A small fraction, however, (about 0.03%), exhibits an apparent MW above 4000 d[70]. Gasparic and Rosengren[70] have also found the peak of the gaussian distribution of Ampholine apparent MW, as determined by gel filtration in

Sephadex G-25 with polyethylene glycols as standards, to be centered around 400 instead of 600 d. Haglund[58], however, thinks this value of the mean to be underestimated and insists on a true mean MW value for Ampholine between 600 and 900 d. Gasparic and Rosengren[71] have also synthesized small MW ampholytes for peptide separations having a distribution curve centered around 300 d, only 0.2% of the total above 1000 d and 1.47% above 600 d. The possibility has also been put forward[72] that there might be Ampholine species with apparent MW's as high as 20 to 30 000 d, but the evidence for it was only inderect. In a recent report, Radola *et al.*[66] have shown the MW of Ampholine to be consistently higher than Servalyt. Moreover, among all the different Ampholine narrow ranges, the alkaline species had a higher MW than the acidic ones. However, the MW of the mean is not given. These authors only state that "experiments with different markers in the range of 300 to > 3000 d suggest that the MW of the major portion of commercial carrier ampholytes exceeds notably the value of several hundreds d" (d = dalton).

Distribution of CA upon focusing

Svensson[19] had demonstrated that, in a region of an electrolysis column where the pH and conductivity courses are linear and constant, the concentration distribution of an isoelectrically focused mono-mono-valent ampholyte was gaussian or nearly so. This concept has been recently challanged by Brown *et al.*[73,74] who found non-gaussian distributions of several small MW ampholytes, especially around neutrality and at high loads. This is quite conceivable, in view of the fact of the unequal buffer capacity of individual components at their pI's and of the non-constancy of conductivity (with a minimum between pH 5.5 and 6.5). However, these authors have also inferred that, following IEF, each ampholyte species is distributed over a wide area of the separation column, usually extending along 7-15% of the column length, and up to 40% at higher loads. This position does not seem tenable, also in view of the fact that it is based on ampholyte patterns resolved by ion-exchange chromatography (which possibly resolves only 10-15% of the total species) and on the IEF distribution of His, which at higher loads, precipitates and aggregates all over the focusing column. Considerable amount of evidence, on IEF of plant pigments[76], of dyes[77], of free amino acids[78], of peptides[79] and detection of focused carrier ampholyte patterns by refractive index gradients[65,55] and by specific stains[66], suggests that indeed even small MW ampholytes focus as very sharp peaks, notwithstanding their higher diffusion coefficients, as compared to proteins, typically extending along 0.1 to 0.5 % of the column length, especially at loads of only a few μg[80].

Optical properties

UV absorbance of Ampholine at 280 nm is generally low, and is usually confined to components of low pI. However, when scanning gels or columns under voltage, where much sharper resolution is obtained and the absorbancy of individual ampholytes may be considerable, it is always best to scan a blank with no sample, in order to subtract spurious peaks due to focused Ampholine[78,81]. UV spectra of different Ampholine ranges[82] have shown that most of them exhibit peak absorbance around 365 nm. The pH range 9-11 also has a peak at 290 nm and the pH range 3.5-5 has a strong, pH dependent, chromophore at 310 nm. This acidic pH range has a marked UV absorbancy and is also distinctly yellow in color as compared to other pH ranges. Some of these chromophores also exhibit a pH-dependent fluorescence spectrum.

Chelating properties

Qualitatively, metal-chelation by Ampholine had been reported in the literature[57]. Galante *et al.*[83] have found that metal chelation by Ampholine usually involves two adjacent nitrogen functions, giving rise to five-membered rings. Chelation maxima occur in the pH region 8.5-10.5, where most of the nitrogens in the polyamine backbone are unprotonated. Mixed type chelation is a possible, though less likely event, since it generates less stable seven-membered rings. Little, if any, chelation occurs solely on carboxyl groups. As compared to EDTA, the chelating power of Ampholine is generally weak. Usually, a given amount of divalent cation (Mg^{++}, Zn^{++}, Ca^{++} or Fe^{+++}) requires a 100 to 500 molar excess of Ampholine for chelation in contrast to EDTA, which usually binds metals in the pH range 3-11 in a 1/1 molar ratio. Carrier ampholytes are, however, as powerful chelators of divalent copper as EDTA, and consequently, many copper-containing proteins may be rapidly inactivated on exposure to ampholytes in IEF. In order to minimize metal chelation, proteins should be run to equilibrium from the anodic side in a pre-focused support, so as to reduce to a minimum the contact time with carrier ampholytes and as close as possible to the expected pI.

Bacto peptone carrier ampholytes

Blanicky and Pihar[84] have prepared ampholytes from bacto peptone by removing proteins by precipitation in 60% ethanol. A mixture of possibly several dozens peptides was obtained, which appeared to be satisfactory, even with conductivity gaps, for IEF in the pH range 3.5-9. The conductivity and buffer capacity in the neutral pH region was improved by adding His. These types of CA's are similar to those described by Vesterberg and Svensson[10] and thus suffer from the same inconveniencies.

Thermal pH gradients

This approach to IEF has been described by Luner and Kolin[85] and is characterized by quick focusing times (15 min for a complete fractionation). Here, the pH determination amounts to a temperature measurement with a thermistor, which gives the pH in a zone to better than 0.01 pH units, from the known pH *vs* temperature curve. Broadly speaking, this approach and the one by Troitzki *et al.*[86] are similar, since they are based on pK variations of buffers either in gradients of organic solvents or in temperature gradients. Lundahl and Hjertén[87] reported that they can only be of limited applicability in IEF, because the pH of ordinary buffers and the pI of proteins do not exhibit sufficiently large differences in temperature coefficients.

Dielectric constant pH gradients

Troitzi *et al.*[86,88] have formed pH gradients by using common buffers in gradients of organic solvents, such as ethanol, dioxane, glycerol or poly-ol gradients, such as mannitol, sucrose, sorbitol. Taking advantage of pH variations of these buffers in different concentrations of these solvents, they were able to generate gradients of *ca* 1.5 pH units in different regions of the pH scale. These pH gradients were stable up to 12 days of IEF under voltages up to 1000 V. They have shown separations of rabbit hemoglobin in pH gradients 7-8.6 formed with borate buffers in a mixed gradient of 0.5% glycerol and 0-30% sucrose and of human serum albumin in a pH gradient 4.5-5.8 formed with acetate buffer in a 0-90% glycerol gradient.

pH gradients obtained with mixtures of acids, bases and ampholytes

It all started in 1969, when Pettersson[89] demonstrated that acidic pH gradients between pH 1 and 3 could be formed by electrolysis of a mixture of acids and ampholytes. He used 6 acids (Nos. 1,2,4,6,8 and 11 in table I) and two ampholytes, glutamic (pI 3.22) and aspartic (pI 2.98) acids, and found that they distributed themselves in order of increasing pK's under an electric field. The strongest acid was closest to the anode, while the weaker acid was almost completely non-ionized in the region of its zone closer to the anodic side. By this method, pepsin (pI 2.5) and seven colored components from red beet sap, with pI's between pH 1.3 and 3.1, were separated. These pH gradients, however, increased stepwise, rather than as a smooth function. A similar approach was used by Stenman and Gräsbeck[90] for the fractionation of the R-type vitamin B_{12}-binding protein. They used a mixture of nine acids (Nos. 3,5,7-12 in table I) and the same two ampholytes. Upon focusing for 5 days at 400 V, in glycerol density gradient, under quasi-equilibrium conditions, they obtained a smooth pH gradient from pH 1.5 to 4.5.

Recently, Chrambach's group, in a series of papers[91-97], has demonstrated that the same focusing results can be obtained in the neutral and basic pH region by using mixtures of non amphoteric, as well as amphoteric, buffers, including mixtures of amino acids. They have used this technique, called buffer focusing (BEF), for the study of pH gradient formation and decay in IEF and for demonstrating a possible parallelism between IEF and ITP. Their results can be summarized as follows:

A) the finding that non-amphoteric buffers, placed in an electric field between acid anolyte and basic catholyte can produce stable pH gradients suggests that focusing is non-isoelectric;

B) this hypothesis, in turns, suggests that focusing is steady-state stacking (ITP) (in order of pK's for non amphoteric and in order of pI's for amphoteric compounds) under conditions such that the buffers used as catholytes and anolytes in ITP are replaced by base and acid, thereby providing a "pH-cage" which prevents the stack from migrating out of the separation column[90].

The spin-offs of their studies are two fold:

1) It is possible to stabilize pH gradients in IEF against decay (cathodic drift)

TABLE I

Acidic and basic buffers and ampholytes used in BEF

ACIDS	pK	BASES	pK	AMPHOLYTES	pK
dichloroacetic	1.48	pyridine	5.5	MES	6.4
phosphoric	2.12	4-picoline	6.2	MOPS	7.2
malonic	2.18	BisTris	6.9	ACES	7.3
monochloroacetic	2.84	Lutidine	7.0	TES	8.0
lactic	3.08	Imidazole	7.5	Tricine	8.6
citric	3.08	HEM	7.2	Bicine	8.7
malic	3.45	NEM	8.0	Taurine	9.7
formic	3.75	TEA	8.4	Glycine	10.5
succinic	4.16	Tris	8.8	GABA	11.3
glutaric	4.34	MEA	10.4	ADA	6.6
acetic	4.75			PIPES	6.8
propionic	4.85			BES	7.15
				HEPES	7.55

Abbreviations: MES = Morpholino ethane sulfonic acid; MOPS = Morpholino-3-propane sulfonic acid; ACES = (N-2-acetamido-2-amino ethane) sulfonic acid; TES = N-tris-(hydroxymethyl) methyl -2-amino propane sulfonate; TRICINE: = N-tris (hydroxymethyl) methyl glycine; BICINE = N,N'-bis (2-hydroxyethyl) glycine; GABA = γ-amino butyric Acid; ADA = N (2-acetamino iminodiacetic acid; BES = N,N-bis(2-hydroxyethyl)-2-amino ethane sulfonic acid; HEPES = N-2-hydroxyethyl piperazine-N'-2-ethane sulfonic acid; PIPES = Piperazine-N,N'-bis(2-ethane sulfonic acid; BisTris = Bis-(2-hydroxyethyl-imino-tris hydroxymethyl) methane; HEM = hydroxyethyl morpholine; NEM = N-ethyl morpholine; TEA = Triethanol amine; Tris = Tris (hydroxymethyl) amino methane; MEA = monoethanol amine.

by equalization of the anolyte pH with the pH of the most acidic amphoteric component in a given Ampholine pH range. In the case of Ampholine pH 6-8, this is achieved by using Thr at the anode and His at the cathode[96];

2) an ITP system with multiple trailing buffer constituents (cascade stack) can be converted into a cascade IEF when the leading and trailing constituents in ITP are replaced with strong acids and bases. Conversely, an IEF system can be transposed into a cascade stack, when the electrolytes are changed into the appropriate leading and terminating buffers of the corresponding ITP system[92]. This is very useful when performing preparative IEF in gel media in electrophoretic apparatuses provided with elution chambers.

It seems to me that these kinds of gradients are only useful for a theoretical, rather than a practical, approach to IEF. Moreover, it remains to be seen whether or not IEF is indeed ITP. Since this parallelism has been drawn mostly by buffer focusing, it must be stressed that, according to Rilbe's theory[45], non-amphoteric BEF can never reach equilibrium conditions, as in IEF, because weak acids and bases, which stop migrating electrophoretically due to total loss of charges, still keep diffusing in strong acids or bases lying on their anodic or cathodic side, respectively. It also remains to be seen whether or not abolition of pH-gradient decay is generally applicable to all pH ranges. However, amphoteric buffers different from Ampholine can be useful additives for the formation of non-linear pH gradients in the new technique "pH gradient modifier IEF" (see ahead).

PREPARATIVE IEF

There are four basic variants of this technique: 1) IEF in liquid density gradients, either stationary or continuously flowing; 2) free-phase IEF, either stationary or in a continuous-flow; 3) IEF in flat beds of porous anticonvective media, either granulated, such as Sephadex, Ultrogel (agarose-acrylamide), Pevikon-Sephadex, or continuously polymerized, such as polyacrylamide; 4) IEF in gel cylinders, again granulated or in continuous matrices. The last two methods differ especially in the sample elution procedure.

IEF in liquid density gradients

One of the earliest methods is the use of columns in which the pH gradient is stabilized against convection by a vertical sucrose density gradient. LKB adapted a design by Svensson[21] and produced two columns of 110 ml (LKB 8100-1) and 440 ml (LKB 8100-2) capacity. Their basic design is shown in Fig. 5. About 5 to 25 mg protein per zone can be focused in the smaller column and *ca* 20-80 mg per zone in the larger. The upper load limit depends on the number of components, their solubility at their pI and the shape of the pH gradient (shallower pH gradients allow higher loads per zone). The density gradient is usually prepared with sucrose solutions. Ideally, the solute should have a high solubility but low viscosity in H_2O. It should be non-ionic and have a high density, a minimum density gradient of 0.12 g/cm^3. The material in the gradient should not react with the sample nor interfere with its subsequent detection. The gradient is prepared with mixers which provide linear density gradients from two equal volumes of light (5%) and dense (50%) sucrose solutions having a density difference of 0.2 g/cm^3. Other solutes, such as sorbitol[26] or glycerol[98,99], may be used instead of sucrose, especially in alkaline regions where sucrose ionizes. Ethylene glycol[100] can be used at concentrations of 60-70% in the dense solution, because of its lower density. Other useful agents include mannitol and polyglycans, such as Dextran[101] and Ficoll[102]. Deuterium oxide has also been suggested as a medium for density gradients[103]. D_2O gradients, which are formed *in situ* by free diffusion of three D_2O solutions for 3 min[100], have so far only been used in the small spectrophotometric cell (1.5 ml) of Jonsson *et al.*[104]. While the density increment obtainable with D_2O is rather small (0.1 g/cm^3) its viscosity at 5°C is, however, about 50% lower than the viscosity of an aqueous sucrose solution of similar density. This means that a more favorable distribution of the field strength along the column and shorter focusing times can be obtained. However, pH readings in D_2O gradients are

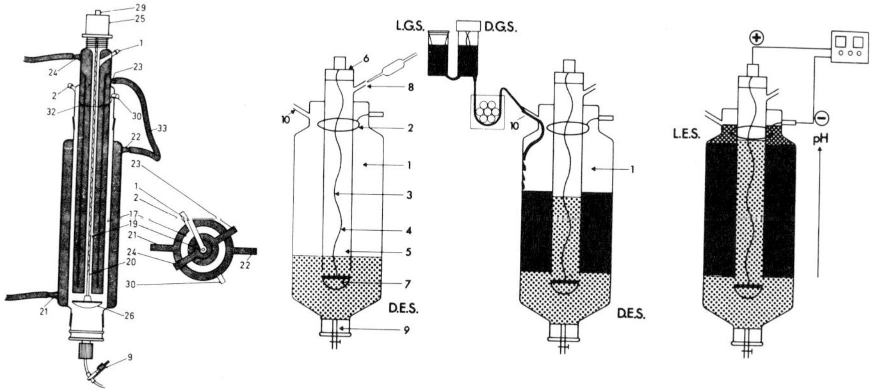

Fig. 5. (Far left): sketch of the LKB 8100 column. (1) gas escape for the lower electrode. (2) gradient and sample inlet. (9) elution capillary. (17) electrofocusing compartment, consisting of a hollow cylinder thermostated on both sides by cooling jackets. (19) lower electrode compartment. (20) teflon support for the lower electrode. (21) and (22) inlet and outlet, respectively, for the outer cooling jacket. (23) and (24) inlet and outlet, respectively, for the inner cooling jacket. (25) setting device for opening and closing valve (26) used to shut off the lower electrode compartment (19) from the electrofocusing chamber (17). (29) and (30) lower and upper electrode terminals respectively, for connection with the power supply. (32) upper platinum electrode. (33) tubing connection between the two cooling jackets. (Center and right): setting up the electrofocusing column. Left: the dense electrode solution (D.E.S.) in layered at the bottom of the column with the valve (7) of the central electrode compartment (5) open. This solution is pipetted in via the gas escape (8), to avoid contamination of the separation chamber. Center: the density gradient is introduced in the column via the sample inlet (10) with the aid of a peristaltic pump. The gradient is best produced with a gradient mixer, shown, schematically, with L.G.S. (light gradient solution) and D.G.S. (dense gradient solution). Right: the light electrode solution (L.E.S.) is now floated on top of the gradient, until the platinum loop (2) is submerged. The electrodes are then connected to a power supply. (Both pictures by courtesy of LKB Produkter AB).

higher than in sucrose gradients, due to shifts in pK_2 values of protolytic groups in proteins and to shifts of the asymmetry potential of the glass electrode in D_2O.

Fig. 5 (right) shows the sequence of operations needed to perform a density gradient IEF experiment . The gradient with Ampholine (1%) and sample is layered on top of a dense cushion containing an electrolyte at the bottom of the column. The second electrolyte is now floated on top of this gradient. Table II lists three types of density gradient stabilization (sucrose, glycerol or ethylene glycol) for the 110 ml and 440 ml columns. Table III gives suggestions for electrode solutions to be used in both columns either when the anode or when the cathode are uppermost. For operation, I am listing some general experimental guidelines.

Load capacity. The carrying capacity of a density gradient rises proportionally with increases in the square on the length of the focused zone (eq. 19). In the LKB 8100-1 column, the maximum theoretical amount of protein in a zone 0.1 cm long is

TABLE II

Density gradient solutions for the 110 ml and 440 ml LKB columns for three types of stabilizing media (sucrose, glycerol and ethylene glycol).

Density gradient solution	110 ml Column LKB 8100-1	440 ml Column LKB 8100-2
Sucrose stabilization		
Dense gradient solution		
Sucrose	27 g	107.5 g
Volume of H_2O + Ampholine (+ sample if added)	37 ml	150 ml
Total volume	54 ml	215 ml
Concentration of sucrose	50% (w/v)	50% (w/v)
Light gradient solution		
Sucrose	2.7 g	10.75 g
Volume of H_2O + Ampholine (+ sample if added)	53 ml	207 ml
Total volume	54 ml	215 ml
Concentration of sucrose	5% (w/v)	5% (w/v)
Glycerol stabilization		
Dense gradient solution		
Glycerol - 87% (v/v)	30 ml	118 ml
Volume of H_2O + Ampholine (+ sample if added)	24 ml	97 ml
Total volume	54 ml	215 ml
Concentration of glycerol	60% (w/v)	60% (w/v)
Light gradient solution		
Total volume of H_2O + Ampholine (+ sample if added)	54 ml	215 ml
Concentration of glycerol	0%	0%
Ethylene glycol stabilization		
Dense gradient solution		
Ethylene glycol - 100% (v/v)	40.5 ml	161 ml
Volume of H_2O + Ampholine (+ sample if added)	13.5 ml	54 ml
Total volume	54 ml	215 ml
Concentration of ethylene glycol	75% (v/v)	75% (v/v)
Light gradient solution		
Total volume of H_2O + Ampholine (+ sample if added)	54 ml	215 ml
Concentration of ethylene glycol	0%	0%

estimated to be 1 mg, while a 5 cm zone would carry 2500 mg. The narrower the pH gradient, the higher the sample load. IEF in pH gradients as narrow as 1 pH unit[105-107] and even 1/2 pH unit[108,109] has been reported. Loads of 1074 mg for sperm-whale myoglobin[49], of 1800 mg for staphylococcal α-toxin[110], of 3870 mg for *clostridium perfringens* α-toxin[111] and of 7300 mg for α-glucosidases[108] have been applied to density gradients.

Protective agents. Oxidation of free -SH groups, with loss of enzyme activity, is one common problem in several preparative separations, including IEF. This can be prevented by using low concentrations (about 10^{-4}M) of thiol compounds such as 2-mercaptoethanol[112], 2, 3 dimercapto propanol[99], dithiothreitol[113] or its isomer, dithioerythritol[99]. Additional problems may arise from the oxidation of cysteine and methionine to cysteic acid and methionine sulphoxide[113]. This can be prevented by performing IEF in the presence of antioxidants, such as thiodiglycol or ascorbic acid (as a protective layer at the anode). The latter seems to be more effective, and also prevents possible modifications of tyrosine and arginine residues[114]. For the use of solubilizing and disaggregating agents, see the Ch. "general experimental aspects".

Electrode polarity. Ideally, conditions should be selected so that components of most interest focus in the lower half of the column. This will minimize diffusion during subsequent elution which is performed in the absence of electric field. However, if precipitates occur, the electrode position should be selected so that the proteins of interest focus away from the flocculi. Also, since the carrier ampholytes show a minimum of conductivity between pH 5.5 and 6.5, the choice of electrode polarity depends on conductivity considerations. One factor which influences the conductivity distribution is the viscosity change along the density gradient, in that increasing viscosity leads to decreasing conductivity. Thus, the polarity is chosen so as to have these two effects counteracting each other, by placing the conductivity minimum at the upper end of the column. Thus, when running pH ranges lower than 6, the cathode should be at the top of the column. On the other hand, when using pH ranges higher than pH 6, the anode should be placed at the column top, thus allowing for higher conductivity in the pH range 6-7.

Sample application and pretreatment. The sample may be incorporated into either the dense or the light sucrose-ampholyte solution, or into both. Since samples are concentrated during the experiment, the volume of sample is not critical and may be as much as 80% of the final column volume. Alternatively, the sample may be introduced as a narrow zone close to the place where it is expected to focus. In this was the run will be shortened, and exposure to pH extremes avoided. The amount of salt and buffer in the sample should be low and should not exceed 0.5 mM for the 110 ml and 1.5 mM for the 440 ml columns. High salt concentrations will create higher current densities at the beginning of the experiment and also generate acid at the anode and base at the cathode. Since most proteins require a certain degree of ionic strength or dipole moment of the solvent in order to form stable solutions, a good practice

TABLE III

Electrode solutions for the 110 ml and 440 ml LKB columns to be used either with the cathode or with anode uppermost.

Electrode solution	110 ml Column LKB 8100-1	440 ml Column LKB 8100-2
Anode at Top of Column		
Cathode solution	15 g sucrose	48 g sucrose
(pH - 11.7 approx.)	+ 10 ml H_2O	+ 30 ml H_2O
	+ 6 ml 1 M NaOH	+ 20 ml 1 M NaOH
Total volume	25 ml	80 ml
Concentration of sucrose	60% (w/v)	60% (w/v)
Anode solution	1.5 ml 1 M H_3PO_4	6 ml 1 M H_2PO_4
	+ 8.5 ml H_2O	+ 34 ml H_2O
Total volume	10.0 ml	40 ml
Cathode at Top of Column		
Cathode solution	25 ml 1 M NaOH	10 ml 1 M NaOH
	+ 7.5 ml H_2O	+ 30 ml H_2O
Total volume	10.0 ml	40 ml
Anode solution	15 g sucrose	48 g sucrose
(pH - 1.2 approx.)	+ 12 ml H_2O	+ 38 ml H_2O
	+ 4 ml 1 M H_3PO_4	+ 12 ml 1 M H_3PO_4
Total volume	25 ml	80 ml
Concentration of sucrose	60% (w/v)	60% (w/v)

is to dialyze against 1% glycine[115,116] or 1% Ampholine (usually the pH 6-8, 7-9 or 8-10 ranges).

Running conditions. Before the advent of the last generation of power supplies (provided with constant current, constant voltage or constant power facilities) it was customary to run the small column at about 2-3W and the 440 ml column at about 6-9W for periods up to five days. Now, with the constant power supplies, the two columns can be run at a maximum of 15W, 1600V and 30W, 2000W, respectively, thus reducing the total experimental time to 15 to 20 hours. Columns with narrow pH ranges or more than 1% Ampholyte usually require considerably larger focusing periods. When eluting the columns, a flow rate of 60 ml/h for the smaller and 240 ml/h for the larger columns are used, while collecting fractions of 1 or 4 ml, respectively. The normal set up for eluting the columns at a constant flow rate is depicted in Fig. 6. In case of sample flocculation, provided the sample of interest focuses away from the precipitate, the column can be emptied from the top, with the aid of a flanged and U-bent capillary tubing, acting as a funnel, as shown in Fig. 6 (center and right). Alternatively, if the sample of interest is precipitated, the ingenious retrieval method of Rathnam and Saxena[117] can be used. In addition to isoelectric precipitation, sample diffusion occurs during elution, since the voltage is disconnected, and some remixing takes place in the annular cross-section leading to a narrow orifice in the elution funnel. Bodding *et al.*[118] have devised a method for eluting the column content under the electric field. Home made columns for density gradient IEF have also been described by Doi and Ohtsuru[119], Azuma *et al.*[120], by Rilbe's group[49], and by Fawcett[121,122] (multiple U-tubes). Commercially, two types of columns,

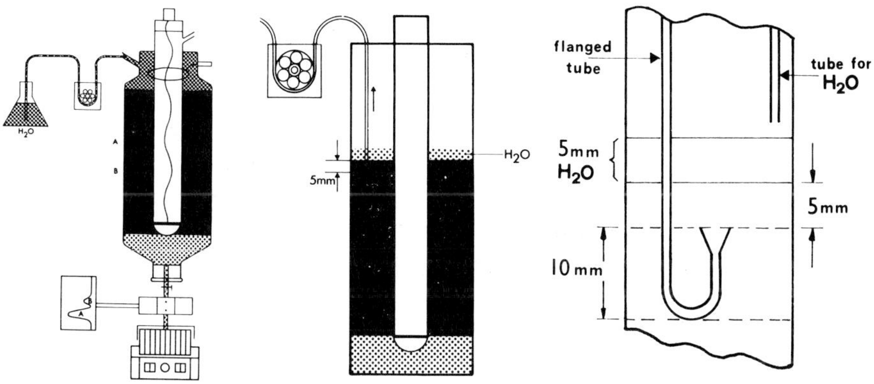

Fig. 6. Left: set-up for elution at a constant flow-rate. The gradient and sample inlet (No.2 in Fig. 5) is plugged with a capillary and H_2O is pumped in and floated on top of the gradient, while emptying the column via the elution capillary (No. 9 in Fig. 5). Center and right: gradient elution in the presence of precipitates. It is assumed that the protein of interest focuses away from the precipitate. The column is then emptied from the top, through nipple (2) (see Fig. 5) by a stepwise procedure. The end of a capillary tube is flanged (60°) and U-bent. This capillary is introduced 5 mm below the surface of the gradient (after removal of the top electrolyte solution) and a 5 mm water layer is layered on top of it with the aid of a second tubing. This layer is then eluted with the aid of a second peristaltic pump operated at flow rate of 100-120 ml/hour. This stepwise procedure is then sequentially repeated. (Courtesy of LKB Produkter AB).

adaptable for IEF (models 212 and 630) are available from Instrumentation Specialty Company (ISCO): they are based on the design of Brakke *et al.*[123]. The electrode chambers are built coaxially around the separation column and are separated from the gradient by two concentric membranes. In the ISCO columns, the experiment progress can be monitored by raising the entire gradient past a densitometer and performing a scan at one or more wavelengths. Their use has been recently reviewed[28]. The only example of continuously-flowing density gradient IEF has been described by Fawcett[121], in a horizontal chamber with 54 inlet and 54 outlet tubes.

On the micro-scale, several "home made" apparatuses have been described. Weller *et al.*[124] have built two small U-tubes of 11 ml and 33 ml total volume. The dummy arm contains only the electrolyte and the other the sucrose density gradient.10 ml J-tubes, fitting standard acrylamide gel electrophoresis apparatus, have been used by Godson[125]. Similar U-tubes systems have been described by Koch and Backx[126]. IEF in 10 ml chromatographic columns has been reported by Kint[127]. Additional microcolumns have been described by Osterman[128] and by Korant and Lonberg[129]. It is also possible to use conventional polyacrylamide gel electrophoresis tubes, plugged either with agar[130] or with polyacrylamide[131] to perform small-scale density gradient IEF. Rilbe has performed IEF in 11 ml parallelepipedic quartz column[31] as well as in a spectrophotometric cell, a flow cuvette code 167-QI from Hellma, of only 1.4 ml volume[65].

Free-phase IEF

Since the stationary (Quast, Ch. 11) as well as continuously-flowing (Just and Werner, Ch. 9) systems have been already detailed in this book, I will only briefly survey them here. In one line of research, Rilbe's group has been active in building and operating multi-compartment electrolysers[31], usually divided into 20 cells by paper or dialysis membranes. Recently, this group has described a host of electrolysers, with open or closed cells, internal or external cooling and with or without stirring [132]. Multi-compartment electrolysers, divided by ceramic diaphragms, are also used by Serva for the preparation of narrow-range carrier ampholytes[56]. Another line of research has developed horizontal troughs having a corrugated bottom, resembling a series of Tiselius cells linked together[67,133,134]. During the run, the liquid is made to overflow the U-tubes, thus bridging the solution and allowing electric conductivity. At the end of the run, the liquid refluxes back in each cell, thus autofractionating the solution. Fawcett[121] has also built a similar apparatus, modified from an earlier design of Kalous and Vacik[135]. On a similar principle is built the sea-serpent apparatus of Rose and Harboe[136]. Essentially the same concept applies to IEF in coils of polyethylene tubing, as described by Macko and Stegemann[137] and Bours[138], in which each turn of the spiral represents one compartment of the Valmet apparatus. An helical glass tube, with each turn provided with a separate outlet, has been recently described by Quast[139]. IEF in free water has also been described by Denckla[140]. A third approach to free- IEF has been attempted by Lundhal and Hjertén[87], using a linear, rotating quartz tube. A final approach to free- IEF is the continuously-flowing thin liquid lamina described by Just and Werner[141,142], in a modified Hanning apparatus[143], for the separation of cells and cell particles.

IEF in flat beds

Radola[50,144-146] has described an excellent method for large scale preparative IEF in throughs coated with a suspension of granulated gels, such as Sephadex G-75 superfine (7.5 g/100 ml) or Sephadex G-200 superfine (4 g/100 ml) or Bio Gel P60, minus 400 mesh (4 g/100 ml). The trough consists of a glass or quartz plate at the bottom of a lucite frame (Fig. 7), with typical dimensions of 40x20 cm or 20x20 cm, with a gel layer thickness of up to 10 mm. The total gel volume in the trough varies from 300 to 800 ml. A slurry of Sephadex, in 1% Ampholine, is poured into the trough and allowed to dry in air to the correct consistency. This is achieved when 25% of the water in the gel has evaporated, and the gel layer does not slide down when the plate is inclined 45°. The plate is run on a cooling block maintained at 2-10°C. The electrical field is applied via flat carbon electrodes or platinum bands in contact with the gel through paper pads soaked in 1 M H_2SO_4 at the anode and 2 M ethylene diamine at the cathode. Initial voltages of 10-15 V/cm and terminal voltages of 20-40 V/cm are applied. As much as 0.05 W/cm^2 are well tolerated in 1 cm thick layers. Samples may be mixed with the gel suspension or added to the surface of the gel as

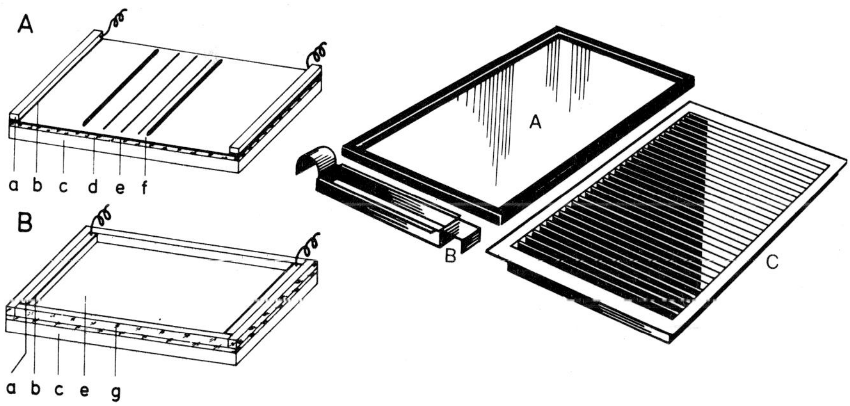

Fig. 7. Left: preparative isoelectric focusing.(A): small scale separations. (B): large scale separations. a: electrode; b: filter paper pad soaked with electrode solution; c: cooling block; d: glass plate; e: gel layer; f: focused proteins; g: trough (from Radola[146]; source: Elsevier Sci. Publ. Co.). Right: trough (A), sample applicator (B) and fractionation grid (C) for preparative IEF in granulated gels using the LKB 2117 Multiphor chamber. After IEF, grid (C) is lowered on trough (A) thus fractionating the gel layer into 30 fractions. Each fraction is scooped up from each channel with a spatula (from Winters *et al.*[147]; source: LKB Produkter AB).

a streak or absorbed into dry gel and poured into a slot in the gel slab. A commercial apparatus, the "Double Chamber", is now available from Desaga, Heidelberg, W. Germany.

After focusing, proteins are located by the paper print technique[50]. Focused proteins in the gel surface are adsorbed onto a strip of filter paper which is then dried at 110°C. After removal of ampholytes by washing in TCA, the paper is stained with low sensitivity dyes such as Light Green SF or Coomassie Violet R-150. Radola's technique offers high resolution, high sample load and easy recovery of focused components. The upper load limit appears to be around 12 mg protein/ml gel suspension. At these high loads, even uncolored proteins are easily detected, since they appear in the gel as translucent zones. Since there is no sieving effect above the Sephadex exclusion limits, the high MW substances, such as virus particles, can be focused without steric hindrance.

Sephadex gels were often plagued by blurred and distorted zones. This was thought to be due to carboxyl groups on the matrix, which Radola[50] suggested might be removed by treatment with propylene oxide. Recently, Winters *et al.*[147,148], have found that most Sephadex batches contain low-MW contaminants which interfere with pH gradient formation. Thus, prior to use in IEF, swollen Sephadex should be washed on a glass filter funnel with 10 gel volumes of distilled H_2O. The gel is then dehydrated by repeated washing in absolute ethanol and dried under vacuum. Winters *et al.*[147] have also described a fractionation grid and a trough to be used in preparative IEF with the LKB Multiphor 2117 chamber (Fig. 7, right). After focusing the grid, lowered on the gel trough, divides the gel bed into 30 fractions (discarding also the edges of the gel, were distortion occurs due to wall effects) which are scraped off with a spatula. The protein is eluted from the gel simply by placing the gel fraction into a syringe plugged with glass-wool and ejecting the eluent with a piston.

Other gel layers described include mixed beds of Pevikon C-870 (a copolymer of polyvinyl chloride and polyvinyl acetate) and Sephadex G-75 superfine, in ratios of *ca* 20:1 (w/w)[149]. Sample recoveries better than 90% were obtained by centrifuging the Pevikon cake. Weise *et al.*[150] have performed IEF in beds of 1.4% agarose added with 0.5% methyl cellulose to increase the viscous drag and reduce electroendoosmosis. The agarose was purified through a strong anion-exchanger (QAE-Sephadex)[151,152]. A similar approach, with the use of 1% liquid non-cross-linked polyacrylamide as an additive to agarose has been describe by Johansson and Hjertén[153] and by Bode[154] (in gel electrophoresis, though). A smart approach to the handling of granulated gel layers, especially for analytical purposes, has been reported by Ziegler and Könler[155]. The samples are focused in a Sephadex layer in the presence of acrylamide monomer (4%, w/v) and a cross linking agent (Bis, 0.13%, w/v). After IEF, the whole layer is polymerized by spraying a solution containing polymerization initiators (1% TEMED and 10% persulphate) in a concentrated buffer onto the gel, under nitrogen atmosphere. One thus has the advantage of focusing very large molecules or even viruses on thin-layers combined with the ease of handling a polymerized gel and the use of a variety of established methods to detect the focused substances. Another approach to the use of granulated layers has been described by Felgenhauer and Pak[156], who perform IEF in Sephadex thin layers followed by electrophoresis at right angles in a polyacrylamide gel gradient for two dimensional size-charge display. On the preparative scale, continuously polymerized slabs of polyacrylamide (2 to 4 mm thick, in 26x17 cm plates) can be used, as reported by Graesslin and Weise[157]. They report loads of up to 40 mg protein and sample recoveries of up to 74% by elution with gentle agitation of excised gel segments.

IEF in gel cylinders

Because of the excellent resolution and load capacity afforded by IEF in polyacrylamide gel cylinders, Righetti and Drysdale[158] have scaled up an analytival apparatus for preparative purposes, by using 16 cm long glass or plastic tubes of 2 cm and 1.2 cm inner diameter. The tubes are held between rubber grommets in a water-tight compartment through which coolant at 2°C is circulated. A similar apparatus has been described by Chrambach *et al.*[159]. For sample recovery, the gel-trapped bands can be excised, squashed and forced electrophoretically into a small cup sealed with dialysis tubing[160,161]. Alternatively, the column can be left intact and the entire content eluted electrophoretically by cascade stacking[91] (*i.e.* by substituting the acid and base at the electrodes in IEF, with appropriate leading and terminating buffers, thus creating and ITP system; see also Ch. 18 in this book). Stathakos[161] has designed a vertical column for multiphasic IEF which consists of a stack of polyacrylamide gel blocks separated by thin liquid interlayers for sample introduction and/or pH monitoring. The idea is similar to the principle of multicompartment electrolysers described by Rilbe *et al.*[132]. It is also possible to perform preparative IEF in granulated gel cylinders, as described by Fawcett[121]. After IEF, the entire Sephadex gel is eluted from the top of the column by pumping dense sucrose solution through the bottom. Gel fractions are collected via a teflon disk cutter which pushes gel segments into a collecting funnel. In another approach, O'Brien *et al.*[162,163] perform IEF in a Sephadex G-15 column plugged at the two extremes with polyacrylamide gel (inmeshed in nylon nets for easy removal). After IEF, the two gel plugs are removed, an elution plunger placed at the column bottom, and the column content recovered by gel filtration with ampholyte-glycerol solution or with distilled water. 96% sample recoveries and loads up to 300 mg per protein component are reported. However, since the small MW carrier ampholytes are retarded by the Sephadex G-15, the protein fractions can be eluted in regions of carrier ampholytes removed from their pI's, thus exhibiting a lower apparent pI.

ANALYTICAL IEF

I will deal here only with two systems, gel tubes and slabs, which are the two most commonly used techniques since most of the know-how and equipment was borrowed from disc-gel electrophoresis. Analytical IEF in small density gradient columns and in granulated gel slabs has been already reviewed in the previous section. The development of analytical techniques for IEF in polyacrylamide gel and its advantages were described in a flurry of papers published independently and almost simultaneously in 1968 (the year of the revolution in most european universities)[164-169]. However, during the transition from IEF in all liquid media to gels, few systematic studies were made to optimize factors such as gel composition and electrolysis conditions. Many of these early systems, particularly in tubes, were plagued by a marked instability in the pH gradients which restricted their general applicability. The first systematic studies on the dos and dont's of gel IEF appeared in 1971[81,170]. Present-day gel IEF techniques are fully standardized and allow the development of sufficiently stable pH gradients to ensure that all molecules of interest reach their pI's.

For versatility, ease of handling and resolving power, gel IEF represents one of the most powerful separation tools today available to protein chemists.

Support media

Since IEF is an equilibrium method, the ideal support should allow virtually unhindered migration of macromolecules to their pI's. By far the best support would be a cellulose acetate membrane, which allows migration of even subcellular particles, requires very little sample and Ampholine and is extremely convenient for staining and destaining, densitometry and storage. Unfortunately, most commercial membranes (Sephraphore III, Separax, Cellogel, Menbranfolien, Oxoid) exhibit too high electroendoosmotic flow and are useless in IEF. Recently, Harada[171] has described the use of an acetate membrane (produced by Fuji Photo Film Co. Ltd, Tokyo, Japan) suitable for IEF. Unfortunately, this author talks just about everything except how the membrane is made or treated. The only information one can squeeze out of his paper is that the membrane contains a surface active agent. At the moment, this method does not seem to have a wide application. The next best is agarose, since, at the usual concentration (1%) allows practically unhindered migration of macromolecules in the multimillion MW range. However, early attempts of IEF in agarose were disastrous since the electroosmotic flow (due to sulphate and carboxyls) was very high and the matrix dissolved at the electrodes, especially at the cathode. Attempts to purify agarose by anion exchangers have been described[151,152]. Grubb[172] has attempted to produce electroendoosmosis-free agarose by covalently bonding to it positively charged groups to counterbalance negatively charged groups, while Låås[173,174] has described the production of virtually charge-free agarose by alkaline desulphatization in the presence of Na borohydride, followed by reduction with Li Al hydride in dioxan. However, present-day IEF in agarose can only be performed in presence of viscous additives[150,153] (methylcellulose, liquid-linear-polyacrylamide, polyethylene oxide with a mean MW of $4x10^6$ daltons) and is stable for only a few hours. The use of granulated gels (Sephadex G-15, G-75 and G-200 superfine, Bio Gel P 60 and Pevikon) has been already described[144-149].

By far, the most popular anticonvective medium appears to be a gel of polyacrylamide. These gels are easily cast in any lab to any desired size, shape and sieving properties. Ideally, the minimum gel strength with a minimum of sieving properties should be used. The lowest acrylamide concentration that will form a satisfactory gel for IEF is about T=3% and C=4%, but these gels are very soft and gluey. Gels of T=5% and C=3%[175] have been found suitable for many purposes, since they allow focusing of proteins up to 500,000 MW in less than 6 hours in 10 cm gels (T=(g acrylamide + g cross linker)/100 ml solution; C=g cross-linker/%T). Originally, a "non-restrictive" gel of 3.125%T and 20%C was described by Davis[176] as the stacking gel in disc electrophoresis. A 5%T, $15\%C_{DATD}$ gel is more mechanically stable and "non-restrictive" for molecules with MW up to 500,000[177]. In fact, Fawcett and Morris[178] and Rodbard *et al.*[179] have demonstrated that as the % cross-linker increases from 10 to 50% the mean pore radii progressively increase and the retardation coefficients of macromolecules of high MW's approach each other and approach zero asymptotically. This has also been nicely demonstrated by Rüchel and Brager[180] with the electron microscope. However, highly cross linked gels with Bis are opaque and brittle and therefore DADT has to be used, since the gels remain optically clear and elastic. A host of cross-linkers, with different properties, are today available. Originally, only N,N'-methylene bisacrylamide (Bis or MBA : $CH_2{=}CH{-}CO{-}NH{-}CH_2{-}NH{-}CO{-}CH{=}CH_2$) was used. Other cross-linkers were then described: ethylene diacrylate (EDA: $CH_2{=}CH{-}CO{-}O{-}CH_2{-}CH_2{-}O{-}CO{-}CH{=}CH_2$)[181] and N,N'-diallyltartardiamide (DADT: $CH_2{=}CH{-}NH{-}CO{-}CH(HO){-}CH(OH){-}CO{-}NH{-}CH{-}CH_2$)[182]. The former, having ester bonds, undergoes base-catalyzed hydrolytic cleavage while the latter, with its 1,2-diol structure, may be selectively oxidized with periodic acid. More recently, other cross-linkers have been synthesized by O'Connel and Brady[183]: N,N',N''-triallyl citric triamide (TACT: $CH_2{=}CH{-}CH_2{-}NH{-}CO{-}CH_2{-}C(HO)(CO{-}NH{-}CH_2{-}CH{=}CH_2){-}CH_2{-}CO{-}NH{-}CH_2{-}CH{=}CH_2$) and N,N'-(1,2-dihydroxyethylene) bisacrylamide (DHEBA:$CH_2{=}CH{-}CO{-}NH{-}CH(OH){-}CH(OH){-}NH{-}CO{-}CH{=}CH_2$) which confer different sieving properties to polyacrylamide gels. Moreover DHEBA gels, in virtue of their 1,2-diol structure, are soluble in dilute periodate while their two amido methylol bonds are more stable to bases than the ester bonds of EDA. Perhaps one of the most interesting developments are gels cross-linked with N,N'-bis-acrylyl cystamine (BAC: $CH_2{=}CH{-}CO{-}NH{-}CH_2{-}CH_2{-}S{-}S{-}CH_2{-}CH_2{-}NH{-}CO{-}CH{=}CH_2$) which can be solubilized under very mild conditions in presence of 2-mercaptoethanol and re-gelled by atmospheric oxidation[184]. Clearly, these types of gels are very useful when high sample recoveries are needed after IEF.

The initiators presently used for polymerization are potassium persulphate (KP), riboflavin (RN) and N,N,N',N'-tetramethylethylene diamine (TEMED). Wyckoff *et al.*[185] suggest a mixture of all three, 0.015% KP, 0.0005% RN and 0.01% (v/v) TEMED. Since RN does not work above pH 7, it is best to substitute it with riboflavin 5'-phosphate. The mixture of monomer and cross-linker should be degassed with a mechanical pump, since oxygen, acting as a free-radical trap, inhibits the polymerization. All reagents should be highly purified; in particular, free acrylic acid should be removed as it will generate fixed charges in the gel causing electroendoosmosis. In addition, unreacted monomeric acrylamide can react with α-amino, sulphydryl and phenolic -OH groups at elevated pH[186]. Acrylamide may be recrystallized from chloroform and Bis from acetone[187]. It must be emphasized that excess persulphate[188-190] or riboflavin[186] generates potentially troublesome free-radicals which have oxidizing properties on thiol groups. Free radicals seem to persist in polyacrylamide gels even after prolonged electrolysis[191] and are only completely removed removed by pre-electrophoresis of 1-5 mM thioglycolate in the gel[186].

IEF in gel cylinders

Most methods for IEF in gel rods use gel dimension similar to those used in analytical disc-gel electrophoresis, *i.e.* 5-15 cm in length and 0.2-0.5 cm in diameter. Righetti and Drysdale[42,81,158] developed an apparatus whose salient features include efficient gel cooling, small electrolyte volumes and close juxtaposition of electrodes and gels (Fig. 8). The gel tubes are held between the electrodes in a separate compartment through which coolant is circulated. The outer wall of the center core extends about 1 cm above the top end plate to form an electrolyte chamber. The lower

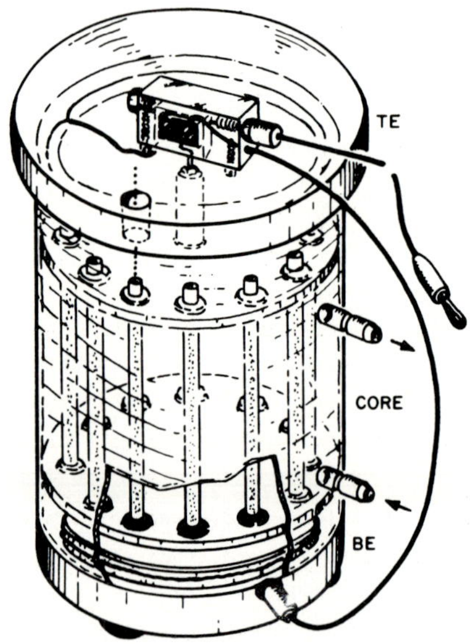

Fig. 8. Schematic drawing of the analytical gel-tube apparatus of Righetti and Drysdale[42,81,158]. TE: top electrode compartment fitted with a safety switch. BE: bottom electrode compartment. The central core is interchangeable with the preparative unit. Available from Medical Research Apparatus (MRA, Boston) (source: North Holland/ American Elsevier) (see ref. 42).

chamber is a separate unit which forms the base of the apparatus. The extremities of the tubes pass through rubber grommets in the plates to make contact with the electrolytes. An aperture in the lid (which rotates freely around a central nylon post) allows samples to be applied directly to the gel surface without raising the lid. This aperture also allows the top compartment to be flushed with a gas if required. The apparatus holds 12 tubes (10 cm long and 0.3 cm inner diameter, volume *ca* 1 ml) but other sizes may be used by changing the rubber grommets. In gel tubes, proteins usually focus as rings rather than uniform disks and consequently are more readily identified by biological activity or staining. Gel tubes are absolutely a must when samples have to be run under strict anaerobic conditions or at subzero temperatures. Except for these particular uses, my feeling is that the "tube technique" will soon belong to the pre-history of IEF.

IEF in gel slabs

For versatility, ease of handling, staining, destaining and storage the slab technique in polyacrylamide gel is by far superior to any other methodology presently available. It is a real bonanza for genetics, evolutionary biology, taxonomical and philogenetic studies, isozyme techniques, clinical chemistry and for any research where several samples have to be run and compared side by side. The gel may be held either vertically or horizontally between cooling plates. I prefer horizontal systems since the gel is subjected to less mechanical stress. Simple electrode arrangements are possible and there is more flexibility in methods for sample application. Separations may be conducted either along the short (10 cm distance) or along the long axis (25 cm). Gels are usually cooled through their bottom surface and the electrolytes are soaked into filter paper strips, placed on the gel surface, on which a cover lid with platinum electrodes is seated. A typical apparatus design for thin layer is shown in Fig. 9. In most systems, the gel is usually cast between two glass plates, separated by a rubber gasket. The plates are clamped together to form a water-tight chamber (Fig. 10, left). One of the two plates may also be used as a template to imprint indentations into the gel for convenient sample application (Fig. 10, right). Since

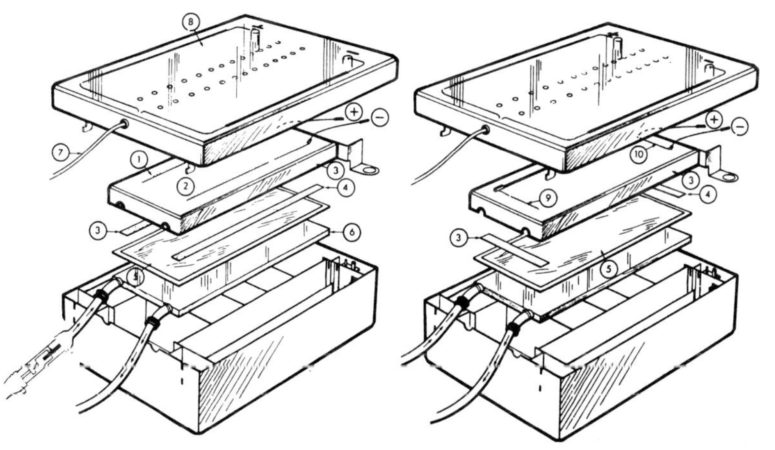

Fig. 9. Drawing of the LKB 2117 Multiphor apparatus. Left: arrangement for IEF in the short side (10 cm). Right: arrangement for IEF in the long side (25 cm). (1) and (2): anodic and cathodic platinum wires, respectively, for runs in the short side; (3): platinum electrode holder and cover lid; (3') and (4): anodic and cathodic filter paper strips, respectively; (5): thin-layer polyacrylamide gel cast on a 1 mm thick gel slab; (6): cooling block; (7): cable connection to power supply; (8): safety cover lid; (9) and (10): anodic and cathodic platinum wires, respectively, for runs on the long side. (source: LKB Produkter AB).

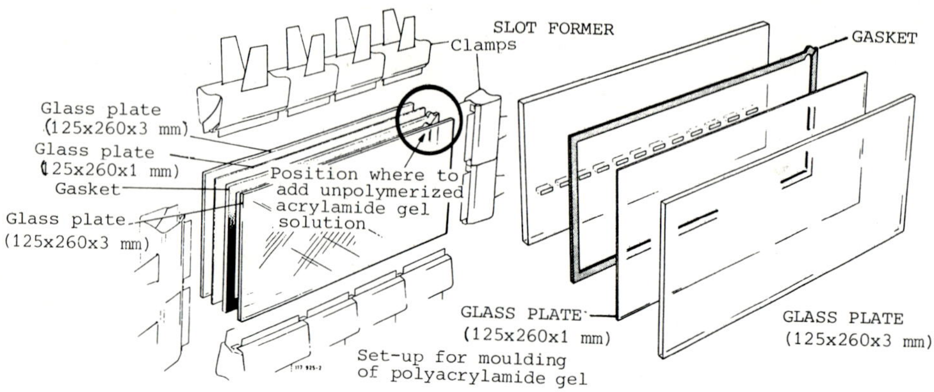

Fig. 10. Polyacrylamide moulding cell for the LKB Multiphor apparatus (left). From Davies (source: Butterworths). Right: same cell, but with a cover lid for pocket formation.

indentations in the gel create conductivity problems and may cause local skewing of the pH gradient, it is best, once the sample has been applied and has migrated into the gel matrix, to fill the troughs with a slurry of Sephadex equilibrated with gel ampholyte. Gels are usually cast in a vertical position by filling the chamber with a syringe trough an open corner in the rubber gasket. (Fig. 11). When the gel has completely polymerized, it should be placed in a refrigerator so that the resulting gel contraction causes a partial separation of the gel from the cover plate and simplifies its removal. For removal of the cover lid, the gel chamber should be placed in a refrigerator so that the resulting gel contraction causes a partial separation of the gel from the cover plate and simplifies its removal. For removal of the cover lid, the gel chamber should be placed on a flat surface and the top plate pried apart from the gel surface by gentle leverage with a spatula against the bottom plate (Fig. 11). The adhesion of the gel to the bottom plate should not be disturbed, as this may cause local distortion in the pH gradient. At this point the electrode strips, soaked in 1M H_3PO_4 and 1M NaOH, are placed on the gel surface, as shown in Fig. 12. It has been suggested (Bio Rad technical bullettin 1030, 1975) to add $Ca(OH)_2$ to the catholyte, so as to precipitate absorbed CO_2 in the NaOH, which would otherwise migrate into the gel. It has also been recommended, when using pH ranges 2.5-6, to utilize as catholyte 1% Ampholine pH 5-7 while using 1% Ampholine pH 7-9 as anolyte in pH ranges 7-5-10.5[192,193]. Even buffers (acetate at pH 4 and ammonia at pH 10) instead of plain acid and base have been suggested[194]. The gel plate is then gently lowered on the cooling block previously wetted with a solution of non-ionic detergent, so as to avoid trapped air bubbles between the two faces (Fig. 12, left). The sample is then applied in a variety of ways, as depicted in Fig. 12 (right). If added on filter paper, it is best to apply it at the anodic side, since at this low pH

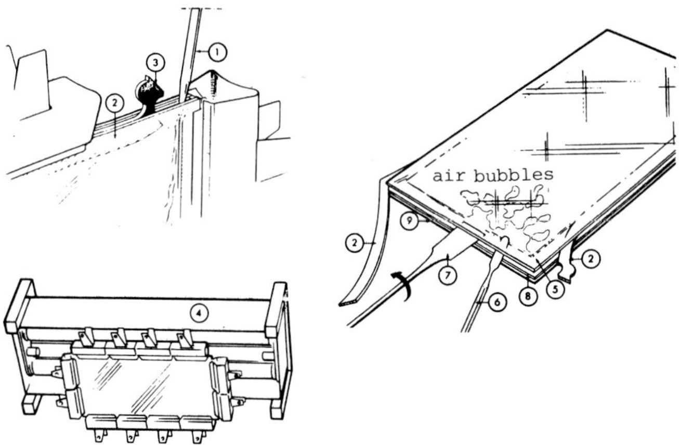

Fig. 11. Sequential procedure for filling a moulding cell (left), photopolymerizing the polyacrylamide gel (center) and removing the cover glass from the thin layer gel (right). This last operation is best done with two spatulas: while slightly pressing down the upper part of the polyacrylamide gel with a thin spatula (6), a broader spatula (7) is slowly twisted as indicated by the arrow. As air bubbles begin to penetrate the space between the upper gel surface and glass plate (5), the broader spatula is gently twisted to a vertical position and the upper glass plate (5) removed. During this operation care should be taken not to loosen the gel from the lower glass plate (8) as this would cause severe band distortion during IEF (source: LKB Produkter AB).

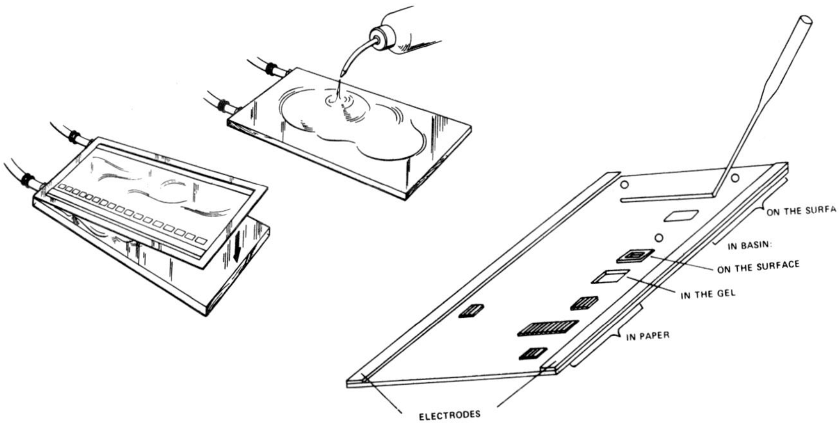

Fig. 12. Left and center: setting of the polyacrylamide gel plate on the cooling block. The block is first wetted with a solution of non-ionic detergent (or kerosene, or light paraffin oil) and then the gel plate is gently lowered on it, so that air bubbles are completely excluded, thus ensuring uniform cooling. Right: different ways of sample application to an IEF gel slab. From left to right: soaked in a square of filter paper; into a rectangle of the same material, and into two squares, one placed close to each electrode; in a basin in the gel; in a basin on the gel surface; as a droplet left on the gel; as a drop spread over a rectangular area; as a streak and finally, as two droplets, one close to each electrode (From Vesterberg[35]; source: LKB Produkter AB).

the carboxyl groups of the paper will be mostly undissociated[79]. Alternatively, cellulose acetate, dessiccated polyacrylamide or Sephadex may be used[195]. Altland[196] has used successfully silicone rubber bands punched with 48 circular holes as sample applicators, pressed on the gel surface. By using a central common electrode, thus dividing the gel into two electrophoretic tracks, 98 samples can be screened simultaneously. A sample applicator made of a plastic frame with rectangular holes (1x0.5 cm), which is fastened by a holder to the gel surface, has also been described[197]. Up to 0.3 ml sample can be applied to each slot. Routinely, we use the following schedule for runs: the constant power supply is set at 13W and a 40 min prerun is performed. After sample application (at the cathode, anode or in the gel middle, as needed) the gel is run for an additional 2 and 1/2 hours (100 V/cm at equilibrium). The cathodic drift begins after the first 4 hours of focusing, thus there is ample time for the components to reach their pI's.

When buying the LKB Multiphor apparatus, the first intelligent thing to do is to throw away the rubber gasket used for the gel sandwich. Given the high cost of reagents (1g Ampholine costs 2.2 times more than 1g of gold; buy gold and save on expensive Ampholine!) and the time needed for staining and destaining, it would be foolish to polymerize a 2mm thick gel. When casting a gel with pockets, I routinely use a 1.05 mm thick gel (a 50 μ thin gel layer on the floor of the chamber is sufficient to seal it properly). When applying the sample on the gel surface, I use 0.7 mm thin gels[79]. This thickness is highly satisfactory because it is thin enough to allow for rapid fixing, quick dye adsorption and rapid leaching out of additives and carrier ampholytes, and thick enough to allow for proper gel handling. A second intelligent move would be to throw away also the cover lid with the electrodes. The platinum wire is very thin and easily bent and dislodged. This results in poor contact with electrolyte solutions and wavy and distorted Ampholine patterns in the gel. From this point of view, the flat platinum bands of the Desaga "Double Chamber" are by far superior.

In a biochemical lab, one often has too few samples to run a gel slab which can easily accomodate 20 to 30 samples. When running only few samples, I polymerize the slab standing on the short side, so that the chamber is filled to the level needed for the sample available. Slabs of only 1/3 or 1/2 the regular length are easily run in the Multiphor chamber. Moreover, when performing a time-course experiment, the slab can be divided into 6-8 strips just before starting the run, leaving a 3 mm channel from strip to strip, thus allowing sequential removal of each strip. This would be the equivalent of having a series of gel tubes, except that handling of the gel and especially radioactivity counting with strip counters (such as the Packard Radiochromatogram Scanner) would be greatly facilitated[75]. Needless to say, this technique also allows the polymerization of juxtaposed layers containing different additives (detergents, urea, formamide, DMSO, etc.) as described by Hobart[198] and by Righetti *et al.*[77]. With all these possible variants, IEF in gel slabs has become a tremendously versatile tool in protein chemistry.

Staining and densitometry

Special procedures have to be used for staining proteins separated by IEF, since the carrier ampholytes form insoluble complexes with many protein stains. This problem can be overcome by first precipitating focused proteins in acid, e.g. 10% TCA, and eluting the acid-soluble ampholytes with extensive washing. After ampholyte removal, proteins can be detected by conventional methods. Today, several direct staining methods have been developed to circumvent this laborious procedure, by solubilizing the dye-ampholyte complexes in alcoholic solvents and/or high temperatures[61]. I will give here examples of the staining methods used:

Fast Green FCF[168]. Immerse gels for 4-8 hours in an aqueous solution of 0.2% dye in 45% ethanol-10% acetic acid. Destain in 10% acetic acid-25% ethanol. This dye tends to fade away with time and has low sensitivity.

Light Green SF[144]. Used by Radola in the paper print technique. After taking the print, the paper is washed 15 min in TCA, rinsed in methanol-H_2O-acetic acid (33:66:10, v/v/v) and stained for 15 min by immersion in the same solution containing 0.2% Light Green SF. For destaining, the above mixture is used in the absence of dye. This stain is 3 to 5 times less sensitive that either Coomassie Brilliant Blue (CBB) R-250 or CBB G-250.

Bromophenol Blue[164]. Immerse gels for 3 hours in an aqueous solution of 0.2% Bromophenol Blue in 50% ethanol-5% acetic acid. Destain in 30% ethanol, 5% acetic acid in H_2O. Gels dehydrate considerably in this solution. Low sensitivity method.

Comassie Brilliant Blue R-250. This is the most popular and sensitive dye, together with CBB G-250. Hayes and Wellner[199] fix the gels in 5% TCA-5% sulphosalicylic acid. The excess acid is washed in H_2O and gels transferred to a solution of 0.066% CBB R-250 in 0.2M Tris-HCl buffer, pH 7.7, for 3 to 4 hours. The background stain is removed in the same, diluted buffer (0.001M) for 24 to 48 hours. According to Spencer and King[200], gels are bathed in 0.01% CBB R-250 in 5% TCA, 5% sulphosalicylic acid and 20% methanol. It is less sensitive than other procedures, but background staining is very low. According to Vesterberg[201], a direct method, without Ampholine interference, consists in heating the gels at 60°C for 15 min in 0.1% dye in 28% methanol, 11% TCA and 3.5% sulphosalicylic acid. Destaining is also affected at 60°C in 28% ethanol-8% acetic acid. Beware, because the combination of alcohol and high temperatures can easily remove the dye from the protein. A modification suggested by Söderholm *et al.*[202] consists in first leaching out Ampholine in 2% sulphosalicylic acid, 11% TCA and 27% methanol at 65°C. After 20 min, the gels are washed in destaining solution (9% acetic acid, 27% ethanol) and then stained in 0.1% dye dissolved in destaining solution. Recently, Vesterberg and Hansén[203] have described four additional variation, using various combinations of TCA, sulphosalicylic acid, perchloric acid (PCA) and urea in the fixing solution. It appears that urea increases the stain sensitivity by unfolding the protein chains. The detection limit appears to be around 0.2 μg protein per band. Righetti and Drysdale[34] have suggested to stain in an acid-alcohol solvent in presence of 01% cupric sulphate to enhance the stain intensity by interaction of Cu^{++} with peptide bonds. Malik and Berrie[204] described a protein staining procedure with CBB R-350 prepared using sulphuric acid, KOH and TCA. This is a rapid staining procedure, but the gel background becomes significantly stained.

Coomassie Brilliant Blue G-250. Diezel *et al.*[205] have introduced the use of CBB G-250 (a dimethyl substitute of the R-250 dye) as a colloidal dispersion in 12.5% TCA. This causes a selective binding to the protein zones and prevents the dye's penetration into the network of the gel, thus avoiding background staining. Upon storage in 5% acetic acid, the bands are intensified as the protein-bound dye becomes soluble, diffuses into the gel and binds again to the interior portions of the protein bands. Reisner *et al.*[206] have found that 0.04% CBB G-250 in 3.5% PCA remains in a leucoform, but changes to the blue form when coupled to proteins. Ampholine does not precipitate, but the gel takes a faint orange to brownish background. By introducing an additional wash in 5% acetic acid after staining, Holbrook and Leaver[207] have claimed a 3 fold increase in sensitivity (but the background becomes heavier). Blakesley and Boezi[208] have described a recent modification which appears to be excellent for IEF gels. To 400 ml of 0.2% dye in H_2O an equal volume of 2N H_2SO_4 is added. After standing for 3 hours, the precipitate is removed through a Whatman No.1 paper. To the clear brown filtrate, 90 ml of 10N KOH are added, producing a dark purple solution. To this, 120 ml of 100% TCA are added, and the greenish solution is ready for use. This dye can be used several times, provided the pH is maintained below 1.0 (in fact for IEF gels, the used dye works much better than the fresh solution, probably because a trace impurity, which tends to precipitate neutral Ampholine, is eliminated; Righetti, unpublished). In 0.7 mm thick slabs, maximum color development is achieved in *ca* 5 hours[79], but the bands are already visible within 15 min. When placed in water, there is a marked color intensification. This dye works wonders for peptide analysis by IEF, allowing fixation and detection of chains as short as 12-15 amino acids[79].

Fluorescent stains. It appears that the ultimate analysis in prestaining has been reached with the introduction of fluorescing stains such as dansyl chloride[209], fluorescamine[210,211] and 2-methoxy-2,4-diphenyl-3(2H)-furanone (MDPF)[212]. Protein quantities as small as 1 ng can be detected with MDPF. However, unless trace labelling is used, it is difficult to conceive that these stains will not alter the protein pI's and introduce an artefactual heterogeneity. These methods thus, as well as covalently bonding of reactive cationic dyes (such as Drimarene Brilliant Blue K-BL)[213] prior to electrophoretic analysis, do not appear at present to be suitable in IEF.

Zymograms and specific stains. Most of the specific stains development for detecting proteins after gel electrophoresis can usually be adapted for use in IEF. Some of these methods depend on a specific stain for a cofactor such as a metal or prostethic groups, e.g. haem, or other component, e.g. carbohydrate, lipid, etc. Others depend on identifying the position of a protein by its biological activity. For tables listing specific stains and zymograms see the book by Righetti and Drysdale[42] and the review of Wadström and Smyth[195]. For zymogram techniques, a gel slab is the matrix of choice, since the different tissue extracts are run side by side and pI alterations of different isozymes are immediately apparent. A slab is also a must when the enzyme substrate is a macromolecule (such as in the case of proteases, DNAases, RNAases, etc.). In this case, an agarose thin slab is polymerized containing the high MW substrate in an appropriate buffer and a contact print is made with the gel slab containing the focused enzyme. After the proper incubation time, both slabs are stained: they appear like the negative and positive in a photographic process[195]. Alternatively, the contact print can be made by utilizing a transparent cellulose acetate sheet impregnated with substrate, which is resistant to tear and easily stained and dried [214].

Autoradiography. This is a highly sensitive method, which allows visualization of trace amount protein, not usually detectable with common protein stains. It has been calculated that this method has at least a 150 fold higher sensitivity than common protein stains[215]. For instance, O'Farrel[216] has detected by autoradiography as many as 1100 protein spots in an *E. coli* cell sap, revealing protein species representing as little as 10^{-4} to 10^{-5} of one percent of the total protein. Again, a flat gel slab is very convenient since the gel plate, containing the TCA-fixed radioactive spots, can be easily overlayered with an X-ray film in an X-ray exposure holder and properly developed and analyzed. The protein can be radiolabelled in vivo or radio-iodinated in vitro with lactoperoxidase or, in the case of antibodies, reacted with a tagged hapten[217] or with radioactive antigens[218]. When enough radioactivity is present, the radioactive gel can also be conveniently analyzed by a spark chamber, such as the LKB 21C5 Radiochromatogram Camera or a similar chamber available from Birchover Instruments Ltd. Recently, it has been demonstrated that the classical strip scanners for thin layers, such as the Packard Radiochromatogram Scanner, are perfectly adeguate for scans of thin layer polyacrylamide gels, provided enough radioactivity (usually at least 0.4 μC) is present in each band[75].

Immuno-detection. Several variants have been described. IEF can simply be followed by immunodiffusion in which case the intact gel is bathed in antiserum[219] or overlayered with antibody-impregnated strips of paper[220] or squashed gel sections are transferred into wells punched in agar gel and subjected to radial diffusion[221]. Alternatively, the focused gel can be subjected to immunoelectrophoresis at right angles. Perhaps the most useful variant is the application of the Laurell's rocket technique[222], by which the focused proteins are forced to migrate electrophoretically into an antibody -impregnated agarose gel. Best results are obtained by the laying-on technique of Söderholm and Smyth[223], instead of embedding the focused acrylamide gel strip into agarose, which causes a strong electroendoosmotic flow at the acrylamide/agarose junction.

Densitometry. Commercial gel scanners developed for cellulose acetate strips or zone and disc electrophoresis were soon found to be unable to keep the pace with the great resolution afforded by IEF, where often few microns thin bands are spaced just few microns apart. It is well known that with most scanners the peak tracing does not go back to baseline if two protein zones are not spaced apart at least one or two mm.

Recently, a soft laser scanner, with a light source of polarized, monochromatic, coherent light of 633 nm has been made available . This radiation is only useful for blue dyes, but will be transparent to red dyes, such as Ponceau S. An additional laser beam at 320 nm for fluorescence and UV absorbtion is also available. The laser beam can be reduced to a microspot having a diameter as narrow as 3-10 μm. It is thus possible to resolve adjacent bands spaced apart of only 80 μm and to resolve them in uniform gaussians, even if the actual band profile is skewed, zigzagged or arc-shaped. The next logical step will be the introduction of tunable laser scanners.

IEF at sub-zero temperature

This could be a useful addition to presently available systems to study reaction intermediates. Often these intermediates have only a short half life at the temperatures commonly used in conventional electrophoresis, but their lifetime could be considerably extended by IEF at sub-zero temperatures, to a point at which their detection becomes feasible. Thus Park[224] has studied the formation of mixed tetramers of HbA-HbS, HbA-HbC, HbA-HbC harlem and HbS-HbC, by developing a system which allows the gel to be run at approximately -10°C. To prevent freezing, gels contain 25% ethylene glycol. In this system, Park[224] has been able to focus CO-hemoglobin partially oxidized with $K_3Fe(CN)_6$ and to resolve them into nine distinct bands, as theoretically predicted. However, a gel matrix cannot be used below 10°C, because the polymer will undergo a glass transition and become opaque and unelastic. Rcently, it has been possible to extend this temperature down to -30°C, by using a mixture of acrylamide with acrylic acid esters in the polymerization step (of the three esters investigated, methyl acrylate, ethyl acrylate and n-butyl acrylate, the first one, in a ratio of about 1:1 with acrylamide, appeared to be the most suitable)[225]. Alcohols of longer aliphatic chains would progressively depress the polymer glass transition but the gel matrix often becomes hard and brittle and not permeable to macromolecules. The best antifreezing solvent appeared to be dimethylsulphoxide (DMSO), at a concentration of about 40%. -30°C appears to be the lowest limit attainable in IEF systems, since focused carrier ampholytes as well as proteins tend to precipitate. By this method, for instance, it was possible to isolate liganded hybrid HbA-HbS molecules, which have a high dissociation rate at 4°C. This method could be very promising for the study of short lived intermediates in enzymatic reactions.

Micro isoelectric focusing

Presently available techniques for IEF in gel tubes or in slabs allow detection of a few μg protein. However, it could be desirable to have systems which allow detection of protein in the nanogram range. Grossbach[226] has developed IEF in capillaries of 50, 100 and 300 μ inner diameter and 65 mm long. 20 to 70 ng of each protein component are detected. The sample is mixed with the monomer solution before polymerization and the quartz tubes are filled by capillary action. After IEF, the gels are pushed out of the capillaries, by means of a tightly fitting steel wire, into a drop of 20% TCA on a depression slide. Gainer[227,228] has described a micromethod capable of detecting proteins at the 10^{-10} to 10^{-9} g level. Glass capillaries 7 cm long and 580 μ inner diameter are used. Capillaries with a too narrow bore might generate a too high electroendoosmotic flow along the glass walls, therefore capillaries having an inner diameter of 500 to 600 μ are preferred. Additional information on micro-IEF is found in the book on micro-methods by Neuhoff[229].

Transient state isoelectric focusing (TRANSIF)

This is Catsimpoolas creature. He has published some 20 papers on the subject (for a review, see ref. 230). TRANSIF involves the study of the kinetic behavior of charged amphoteric molecules during their transport by the electric field in an IEF system. TRANSIF is thus a kinetic method which is assumed to consists of three stages: initial focusing (IF), defocusing (DF) and refocusing (RF). The first stage involves the electrophoretic migration of amphoteres to their pI with the attainment of steady-state focusing. The second stage (DF) is concerned with analysis of diffusion in the absence of current and the third stage (RF) studies the reapproach to the steady state when the electrical field is reapplied. TRANSIF gives informations on such parameters as minimal focusing time, segmental pH gradient d(pH)/dx, and apparent physical constants of proteins, such as pI, diffusion coefficient D, retardation coefficient C_R and the slope of the pH-mobility curve at pI, dM/d(pH). Unfortunately, measurements of D and C_R appear to be unrealistically higher than expected. Measurements of true physicochemical parameters must thus await a second generation of ampholytes exhibiting the same conductivity, buffer capacity and viscosity, at their pI, all along the separation column.

Two-dimensional techniques

There are at least six variants utilizing IEF as one of the two dimensions (IEF-SDS gel electrophoresis or IEF-SDS gel gradient electrophoresis; IEF-gel electrophoresis; IEF-gel gradient electrophoresis; IEF-immunoelectrophoresis; IEF-isotachophoresis (ITP) and IEF-electrophoresis in a stationary pH gradient). These techniques are reviewed in detail in this book (see Ch. 16 by Gianazza and Righetti). I will briefly comment here on the last two methods. IEF-ITP[231] should allow an investigation of the differences between the separation principles of these two methods. In ITP, the degree of protein separation can be varied by changing the mobility of the leading ion and in IEF the separation of proteins occurs until the mobility is zero, a case of ITP at zero velocity. A combined IEF-ITP separation therefore should allow separation, by IEF, of proteins of similar mobility under ITP and separation, by ITP, of proteins having similar isoelectric point under IEF. The last method, recently introduced by Rosengren *et al.*[232], is a real innovation in IEF. A gel slab is first focused without sample. After IEF, the gel is turned 90°, the sample applied in a narrow slot going from anode to cathode at the center of the gel, and electrophoresis performed at right angles to IEF. Since all the carrier ampholytes are focused and practically stationary during the electrophoretic step, the resulting, sigmoidal-shaped protein pattern is indeed the protein titration curve in the pH range 3-10. The pI is the intersection point between the application slot and the protein curve. This very elegant method appears to hold tremendous future applications.

GENERAL EXPERIMENTAL ASPECTS

I will discuss here briefly some experimental aspects common to all the possible variants of IEF, preparative and analytical, in tubes or slabs, in granulated layers as well as continuous matrices. They include the use of additives, pH measurements, removal of carrier ampholytes, pH gradient decay and possible artefacts.

Additives

I have already discussed under the heading "protective agents" the use of thiol-protecting agents and of reducing compounds, such as ascorbic acid, for prevention of anodic oxidation. One of the main uses of additives is in preventing or lessening isoelectric precipitation of macromolecules. In addition to stabilizing or solubilizing focused bands, some additives have the added advantage of reducing differences in osmolarity along the gradient caused by focused ampholytes. Also, by increasing the osmolarity, additives reduce the risk of an unacceptably large surface film of water. Non-ionic detergents have been successfully used in concentrations ranging from 0.1 to 5%. They have proven especially useful for maintaining the solubility of focused membrane proteins. Many such detergents are now available under various trade names, e.g.: Tween 80, Emasol, Brij 39, Triton X-100, Nonidet P-40 (NP40). In addition to non-ionic, also zwitter-ionic surfactants, such as the alkylbetaines: $R\text{-}\overset{+}{N}(CH_3)_2\text{-}CH_2\text{-}COO^-$ ($R=C_{10}$ to C_{16}) (Empigen BB) or the sulphobetaines: $R\text{-}\overset{+}{N}(CH_3)_2\text{-}CH_2\text{-}CH_2\text{-}SO_3^-$ ($R=C_{12}$) (sulphobetaine DLK) can be effectively used in IEF[233]. Alper *et al.*[234] have used a similar approach by incorporating in the focusing system 0.2M taurine ($^+NH_3\text{-}CH_2\text{-}CH_2\text{-}SO_3^-$) which is isoelectric over a wide pH range and serves to raise the medium osmotic pressure without contributing to its electrolyte concentration or to the viscosity of the liquid phase. Particularly attractive seems to be a combination of 9M urea and 2% NP40 in the focusing medium[216,235]. Urea dissociates aggregates and subunit assemblies and unfolds polypeptide chains, thereby reducing the influence of protein conformation on the focusing pattern and releasing possible non-covalently bound cofactors, while NP40 helps disaggregating and solubilizing membrane proteins. In this system, multiple banding patterns or artifactual bands described to occur in IEF in absence of urea and detergents[236,237], are not observed. Furthermore, there is no evidence that carbamylation of proteins by cyanate present in the concentrated urea solution takes place, probably because carrier ampholytes provide protection against potential carbamylation. When using a slightly modified medium (8M urea and 3% NP40) we have found a very interesting phenomenon[238]. It turns out that NP40 not only binds to carrier ampholytes[79] (when the gel slab, after IEF, is bathed in TCA, the NP40 is seen condensed along the ampholyte refractile lines, and not uniformly distributed in the gel; Righetti, unpublished) but also to proteins. Surprisingly, when focusing β and γ chains of hemoglobin (Hb), we found that bound NP40 altered the pI's of these two chains. In absence of NP40 the pI differences among the chains was only *ca.* 0.06 pH units, and β and γ polypeptides were barely 1 mm apart in the gel. In presence of NP40 the pI of the β chain was increased by about 0.1 pH units while the pI of the γ chain was concomitantly lowered to the same extent and now the two chains were 5-6 mm apart in the gel. I don't have a ready explanation for this phenomenon, but it is certainly something worth exploring and exploiting.

In the case of tenacious complexes, other disaggregating agents have been successfully tried: 50% DMSO, 50% dimethyl formamide (DMF), 50% tetramethyl urea (TMU) and up to 90% formamide (FA)[77]. Except for FA, the concentration limit in IEF appears to be about 50%, as higher levels will precipitate Ampholine. If they are added to a gel, DMF and TMU should be used at levels of about 30-35%, as higher levels hamper gel polymerization. In the case of Ampholine-dye complexes, the most powerful disaggregating agents are DMF and TMU and the weakest FA (on an equimolar basis). However, since FA can be used at levels up to 90-95%, without precipitating Ampholine or inhibiting gel polymerization, its use in IEF appears to be very promising. Unfortunately, I have found that it is rapidly hydrolyzed at the cathode (even when 1M NaOH is substituted with weaker bases) giving free ammonia and formic acid, which severely interfere with pH gradient formation and sample focusing.

pH measurements

The pI of a protein determined by IEF also represents its isoionic point in the absence of complex-forming ions. By definition, the isoionic point is a measure of the intrinsic acidity of a pure protein, as it is defined as that pH which does not change on addition of a small amount of pure protein[239]. This definition is also applicable to a protein in IEF, as the pH of the isoelectric zone does not alter on addition of more protein; pI values estimated by IEF are temperature dependent and usually decrease with increasing temperature[10]. The difference in pI for the same protein, measured at 25° and 4°C, could be as high as 0.5 pH units, the higher value being obtained at the lower temperature. This difference is usually more pronounced in alkaline regions and when a protein has a pI value close to the pK of some of its functional groups. In fact, apart from carboxyls, the protolytic groups in proteins have rather large standard heats of ionization (6-14 Kcal/mole) and therefore neutral and basic proteins will exhibit large dpI/dT increments. As shown by Fredriksson[240], in fact, the dpH/dT and dpI/dT curves for carrier ampholytes and proteins, respectively, coincide only below pH 5 and greatly diverge in the pH range 5.5 to 9, especially as a function of the imidazole group content of a protein.

Ideally, a good mixture of pH markers, of known pI's at 4 and 25°C, could avoid tedious pH measurements after IEF. Tables on the use of protein and dye pH markers have been reported[40,37,144,241,242]. Protein markers should not be heterogeneous or at least have a major, easily identifiable, band in order to avoid confusion in pH assessments. Dyes would appear to be very promising, except that most of them, especially sulphonated derivatives, interact strongly with carrier ampholytes, originating a multitude of bands representing specific ampholyte-dye complexes[77]. For pH measurements after IEF in density gradients, Jonsson *et al.*[243] Secchi[244] and Strongin *et al.*[245] have described flow cells which allow continuous pH monitoring of the eluate. The pH curve can be overimposed to the UV profile in the same chart, thus providing easy reference for pH assessments[244]. Gelsema and DeLigny[246-248] have stressed that, when using additives (especially in gradients) several correction factors should be introduced to correct for the apparent isoelectric point (pI_{app}) determined. Their conclusions can be summarized as follows:

a) pI_{app} -pI values, *i.e.* the errors made in the measurement of isoelectric points by the conventional execution of density gradient IEF, depend upon the acidity, basicity and the chemical type of the ampholyte and upon the solvent composition;

b) these errors are within ± 0.1 pH unit for Ampholine (and presumably proteins) having $3 < pI < 9$ in gradients of 0-60% sucrose;

c) the temperature coefficient of pI values of Ampholine (and presumably proteins) is in good agreement with that of pK values of carboxylic and amino groups;

d) the primary medium effect (pI -pI) of sucrose, glycerol and ethylene glycol on pI values of Ampholine is in good agreement with that (pK -pK) on pK values of carboxylic acids and alkyl substituted ammonium ions;

e) pI_{app}-pI values holding for Ampholine can be used for the correction of pI_{app} values of proteins determined by density gradient IEF.

When using 6M urea as an additive in sucrose density gradient IEF, Ui[249] suggests an overall correction factor of 0.42 pH unit, to be subtracted to the apparent pI of the protein in this medium. A correction factor ranging from 0.7 to 0.9 pH unit has been recommended by Josephson *et al.* in 7M urea solutions[250], while Gianazza *et al.*[251] find that the urea effect varies with the various Ampholine pH ranges, from 5 to 7.5 hundreths of a pH unit/unit of urea molarity in going from acidic to alkaline pH ranges. In agreement with these last findings, Gelsema and DeLigny[246] report that the use of a unique correction term irrespective of the pI values of the ampholytes studied is incorrect. For pH measurements in gel tubes, it was customary to section the gel in 2 to 5 mm segments, which were eluted in 200 to 300 µl of 10 mM KCl prior to pH readings. Recently, this operation has been automated with continuous pH readings on an intact gel rod[252]. A contact microelectrode (Ingold No. E547310--1), mounted on a mechanical drive, moves along the surface of a gel cylinder resting on a trench in a plastic block. The gel is scanned at a rate of 1 cm/1.5 min and the pH is automatically recorded on a chart. In gel slabs, the pH can be measured directly on the gel surface with a flat membrane electrode, such as LOT type 403-30-M8 from Ingold (with a 6 mm membrane diam.) while punching 1-2 mm reference holes on the gel as markers for pH readings[40]. The original antimony electrode of Beeley *et al.*[253] has been mostly abandoned because of its low reproducibility and lack of nernstian properties. Recently, a microelectrode of iridium (a needle, with a 0.2 to 0.5 mm tip) has been described[251,254]. This is a true nernstian electrode, with linear pH/volts and pH/temperature responses and seems to be the only microelectrode able to perform pH readings at 4°C, the usual temperature for IEF experiments (with all other microelectrodes, the gel has to be warmed up to room temperature before pH readings).

Removal of carrier ampholytes

After IEF, it is often desirable to remove the carrier ampholytes from the protein fraction. Since most ampholytes have a mean MW between 600 and 900 d[58], it would appear to be a simple matter to remove them by dialysis. However, working with ^{14}C-Ampholine, Vesterberg[255] found that a complete removal by dialysis requires at least 32 hours, even when dialysing against 0.1M phosphate buffer, pH 7, in 0.5M NaCl. Nilson *et al.*[256] and Li and Li[257] obtained a very efficient removal of ampholytes by precipitation and subsequent washings of the protein with ammonium sulphate, in concentrations ranging from 60 to 100%. The protein precipitate is then washed three times with the same $(NH_4)_2SO_4$ solution. It appears that all ampholytes are soluble even at 100% salt saturation, so that their removal is quantitative. Quast and Vesterberg[258] obtained a good separation of proteins from ampholytes by gel filtration in Sephadex G-50 fine. The separation is performed in columns equilibrated with 0.1M phosphate buffer, containing 0.5M NaCl, or 0.1M ammonium bicarbonate or 0.1M imidazole acetate. These buffers are usually kept at pH of about 7. The last two buffers have the advantage of being volatile, which allows them to be removed by lyophilization. Whichever method is used, it is important to have a high ionic strength in the buffer (0.1M, or 0.5M salt) to disrupt weak complexes which might form between Ampholine and proteins by electrostatic interaction. Alternatively, Brown and Green[259] have proposed the use of a mixed-bed ion-exchange resin (Bio Rad AG 501-X8) for separation of proteins and carrier ampholytes. This seems to be the method of choice, as also confirmed by Baumann and Chrambach[260], who found, by using ^{14}C-ampholyte, that less than 0.005 moles of Ampholine would remain bound to 1 mole of protein after chromatography on the bifunctional resin. This tantamounts to quantitative removal. Of course, electrophoretic removal, as described by Suzuki *et al.*[160] and by Stathakos[161] remains always a valid alternative.

pH gradient decay

The "cathodic drift"[81] or "plateau phenomenon"[159] seems incompatible with the concept of IEF, yet is an ever-present hazard that the users of the technique must learn to cope with. The pH drift during IEF seems to have a multifactorial origin[58,121,261]. Seven hypothesis have been advanced to explain pH gradient decay: 1) electrophoretic migration of isoelectric Ampholine; 2) electroendoosmosis; 3) formation of a zone of pure water in the neutral region of the pH gradient; 4) "isoelectric focusing" of water (considered to be an ampholyte) with resulting accumulation of water in the neutral region and backflow towards the periphery of the gel; 5) selective deficiency or progressive destruction of neutral ampholytes; 6) progressive gain or loss of charged ligands by Ampholine; 7) diffusion of Ampholine out of and/or electrolytes into the gel. From experiments with ^{14}C-Ampholine and $^{3}H_2O$ it has been concluded that the major force involved in pH gradient decay appears to be electrophoresis of Ampholine, resulting in its depletion from the gel. It is possible that the movement takes place "en bloc" especially in the cathodic direction, possibly due to chemical interaction among Ampholine species or to their modification by reaction with acid or base. Electroendoosmosis is a possible, albeit minor, force in pH gradient decay, and could play a major role in the predominantly unidirectional (cathodic) drift of Ampholine. On the other hand, according to Rilbe[262], it is possible to explain the plateau phenomenon and consequent cathodic drift as effects of electroosmosis and hydration water transport. The general remedy applied is an increase in viscosity of the medium and increase in carrier ampholyte concentration. High voltage techniques favor the cathodic drift, thus the experiment should be terminated within 2 hours in gel slabs (when applying more than 100 V/cm) to avoid excessive decay of the pH gradient. Recently, Nguyen and Chrambach[96] have found that it is possible to stabilize pH gradients against decay by equalization of the anolyte pH with the pH of the most acidic amphoteric component in a given Ampholine pH range. In the case of pH range 6-8, this is achieved by using Thr at the anode and His at the cathode. Oddily, the choice of catholyte does not modify the cathodic drift. It remains to be seen whether or not this is a generally applicable to any pH range.

"pH gradient-modifier" IEF

Introduced by Brown *et al.*[74,263] with the name of "separator IEF" this technique has been further studied and exploited by Beccaria *et al.*[264]. It consists in modifying and flattening an Ampholine pH gradient by addition of one or more amphoteric compounds which, by condensing at their pI's, will alter the pH course at or near their pI. This results in an improved separation for the proteins focusing in the flattened pH region. The following compounds have been used as "modifiers": tetraglycine (pI 5.2), Pro (pI 6.3), Thr (pI 6.5), β-alanine (pI 6.9), 5-amino valeric acid (pI 7.5), 4-amino butyric acid, 7-amino caprylic acid, His (pI 7.6) and 6-amino caproic acid (pI 8.0). All of them, except His, are poor carrier ampholytes (pI-pK values of 3 to 4) and thus focus as a broad plateau in the Ampholine gradient. Therefore, in order to be able to modify the pH gradient, they have to be used at rather high concentrations, usually about 0.3 to 0.6M. On the other hand, good ampholytes, such as His (or the dipeptide His-Gly, successfully used for the separation of HbA from HbA_{IC}, a minor glycosylated Hb)[264] are already effective at concentrations of 20 to 50 mM and probably act by interposing themselves between the two species to be separated. pH gradient modifier IEF adds another dimension to IEF by further improving separations which are at the resolution limit of the technique.

Artefacts

In many cases, proteins homogenoeus by previous criteria, have been found heterogeneous when examined by IEF. Many of the possibilities for charge heterogeneity have been discussed by Epstein and Schechter[265], Kaplan[266] and Williamson *et al.*[267]. Perhaps the most intriguing question is whether or not the microheterogeneity so often detected, rather than being inherent to the sample being examined, is artefactually elicited by interaction of the protein with the carrier ampholytes. The results reported so far have been conflicting. Thus, while Frater[236] has reported binding of Ampholine to acidic wool proteins and Kaplan and Foster[268] and Wallevik[269] complex formation between Ampholine and bovine serum albumin, these interactions have been excluded by Dean and Messer[270] and Baumann and Chrambach[72]. Recently, however, a whole set of artefacts has been discovered by my reasearch group[77,271-274]. They include binding among Ampholine and nucleic acids (notably t-RNA), sulphonated dyes, heparin and practically all polyanions (polyphosphates, polysulphates and polycarboxylates, including polyglutamate). These cases are only apparently unrelated, infact they belong to the same category of phenomena. In all these cases, binding is strongest in the pH range 3.2-4.5, where the Ampholine polyamine backbone exhibits a ordered sequence of 4 to 5 positive charges, while the carboxyl groups are mostly protonated. Binding is affected by the type of negative charge, its density and spatial orientation on the polyanion. On the basis of charge type, the binding strength decreases in the order: polyphosphate > polysulphate > polycarboxylate. The binding is pH-dependent: it is very strong at pH 3 (where Ampholine behaves as a polycation), rather weak at pH 5 and abolished at pH 7 (where Ampholine reverses its charge and behaves more as a polyanion). Moreover, upon binding, a homogeneous species is not just resolved into two peaks (the unbound versus the bound species) but into a multitude of bands, each one representing a specific Ampholine-macromolecule complex. Thus, it is typical to see more than 20 bands resolved in a range of abour one pH unit, demonstrating that there are at least 20 carrier ampholytes able to form complexes with a given macromolecule in one pH unit range. This is in contrast with theoretical considerations by Cann's group[275-277], who have demonstrated, in the three possible cases investigated (carrier ampholyte-induced macromolecular isomerization, pH dependent conformational transitions and carrier ampholyte-induced macromolecular association or dissociation into subunits) bimodal transient and equilibrium IEF patterns, with one of the two peaks growing at the expenses of the other, in the third case. Recently, however, Cann (see Ch. 19 in this book) has theoretically demonstrated that, given *n* carrier ampholytes interacting with a single macromolecule, *n+1* peaks will be generated, thus fully supporting our experimental evidence. On the opposite extreme of the pH scale, we have demonstrated that the same binding phenomena can be elicited by polycations (notably polyarginine and polylysine)[278].

Recently, some more evidence has been brought forward on the binding of particular proteins, such as lipoprotein lipase[279] and rat incisor phosphoprotein[280] to carrier ampholytes. While these might just be isolated cases, it is always best to check for artefacts when dealing with a new protein in IEF (focusing in presence of disaggregating agents, sample application at the anode and cathode, changing the ratio Ampholine/protein, etc.)[42]. My feeling is that, when dealing with IEF of proteins, in the vast majority of the cases the microheterogeneity found is inherent to the sample and not artefactually elicited by the IEF technique.

Chromatofocusing

This new technique, described by Sluytermann and Wijdenes[281-283], is a kind of isoelectric focusing in the absence of an externally applied potential. The experiment is run as follows: an anion or cation exchange resin is buffered at a proper pH and the sample applied. Elution is performed by using a buffer at another, appropriate pH, or a suitable mixture of Ampholine in a given pH range. This creates a pH gradient which is "internally" generated, as opposed to "externally" generated pH gradients made by mixing two buffers of different pH's in a mixing chamber, prior to the column application. Within this pH gradient proteins are eluted according to their pI's and focusing effects can be demonstrated by showing that the same protein, applied to the column at different time intervals during elution, condenses or focuses within the same zone. A similar approach, called "ampholyte displacement chromatography", has been described by Leaback and Robinson[284]. Here the proteins are displaced from an ion exchange adsorbent using a mixture of carrier ampholytes of closely similar isoelectric points. A spin-off of the IEF research is also a new chromatographic technique, called "isoelectric chromatography", recently described by Petrilli *et al.*[285]. The idea here is to run the unpurified protein suquentially through an anion and a cation exchanger, both buffered at the isoelectric pH of the protein to be purified. In this way all the impurities will be absorbed while the required protein will flow through unabsorbed. The resolution here is given around 0.2 to 0.3 pH units in pI differences among the species to be separated.

FUTURE TRENDS AND DEVELOPMENTS

This early 1978 panorama of IEF allows us to make some predictions. Drift-free IEF systems appear to be now in sight[196,262], and this will solve a bad nuisance of present-day IEF. Excellent staining systems have been described[79,206-208], which allow quick gel staining with no background coloring. Coupled to laser scanners and computers, this should allow in the near future rapid collection, computation and storage of data, particularly important when running routine analysis and large screening programs. To this purpose, the recent introduction in the market of electroendoosmosis-free cellulose acetate (in which the -COOH groups have been extensively methylated, as described by Ambler[286]) appears to be a great step ahead for quick focusing, quick staining and rapid data collection. Also, the recently introduced focusing technique at -20 to -30°C could open up a completely new field of protein

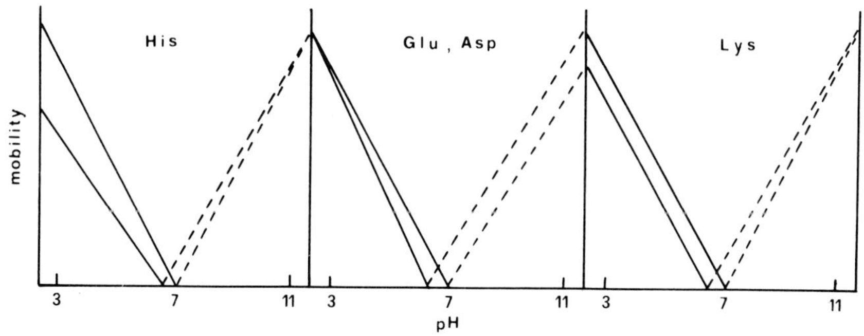

Fig. 13. Theoretical titration curves of normal and mutant hemoglobins. For simplicity, straight lines instead of sigmoidal curves have been drawn. The observations are made during the transient state from both, above (broken lines) and below (solid lines) the isoelectric plane. Only the cases regarding His, Glu or Asp and Lys mutants are shown (Righetti, P.G., unpublished).

and enzyme chemistry, particularly for the study of too short lived reaction intermediates.

Finally, I should like to stress that some useful information could be obtained not only by equilibrium IEF, but also by transient state IEF. For instance, by following the rate of migration of a protein and of its genetic mutants, it is possible to determine which charged amino acid has been substituted. I am giving here a theoretical example for His, Glu or Asp and Lys mutants calculated for hemoglobins. By following the rate of migration of Hb and its mutants, from below and above the isoelectric plane, and plotting it against the pH of the zone in which the protein is travelling, one obtains a titration curve. When examining the shape of the two half-legs of the titration curve, it can be seen that, in the case of a His mutant, the two curves will be converging toward the pI from below the pI plane and diverging toward the pI from above the pI plane. In the case of Glu or Asp mutants (at the moment, I think it will be difficult to distinguish among the two) the two curves will be diverging toward the pI from below the pI plane and parallel above pH 7. Exactly the opposite behaviour can be predicted for Lys mutants, while in the case of Arg mutants both curves will be parallel from both, above and below the pI plane (not shown). I think it will also be possible to predict mixed mutations (e.g. Glu→Lys mutants, etc.). Clearly, the horizon of IEF is in full expansion.

ACKNOWLEDGEMENTS

Supported in part by grants CT76.00311.06, 76.01488.04, CT77.01471.04 and CT77.00278.06 from Consiglio Nazionale delle Ricerche (CNR) Roma, Italy. I am indebted to my colleagues and former students (E. Galante, E. Gianazza, O. Brenna, M. Pagani and T. Caravaggio) who have made possible through their enthusiasm and dedication our own research in IEF.

REFERENCES

1. Catsimpoolas, N. (ed.)(1973) Isoelectric Focusing and Isotachophoresis, Ann. N.Y. Acad. Sci. vol. 209
2. Allen, R.C. and Maurer, H.R. (eds.)(1974) Electrophoresis and Isoelectric Focusing in Polyacrylamide Gels, W. de Gruyter, Berlin
3. Arbuthnott, J.P. and Beeley, J.A. (eds.)(1975) Isoelectric Focusing, Butterworths, London
4. Righetti, P.G. (ed.)(1975) Progress in Isoelectric Focusing and Isotachophoresis, North Holland/American Elsevier, Amsterdam and New York
5. Radola, B.J. and Graesslin, D. (eds.)(1977) Electrofocusing and Isotachophoresis, W. de Gruyter, Berlin
6. Matuo, Y. and Horio, T. (1967) Prot. Nucl. Acid Enz. 12, 737-748
7. Svensson, H. (1967) Svensk Kem. Tidskr. 79, 654-659
8. Kolin, A. (1970) in "Methods in Medical Research" (Olson, R.E., ed.) vol. 12, Year Book Med. Publ., Inc., Chicago, pp. 326-358
9. Righetti, P.G., Gianazza, E. and Bianchi Bosisio, A. (1976) Ital. J. Clin. Chem. 1, 11-46
10. Vesterberg, O. and Svensson, H. (1966) Acta Chem. Scand. 20, 820-834
11. Allen, R.C., Harley, R.A. and Talamo, R.C. (1974) Amer. J. Clin. Pathol. 62, 732-739
12. Ikeda, K. and Suzuki, S. (1912) U.S. Patent No. 1015891
13. Williams, A.R. and Waterman, R.E. (1929) Proc. Soc. Exp. Biol. Med. 27, 56-59
14. Tiselius, A. (1941) Svensk Kem. Tidskr. 53, 305-312
15. Kolin, A. (1977) in ref. 5, pp. 3-33
16. Kolin, A. (1954) J. Chem. Phys. 22, 1628-1629
17. Kolin, A. (1955) Proc. Natl. Acad. Sci. 41, 101-110
18. Kolin, A. (1958) in "Methods Biochem. Anal." (Glick, D., ed.) vol. 6, Wiley, Interscience, New York, pp. 259-288
19. Svensson, H. (1961) Acta Chem. Scand. 15, 325-341
20. Svensson, H. (1962) Acta Chem. Scand. 16, 456-466
21. Svensson, H. (1962) Arch. Biochem. Biophys., Suppl. 1, 132-140
22. Vesterberg, O. (1969) Acta Chem. Scand. 23, 2653-2666
23. Rilbe, H. (1976) Sci. Tools 23, 18-22
24. Haglund, H. (1971) in "Methods Biochem. Anal." (Glick, D., ed.) vol. 19, Wiley, Interscience, New York, pp. 1-104
25. Catsimpoolas, N. (1973) Separ. Sci. 8, 71-121; (1975) ibid. 10, 55-76
26. Vesterberg, O. (1971) in Methods Enzymol. (Jackoby, W.B. ed.) vol. 22, Academic Press, New York, pp. 389-412
27. Vesterberg, O. (1973) in "VII Symposium Chromatogr. Electrophor.", Publ. Press Acad. Europ., Bruxelles, pp. 81-98
28. Leaback, D.H. and Wrigley, C.W. (1976) in "Chromatographic and Electrophoretic Techniques" (Smith, I., ed.) vol II, Heinemann, London, pp. 272-320
29. Morris, C.J.O.R. and Morris, P. (1976) Separation Methods in Biochemistry, 2nd ed., Pitman Publ., London, pp. 853-876
30. Strickland, R.D. (1976) Anal. Chem. 48, 39R-44R
31. Rilbe, H. (1970) in Prot. Biol. Fluids (Peeters, H., ed.) Pergamon Press, London

17, 369-382.

32. Righetti, P.G. (1975) Separ. Purif. Methods 4, 23-72
33. Wrigley, C.W. (1971) in Methods Enzymol. (Jakoby, W.B., ed.) vol. 22, Academic Press, New York, pp. 559-564
34. Righetti, P.G. and Drysdale, J.W. (1974) J. Chromatogr. 98, 271-321
35. Vesterberg, O. (1973) Sci. Tools 20, 22-29
36. Wadström, T. and Smyth, C.J. (1973) Sci. Tools 20, 17-21
37. Bours, J. (1973) Sci. Tools 20, 29-34
38. Williamson, A.R. (1973) in "Handbook Exptl. Immunol." (Weir, D.M., ed.) Blackwell Sci. Publ., Oxford, Ch. 8, pp. 1-23
39. Latner, A.L. (1975) in "Advances in Clinical Chemistry" (Bodansky, O. and Latner, A.L., eds.) vol. 17, Academic Press, New York, pp. 193-250
40. Righetti, P.G. and Caravaggio, T. (1976) J. Chromatogr. 127, 1-28
41. Catsimpoolas, N. (ed.) (1976) Isoelectric Focusing, Academic Press, New York
42. Righetti, P.G. and Drysdale, J.W. (1976) Isoelectric Focusing, North Holland, American Elsevier, Amsterdam and New York
43. Rilbe, H. (1973) Ann. N.Y. Acad. Sci. 209, 11-22
44. Rilbe, H. (1971) Acta Chem. Scand. 25, 2768-2769
45. Rilbe, H. (1976) in ref. 41, pp. 14-52
46. Almgren, M. (1971) Chem. Scripta 1, 69-75
47. Giddings, J.C. and Dahlgren, (1971) Separ. Sci. 6, 345-356
48. Svensson, H. (1967) Prot. Biol. Fluids (Peeters, H. Ed.) Pergamon Press, London 15, 515-522
49. Rilbe, H. and Pettersson, S. (1975) in ref. 3, pp. 44-57
50. Radola, B.J. (1975) in ref. 3, pp. 182-197
51. Svensson, H. (1966) J. Chromatogr. 25, 266-273
52. Vesterberg, O. (1973) Ann. N.Y. Acad. Sci. 209, 23-33
53. Vesterberg, O. (1976) in ref. 41, pp. 53-76
54. Vinogradov, S.N., Lowenkron, S., Andonian, H.R., Bagshaw, J., Felgenhauer, K. and Pak, S.J. (1973) Biochem. Biophys. Res. Commun. 54, 501-506
55. Righetti, P.G., Pagani, M. and Gianazza, E. (1975) J. Chromatogr. 109, 341-356
56. Grubhofer, N. and Borja, C. (1977) in ref. 5, pp. 111-120
57. Davies, H. (1970) in Prot. Biol. Fluids (Peeters, H., ed.) Pergamon Press, London vol. 17, pp. 389-396
58. Haglund, H. (1975) in ref. 3, pp. 3-22
59. Fredriksson, S. (1977) J. Chromatogr. 135, 441-446
60. Fawcett, J.S. (1977) in ref. 5, pp. 59-69
61. Righetti, P.G., Balzarini, L. Gianazza, E. and Brenna, O. (1977) J. Chromatogr. 134, 279-284
62. Felgenhauer, K. and Pak, S.J. (1973) Ann. N.Y. Acad. Sci. 209, 147-153
63. Frater, R. (1970) J. Chromatogr. 50, 575-585
64. Brown, R.K., Lull, J.M., Lowenkron, S., Bagshaw, J.C. and Vinogradov, S.N. (1976) Anal. Biochem. 71, 325-332
65. Rilbe, H. (1973) Ann. N.Y. Acad. Sci. 209, 80-93

66. Radola, B.J., Tschesche, H. and Schuricht, H. (1977) in ref. 5, pp. 97-110

67. Valmet, E. (1969) Sci. Tools 16, 8-13

68. Gianazza, E.,Pagani, M., Luzzana, M. and Righetti, P.G. (1975) J. Chromatogr. 109, 357-364

69. Fawcett, J.S. (1973) Ann. N.Y. Acad. Sci. 209, 112-126

70. Gasparic, V. and Rosengren, A. (1975) in ref. 3, pp. 178-181

71. Gasparic, V. and Rosengren, A. (1974) Sci. Tools 21, 1-2

72. Baumann, G. and Chrambach, A. (1975) Anal. Biochem. 64, 530-536

73. Brown, R.K., Gaspers, M.L., Lull, J.M., Vinogradov, S.N., Felgenhauer, K. and Nekic, M. (1977) J. Chromatogr. 131, 223-232

74. Brown, R.K., Caspers, M.L. and Vinogradov, S.N. (1977) in ref. 5, pp. 87-96

75. Righetti, P.G. and Chrambach, A. (1978) Anal. Biochem. 90, 633-643

76. Jonsson, M. and Pettersson, E. (1968) Sci. Tools 15, 2-5

77. Righetti, P.G., Gianazza, E., Brenna, O. and Galante, E. (1977) J. Chromatogr. 137, 171-181

78. Catsimpoolas, N. (1973) Ann. N.Y. Acad. Sci. 209, 65-79

79. Righetti, P.G. and Chillemi, F. (1978) J. Chromatogr. 157, 243-251

80. Righetti, P.G. (1977) J. Chromatogr. 138, 213-215

81. Righetti, P.G. and Drysdale, J.W. (1971) Biochim. Biophys. Acta 236, 18-28

82. Righetti, P.G., Righetti, A.B.B. and Galante, E. (1975) Anal. Biochem. 63, 423-432

83. Galante, E., Caravaggio, T. and Righetti, P.G. (1975) in ref. 4, pp. 3-12

84. Blanicky, P. and Pihar, O. (1972) Coll. Czech. Chem. Commun. 37, 319-325

85. Luner, S.J. and Kolin, A. (1970) Proc. Natl. Acad. Sci. 6, 898-903

86. Troitzi, G.V., Savialov, V.P. and Abramov, V.M. (1974) Dokl. Akad. Nauk USSR 214, 955-958

87. Lundahl, P. and Hjertén, S. (1973) Ann. N.Y. Acad. Sci. 209, 94-111

88. Troitski, G.V., Savialov, V.P., Kirjukhin, I.F., Abramov, V.M. and Agitsky, G.J. (1975) Biochim. Biophys. Acta 400, 24-31

89. Pettersson, E. (1969) Acta Chem. Scand. 23, 2631-2635

90. Stenmann, U.K. and Grasbeck, R. (1972) Biochim. Biophys. Acta, 286, 243-251

91. Nguyen, N.Y. and Chrambach, A. (1976) Anal. Biochem. 74, 145-153

92. Nguyen, N.Y., Rodbard, D., Svendsen, P.J. and Chrambach, A. (1977) Anal. Biochem. 77, 39-55

93. Nguyen, N.Y., Salokangas, A. and Chrambach, A. (1977) Anal. Biochem. 78, 287-294

94. Nguyen, N.Y. and Chrambach, A. (1977) Anal. Biochem. 79, 462-469

95. Caspers, M.L. and Chrambach, A. (1977) Anal. Biochem. 81, 28-39

96. Nguyen, N.Y. and Chrambach, A. (1977) Anal. Biochem. 81, 54-62

97. Nguyen, N.Y. and Chrambach, A. (1977) Anal. Biochem. 82, 226-235

98. Vesterberg, O. and Berggren, B. (1966) Arkiv fur Kemi 27, 119-125

99. Wadström, T. and Hisatsune, K. (1970) Biochem. J., 120, 725-731

100. Kostner, G., Albert, W. and Holasek, A. (1969) Hoppe-Seyler's Z. Physiol. Chem. 350, 1347-1352

101. Leise, E.M. and LeSane, T. (1974) Prep. Biochem. 4, 395-400

102. Sherbet, G.V. and Lakshmi, M.S. (1973) Biochim. Biophys. Acta 298, 50-58

103. Fredriksson, S. (1975) J. Chromatogr. 108, 153-167

104. Jonsson, M., Pettersson, E. and Rilbe, H. (1973) Anal. Biochem. 51, 557-576

105. Berghem, L. and Pettersson, L. (1973) Eur. J. Biochem. 37, 21-26

106. Schleger, M. and Christen, P. (1974) Biochem. Biophys. Res. Commun. 61, 117-122

107. Wadström, T. and Vesterberg, O. (1971) Acta Path. Microbiol. Scand. 79, 248-255

108. Janson, J.C. (1972) Ph.D. Thesis, Acta Universitatis Upsaliensis, Uppsala

109. Nishimune, T. and Hayashi, R. (1973) Bull. Yamaguchi Med. School 20, 9-16

110. Goode, R.L. and Balwin, J.N. (1973) Prep. Biochem. 3, 349-361

111. Smyth, C.J. and Arbuthnott, J.P. (1974) Med. Microbiol. 7, 41-66

112. Catsimpoolas, N. (1969) FEBS Letters 4, 259-261

113. Jacobs, S. (1971) in Prot. Biol. Fluids (Peeters, H., ed.) Pergamon Press, 19, 499-502

114. Jacobs, S. (1973) Analyst 98, 25-33

115. Vesterberg, O., Wadström, T., Vesterberg, K., Svensson, H. and Malmgren, B. (1967) Biochim. Biophys. Acta 133, 435-445

116. Wadström, T. (1967) Biochim. Biophys. Acta 147, 441-449

117. Rathnam, P. and Saxena, B.B. (1970) J. Biol. Chem. 245, 3725-3731

118. Boddin, M., Hilderson, J., Lagrou, A. and Dierick, W. (1975) Anal. Biochem. 64, 293-296

119. Doi, E. and Ohtsuru, C. (1974) Agr. Biol. Chem. 38, 1747-1748

120. Azuma, J.I., Kashimura, N. and Komano, T. (1977) Anal. Biochem. 81, 454-457

121. Fawcett, J.S. (1975) in ref. 3, pp. 23-43

122. Fawcett, J.S. (1975) in ref. 4, pp. 25-37

123. Brakke, M.K., Allington, R.W. and Langille, F.A. (1968) Anal. Biochem. 25, 30-39

124. Weller, D.L., Heaney, A. and Sjogren, R.E. (1968) Biochim. Biophys. Acta 168, 386-388

125. Godson, G.N. (1970) Anal. Biochem. 35, 66-76

126. Koch, H.J.A. and Backx, J. (1969) Sci. Tools 16, 44-47

127. Kint, J.A. (1975) Anal. Biochem. 67, 679-683

128. Osterman, L. (1970) Sci. Tools 17, 31-33

129. Korant, B.D. and Lonberg, K. (1974) Anal. Biochem. 59, 75-82

130. Massey, H.T. and Deal Jr., W.C. (1973) J. Biol. Chem. 248, 56-62

131. Behnke, J.N., Dagher, S.M., Massey, T.M. and Deal Jr., W.C. (1975) Anal. Biochem. 69, 1-9

132. Rilbe, H., Forchheimer, A., Pettersson, S. and Jansson, M. (1975) in ref. 4, pp. 51-63

133. Valmet, E. (1970) in Prot. Biol. Fluids (Peeters, H., ed.) Pergamon Press, Oxford, 17, 401-407

134. Talbot, P. and Caie, I.S. (1975) in ref. 3, pp. 74-77

135. Kalous, V. and Vacik, J. (1959) Chem. Listy 53, 35-37

136. Rose, C. and Harboe, N.M.G. (1970) in Prot. Biol. Fluids (Peeters, H., ed.) Pergamon Press, Oxford 17, 397-400

137. Macko, V. and Stegemann, H. (1970) Anal. Biochem. 37, 186-190

138. Bours, J. (1973) Exp. Eye Res. 16, 501-515

139. Quast, R. (1977) in ref. 5, pp. 455-462

140. Denkla, W.D. (1977) in ref. 5, pp. 423-432

141. Just, W.N., Leon-V, J.O. and Werner, G. (1975) in ref. 4, pp. 265-280

142. Just, W.W. and Werner, G. (1977) in ref. 5, pp. 481-494

143. Hanning, K. (1961) Z. Anal. Chem. 181, 244-254

144. Radola, B.J. (1973) Biochim. Biophys. Acta 295, 412-428

145. Radola, B.J. (1973) Ann. N.Y. Acad. Sci. 209, 127-143

146. Radola, B.J. (1974) Biochim. Biophys. Acta 386, 181-195

147. Winters, A., Perlmutter, H. and Davies, H. (1975) LKB Application Note No. 198

148. Winters, A. (1977) in ref. 5, pp. 433-442

149. Otavsky, W.I., Bell, T., Saravis, C. and Drysdale, J.W. (1977) Anal. Biochem. 78, 302-307

150. Weise, H.C., Graesslin, D. and Glismann, D. (1975) in ref. 4, pp. 93-98

151. Johansson, B.G. and Stanflo, J. (1971) Anal. Biochem. 40, 232-238

152. Quast, R. (1971) J. Chromatogr. 54, 405-412

153. Johansson, B.G. and Hjertén, S. (1974) Anal. Biochem. 59, 200-213

154. Bode, H.J. (1977) Anal. Biochem. 83, 204-210

155. Ziegler, A. and Kohler, G. (1976) FEBS Letters 64, 48-51

156. Felgenhauer, K. and Pak., S.J. (1975) in ref. 4, pp. 115-120

157. Graesslin, D. and Weise, H.C. (1974) in ref. 2, pp. 199-205

158. Righetti, P.G. and Drysdale, J.W. (1973) Ann. N.Y. Acad. Sci. 209, 163-186

159. Chrambach, A., Doerr, P., Finlayson, G.R., Miles, L.E.M., Sherins, R. and Rodbard, D. (1973) Ann. N.Y. Acad. Sci. 209, 44-64

160. Suzuki, T., Benesch, R.E., Yung, S. and Benesch, R. (1973) Anal. Biochem. 55, 249-254

161. Stathakos, D. (1975) in ref. 4, pp. 65-75

162. O'Brien, T.J., Liebke, H.H., Cheung, H.S. and Johnson, L.K. (1976) Anal. Biochem. 72, 38-44

163. Johnson, L.K., Liebke, H.H. and O'Brien, T.J. (1976) Anal. Biochem. 76, 311-320

164. Awdeh, Z.L., Williamson, A.R. and Askonas, B.A. (1968) Nature 219, 66-67

165. Dale, G. and Latner, A.L. (1968) Lancet, April 20, 847-848

166. Fawcett, J.S. (1968) FEBS Letters 1, 81-82

167. Leaback, D.H. and Rutter, A.C. (1968) Biochem. Biophys. Res. Commun. 32, 447-453

168. Riley, R.F. and Coleman, M.K. (1968) J. Lab. Clin. Med. 72, 714-720

169. Wrigley, C. (1968) J. Chromatogr. 36, 362-365

170. Finlayson, G.R. and Chrambach, A. (1971) Anal. Biochem. 40, 292-311

171. Harada, S. (1975) Clin. Chim. Acta 63, 275-283

172. Grubb, A. (1973) Anal. Biochem. 55, 582-592

173. Låås, T. (1972) J. Chromatogr. 66, 347-355

174. Lönnerdal, B. and Låås, T. (1976) Anal. Biochem. 72, 527-532

175. Vesterberg, O. (1971) Biochim. Biophys. Acta 243, 345-348

176. Davis, B.J. (1964) Ann. N.Y. Acad. Sci. 121, 404-427

177. Baumann, G. and Chrambach, A. (1976) Anal. Biochem. 70, 32-40
178. Fawcett, J.S. and Morris, C.J.O.R. (1966) Separ. Sci. 1, 9-26
179. Rodbard, D., Chrambach, A. and Weiss, G.H. (1972) Separ. Sci. 7, 705-723
180. Rüchel, R. and Brager, M.D. (1975) Anal. Biochem. 68, 415-428
181. Paus, P.N. (1971) Anal. Biochem. 42, 372-376
182. Anker, H.S. (1970) FEBS Letters 7, 293-295
183. O'Connel, P.H.B. and Brady, C.J. (1976) Anal. Biochem. 76, 63-73
184. Hansen, J.N. (1976) Anal. Biochem. 76, 37-44
185. Wyckoff, M., Rodbard, D. and Chrambach, A. (1977) Anal. Biochem. 78, 459-482
186. Dirksen, M.L. and Chrambach, A. (1972) Separ. Sci. 7, 747-772
187. Loening, U.E. (1967) Biochem. J. 102, 251-257
188. Brewer, J.M. (1967) Science, 156, 256-257
189. Fantes, K.H. and Furminger, I.G.S. (1967) Nature 215, 750-751
190. Mitchell, W.M. (1967) Biochim. Biophys. Acta 147, 171-174
191. Peterson, R.F. (1971) J. Agr. Food. Chem. 19, 595-599
192. Davies, H. (1975) in ref. 3, pp. 97-113
193. Vesterberg, O. (1975) in ref. 3, pp. 78-96
194. Swanson, M.J. and Sunders, B.E. (1975) Anal. Biochem. 66, 520-524
195. Wadström, T. and Smyth, C. (1975) in ref. 3, pp. 152-177
196. Altland, K. (1977) in ref. 5, pp. 295-301
197. Chuat, J.C. and Pilon, C. (1977) Anal. Biochem. 82, 258-261
198. Hobart, M.J. (1975) in ref. 3, pp. 275-280
199. Hayes, M.B. and Wellner, D. (1969) J. Biol. Chem. 244, 6636-6644
200. Spencer, E.M. and King, T.P. (1971) J. Biol. Chem. 246, 201-208
201. Vesterberg, O. (1972) Biochim. Biophys. Acta 257, 11-19
202. Söderholm, J., Allestam, P. and Wadström, T. (1972) FEBS Letters 24, 89-92
203. Vesterberg, O. and Hansén, L. (1977) in ref. 5, pp. 123-133
204. Malik, N. and Berrie, A. (1972) Anal. Biochem. 49, 173- 176
205. Diezel, W., Kopperschläger, G. and Hofmann, E. (1972) Anal. Biochem. 48, 617-620
206. Reisner, A.H., Nemes, P. and Bucholtz, C. (1975) Anal. Biochem. 64, 509-516
207. Holbrook, I.B. and Leaver, A.G. (1976) Anal. Biochem. 75, 634-636
208. Blakesley, R.W. and Boezi, J.A. (1977) Anal. Biochem. 82, 580-582
209. Talbot, D.N. and Yaphantis,, D.A. (1971) Anal. Biochem. 44, 246- 254
210. Stein, S., Böhlen, P., Stone, J., Dairman, W. and Udenfriend, S. (1973) Arch. Biochem. Biophys. 155, 203- 208
211. Handschin, U.E. and Ritschard, W.J. (1976) Anal. Biochem. 71, 143- 148
212. Barger, B.O., White, F.C., Pace, J.L., Kemper, D.L. and Ragland, W.L. (1976) Anal. Biochem. 70, 327- 335
213. Bosshard, H.F. and Datyner, A. (1977) Anal. Biochem. 82, 327-333
214. Righetti, P.G., Molinari, B.M. and Molinari, G. (1977) J. Dairy Res. 44, 69-72
215. Weinstein, J.D., Obrig, T.G. and Weiggle, J.B. (1978) J. Chromatogr., in press
216. O'Farrel, P. (1975) J. Biol. Chem. 250, 4007-4021

217. Williamson, A.R. (1971) Eur. J. Immunol. 1, 390-394

218. Keck, K., Grossberg, A.L. and Pressman, D. (1973) Eur. J. Immunol. 3, 99-102

219. Carrel, S., Theilkaes, L., Skvaril, S. and Barandun, S. (1969) J. Chromatogr. 45, 483-486

220. Cotton, R.G.H. and Milstein, C. (1973) J. Chromatogr. 86, 219-221

221. Catsimpoolas, N. (1973) Ann. N.Y. Acad. Sci. 209, 144-146

222. Laurell, C.B. (1966) Anal. Biochem. 15, 45-52

223. Söderholm, J. and Smyth, C. (1975) in ref. 4, pp. 99-114

224. Park, C.M. (1973) Ann. N.Y. Acad. Sci. 209, 237-257

225. Perrella, M., Heyda, A., Mosca, A. and Rossi Bernardi, L. (1978) Anal. Biochem. 88, 212-224

226. Grossbach, U. (1972) Biochem. Biophys. Res. Commun. 49, 667-672

227. Gainer, H. (1971) Anal. Biochem. 44, 589-605

228. Gainer, H. (1973) Anal. Biochem. 51, 646-650

229. Neuhoff, V. (1973) Micromethods in Molecular Biology, Springer Verlag, Berlin, pp. 49-56

230. Catsimpoolas, N. (1976) in ref. 41, pp. 229-258

231. Brogen, C.H. (1977) in ref. 5, pp. 549-558

232. Rosengren, A., Bjellqvist, B. and Gasparic, V. (1977) in ref. 5, pp. 165-171

233. Allen, J.C. and Humphries, C. (1975) in ref. 3, pp. 347-354

234. Alper, C.A., Hobart, M.J. and Lachmann, P.J. (1975) in ref. 3, pp. 306-312

235. Zechel, K. (1977) Anal. Biochem. 83, 240-251

236. Frater, R. (1970) J. Chromatogr. 50, 469-474

237. Yapa, P.A.J. (1973) J. Chromatogr. 87, 311-313

238. Righetti, P.G., Gianazza, E., Gianni, A.M., Comi, P., Giglioni, B. and Ottolenghi, S. (1978) Blood, in press

239. Sorensen, S.P.L., Linderstrøm-Lang, K. and Lund, E. (1926) C.R. Trav. Lab. Carlsberg 16, 5-15

240. Fredriksson, S. (1977) in ref. 5, pp. 71-83

241. Conway-Jacobs, A. and Lewin, L.M. (1971) Anal. Biochem. 43, 394-400

242. Nakhleh, E.T., Samra, S.A. and Awdeh, Z.L. (1972) Anal. Biochem. 49, 218-224

243. Jonsson, M., Pettersson, E. and Rilbe, H. (1969) Acta Chem. Scand. 23, 1553-1559

244. Secchi, C. (1973) Anal. Biochem. 51, 448-455

245. Strongin, A.J.A., Balduev, A.P. and Levin, Z.D. (1973) Sci. Tools 20, 34-35

246. Gelsema, J.W. and DeLigny, C.L. (1977) J. Chromatogr. 130, 41-50

247. Gelsema, J.W., DeLigny, C.L. and Van Der Veen, N.G. (1977) J. Chromatogr. 140, 149-155

248. Gelsema, J.W., DeLigny, C.L. and Van Der Veen, N.G. (1978) J. Chromatogr., 154, 161-174.

249. Ui, N. (1971) Biochim. Biophys. Acta 229, 567-581

250. Josephson, R.V., Maheswaran, S.K., Morr, C.V., Jenness, R. and Lindorfer, R.K. (1971) Anal. Biochem. 40, 476-482

251. Gianazza, E., Righetti, P.G., Bordi, S. and Papeschi, G. (1977) in ref. 5, pp. 173-179

252. Chidakel, B.E., Nguyen, N.Y. and Chrambach, A. (1977) Anal. Biochem. 77, 216-225

253. Beeley, J.A., Stevenson, S.M. and Beeley, J.G. (1972) Biochim. Biophys. Acta 285, 293-300

254. Papeschi, G., Bordi, S., Beni, C. and Ventura, C. (1976) Biochim. Biophys. Acta 453, 192-199

255. Vesterberg, O. (1970) in Prot. Biol. Fluids (Peeters, H., ed.) Pergamon Press, London, 17, 383-387

256. Nilsson, P., Wadström, T. and Vesterberg, O. (1970) Biochim. Biophys. Acta 221, 146-148

257. Li, Y.T. and Li, S.C. (1973) Ann. N.Y. Acad. Sci. 209, 187-197

258. Quast, R. and Vesterberg, O. (1968) Acta Chem. Scand. 22, 1499-1508

259. Brown, W.D. and Green, S. (1970) Anal. Biochem. 34, 593-595

260. Baumann, G. and Chrambach, A. (1975) Anal. Biochem. 69, 649-651

261. Baumann, G. and Chrambach, A. (1975) in ref. 4, pp. 13-23

262. Rilbe, H. (1977) in ref. 5, pp. 35-50

263. Caspers, M.L., Posey, Y. and Brown, R.K. (1977) Anal. Biochem. 79, 166-180

264. Beccaria, L., Chiumello, G., Gianazza, E., Luppis, B. and Righetti, P.G. (1978) Amer. J. Hematol. 4, 367-374

265. Epstein, C.J. and Schechter, A.N. (1968) Ann. N.Y. Acad. Sci. 151, 85-101

266. Kaplan, N.O. (1968) Ann. N.Y. Acad. Sci. 151, 382-399

267. Williamson, A.R., Salaman, M.R. and Kreth, W.H. (1973) Ann. N.Y. Acad. Sci. 209, 210-224

268. Kaplan, L.J. and Foster, J.F. (1971) Biochemistry, 10, 630-636

269. Wallevik, K. (1973) Biochim. Biophys. Acta 322, 75-87

270. Dean, R.T. and Messer, M. (1975) J. Chromatogr. 105, 353-358

271. Galante, E., Caravaggio, T. and Righetti, P.G. (1976) Biochim. Biophys. Acta 442, 309-315

272. Righetti, P.G. and Gianazza, E. (1978) Biochim. Biophys. Acta 532, 137-146

273. Gianazza, E. and Righetti, P.G. (1978) Biochim. Biophys. Acta, 540, 357-364

274. Righetti, P.G., Brown, R. and Stone, A.L. (1978) Biochim. Biophys. Acta, 542, 232

275. Cann, J.R. and Stimpson, I.D. (1977) Biophys. Chem. 7, 103-114

276. Stimpson, I.D. and Cann, J.R. (1977) Biophys. Chem. 7, 115-119

277. Cann, J.R., Stimpson, D.I. and Cox, D.J. (1978) Anal. Biochem., in press

278. Gianazza, E. and Righetti, P.G. (1978) J. Chromatogr., in press

279. Bengtsson, G. and Olivecrona, T. (1977) in ref. 5, pp. 189-195

280. Jonsson, M., Fredriksson, S., Jontell, M. and Linde, A. (1978) J. Chromatogr., 157, 235-241

281. Sluyterman, L.A.A.E. and Wijdenes, J. (1977) in ref. 5, pp. 463-466

282. Sluyterman, L.A.A.E. and Elgersma, O. (1978) J. Chromatogr., 150, 17-30

283. Sluyterman, L.A.A.E. and Wijdenes, J. (1978) J. Chromatogr., 150, 31-44

284. Leaback, D.H. and Robinson, H.K. (1975) Biochem. Biophys. Res. Commun. 67, 248-254

285. Potrilli, P., Sannia, C. and Marino, C. (1977) J. Chromatogr. 135, 511-513

286. Ambler, J. (1978) Clin. Chim. Acta, 85, 183-191

Electrokinetic Separation Methods,
P.G. Righetti, C.J. van Oss and J.W. Vanderhoff

ISOTACHOPHORESIS

MILAN BIER and THOMAS T. ALLGYER
Biophysics Technology Laboratory, University of Arizona, Tucson, AZ 85721

INTRODUCTION

Isotachophoresis is an electrophoretic technique which separates and orders ions by virtue of their differences in mobility. The name reflects a distinguishing characteristic of the method, to wit, equal (iso) velocities (tacho) of the ion zones in the steady state. The technique of isotachophoresis encompasses in earlier terminology the ionic migration method, moving-boundary analysis, displacement electrophoresis, steady-state stacking, cons electrophoresis, transphoresis, and ionophoresis. This article is intended to provide the reader with an understanding of the fundamental theory of the technique and to illustrate its analytical and preparative applications. It is not our intent to provide an exhaustive treatment of the theory nor a comprehensive review of the literature. The theory of isotachophoresis has been developed in detail elsewhere[1-4], the instrumentation and application of analytical isotachophoresis have been recently reviewed[4], and a discussion of some preparative applications of the method is presented elsewhere in this volume[5].

For isotachophoresis to occur, some stringent requirements have to be met. A discontinuous electrolyte system is used, comprising a leading and a terminating electrolyte, and the sample ions are introduced as a zone at their interface. The leading electrolyte must contain only one ion species of the same charge as the sample ions to be separated, and it must have a net mobility higher than that of any of the sample ions. Net mobility is defined as the mean velocity in unit d.c. electric field; for weak electrolytes, it is the product of absolute ion mobility and degree of ionization. The terminating electrolyte must also contain only one co-ion species, but of net mobility lower than that of any of the sample ions. A common counterion should be used throughout the system. If these mobility requirements are not met, sample ions which have net mobilities greater than that of the leading ion or smaller than that of the terminating ion will migrate zone electrophoretically within the leading or terminating electrolyte zones.

The electric field is applied so that the direction of migration is toward the leading electrolyte. After a sufficient migration time, the sample ions will arrange in contiguous zones in order of their mobilities and, provided the current is constant and the separation apparatus is of constant cross-sectional area, all zones will continue migrating at constant velocity without further changes. This is the isotachophoretic steady state. The velocity of each zone is then equal to the velocity of the leading ion zone and the concentration of each ion species is self-adjusted, so that the electric field strength in each zone is inversely proportional to the mobility of that zone's constituent ion. In accordance with the principle of electroneutrality, the counterion concentration in each zone equals the co-ion concentration.

The process of an isotachophoretic separation is shown graphically in Figure 1.

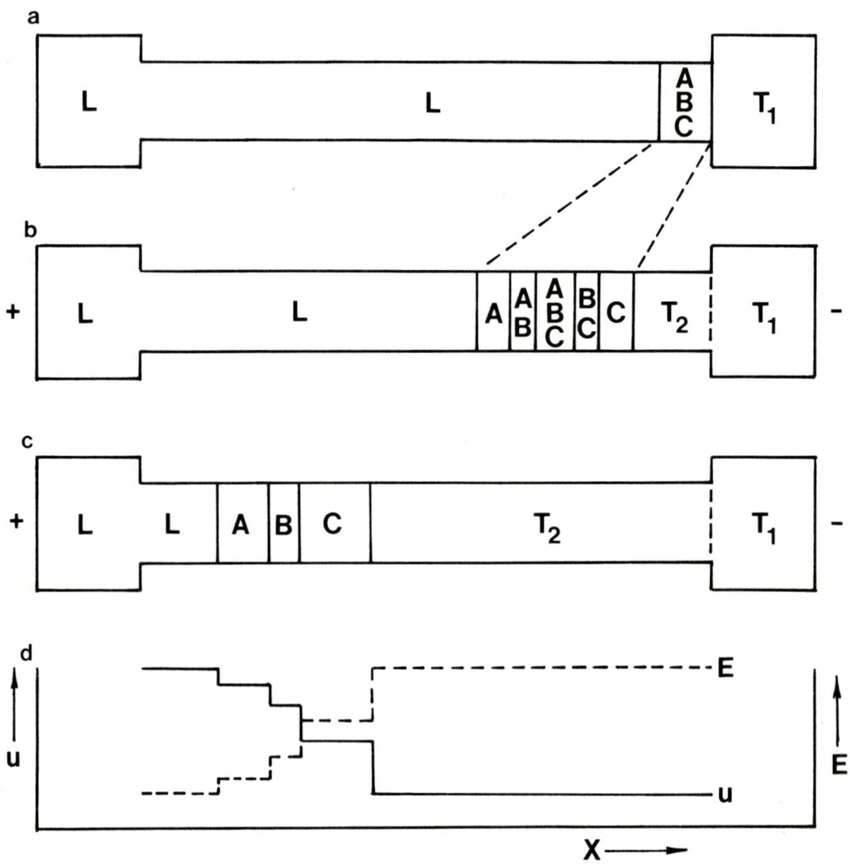

Fig. 1. An isotachophoretic separation. The sample anions (A, B, and C) are introduced between the leading anionic species L and the terminating anionic species T. A common counterion is present throughout. (a) Conditions prevailing after the introduction of the sample ions prior to the application of current. (b) After some period of current application, prior to attaining steady state. (c) Isotachophoretic steady state: all anionic species are separated into discrete and contiguous zones which migrate with identical velocities. (d) The electric field strength E and mobility u profiles of the steady state. T_1 and T_2 represent the initial and Kohlrausch-adjusted concentrations of the terminator electrolyte, and X represents the position along the separation column.

THEORY

Isotachophoresis is a specific application of electrophoretic theory based upon three theoretical observations made by Kohlrausch[6] in 1897. He first demonstrated theoretically that the passage of an electric current through a homogenous electrolyte solution does not produce any changes in concentration or composition except in the immediate vicinity of the electrodes. Since the passage of a current by electrolytic conduction accompanies the transfer of mass, the conductance of a solution is a function of the ionic mobilities of its constituents. The specific conductance of the solution is given by

$$\kappa = 96.5\sum^{k}\sum^{n} u_{j(i)} c_{j(i)} |i| \tag{1}$$

where k is the number of constituents in the solution, n is the number of subspecies of constituent j, u_j is the ion mobility of the ion subspecies of constituent j, and $c_{j(i)}$ is the concentration of subspecies of constituent j with valence i. Both the mobility and concentration terms are signed quantities, positive for cations and negative for anions. The electric field strength E is constant throughout the solution and is given by

$$E = I/\kappa \tag{2}$$

where I is the current density, i.e., the current per unit cross-sectional area. The fractional current carried by a particular ion species is given by $u_j c_j/\kappa$, this quantity is also known as the transference number. The principle of electroneutrality applies, i.e., the concentrations of positive and negative ions of any given volume element must be identical.

Kohlrausch's second observation was that passage of a current through a solution in which the only discontinuity is a concentration boundary, i.e., the boundary between different concentrations of the same electrolyte, does not produce a migration of the boundary. In this case two homogeneous phases are considered each of which can be described by the above equations. In addition, it is convenient to specify relative mobilities r_j which may be obtained by dividing the set of mobilities by any single member of the set or by the mobility of a suitable reference ion. The relative mobilities are essentially insensitive to concentration and temperature, and therefore are assumed to be constant throughout the system. Equation 1 may be restated in terms of relative mobilities to give the relative conductance, σ

$$\sigma = 96.5\sum^{k}\sum^{n} r_{j(i)} c_{j(i)} |i| \tag{3}$$

In an isotachophoretic separation a boundary of this nature is formed between the initial concentration of terminator electrolyte and the Kohlrausch-adjusted concentration of terminator electrolyte which migrates behind the sample ion zones. In reality this concentration boundary is not truly immobile: its slight migration results from the approximate nature of the assumption that relative mobilities are independent of concentration.

Kohlrausch's third observation was that passage of a current through a solution which contains a species discontinuity, i.e., a boundary which separates zones of different ions does produce migration of the boundary. The simplest system that demonstrates a species discontinuity contains three constituent ions comprising two zones, i.e., two ions separated by a boundary with a common counterion. The boundary will be maintained only if the net mobility of the ion species in the trailing zone is smaller than the net mobility of the ion species in the leading zone. Since there is no background electrolyte to support the conductance of current between the zones, the ion zones must remain contiguous and must migrate with identical velocities. Kohlrausch developed a mathematical treatment of this phenomenon assuming constant ion mobilities and demonstrated that the concentration of ions behind the boundary are specifically established or regulated by the zone preceding the boundary. The mathematical description of this phenomenon, termed the Kohlrausch regulating function, is essentially a statement of the balance of electric current, i.e., the current density in all zones is identical. The function was originally derived by Kohlrausch in terms of constant ion mobilities, and as such it could be rigorously applied to only strong electrolytes at high dilution. Dole[7] redefined the function in terms of relative ion mobilities, and Dismukes and Alberty[8] derived the function for weak electrolytes in terms of relative mobilities. This form is given by

$$\overline{\omega} = \sum^{k} \overline{c}_j / r_j \tag{4}$$

where $\overline{c}_j$ is the total concentration of constituent j including all its equilibrium forms. The function $\overline{\omega}$ has the same value in the two zones separated by a moving boundary. Inasmuch as a multicomponent isotachophoretic separation may be regarded as a series of such zones, $\overline{\omega}$ is constant for all zones. Thus each zone in an isotachophoretic train is uniquely defined by the zone into which its frontal boundary migrates. Kohlrausch proposed an alternate form of the regulating function, which has recently been derived anew by Routs[1] and is perhaps more obviously applicable to two zones separated by a single boundary:

$$c_b/c_a = u_b(u_a + u_c)/u_a(u_b + u_c) \tag{5}$$

where c_a and c_b are the concentrations of ion a in the leading zone and ion b in the trailing zone, respectively, u_a and u_b are their respective net mobilities, and u_c is the mobility of the common counterion.

Another expression of the relationship between the compositions of two adjacent zones is the moving boundary equation. This equation, which is essentially a statement of the conservation of mass, was derived by Dole[7] using the relative mobilities of strong electrolytes and later by Alberty[9] for weak electrolytes in the form

$$\overline{u}_j^{\alpha} \overline{c}_j^{\alpha} / \kappa^{\alpha} - \overline{u}_j^{\beta} \overline{c}_j^{\beta} / \kappa^{\beta} = v^{\alpha\beta}(\overline{c}_j^{\alpha} - \overline{c}_j^{\beta}) \tag{6}$$

where $\overline{u}_j$ is the net mobility of constituent j including all its equilibrium forms, and the superscripts refer to the zone (α or β) to which the quantity applies. This is relevant to a constituent j in both zones separated by a moving boundary with displacement $v^{\alpha\beta}$, and thus may be applied to the common counterion.

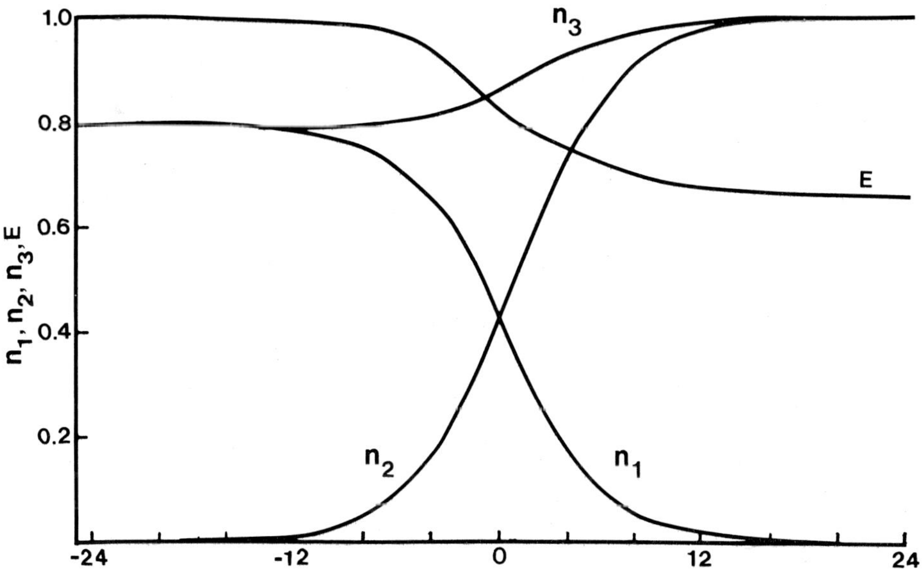

Fig. 2. Concentration and electric field profiles. The ionic components are represented as sodium n_1, potassium n_2, and chloride n_3; their assumed mobilities were respectively 35.98 x 10^{-5}, 54.40 x 10^{-5}, and 55.25 x 10^{-5} $cm^2/V.sec$. The temperature was 283°K and the applied field strength 250 V/cm. The horizontal axis represents the dimensionless distance along the axis of the separating column referred to a moving coordinate system with velocity equal to boundary displacement (From Coxon and Binder (10).

Diffusion of constituents from both zones tends to degrade the boundary between adjacent zones; in isotachophoresis the boundary attains an equilibrium between the broadening effect of diffusion and the sharpening effect of the electric field gradient expressed across the boundary. If an ion from the leading zone diffuses into the trailing zone it encounters a higher electric field strength and thus experiences an increase in velocity which tends to return it to its proper zone. Likewise if an ion from the trailing zone diffuses ahead of the boundary, it will experience a decrease in velocity in response to the lower field strength encountered. This self-restorative nature of the boundary reflects the applied electric field strength, i.e., the thickness of the boundary decreases with increasing field strength. Coxon and Binder[10] have given a mathematical model for a one-dimensional solution interface which allows for the presence of a counterion, the effects of diffusion, and the existence of a non-zero sum of charges in the frontal region in violation of the principle of electroneutrality. The concentration and electric field profiles typical of their analysis are illustrated in Figure 2 for the NaCl-KCl aqueous system. This depicts the form of the concentration changes that occur at the boundary and demonstrates that the change in electric field strength occurs by a gradient rather than a discontinuity. As a point of interest, an interface thickness of 0.027 mm was calculated for the conditions given.

The boundary illustrated in Figure 2 is ideal in that no disturbing influences other than diffusion were considered. In real systems the sharpness of the boundary is compromised by the superimposable effects of electroosmosis, radial and axial temperature gradients, pH phenomena, and gravity-dependent phenomena. In order to gain an appreciation of the complexity of the technique, these phenomena will be discussed briefly.

Electroosmosis. Electroosmosis is a common disturbing factor in many forms of electrophoresis and is due to the zeta potential exhibited by the walls of the apparatus. In isotachophoresis the phenomenon is more complicated than in zone electrophoresis, as the isotachophoretic zones have different ionic compositions and different field strengths. In practice, it is essential that electroosmosis be controlled, otherwise sharply parabolic boundary profiles are observed. In preparative isotachophoresis, electroosmosis is abolished by the use of solid support media, either gel or granular material. In free solutions, mainly used for analytical purposes, the addition of uncharged hydrophilic polymers, such as hydroxyethyl cellulose, or non-ionic surfactants, is effective for control of electroosmosis. Excellent anti-osmotic coatings for glass walls have been developed[11,12].

Radial and axial temperature gradients. Isotachophoretic separations are characterized by ionic zones of increasing field strengths and correspondingly increasing Joule heating from leader to terminator. The differential heating of ion zones produces strong axial temperature gradients at the boundaries. In analytical systems these gradients are exploited for the detection of the boundary position. In addition, the imposition of external cooling (or lack of insulation) produces radial temperature gradients with the temperature being highest at the axis of the column and declining toward the periphery. Temperature gradients exert a direct influence on an isotachophoretic separation primarily as a result of the temperature dependence of electrophoretic mobility (increasing about 2%/^{o}K).

Martin and Everaerts[13] have calculated the axial temperature profile across an idealized separation boundary assuming the radial temperature variation to be negligible. The effect of an axial temperature gradient is to decrease the difference in mobilities of a component on either side of the boundary, thereby compromising an element of boundary sharpening and increasing the boundary thickness. The conventional method of reducing axial temperature gradients is to use efficient cooling, a minimal leader-terminator mobility difference, and the smallest applied voltage that is consistent with reasonable separation time[14].

The radial temperature variation and the resultant variation in mobility have been considered theoretically in regions distant from the boundaries where axial temperature differences are negligible. Hjerten[11] derived an expression for the mobility variation resulting from the radial temperature gradient with a constant inner wall temperature. The resultant boundary velocity profiles are parabolic and progressively more bowed from leader to terminator. Brown and Hinckley[15-17] considered the effects of finite wall thickness and obtained temperature profiles in the solution and column wall assuming a constant outer wall temperature. The transient heat conduction problem for isotachophoresis columns of circular and rectangular cross-section has been treated by Coxon and Binder[18,19]. Since the electrical field strength is not arbitrary, it has been shown that there is a maximum permissible radial size of a column of circular cross-section to maintain an adequate balance between heat generated and heat dissipated. Similar considerations have established a one-dimensional limit to columns of rectangular cross-section; however, in principle the second dimension of the cross-section may be extended indefinitely without disturbing the thermal balance. This may be pertinent to the design of preparative apparatus.

Gravity-dependent phenomena. Solution density variations arising from thermal and concentration gradients may cause convection under the influence of gravity with a resultant distortion of zone boundaries. With particulate suspensions (e.g. cell suspensions) gravity may also cause particle sedimentation. Density differences between solutions of small ions at high dilution are generally not large, but differences between protein and protein-free solutions may be quite large and become critical even at low ionic concentrations. The convective effects may be reduced by imposing a continuous (e.g. polyacrylamide) or discontinuous (e.g. Sephadex) solid support medium; however, the use of a solid support introduces differential mobility restrictions as a result of chromatographic and adsorptive effects. Convective flows may also be counteracted (equalized in two dimensions) by rotation of a separation column about its long axis[11,20]. Total suppression of gravitational effects may be achieved by retreat to a zero-gravity environment. Research of this alternative has been sponsored and conducted by the National Aeronautics and Space Administration to establish the feasibility of a zero-gravity electrophoretic facility in the Space Shuttle[21-23].

pH Phenomena. Isotachophoresis becomes increasingly ineffective in pH ranges below pH 4 or above pH 10.5, as a result of the relatively large part of the current carried by the highly mobile hydrogen and hydroxyl ions. In the intermediate pH region, the concentrations of H^+ and OH^- ions can be neglected only when all sample ions are strong electrolytes, i.e. fully ionized. When one of the sample ions is a weak electrolyte, it influences the pH of adjacent zones through the dissociation equilibria expressed at its zone boundaries. In order to eliminate such undesireable effects, a counterion with buffering capacity is commonly used. Since this buffer is a constituent in terms of the Kohlrausch regulating function, the concentrations of its equilibrium species are specified in each ion zone. In anionic systems this

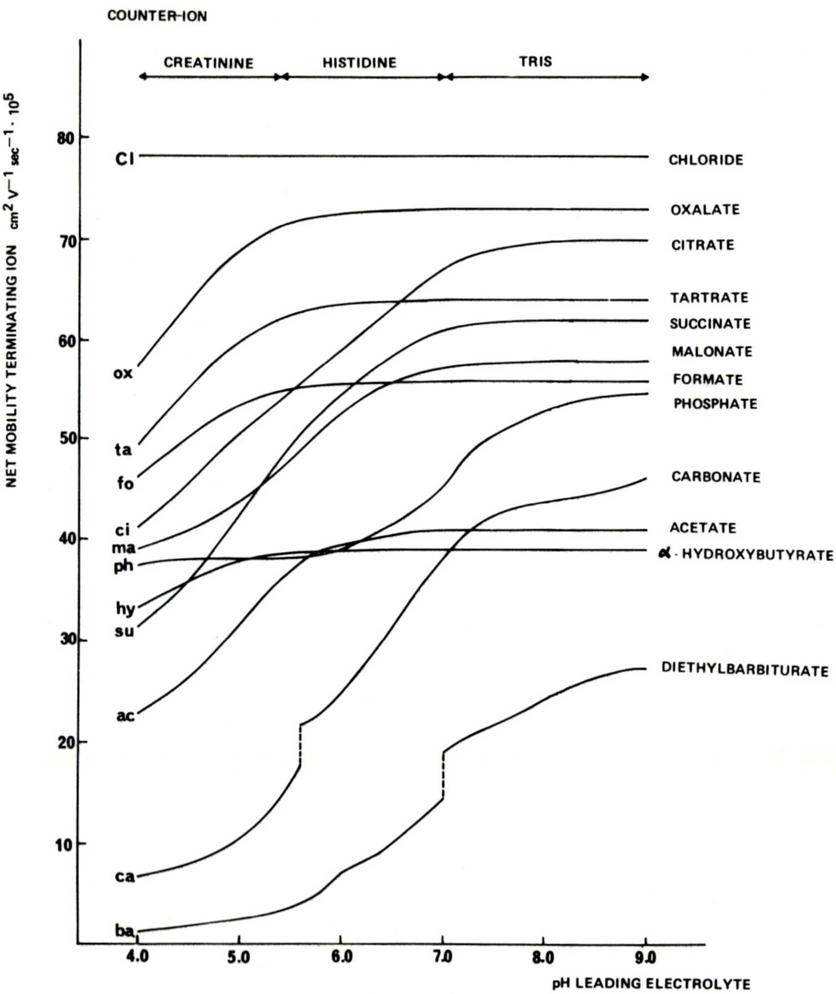

Fig. 3 The net mobilities of several anions are plotted as a function of the pH of the leading electrolyte. The leading ion is 0.01 M HCl, buffered with counterions as indicated (From Routs[1]).

results in a stepwise pH increase through the ion zones from leader to terminator. In cationic systems, the pH decreases from leader to terminator. The net mobility of a sample ion is dependent upon the pH of its zone and is therefore a function of the pH and concentration of the leading electrolyte. The order of net mobilities of various ions may be altered by changing the pH of the leading electrolyte. For instance, Routs[1] demonstrated that by changing the counterion and the pH of the leading electrolyte (chloride ion) citrate could be made to migrate either before, behind, or with the tartrate ion. Figure 3, taken from Routs[1], illustrates this and presents the plot of the net mobilities of various anions as a function of the pH of the leading electrolyte. This emphasizes the importance of choosing the correct pH for the leading electrolyte when separating ions which may be incompletely ionized. If the differences in absolute mobilities of the ions of interest are adequate for separation, the pH of the leading electrolyte may be established such that the ions are completely ionized, while if the ions of interest have approximately identical absolute mobilities, the pH of the leading electrolyte should be chosen to maximize the differences in net mobilities so that the separation is optimized[24]. In the case of sample ions of known mobilities, the computer programs of Routs[1] and Jovin[25,26] may be utilized to simplify the task of selection of appropriate leading and terminating electrolyte systems. The selection of electrolyte systems for samples of unknown composition (or unknown constituent mobility) must be based partly upon trial and error. Initially a leader-terminator pair with a wide mobility interval is used. Introduction of a co-ion of known mobility and pH characteristics may give sufficient information of the mobility range of sample components to allow a more judicious secondary selection of leading and terminating electrolytes. The library of 45 buffers and ions used by Jovin to generate discontinuous systems for multiphasic zone electrophoresis (Figure 4 of reference 26) provides pK and relative mobility data suitable for establishing electrolyte systems for both anionic and cationic isotachophoretic separations.

INSTRUMENTATION

Analytical Methods. Analytical isotachophoresis is usually carried out in free solution in thin-wall, narrow-bore teflon capillaries. Convection and electroosmosis are minimized by thermostating the capillary and by the addition of 0.4% of hydroxypropylmethylcellulose to the buffers. The boundaries are allowed to migrate past stationary boundary detectors. The only commercially available apparatus of this type is the LKB 2127 Tachophor, illustrated schematically in Figure 4. The coiled teflon capillaries are only 0.5 mm in internal diameter and up to 36 cm in length. Sample application is by syringe injection (0.1 to 50 µl volumes) at the interface between the terminating electrolyte reservoir and the leader-filled capillary. The thermal and ultraviolet (UV) detectors are located at the other end of the capillary, close to the leading electrolyte reservoir. The high-voltage power supply is current stabilized.

The main element of analytical instrumentation is the boundary detection system. Everaerts and his associates have done most of the work in this field, and have provided an extensive review[4]. Three detection methods are in general use and are designed to measure changes in conductivity, temperature or UV absorption across ion boundaries. In principle, only the first two methods are universally applicable,

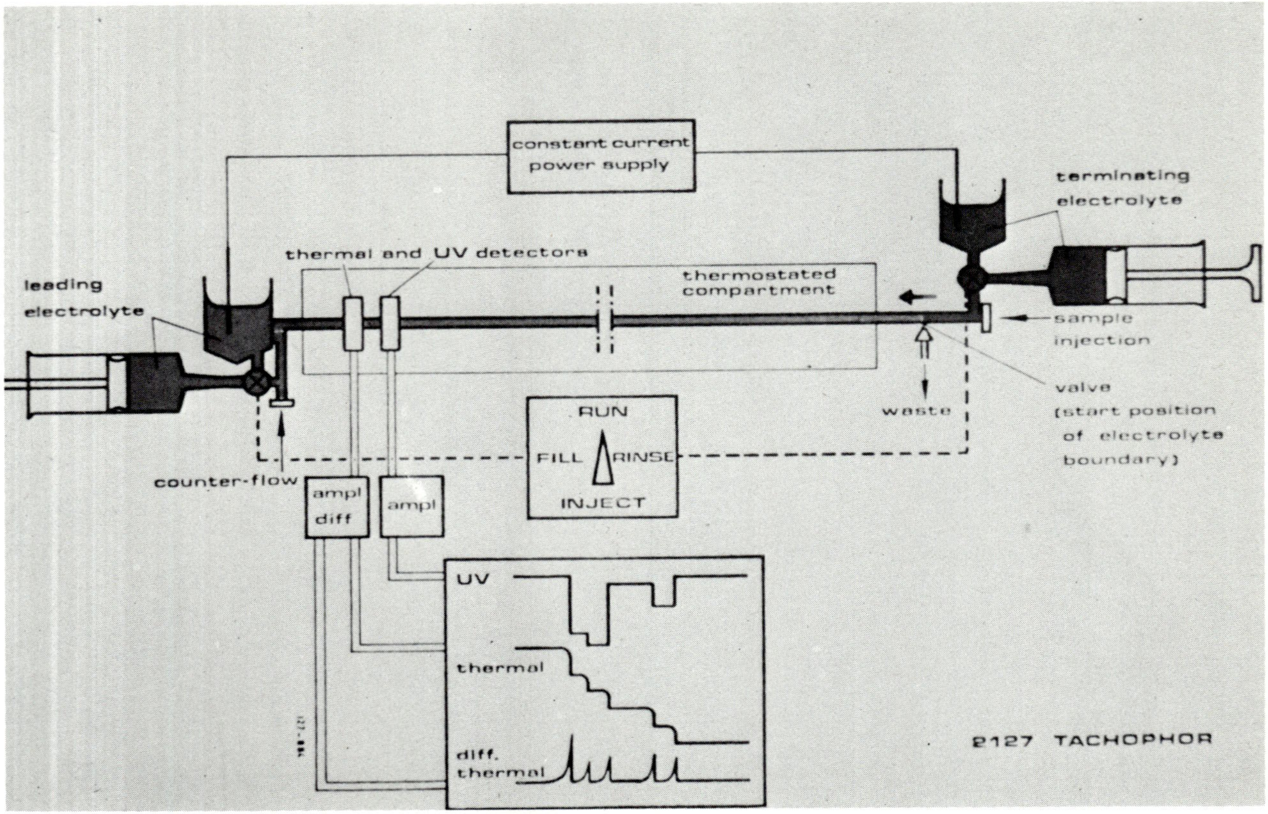

Fig. 4. Schematic presentation of the LKB Tachophor 2127 apparatus and its recorder tracings (From LKB-Produkter AB, Bromma, Sweden).

while the third method depends on the presence of UV-absorbing compounds. In practice, however, even when such components are not willfully included, there are unavoidable UV-absorbing trace impurities which concentrate at zone interfaces and are useful markers of zone length. It is also possible to use a UV-absorbing counterion, as its concentration will vary from zone to zone.

Thermistors are used for the detection of the temperature gradients, which can be recorded in integral or differential mode. The height of each boundary step in the integral curve is a function of the mobility difference across the boundary and, thus, can be used to identify the ion species. The distance between the boundaries is a measure of the length of the column occupied by each ion zone and, since the concentration of the ion is constant within the zone, can be used for the quantitation of the given ion. As the concentration of the Kohlrausch-adjusted zones is dependent on the leader concentration and the ratio of mobilities of leader and sample ions, great care must be exerted in the calibration of the system.

Conductometric boundary detection is carried out by means of microelectrodes. Everaerts et al.[4] have shown that the resolving power of such microelectrodes is of the order of 0.05 mm, about two orders of magnitude better than the thermal detectors. Thus, the conductometric method permits the use of smaller samples and shorter analysis times. As with the thermal method, conductometric detection permits identification and quantitation of ion species.

Ultraviolet sensors do not permit the identification of the sample ions, as UV absorption is a function of molecular structure rather than ion mobility. Nevertheless, it is particularly useful for the determination of zone length, i.e. for quantitation, because of the aforementioned trace impurities. Since most biological samples are UV-absorbing, it is common practice to record simultaneously the UV absorption tracing and either the thermal or conductometric output for every run. The resolution of UV detection is comparable to that of the conductometric method.

Counterflow of leader has been suggested as a means of correcting thermal and electroosmotic boundary distortion. It may also be used to effectively extend the length of the migration path during separation. This is of particular advantage when the mobility difference between ions to be separated is small. Preetz[27] has given a theoretical treatment of counterflow in isotachophoresis, and Everaerts et al.[4] have discussed a variety of devices for its regulation. With judicious choices of leader concentration and sample size, counterflow may be utilized to virtually immobilize the migrating zones. The sample ion zones then occupy most of the separation column and little of the applied potential gradient is spent on leader and terminator zones. In principle, counterflow can be used with both preparative and analytical techniques.

Other analytical electrophoretic methods frequently use solid support media, such as polyacrylamide gels, thin layers or filter paper. Isotachophoresis is compatible with these techniques, but the need for color development for zone visualization results in loss of the sharpness of isotachophoretic boundaries due to diffusion. This may explain the lack of general interest in these procedures, though they may find application for some specific purposes.

Convective mixing of isotachophoretic boundaries in free solutions can also be prevented using the principle of rotational stabilization first proposed by Hjerten[11]. Separation is carried out in a thermostated, narrow-bore glass tube slowly rotating around its horizontal axis. As the direction of the gravity vector is continuously changing with reference to the fluid column, gross convection is prevented. This principle of stabilization is of interest mainly for the separation of particulate matter, such as bacteria or blood cells.

Preparative Methods. Preparative isotachophoresis is commonly carried out in thermostatted vertical columns, such as the LKB 7900 Uniphore column electrophoresis apparatus. This apparatus permits easy application of sample and terminator on top of precast solid support media. The lower end of the glass column contains an 0.3 ml elution chamber, bounded by membranes, with a slow radial flow of eluting (leading) electrolyte. Large electrolyte reservoirs are separated from the column by membranes which prevent bulk liquid flow.

Sucrose density gradients are not effective for the stabilization of the steep concentration gradients characteristic of isotachophoresis. Solid support media are therefore necessary. Polyacrylamide gels were generally used for this purpose, but the steep osmotic and electroosmotic gradients associated with the concentration gradients across advancing boundaries frequently cause gel waisting, resulting in various boundary anomalies[28]. Bier et al.[29,30] have found that these boundary anomalies can be completely avoided by the use of Sepahdex G-200 as the anticonvective support medium. Sephadex columns are easier to pour and are free of the catalysts used in the polymerization of the acrylamide-based gels. Due to their molecular sieving, they will, however, exhibit different spacing characteristics for low and high molecular weight ions.

In either case, the leading electrolyte is incorporated into the gel or Sephadex matrix on top of which sample and terminator solutions are carefully layered. Layering is facilitated by increasing the density of the sample through addition of sucrose. To minimize the danger of thermally induced convection, separation is started at low voltage until the sample has fully migrated into the gel. The boundaries of even colorless proteins are usually clearly visible due to the sharp refraction gradients of the concentrated zones. The profile of the eluting proteins can be determined by the usual continuous-flow UV monitors, and eluate fractions collected. Sephadex columns may also be fractionated by extruding them with a tightly fitting piston, the extruded gel being cut into 0.5-1 cm lengths from which the sample proteins can be eluted by the addition of a suitable buffer[30]. This procedure can save time, since extrusion of fractions can be carried out as soon as the steady state is obtained without the need for waiting until the trailing components migrate to the elution chamber.

Separations in free solution are possible in the rotationally stabilized apparatus of Hjerten[11], which has been modified for preparative work by Bier[20]. A schematic drawing of this apparatus is shown in Figure 5. The apparatus consists of a thick-wall horizontally rotating glass tube, a co-rotating electrode compartment and a stationary electrode compartment. The rotating electrode compartment is separated from the rotating tube by means of a sliding gate, which permits separate filling of the tube, and also serves for the introduction of the sample. The interface between the rotating tube and the stationary electrode constitutes a frictional seal, with provisions for elution of migrating zones by means of a radially directed stream of eluting buffer. In isotachophoresis, the rotating electrode chamber is filled with terminating electrolyte, the tube and stationary electrode chamber being filled with the leader.

It is also of historic interest to note that the first application of isotachophoresis was carried out by Kendall and Crittenden[31] using an apparatus consisting of a series of 3 ft long sections of gel-filled glass tubes, 1.5 inches in internal diameter. The boundary advancement was followed by manual detection of the temperature gradients. As the boundary advanced to the end of one tube, the rear tube

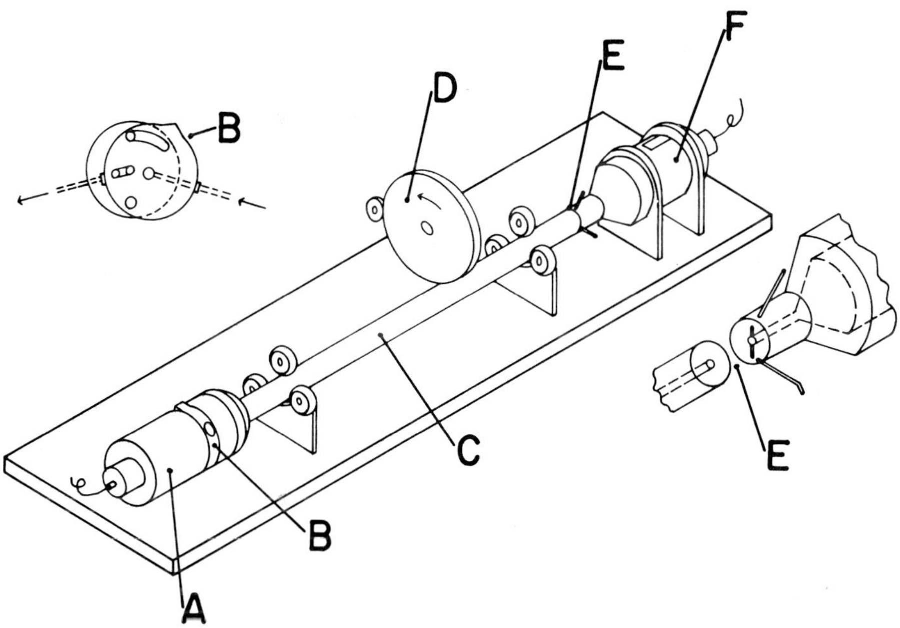

Fig. 5. Schematic presentation of the rotationally stabilized apparatus for cell electrophoresis. A - rotating electrode chamber; B - sliding gate for sample introduction; C - rotating glass tube; D - driving mechanism; E - frictional seal for fraction elution; F - stationary electrode (From Bier[20]).

section was removed, and a new tube inserted ahead of the boundary. This illustrates the potential of isotachophoresis for large scale fractionations.

APPLICATIONS

The theory of isotachophoresis is easiest to apply to relatively simple ionic species, such as monovalent or divalent ions. Early studies in isotachophoresis were therefore mainly aimed at separation of simple inorganic or organic ions, and the determination of their transference numbers. Good agreement with theory was generally observed[1,4]. The application of the method to more complex systems is often complicated by the lack of detailed information on the mobility of various ionic species. This applies particularly to proteins and other compounds of biological interest. Systematic study of such systems has been carried out only more recently.

Simple Cations and Anions. A detailed review of all systems investigated is beyond the scope of the present article. It may suffice to show some significant examples reported by Everaerts et al[4], who have carried out the most systematic study of simple ion mixtures and have established several operational systems for their resolution. These examples were chosen mainly to illustrate the versatility of the method and the applicability of mixed solvents. Mixed solvents can be used either to increase the solubility of sample ions or to increase the differences in their mobility through manipulation of the dielectric constant of the medium.

Mixtures of cations including barium, sodium, tetramethylammonium, lead, histidine, creatinine, benzidine, ε-aminocaproic acid, and γ-aminobutyric acid were separated with 0.01 M potassium as the leading ion at pH 5.4, Tris as the terminating ion, and acetate as the counterion. The separation was conducted in both H_2O and D_2O.

The alkali metals cesium, rubidium, and potassium can not be easily separated in water because the differences in their net mobilities are too small (less than 10%). The differences in their mobilities are larger in 95% methanol, and an effective separation was carried out in this solvent using hydrogen as the leading ion and copper as the terminating ion. Since in this case a buffering counterion is not required, chloride was used as the counterion species. The current was stabilized at 70 μA and detection was thermometric.

The oxidation products of acetylsalicylic acid were separated as anions using 0.01 M chloride as the leading ion at pH 3.2, propionate as the terminating ion, and β-alanine as the counterion. The current was stabilized at 80 μA and detection was accomplished by combination of conductivity and UV-absorption measurements. Clear separation of phosphate, salicylate, acetylsalicylate, and acetate was obtained.

Fatty acids, especially the higher fatty acids, are too insoluble in water for this solvent to be used for their separation. Effective separation was obtained in 95% methanol with 0.01 M chloride as leading ion at pH 9.0, cacodylic acid as the terminating ion, and Tris as the counterion. The current was stabilized at 70 μA, and detection was thermometric. Formate (C_1), acetate (C_2), butyrate (C_4), n-caproate (C_6), n-caprylate (C_8), n-caprate (C_{10}), n-laurate (C_{12}), n-myristrate (C_{14}), n-palmitate (C_{16}), and n-stearate (C_{18}) were resolved.

Amino acids and peptides. The analysis of amino acids is interesting because in principle they can be separated either as anions or as cations. Kopwillem and Lundin[32] used alkaline electrolyte systems to separate the acidic and neutral amino acids as anions and the basic amino acids as cations. The use of three different electrolyte systems permitted analysis of all common amino acids with the exception of leucine and isoleucine which formed a mixed zone. Equimolar mixtures were resolved with 20-40 minutes. Everaerts et al.[4] pursued the separation of amino acids at both low and high pH, attempting to demonstrate resolution of all common amino acids with a single electrolyte system. Several systems were tested which relied upon metal-ion complex formation and combination of solvents to improve resolution; however, they were not able to establish a single operational system which effectively resolved all common amino acids.

Kopwillem et al.[33] have monitored by isotachophoresis the purification of peptides synthesized by the solid phase method. The purifications of somatostatin (a cyclic tetradecapeptide), a decapeptide, and an undecapeptide were followed using thermometric and UV-absorption detection. Trace UV-absorbing impurities acted as markers for the zone boundaries of the non-UV-absorbing peptides. The leading electrolyte was 0.005 M hydrogen chloride, 0.006 to 0.020 M Tris and 0.5% methylcellulose. Depending on the Tris concentration, the pH of the leading electrolyte was varied from 7.2 to 8.6. The terminating electrolyte was valine. A current of 40 μA gave an analysis time of 30 minutes. This work proved the usefulness of isotachophoresis for the analysis of this important group of biochemicals, which is difficult to accomplish by other means. Miyazaki and Katoh[34] analyzed mixtures of plasma kinins, human angiotensins I and II, reduced and oxidized glutathione, and di- and tripeptides with the same amino acid residues. Six different electrolyte systems were utilized in the pH range of 6.5 to 9.6 in order to optimize separations.

Nucleotide Analysis. Exact data on pK's and absolute mobilities of nucleotides not being available, Everaerts et al.[4] conducted preliminary experiments in the pH range 3 to 7 to determine the pH at which the net mobility differences were maximized. It was found that complex mixtures of the mono-, di-, and triphosphates of adenosine, cytidine, guanosine, and uridine could be separated in the pH range 3 to 4. At pH 3.7, 0.01 M chloride was used as the leading ion, caproate the terminating ion, and α-naphthylamine as the counterion. The time of analysis was approximately 30-45 minutes using thermometric detection and approximately 15 minutes using high-resolution UV-absorption and conductivity sensors.

Applications in Enzymology. Bocek et al.[35] analyzed a mixture of typical Krebs cycle acids. The selection of suitable separation conditions was performed experimentally by measuring the dependence of the relative mobilities on the pH of the leading electrolyte. Use of 0.011 M chloride as leading ion at pH 3.8, acetate as terminating ion, and β-alanine as counterion permitted the resolution of nine Krebs cycle acids in less than four minutes.

Kopwillem[36] demonstrated the feasibility of separating the essential ion species of a mitochondrial incubation mixture without pretreatment of the sample. The reaction studied was the transformation of pyruvate to succinate by calf-heart mitochondrial enzymes. The leading electrolyte contained 0.01 M chloride with 0.032 M β-alanine as the counterion giving a pH of 3.9. The terminating ion was caproic acid. The combination of thermal and UV detectors permitted the identification and quantitation of the reaction components.

Everaerts and Verheggen[37] followed the conversion of glucose and fructose into glucose-6-phosphate and fructose-6-phosphate by the enzyme hexokinase. The hexokinase reaction represents a class of enzymatic reactions which have been analyzed conventionally by coupling to a second reaction in which one of the reactant-product pairs (e.g. NADP-NADPH) has a UV-absorption differential permitting its spectrophotometric quantitation. Isotachophoresis permitted the direct analysis of the reaction mixture without recourse to a coupling reaction system; however, the reaction cannot be followed continuously. For kinetic measurements aliquots of the reaction mixture must be withdrawn at given times, quenched, and analyzed independently. Analysis was conducted using 0.01 M chloride as leading ion at pH 3.88, caproic acid as terminating ion, and β-alanine as counterion.

Willemsen[38] has analyzed the conversion of pyruvate to lactate by the enzyme lactate dehydrogenase. NAD is a cofactor in the reaction and presents a technical difficulty because of its very low net mobility. A terminating electrolyte of lower net mobility must be used, and purity of the terminating electrolyte is critical because impurities are likely to migrate with a higher velocity and disrupt the analysis. Morpholinoethanesulphonic acid (MES) was selected as a terminator, and the reaction components were separated at pH 4.75 using 0.01 M chloride as the leading ion and histidine as the counterion.

Biomedical Applications. Sjodin, Kopwillem, and Karlsson[39] have demonstrated that tissue metabolites can be analyzed both qualitatively and quantitatively by isotachophoresis. A water-soluble extract was prepared by either perchloric acid or ethanol treatment of tissues obtained by muscle biopsy. Detection was by thermometric and UV-absorption systems. Identification of the different zones of the separated sample was performed by adding selected individual ions to the sample extract; subsequent isotachophoresis then demonstrated elongation of the corresponding zone. By this method the following ions zones could be identified: adenosine triphosphate, adenosine diphosphate, adenosine monophosphate, creatine phosphate, inorganic phosphate, glucose-6-phosphate, lactate, and both the oxidized and reduced forms of nicotinamide adenine dinucleotide. Quantitative estimations were performed on metabolites present in excess of 0.1 mmole/Kg wet muscle tissue. Accuracy, sensitivity, and reproducibility were comparable to conventional methods. The time required for analysis was less than one hour.

Uyttendaele et al.[40] have conducted analysis of the proteins in human sweat and urine by isotachophoresis on agarose slab gels. Ampholines were added to the samples before analysis to allow the protein zones to separate during migration. Detection of the proteins was by staining with Coomassie brilliant blue, and identification of the zones was accomplished by immunological analysis with specific antisera. Urine samples from patients with kidney disease were classified according to the type of proteinuria. This method of analysis is applicable to biological fluids of very low protein concentrations with no requirement for pre-concentration of the samples. A detection limit of 50 ng/ml was obtained for albumin.

Kopwillem et al.[41] have demonstrated the application of analytical isotachophoresis to the detection of phenylketonuria by quantitative analysis of serum phenylalanine.

Kjellin et al.[42] analyzed the proteins of cerebrospinal fluid. Very small volumes (15-30 μl) of unconcentrated cerebrospinal fluid were examined effectively. Albumin and γ-globulin regions were identified in samples from normals and patients with multiple sclerosis and meningoencephalitis.

Everaerts et al.[4] have analyzed normal and pathological human sera by isotachophoresis with and without Ampholine. The protein zones were not resolved by the detectors in the absence of Ampholine. With Ampholine both the UV and conductivity tracings of the separations were extremely complex and demonstrated reproducible differences between samples; however, assignment of the differences was not possible.

Proteins. The application of isotachophoresis to the separation of proteins presents some uniquely appealing aspects. The method is attractive because of the sharpness of zone boundaries, the stability of the zones in the steady state, and the self-concentrating nature of the process. As a result of the latter, even at low leader concentration, protein concentrations of the order of 10% are readily obtainable. As separation is not carried out at the isoelectric point, precipitation of proteins is generally avoided, which is a distinct advantage over isoelectric focusing. Thus, the method has the potential of relatively large scale protein fractionation.

The method also has some unique problems. The separated zones remain contiguous, and assignment of individual proteins to the observed peaks is impossible in analytical applications. The same problem plagues preparative procedures, where physical separation of the contiguous zones is desired. A further complication is that the isotachophoretic theory has not been rigorously extended to such complex polyelectrolytes as proteins. The dependence of net protein mobility on leader pH and concentration cannot be theoretically predicted. This limits the applicability of the computer programs of Routs[1] and Jovin[25,26], and requires considerable experimentation to optimize a specific separation.

Contiguous isotachophoretic zones can be separated from each other by the use of spacer ions of intermediate net mobilities, as first suggested by Kendall[43] in 1928. Ampholine is most commonly used for this purpose. Ampholine (trademark of LKB Produkter, Bromma, Sweden) is a random product of polymerization of acrylic acid and polyamines, developed by Vesterberg[44] for isoelectric focusing. This product contains a great number of isomers and homologues of polyamino-polycarboxylic acids with a nearly continuous distribution of isoelectric points and net mobilities. As a result, Ampholine not only provides appropriate spacers for separation of contiguous zones, but also dilutes each zone and broadens its boundaries. Addition of Ampholine also increases the time necessary for the establishment of the steady state. Thus, there is a practical limit to the concentration of Ampholine that can be added, as beyond this, it becomes counter-productive.

Most protein separations are carried out in anionic systems, i.e. above the isoelectric points of the proteins. Occasionally, cationic systems may be useful, as, for instance, in the isolation of a protein with a high isoelectric point. In such a case, all proteins with isoelectric points below the pH of the leading electrolyte will not migrate in the direction of the leading electrolyte, but will migrate zone electrophoretically into the terminator.

Svendsen and Rose[45] were the first to employ Ampholine for the separation of serum proteins, and its use was further studied by Catsimpoolas and coworkers[46,47]. The latter authors studied the effects of the pH range and concentration of Ampholine, leading electrolyte pH and concentration, and gel acrylamide concentration. The separation patterns obtained were in general superior to those of simple zone electrophoresis but inferior to those obtained by disc electrophoresis. The lack of better resolution was attributed to the inherent continuous-mobility characteristic of Ampholine, which does not permit spacing without dilution. Experiments by Chrambach et al.[48] also indicated the inadequacy of Ampholine for the resolution of multicomponent systems such as serum. Despite these shortcomings, isotachophoresis with Ampholine has been applied with varying success to both analytical and preparative protein separations.

Preparative isotachophoresis has been successfully employed in the purification of a variety of proteins which have been difficult to isolate by other methods. These include D- and E-antigenic fibrinogen breakdown products[49], human IgD myeloma protein[50], human bronchogenic carcinoma antigen[51], _Escherichia coli_ enterotoxin[52], human α_1-antitrypsin[53], two components of rat α_1-macroglobulin[54], and prolactin from canine pituitary extracts[55].

Several investigators have compared the efficacy of isotachophoresis with Ampholine to that of isoelectric focusing. Routs[1] discussed differences in the two techniques and compared the separations of chicken hemoglobin by both methods. He also examined the isotachophoretic separation of human serum using two different leading ions, 0.011 M N-Tris(hydroxymethyl)methyl-2-amino ethanesulfonic acid (TES) at pH 7.65 and 0.013 M cacodylic acid at pH 7.02. Both electrolyte systems employed the same counterion (Tris) and terminating ion (β-alanine).

Arlinger[56] has used analytical isotachophoresis in the study of hemoglobin and hemoglobin cyanide. The leading ion used was 0.005 M chloride at pH 7.45, Tris was the counterion, and phenol the terminating ion. Hemoglobin cyanide was fractionated by isoelectric focusing and the isoelectric fractions were shown to be isotachophoretically heterogeneous.

Yotis and Catsimpoolas[57] characterized highly purified Clostridium perfringens type A enterotoxin by scanning isoelectric focusing and isotachophoresis on polyacrylamide gels. Phosphate (0.023 M) at pH 6.0 was used as the leading ion, glycine as the terminating ion, and Tris as the counterion.

Ziegler and Kohler[58] compared the resolving power of isotachophoresis and isoelectric focusing for immunoglobulins secreted by myeloma cell lines and myeloma-spleen hybrids. They utilized polyacrylamide slab gels and concluded that isotachophoresis could resolve proteins in the presence of Ampholine which could not be analyzed by isoelectric focusing because of their sensitivity to low ionic strength and/or their tendency to precipitate at a pH near their isoelectric point.

Van Kleef et al.[59] have considered the analytical and preparative separation of the closely related subunits of bovine α-crystallin in 6 M urea. They used cacodylic acid as the leading ion, β-alanine as the terminating ion, and Tris as the counterion. The effects of the amount and pI range of Ampholine, and the leading electrolyte pH on the analytical separation were examined prior to scale enlargement to the preparative level. The fractions eluted from the preparative separation were characterized by isoelectric focusing. Although fractions demonstrated differential enrichments, none of the subunits were isolated in pure form.

Lancaster and Sprouse[60] have optimized conditions for separation of partially purified histoplasmin derivative components by analytical isotachophoresis, and successfully applied those conditions preparatively using 0.015 M acetate at pH 4.7 as the leading ion, glycine and Tris as terminating and counterions, respectively. They concluded that isotachophoresis is particularly suitable for the isolation of relatively small protein components since these demonstrate rapid diffusion and concomitantly reduced resolution with other electrophoretic methods.

Bog-Hansen et al.[61] described the isotachophoretic preparative fractionation of membrane proteins solubilized by non-ionic detergents. Separations of proteins from human erythrocyte membranes and from bovine milk fat-globule membranes were demonstrated, and the possibility of introducing biospecific interacting components (e.g. antibodies) in the isotachophoretic system was indicated.

Brogren and Svendsen[62] used preparative isotachophoresis in the purification of human serum cholinesterase. Specific impurities (haptoglobin and albumin) were removed in an intermediate step by binding to hemoglobin and anti-albumin respectively, and the mobility of the cholinesterase (a sialoglycoprotein) was altered by neuraminidase treatment. The isolated proteins, both native and asialo-cholinesterase, were immunochemically pure.

Above examples illustrate the potential applicability of isotachophoresis to many protein systems, and the usefulness of Ampholine as a universal spacer. Nevertheless, in truly complex mixtures such as serum, Ampholine alone does not permit separation of the individual components in pure form and considerable overlap has been routinely observed. On the basis of the data available, it was not possible to state unequivocally that isotachophoresis is truly a high resolution method for protein separation, comparable to isoelectric focusing or disc electrophoresis.

To determine if isotachophoresis is truly a high resolution method, Bier and coworkers[28-30] have developed an alternative approach, using discrete spacers for both analytical and preparative isotachophoresis. Ideally, the use of _n_ spacers with well-defined intermediate net mobilities should separate a complex protein mixture into _n + 1_ subfractions with no overlap or cross-contamination of proteins across the spacer zones. Unlike Ampholine, discrete spacers should not result in dilution of the protein zones, and should not compromise boundary sharpness. The length of the spacer zones can be extended at will to facilitate preparative recovery of the subfractions. In principle, if the isolation of a single protein is desired, a judicious choice of only two bracketing discrete spacers should suffice.

The requirement for such discrete spacers is that they have net mobilities comparable to those of the proteins in a given system: they need not be amphoteric. Nevertheless, amino acids and various simple peptides were chosen for above study, as Vestermark[63] had previously shown that they have mobilities comparable to those of serum proteins. The study was carried out using a serum protein preparation impoverished of albumin through ammonium sulfate precipitation, as untreated serum forms an excessively long albumin zone.

Analytical isotachophoresis has clearly shown that the use of four spacers separates the isotachophoretic train into five distinct groups as shown in Figure 6. Best resolution of components was obtained by the simultaneous use of Ampholine and discrete spacers, the discrete spacers separating the proteins into five major groups, the Ampholine further separating the components of each group. With a partially automated Tachophor apparatus, Delmotte[64] has utilized such mixtures of discrete and continuous spacers for an analytical survey of serum proteins in clinical abnormalities.

Comparable separation of the serum preparation using discrete spacers was also obtained in preparative runs on polyacrylamide and Sephadex gels. The advantages of this latter support medium were previously mentioned. The spacer-separated subfractions were clearly visible and were exquisitely sharp, as shown in previously published photographs[28,29]. The spacer zones could be made sufficiently long (of the order of 3 cm) to permit easy separation of each subfraction without risk of cross-contamination. The protein content of each fraction was analyzed by quantitative

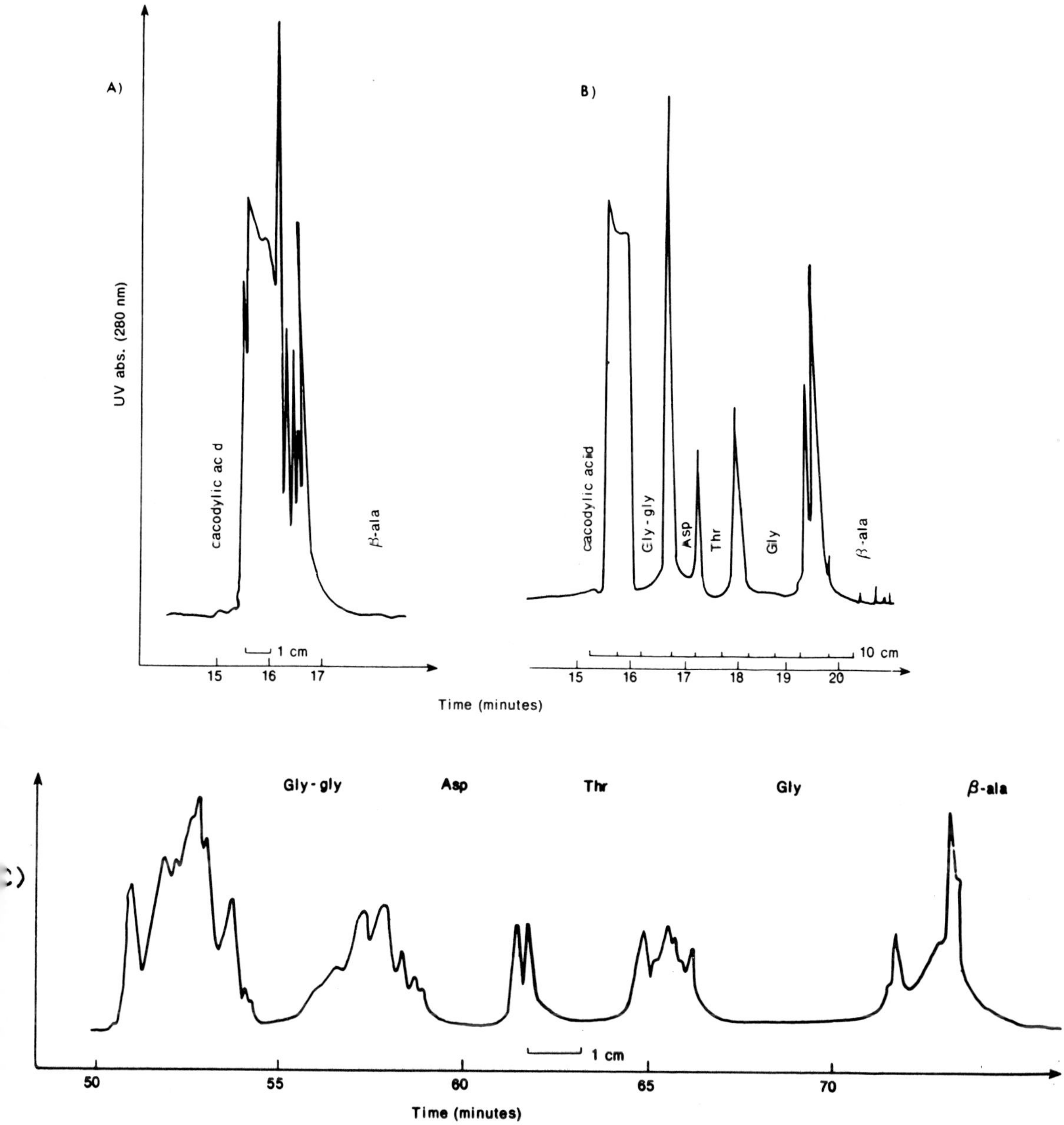

Fig. 6. Comparison of isotachophoretic patterns of serum proteins obtained with the Tachophor. Without spacers a complex pattern of contiguous peaks is obtained (A), whereas the addition of four discrete spacers (glycylglycine, asparagine, threonine, and glycine) separates the pattern into five distinct subgroups (B). Best resolution is obtained with the combined use of Ampholine and discrete spacers (C). Leading ion was 5 mM cacodylic acid, pH 7.5, and terminating ion was β-alanine. (From Kopwillem et al.[30]).

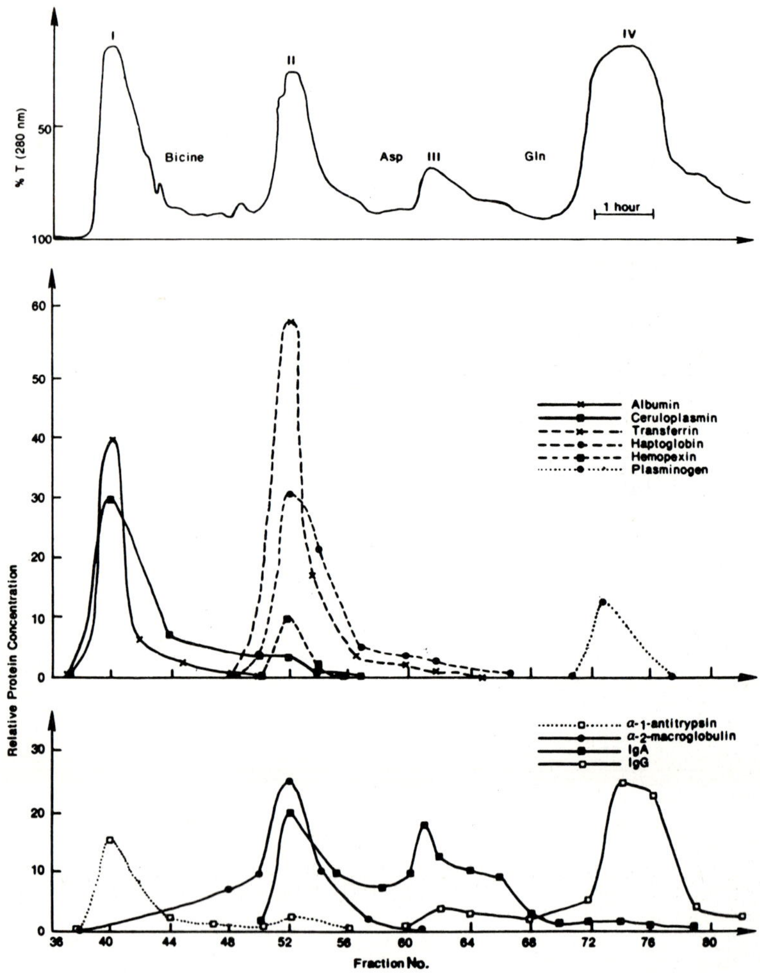

Fig. 7. Isotachophoretic separation of 180 mg of a plasma proteins preparation using 240 μmoles of bicine, asparagine, and glutamine as spacers. Sephadex G-200 was used as the support medium with 20 mM cacodylic acid, pH 7.1, as leading electrolyte and β-alanine as terminator. The eluted proteins were quantitated by immunoelectrophoresis. (From Bier et al.[28]).

immunoelectrophoresis or immunodiffusion, and the distribution of ten specific proteins in each fraction determined. A typical result is shown in Figure 7. As anticipated, most of the proteins were sharply resolved by the discrete spacers into the individual subfractions. The distribution of individual proteins in the various subfractions could be manipulated by choice of spacers used and the leader pH. The net mobility of amino acids is more pH-dependent than that of proteins, and this adds

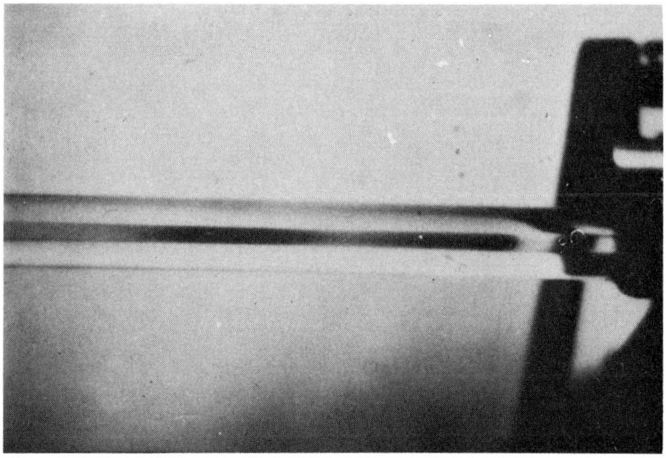

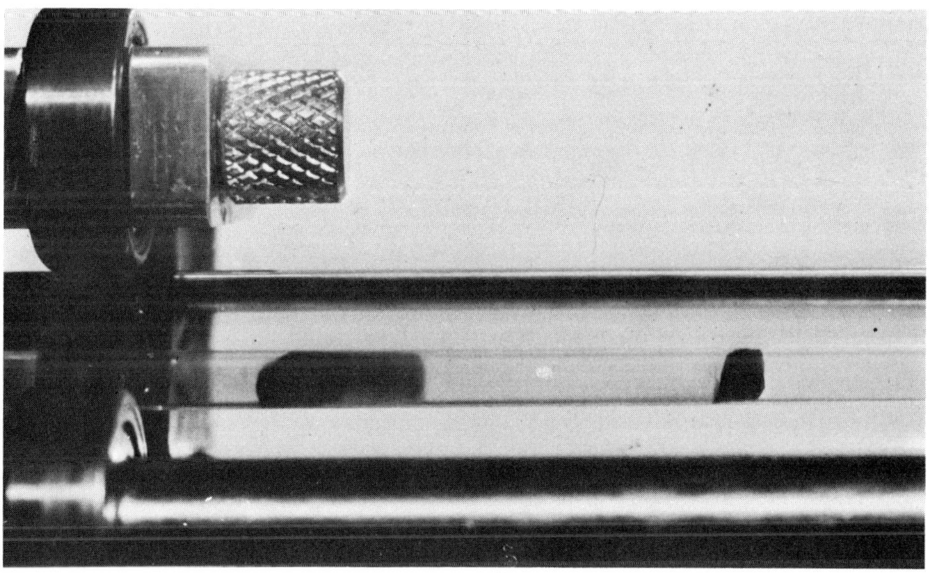

Fig. 8. Comparison of zone electrophoretic and isotachophoretic separation of erythrocytes of cat and sheep. Zone electrophoresis (top photograph) easily separates the two species, but gives diffuse boundaries. Isotachophoresis (bottom photograph) gives sharp and self-adjusting boundaries, but no spacers were found giving consistently good results. The separation was carried out in the Bier[20] rotationally stabilized apparatus.

to the flexibility of the method. However, in no case was it possible to isolate one protein only in a subfraction. Moreover, in each system tried some proteins were not sharply resolved by the spacers, but were found in two or more subfractions.

This study has shown that isotachophoresis is indeed capable of sharp resolution, but that in some instances protein-protein interactions or other causes may result in protein overlap. The use of Sephadex gels precludes the possibility that these overlaps resulted from boundary distortions, often observed in polyacrylamide gels, but consistently absent in Sephadex.

Particle and Cell Separations. Satellite tobacco necrosis virus has been fractionated isotachophoretically by Hjerten in his rotationally stabilized apparatus[65]. The preparative modification [20] of the Hjerten apparatus has been used by Bier et al.[66] to study the feasibility of isotachophoretic separation of erythrocytes of various animal species. Figure 8 compares the separation of a mixture of cat and sheep erythrocytes by isotachophoresis and zone electrophoresis. Erythrocytes were found to behave as an 'ionic' species in terms of isotachophoresis: their concentration was self-adjusting in compliance with the Kohlrausch regulating function. The sharp isotachophoretic boundaries contrast with the more diffuse zone electrophoretic boundaries. Unfortunately, extensive investigations of the separability of erythrocytes either in the rotating apparatus or the analytical LKB apparatus have failed to produce good spacers for such cell mixtures. Neither Ampholine nor a variety of discrete spacers acted in a satisfactory and predictable manner. Thus, the possibility of using isotachophoresis for cell separations remains an open question.

CONCLUSION

The feasibility of using isotachophoresis as a means of analyzing a variety of low molecular weight sample systems has been demonstrated. Application of isotachophoresis for these purposes has been limited as other better established methods generally offer comparable or superior sensitivity and reproducibility. However, isotachophoresis provides a new dimension in the electrophoretic analysis and fractionation of peptides and proteins, and the technique is enjoying expanding applications in this area. The multiplicity of significant factors confers a high degree of flexibility to the method but it also demands that applications be approached on an individual problem basis. The main reason for the lack of more widespread use of the technique is probably the complexity of the method which demands a great deal of expertise. Other electrophoretic techniques are easier to apply and do not require a detailed understanding of the underlying theories. In specific instances when the objectives of the application are clearly defined, isotachophoresis may possess important advantages. This is the case, for example, with small peptides, which are difficult to analyze by isoelectric focusing or other electrophoretic methods, as they may not be UV-absorbing, are difficult to fix and stain on gels, and diffuse rapidly. The speed of analytical isotachophoresis is another advantage, coupled with its high resolution and sensitivity, and ready adaptability to highly dilute solutions or small sample volumes. The isolation of a specific protein may be accomplished if it can be suitably bracketed by discrete spacers or its mobility can be influenced by experimental conditions so that it is either the fastest or the slowest component in a mixture. Under such circumstances, specific optimization of the method is obviously possible.

ACKNOWLEDGEMENTS

This research was supported in part by NASA grant NSG-7333 and NASA contract NAS8-29566.

REFERENCES

1. Routs, R. J. (1971) Electrolyte Systems in Isotachophoresis and their Application to some Protein Separations, Thesis, University of Technology, Eindhoven.
2. Jovin, T. M. (1973) Biochemistry 12, 871 879.
3. Nee, T. W. (1974) J. Chromatogr. 93, 7-15.
4. Everaerts, F. M., Beckers, J. L. and Verheggen, Th. P.E.M. (1976) Isotachophoresis: Theory, Instrumentation and Applications, Elsevier Scientific, Amsterdam, pp. 1418.
5. Chrambach, A. and Nguyen, N. Y. (1978) This volume.
6. Kohlrausch, F. (1897) Ann. Physik. 62, 209-239.
7. Dole, V. P. (1945) J. Amer. Chem. Soc. 67, 1119-1126.
8. Dismukes, E. B. and Alberty, R. A. (1954) J. Amer. Chem. Soc. 76, 191-197.
9. Alberty, R. A. (1950) J. Amer. Chem. Soc. 72, 2361-2367.
10. Coxon, M. and Binder, M. J. (1974) J. Chromatogr. 95, 133-145.
11. Hjerten, S. (1967) Free Zone Electrophoresis, Thesis, Institute of Biochemistry, Uppsala.
12. Patterson, W. J. (1976) NASA Technical Memorandum TM X-73311.
13. Martin, A. J. P., and Everaerts, F. M. (1970) Proc. Roy. Soc. Lond., Ser. A, 316, 493-514.
14. Hinckley, J. O. N. (1974) Clin. Chem. 20, 973-991.
15. Hinckley, J. O. N. (1975) J. Chromatogr. 109, 209-217.
16. Brown, J. F. and Hinckley, J. O. N. (1975) J. Chromatogr. 109, 218-224.
17. Brown J. F. and Hinckley, J. O. N. (1975) J. Chromatogr. 109, 225-231.
18. Coxon, M. and Binder, M. J. (1974) J. Chromatogr. 101, 1-16.
19. Coxon, M. and Binder, M. J. (1975) J. Chromatogr. 107, 43-50.
20. Bier, M. (1977) U. S. Patent 4,040,940.
21. Bier, M., Hinckley, J. O. N., Smolka, A. J. K. and Snyder, R. S. (1974) Protides Biol. Fluids, Proc. Colloq. 22, 673-678.
22. Bier, M. (1977) in Materials Sciences in Space with Applications to Space Processing (L. Steg, ed.), American Institute of Aeronautics and Astronautics, New York, pp. 41-55.

23. Snyder, R. S. and Allen, R. E. (1977) in Materials Sciences in Space with Application to Space Processing (L. Steg, ed.), American Institute of Aeronautics and Astronautics, New York, pp. 399-410..

24. Beckers, J. L. and Everaerts, F. M. (1972) J. Chromatogr. 69, 165-179.

25. Jovin, T. M., Dante, M. L. and Chrambach, A. (1970) Multiphasic Buffer Systems Output PB 196085 to 196092 and 203016, Nat. Tech. Inform. Serv., Springfield, Va.

26. Jovin, T. M. (1973) Ann. N. Y. Acad. Sci. 209, 477-496.

27. Preetz, W. (1966) Talanta 13, 1649-1660.

28. Bier, M., Cuddeback, R. M. and Kopwillem, A. (1977) J. Chromatogr. 132, 437-450.

29. Bier, M. and Kopwillem, A. (1977) in Electrofocusing and Isotachophoresis (Radola, B. J. and Graesslin, D., eds.), Walter de Gruyter, Berlin, pp. 567-576.

30. Kopwillem, A., Merriman, W. G., Cuddeback, R. M., Smolka, A. J. K., and Bier, M. (1976) J. Chromatogr. 118, 35-46.

31. Kendall, J. and Crittenden, E. D. (1923) Proc. Nat. Acad. Sci. USA 9, 75-78.

32. Kopwillem, A. and Lundin, H. (1974) LKB Application Note 183.

33. Kopwillem, A., Moberg, U., Westin-Sjodahl, G., Lundin, R., and Sievertsson, H. (1975) Anal. Biochem. 67, 166-181.

34. Miyazaki, H. and Katoh, K. (1976) J. Chromatogr. 119, 369-383.

35. Bocek, P., Lekova, K., Deml, M. and Janak, J. (1976) J. Chromatogr. 117, 97-104.

36. Kopwillem, A. (1973) J. Chromatogr. 82, 407-409.

37. Everaerts, F. M. and Verheggen, Th. P. E. M. (1974) J. Chromatogr. 91, 837-851.

38. Willemsen, A. J. (1975) J. Chromatogr. 105, 405-410.

39. Sjodin, B., Kopwillem, A. and Karlsson, J. (1975) Scand. J. Clin. Lab. Invest. 35, 699-703.

40. Uyttendaele, K., de Groote, M., Blaton, V., Peeters, H. and Alexander, F. (1977) J. Chromatogr. 132, 261-266.

41. Kopwillem, A., Lundin, H., Righetti, A. B. B. and Righetti, P. G. (1974) Protides Biol. Fluids, Proc. Colloq. 22, 737-742.

42. Kjellin, K. G., Moberg, U. and Hallander, L. (1975) Sci. Tools 22, 3-7.

43. Kendall, J. (1928) Science 67, 163-167.

44. Vesterberg, O. (1969) Acta Chem. Scand. 23, 2653-2666.

45. Svendsen, P. J. and Rose, C. (1970) Sci. Tools 17, 13-17.

46. Griffith, A., Catsimpoolas, N. and Kenney, J. (1973) Ann. N. Y. Acad. Sci. 209, 457-468.

47. Catsimpoolas, N. and Kenney, J. (1972) Biochem. Biophys. Acta 285, 287-292.

48. Chrambach, A., Kapadia, G. and Cantz, M. (1972) Separ. Sci. 7, 785-816.

49. Clemmensen, I. and Svendsen, P. J. (1973) Sci. Tools 20, 5-6.

50. Jefferis, R. and Butwell, A. J. (1975) Sci. Tools 22, 1-3.

51. Frost. M. J., Rogers, G. T. and Bagshawe, K. D. (1975) Br. J. Cancer 31, 379-386.

52. Dorner, F., Jaksche, H. and Stockl, W. (1976) J. Infect. Diseases 133, S142-156.

53. Clemmensen, I. and Christensen, F. (1976) Biochem, Biophys. Acta 429, 591-599.

54. Gauthier, F., Gutman, N., Muh, J. P. and Mouray, H. (1976) Anal. Biochem. 71, 181-185.

55. Knight, P. J., Gronow, M. and Hamilton, J. M. (1976) J. Endocr. 69, 127-132.

56. Arlinger, L. (1975) Biochim. Biophys. Acta 393, 396-403.

57. Yotis, W. W. and Catsimpoolas, N. (1975) J. Appl. Bact. 39, 147-156.

58. Ziegler, A. and Kohler, G. (1976) FEBS Letters 71, 142-146.

59. Van Kleef, F. S. M., Peeters, M. and Hoenders, H. J. (1977) Anal. Biochem. 77, 122-129.

60. Lancaster, M. V. and Sprouse, R. F. (1977) Anal. Biochem. 77, 158-167.

61. Bog-Hansen, T. C., Svendsen, P. J., Bjerrum, O. J., Nielsen, C. S. and Ramlau, J. (1974) Protides Biol. Fluids, Proc. Colloq. 22, 679-684.

62. Brogren, C. H. and Svendsen, P. J. (1974) Protides Biol. Fluids, Proc. Colloq. 22, 685-689.

63. Vestermark, A. (1966) Cons Electrophoresis, University of Stockholm, Stockholm, pp. 5-46.

64. Delmotte, P. (1978) Electrophoresis '78 Symposium, MIT, Cambridge.

65. Hjerten, S. (1974) Protides Biol. Fluids, Proc. Colloq. 22, 669-672.

66. Bier, M., Smolka, A. J. K. and Hinckley, J. O. N. Unpublished data.

SUBJECT INDEX